Lecture Notes in Electrical Engineering

Volume 27

Lecture Notes in Electrical Engineering

(Continued after index)

Nikos Mastorakis • Valeri Mladenov •
Vassiliki T. Kontargyri

Editors

Proceedings of the European Computing Conference

Volume 1

 Springer

Editors

Nikos Mastorakis
Hellenic Naval Academy
Military Institutes of University
   Education
Piraeus, Greece
mastor@hna.gr

Valeri Mladenov
Technical University of Sofia
Sofia, Bulgaria
valerim@tu-sofia.bg

Vassiliki T. Kontargyri
National Technical
   University of Athens
Athens, Greece
vkont@central.ntua.gr

ISSN 1876-1100
ISBN 978-1-4899-7769-4
DOI 10.1007/978-0-387-84814-3

ISSN 1876-1119 (eBook)
ISBN 978-0-387-84814-3 (eBook)

Printed on acid-free paper

springer.com

Contents

Contributors

J. El Abbadi Lab d'Electronique et de Télécommunication, EMI, Rabat, Morocco

R. Alesii Università degli Studi dell'Aquila, Centro di Eccellenza DEWS, 67100 Poggio di Roio (AQ), Italy, alesii@ing.univaq.it

Manuel de Jesús Nandayapa Alfaro Universidad Autónoma de Ciudad Juárez (UACJ), Avenida del Charro 450 Norte Ciudad Juárez Chihuahua México, mnandaya@uacj.mx

Jesús B. Alonso Centro Tecnológico para la Innovación en Comunicaciones (CETIC), Dpto. de Señales y Comunicaciones, Universidad de Las Palmas de Gran Canaria, Campus de Tafira, E35017, Las Palmas de Gran Canaria, Spain, jalonso@dsc.ulpgc.es

Antonios S. Andreatos Division of Computer Engineering and Information Science, Hellenic Air Force Academy, Dekeleia, Attica, TGA-1010, Greece, aandreatos@hafa.gr, aandreatos@gmail.com

Youssef Balouki Department of Mathematics and Computer Science, Mohammed V University, B.P 1014, Av. Ibn Bettouta, Rabat Morocco, balouki@fsr.ac.ma

Celia A. Zorzo Barcelos Federal University of Uberlândia 38.400-902, Uberlândia, MG, Brasil, celiazb@ufu.br

Marcos Aurélio Batista Federal University of Goiás 75.701-970, Catalão, GO, Brasil, marcos@catalao.ufg.br

Afrodita Balasa Boldea Department of Computer Science, West University of Timisoara, 4,Vasile Parvan, Timisoara1900, Romania, alinusha_b@yahoo.com

Costin-Radu Boldea Department of Computer Science, West University of Timisoara, 4,Vasile Parvan, Timisoara1900, Romania, cboldea@info.uvt.ro

Noura Boudiaf Computer Science Department, University of Oum El Bouaghi 04000 Oum El Bouaghi, Algeria, boudiafn@yahoo.com

Mohamed Bouhdadi Department of Mathematics and Computer Science, Mohammed V University, B.P 1014, Av. Ibn Bettouta, Rabat, Morocco, bouhdadi@fsr.ac.ma

Mohamed Salim Bouhlel Research Unit, Sciences and Technologies of Image and Telecommunications (SETIT), Higher Institute of Biotechnology of Sfax, Tunisia, medsalim.bouhlel@enis.rnu.tn

S. Bunluechokchai Department of Industrial Physics and Medical Instrumentation, Faculty of Applied Science, King Mongkut's Institute of Technology, North Bangkok, 1518 Pibulsongkram Road, Bangsue, Bangkok 10800 Thailand

Rui Camacho DEEC & LIACC, Universidade do Porto, R. Dr Roberto Frias, s/n, 4200-465 Porto, Portugal, rcamacho@fe.up.pt

E. Camahort Department of Computer Systems, Polytechnic University of Valencia, Valencia, Spain

Leonarda Carnimeo Dipartimento di Elettrotecnica ed Elettronica, Politecnico di Bari, Via Orabona, 4 70125 BARI – Italy, carnimeo@deemail.poliba.it

Hsueh-Yung (Robert) Chao Department of Communication Engineering, National Chiao Tung University, 1001 Ta-Hsueh Road, Hsinchu 300, Taiwan

Hassan Charaf Department of Automation and Applied Informatics, Budapest University of Technology and Economics, H-1111 Budapest, Goldmann György tér 3., Hungary

Chokri Chemak Research Unit, Sciences and Technologies of Image and Telecommunications (SETIT), Higher Institute of Biotechnology of Sfax, Tunisia, cchemak@univ-fcomte.fr

M. Chover Department of Computer Systems, Jaume I University, Castellón, Spain

Zygmunt Ciota Department of Microelectronics and Computer Science, Technical University of Lodz, Poland

Kelly Connolly The MITRE Corporation, 202 Burlington Road, Bedford, MA 01730-1420, USA

Douglas Farias Cordeiro Federal University of Uberlândia 38.400-902, Uberlândia, MG, Brasil, douglasf@pos.facom.ufu.br

David Wolfe Corne School of Mathematical and Computer Sciences, Heriot-Watt University, Edinburgh EH14 4AS, UK

Ciprian Cuzmin Department of Manufacturing Engineering, "Dunărea de Jos" University of Galaţi, Galaţi, 111 Domnească Street, zip code 800201, Romania

Krzysztof A. Cyran Institute of Informatics, Silesian University of Technology, Gliwice, Poland

Monica Dascălu IMT Bucharest, Erou Iancu Nicolae 32B Street, Romania

Nor Ridzuan Daud Faculty of Computer Science and Information Technology, University of Malaya, 50603 Kuala Lumpur, Malaysia

James DeArmon The MITRE Corporation, 202 Burlington Road, Bedford, MA 01730-1420, USA

C. Ray Diez Department of Technology, The University of North Dakota, Grand Forks, North Dakota, USA, Clayton.Diez@mail.business.und.edu

Walter Dosch Institute of Software Technology and Programming Languages, University of Lübeck, Lübeck, Germany

Daniel Dumitriu Institute of Solid Mechanics, Romanian Academy, Bucureşti, Romania, dumitri04@yahoo.com

Lambros Ekonomou Hellenic American University, 12 Kaplanon Str., 106 80 Athens, Greece, leekonom@gmail.com

Alexandru Epureanu Department of Manufacturing Engineering, "Dunărea de Jos" University of Galaţi, Galaţi, 111 Domnească Street, zip code 800201, Romania

Muhammad Abuzar Fahiem University of Engineering and Technology, Lahore, Pakistan; Lahore College for Women University, Lahore, Pakistan, abuzar@uet.edu.pk

Rong-Jyue Fang Department of Information Management, Southern Taiwan University of Technology, Taiwan, ROC, rxf26@mail.stut.edu.tw

Miguel A. Ferrer Centro Tecnológico para la Innovación en Comunicaciones (CETIC), Dpto. de Señales y Comunicaciones, Universidad de Las Palmas de Gran Canaria, Campus de Tafira, E35017, Las Palmas de Gran Canaria, Spain, mferrer@dsc.ulpgc.es

Bertalan Forstner Department of Automation and Applied Informatics, Budapest University of Technology and Economics, H-1111 Budapest, Goldmann György tér 3., Hungary, bertalan.forstner@aut.bme.hu

Georgios P. Fotis National Technical University of Athens, 9 Iroon Politechniou Str., 157 80 Athens, Greece, gfotis@gmail.com

Eduard Franti IMT Bucharest, Erou Iancu Nicolae 32B street, Romania, edif@atlas.cpe.pub.ro

A. Garcés Department of Computer Systems, Jaume I University, Castellón, Spain

F. D. Lorenzo-García Dpto. de Señales y Comunicaciones, Universidad de Las Palmas de Gran Canaria, Campus de Tafira, E35017, Las Palmas de Gran Canaria, Spain

A. G. Ravelo-García Dpto. de Señales y Comunicaciones, Universidad de Las Palmas de Gran Canaria, Campus de Tafira, E35017, Las Palmas de Gran Canaria, Spain, aravelo@dsc.ulpgc.es

G. Gargano Università degli Studi dell'Aquila, Centro di Eccellenza DEWS, 67100 Poggio di Roio (AQ), Italy, gino.gargano@email.it

S.A.M. Gilani Faculty of Computer Science and Engineering, Ghulam Ishaq Khan Institute of Engineering Sciences and Technology, Topi, Pakistan, asif@giki.edu.pk

E. Glavas A.T.E.I. of Epirus, Department of Teleinformatics and Management, GR-471 00 Arta, Greece, eglavas@teiep.gr

Bindu Goel University School of Information Technology, Guru Gobind Singh Indraprastha University, Kashmere Gate, Delhi -06, India, bindu_delus@yahoo.com

F. Graziosi Università degli Studi dell'Aquila, Centro di Eccellenza DEWS, 67100 Poggio di Roio (AQ), Italy, graziosi@ing.univaq.it

Eugene Grichuk Moscow Engineering Physics Institute, Kashirskoe Schosse, 115409 Moscow, Russia, es@t-25.ru

M. Amac Guvensan Department of Computer Engineering, Yildiz Teknik University, Besiktas, Istanbul, Turkey, amac@ce.yildiz.edu.tr

Wen-Hui Han Department of Nursing, Yuanpei University, No.306, Yuanpei Street, Hsin Chu 30015, Taiwan, wenhui@mail.ypu.edu.tw

Shaiq A. Haq University of Engineering and Technology, Lahore, Pakistan

Yasumichi Hasegawa Gifu University, 1-1 Yanagido, Gifu, 501-1193 Japan, yhasega@gifu-u.ac.jp

Takeshi Hashimoto Department of Electrical and Electronics Engineering, Shizuoka University, 5-1, 3-chome Johoku, 432-8561 Hamamatsu, Japan

See-Chien Hou National Chung Cheng University, Taiwan, ROC

Mei-Huang Huang Institute of Technological and Vocational Education, National Pingtung University of Science and Technology, Taiwan

Tzai-Hung Huang Institute of Technological and Vocational Education, National Pingtung University of Science and Technology, Taiwan

J. Huerta Department of Computer Systems, Jaume I University, Castellón, Spain

Anca Iordan Engineering Faculty of Hunedoara, Technical University of Timişoara, Revolutiei 5, 331128 Hunedoara, Transylvania, Romania, anca.iordan@fih.upt.ro

John James The MITRE Corporation, 202 Burlington Road, Bedford, MA 01730-1420, USA

A. Jraifi Groupe CPR, Canal, Radio & Propagation, Lab/UFR-PHE, Faculté des Sciences de Rabat, Morocco

Arbana Kadriu CST Department, SEE-University, b.b. 1200 Tetovo, FYROM, a.kadriu@seeu.edu.mk

G. Karangelis Oncology Systems Ltd., Shrewsbury, SY1 3AF, UK

Alireza Kargar Department of Electronics and Engineering, School of Engineering, Shiraz University, Shiraz, Iran; Shiraz Nanotechnology Research Institute, Shiraz, Iran

V.N. Kasyanov A.P. Ershov Institute of Informatics Systems/Novosibirsk State University, Novosibirsk, 630090, Russia

E.V. Kasyanova A.P. Ershov Institute of Informatics Systems/Novosibirsk State University, Novosibirsk, 630090, Russia

Sokratis K. Katsikas Department of Technology Education and Digital Systems, University of Piraeus, 150 Androutsou St., Piraeus, 18532, Greece, ska@unipi.gr

Imre Kelényi Department of Automation and Applied Informatics, Budapest University of Technology and Economics, H-1111 Budapest, Goldmann György tér 3., Hungary

A. El Khafaji Lab d'Electronique et de Télécommunication, EMI, Rabat, Morocco

Amin Khansefid Electrical and Computer Engineering Department, University of Tehran, Campus #2, University of Tehran, North Kargar Ave., Tehran, Iran, a.khansefid@ece.ut.ac.ir

Zoheir Kordrostami Department of Electrical Engineering, School of Engineering, Shiraz University, Shiraz, Iran; Shiraz Nanotechnology Research Institute, Shiraz, Iran, zkrostami@shirazu.ac.it

Stavroula Kourtesi Hellenic Public Power Corporation S.A., 22 Chalcocondyli Str., 104 32 Athens, Greece, stavriani_kourtesi@yahoo.gr

Guennadi Kouzaev Department of Electronics and Telecommunications, Norwegian University of Science and Technology, Trondheim, Norway

Yi-Ting Kuo Department of Communication Engineering, National Chiao Tung University, 1001 Ta-Hsueh Road, Hsinchu 300, Taiwan

Margarita Kuzmina Keldysh Institute of Applied Mathematics RAS, Miusskaya Sq. 4, 125047 Moscow, Russia, kuzmina@spp.keldysh.ru

Jean Christophe Lapayre Computer Laboratory of Franche-Comte (L.I.F.C.), Franche-Compte University of Sciences and Techniques, France, Jean christophe.lapayre@univ-fcomte.fr

Chi-Jen Lee Department of Industrial Technology Education, National Kaohsiung Normal University, Taiwan, mickeylee@gmail.com

Chung-Ping Lee Graduate Institute of Educational Entrepreneurship and Management, National University of Tainan, Taiwan, cl87369@yahoo.com.tw

T. Leeudomwong Department of Industrial Physics and Medical Instrumentation, Faculty of Applied Science, King Mongkut's Institute of Technology, North Bangkok, 1518 Pibulsongkram Road, Bangsue, Bangkok 10800 Thailand

Yiming Li Department of Communication Engineering, National Chiao Tung University, 1001 Ta-Hsueh Road, Hsinchu 300, Taiwan

Shi-Jer Lou Institute of Technological and Vocational Education, National Pingtung University of Science and Technology, Taiwan, ROC, lou@mail.npust.edu.tw

Yi-Hui Lui Institute of Technological and Vocational Education, National Pingtung University of Science and Technology, Taiwan

Demétrio Renó Magalhães Depto de Exatas & CSI, UnilesteMG, Av Pres. Tancredo Neves, 3500, 35170-056, C. Fabriciano, MG, Brasil, reno@uniliestemg.br

Paula Mahoney The MITRE Corporation, 202 Burlington Road, Bedford, MA 01730-1420, USA

G.A. Mallah National University of Computer and Emerging Sciences, ST-4, Sector-17/D, Shah Latif Town, Karachi, Pakistan

Eduard Manykin Russian RC "Kurchatov Institute", Kurchatov Sq. 1, 123182 Moscow, Russia, edmany@isssph.kiae.ru

Goran Martinovic Faculty of Electrical Engineering, Josip Juraj Strossmayer University of Osijek, Kneza Trpimira 2b, Croatia, goran.martinovic@etfos.hr

N.E. Mastorakis WSEAS Headquarters, Agiou Ioannou Theologou 17-23, 15773, Zografou, Athens, Greece

T. Melesanaki Division of Telecommunications, Microwave Communications and Electromagnetic Applications Laboratory, Department of Electronics, Technological Educational Institute (T.E.I.) of Crete, Chania Branch, Romanou 3, Chalepa, 73133 Chania Crete, Greece, ivardia@chania.teicrete.gr

Nisar Ahmed Memon Faculty of Computer Science and Engineering, Ghulam Ishaq Khan Institute of Engineering Sciences and Technology, Topi, Pakistan, memon_nisar@yahoo.com

Marian Cristian Mihăescu Software Engineering Department, University of Craiova, Craiova, Romania

Jamie D. Mills Department of Educational Studies, University of Alabama, 316-A Carmichael Hall, Box 870231, Tuscaloosa, AL 35487-0231, USA, jmills@bamaed.ua.edu

Gabriela Mircea Faculty of Economic Sciences, West University of Timisoara, Romania, gabriela.mircea@fse.uvt.ro

Anwar Majid Mirza Faculty of Computer Science and Engineering, Ghulam Ishaq Khan Institute of Engineering Sciences and Technology, Topi, Pakistan, anwar.m.mirza@gmail.com

Valeri Mladenov Department of Theoretical Electrical Engineering, Technical University of Sofia, 8 Kliment Ohridski St., Sofia 1000, Bulgaria, valerim@tu-sofia.bg

Payman Moallem Faculty of Engineering, University of Isfahan, Isfahan, Iran, p_moallem@eng.ui.ac.ir

K.S. Mohammed Electronic Department, National Telecommunication Institute, Cairo, Egypt

S. Amirhassan Monadjemi Faculty of Engineering, University of Isfahan, Isfahan, Iran, monadjemi@eng.ui.ac.ir

John Morris The MITRE Corporation, 202 Burlington Road, Bedford, MA 01730-1420, USA, jemorris@mitre.org, www.mitre.org

Vassilios C. Moussas School of Technological Applications (S.T.E.F.), Technological Educational Institution (T.E.I.) of Athens, Egaleo, Greece, vmouss@teiath.gr

Mihaela Muntean Faculty of Economic Sciences, West University of Timisoara, Romania, mihaela.muntean@fse.uvt.ro

Angelos Nakulas National and Kapodistrian University of Athens, 11 Asklipiu Str., 153 54 Athens, Greece, aaatos@gmail.com

J. L. Navarro-Mesa Dpto. de Señales y Comunicaciones, Universidad de Las Palmas de Gran Canaria, Campus de Tafira, E35017, Las Palmas de Gran Canaria, Spain

Adrian-Ioan Niculescu Institute of Solid Mechanics, Romanian Academy, Bucureşti, Romania, adrian_ioan_niculescu@yahoo.com

Wodzimierz Ogryczak Institute of Control and Computation Engineering, Warsaw University of Technology, 00-665 Warsaw, Poland, wogrycza@ia.pw.edu.pl

Ali Olfat Electrical and Computer Engineering Department, University of Tehran, Campus #2, North Kargar Ave., Tehran, Iran, aolfat@ut.ac.ir

R. Orzechowski Chair of Wireless Communication, Poznan University of Technology, Piotrowo 3a, 60-965 Poznan, Poland

Manuela Pănoiu Engineering Faculty of Hunedoara, Technical University of Timişoara, Revolutiei 5, 331128 Hunedoara, Transylvania, Romania

Caius Pănoiu Technical University of Timişoara, Engineering Faculty of Hunedoara, Revolutiei 5, 331128 Hunedoara, Transylvania, Romania

Stylianos Sp. Pappas Department of Information and Communication Systems Engineering, University of the Aegean, 83200 Karlovassi Samos, Greece, spappas@aegean.gr

Sonja Petrovic-Lazarevic Department of Management, Monash University, 26 Sir John Monash Drive, Caulfield East, VIC 3145, Australia, sonja.petrovic-lazarevic@buseco.monash.edu.au

L. Pomante Università degli Studi dell'Aquila, Centro di Eccellenza DEWS, 67100 Poggio di Roio (AQ), Italy, pomante@ing.univaq.it

Andreas Pomportsis Department of Informatics, Aristotle University of Thessaloniki, 54006, Thessaloniki, Greece

Elvira Popescu Software Engineering Department, University of Craiova, Romania; Heudiasyc UMR CNRS 6599, Université de Technologie de Compiègne, France

George Popov Department of Computer Science, Technical University, Sofia, 8 Kliment Ohridski St., Sofia 1000, Bulgaria, popovg@tu-sofia.bg

Mircea Preda Computer Science Department, University of Craiova, Romania

R. Quirós Department of Computer Systems, Jaume I University, Castellón, Spain

Tomáš Rebok Faculty of Informatics, Masaryk University, Botanická 68a, 602 00 Brno, Czech Republic, xrebok@fi.muni.cz

Francisco Reinaldo DEEC and LIACC, Universidade do Porto, R. Dr Roberto Frias, s/n, 4200-465 Porto, Portugal, reifeup@fe.up.pt

Luís P. Reis DEEC and LIACC, Universidade do Porto, R. Dr Roberto Frias, s/n, 4200-465 Porto, Portugal, lpreis@fe.up.pt

P. Remlein Chair of Wireless Communication, Poznan University of Technology, Piotrowo 3a, 60-965 Poznan, Poland

C. Rinaldi Università degli Studi dell'Aquila, Centro di Eccellenza DEWS, 67100 Poggio di Roio (AQ), Italy, rinaldi@ing.univaq.it

G.E. Rizos Department of Computer Science and Technology, University of Peloponnese, GR-221 00 Tripolis, Greece, georizos@uop.gr

Carlos F. Romero Centro Tecnológico para la Innovación en Comunicaciones (CETIC), Dpto. de Señales y Comunicaciones, Universidad de Las Palmas de Gran Canaria, Campus de Tafira, E35017, Las Palmas de Gran Canaria, Spain, fabian_romero@ciudad.com.ar

András Rövid Department of Chassis and Lightweight Structures, Budapest University of Technology and Economics, Bertalan Lajos u. 2. 7. em., 1111 Budapest, Hungary

Hajar Sadeghi Faculty of Engineering, University of Isfahan, Isfahan, Iran, hsadeghi@eng.ui.ac.ir

Lakhdar Sahbi Computer Science Department, University of Oum El Bouaghi, 04000 Oum El Bouaghi, Algeria, sahbi_lak@yahoo.fr

E.H Saidi VACBT, Virtual African Center for Basic Science and Technology, Focal Point Lab/UFR-PHE, Fac Sciences, Rabat, Morocco

Farhat Saleemi Lahore College for Women University, Lahore, Pakistan

Ahmed Sameh Department of Computer Science, The American University in Cairo, 113 Kaser Al Aini Street, P.O. Box 2511, Cairo, Egypt, sameh@aucegypt.edu

Vianey Guadalupe Cruz Sánchez Centro Nacional de Investigación y Desarrollo Tecnológico (CENIDET), Interior Internado Palmira s/n, Col. Palmira, Cuernavaca, Morelos, México, vianey@cenidet.edu.mx

George Savii Mechanical Engineering Faculty, Technical University of Timişoara, Mihai Viteazu 1, 300222 Timişoara, Romania

France Sevšek University of Ljubljana, University College of Health Studies, Poljanska 26a, 1000, Ljubljana, Slovenia, france.sevsek@vsz.uni-lj.si

N.A. Shaikh National University of Computer and Emerging Sciences, ST-4, Sector-17/D, Shah Latif Town, Karachi, Pakistan

Z.A. Shaikh National University of Computer and Emerging Sciences, ST-4, Sector-17/D, Shah Latif Town, Karachi, Pakistan

Mohammad Hossein Sheikhi Department of Electrical Engineering, School of Engineering, Shiraz University, Shiraz, Iran; Shiraz Nanotechnology Research Institute, Shiraz, Iran

Yogesh Singh University School of Information Technology, Guru Gobind Singh Indraprastha University, Kashmere Gate, Delhi -06, India, ys66@rediffmail.com

Tudor Sireteanu Institute of Solid Mechanics, Romanian Academy, Bucureşti, Romania, siret@imsar.bu.edu.ro

Tomasz Śliwiński Institute of Control and Computation Engineering, Warsaw University of Technology, 00-665 Warsaw, Poland, tsliwins@ia.pw.edu.pl

E. Stergiou A.T.E.I. of Epirus, Department of Informatics and Telecommunications Technology, GR-471 00 Arta, Greece, ster@teiep.gr

Tatsuo Suzuki Gifu University, 1-1 Yanagido, Gifu, 501-1193, Japan, tasuzuki@gifu-u.ac.jp

Huma Tauseef Lahore College for Women University, Lahore, Pakistan

Z. Cihan Taysi Department of Computer Engineering, Yldz Technical University, 34349 Besiktas / Istanbul, Turkey, cihan@ce.yildiz.edu.tr

Virgil Teodor Department of Manufacturing Engineering, "Dunărea de Jos" University of Galaţi, Galaţi, 111 Domnească Street, zip code 800201, Romania, virgil.teodor@ugal.ro

Drazen Tomic Faculty of Electrical Engineering, Josip Juraj Strossmayer University of Osijek, Kneza Trpimira 2b, Croatia, drazen.tomic@etfos.hr

P. Tosaranon Department of Industrial Physics and Medical Instrumentation, Faculty of Applied Science, King Mongkut's Institute of Technology, North Bangkok, 1518 Pibulsongkram Road, Bangsue, Bangkok 10800 Thailand

Carlos M. Travieso Centro Tecnológico para la Innovación en Comunicaciones (CETIC), Dpto. de Señales y Comunicaciones, Universidad de Las Palmas de Gran Canaria, Campus de Tafira, E35017, Las Palmas de Gran Canaria, Spain, ctravieso@dsc.ulpgc.es

Philippe Trigano Heudiasyc UMR CNRS 6599, Université de Technologie de Compiègne, France

Hua-Lin Tsai Department of Industrial Technology Education, National Kaohsiung Normal University, Taiwan, kittyhl@gmail.com

Tien-Sheng Tsai Department of Industrial Technology Education, National Kaohsiung Normal University, Taiwan, mullis@mail.joseph.org.tw

Kuo-Hung Tseng Meiho Institute of Technology, Taiwan, ROC, ken@meiho.edu.tw

N. Tzioumakis Division of Telecommunications, Microwave Communications and Electromagnetic Applications Laboratory, Department of Electronics,

Technological Educational Institute (T.E.I.) of Crete, Chania Branch, Romanou 3, Chalepa, 73133 Chania, Crete, Greece, ivardia@chania.teicrete.gr

W. Ussawawongaraya Department of Industrial Physics and Medical Instrumentation, Faculty of Applied Science, King Mongkut's Institute of Technology, North Bangkok, 1518 Pibulsongkram Road, Bangsue, Bangkok 10800 Thailand

I. O. Vardiambasis Division of Telecommunications, Microwave Communications and Electromagnetic Applications Laboratory, Department of Electronics, Technological Educational Institute (T.E.I.) of Crete, Chania Branch, Romanou 3, Chalepa, 73133 Chania, Crete, Greece, ivardia@chania.teicrete.gr

Péter Várlaki Department of Chassis and Lightweight Structures, Budapest University of Technology and Economics, Bertalan Lajos u. 2. 7. em., 1111 Budapest, Hungary

D.C. Vasiliadis Department of Computer Science and Technology, University of Peloponnese, GR-221 00 Tripolis, Greece, dvas@uop.gr

Andreas Veglis Department of Journalism and MC, Aristotle University of Thessaloniki, 54006, Thessaloniki, Greece

Osslan Osiris Vergara Villegas Universidad Autónoma de Ciudad Juárez (UACJ), Avenida del Charro 450 Norte Ciudad Juárez Chihuahua, México, overgara@uacj.mx

Pofen Wang Department of Information Management, Southern Taiwan University of Technology, Taiwan, rxf26@mail.stut.edu.tw, pohfen@ma.ks.edu.tw

P. Woraratsoontorn Department of Industrial Physics and Medical Instrumentation, Faculty of Applied Science, King Mongkut's Institute of Technology, North Bangkok, 1518 Pibulsongkram Road, Bangsue, Bangkok 10800, Thailand

Ming-Chang Wu National United University, Taiwan, ROC

Hsieh-Hua Yang Department of Health Care Administration, Oriental Institute of Technology, 58, sec. 2, Szechwan Rd., Banciao City, Taipei Country 220, Taiwan, yansnow@gmail.com, FL008@mail.oit.edu.tw

Hung-Jen Yang Department of Industrial Technology Education, National Kaohsiung Normal University, Kaohsiung County, Taiwan, ROC, hjyang@nknucc.nknu.edu.tw, Hungjen.yang@gmail.com

A. Gokhan Yavuz Department of Computer Engineering, Yldz Technical University, 34349 Besiktas, Istanbul, Turkey, gokhan@ce.yildiz.edu.tr

Chia-Hung Yen Institute of Technological and Vocational Education, National Pingtung University of Science and Technology, Taiwan

Jui-Chen Yu National Science and Technology Museum, #720 Ju-Ru 1st Rd., Kaohsiung, Taiwan, ROC, raisin@mail.nstm.gov.tw

Adrian Zafiu Department of Electronics, Communications and Computers, University of Pitesti, Str. Targul din Vale, nr.1, Pitesti, Arges, Romania, adrian_zafiu@yahoo.com

Annely-Mihaela Zafiu CAS Arges, Aleea Spitalului Nr.1, Pitesti, Romania, annelyzafiu@yahoo.com

Drago Zagar Faculty of Electrical Engineering, Josip Juraj Strossmayer University of Osijek, Kneza Trpimira 2b, Croatia, drago.zagar@etfos.hr

Krista Rizman Žalik Faculty of Natural Sciences and Mathematics, University of Maribor, Slovenia

Jian Ying Zhang Department of Management, Monash University, 26 Sir John Monash Drive, Caulfield East, VIC 3145, Australia, jian.zhang@buseco.monash.edu.au

X.D. Zhuang WSEAS Headquarters, Agiou Ioannou Theologou 17-23, 15773, Zografou, Athens, Greece

S. Zimeras Department of Statistics and Actuarial-Financial Mathematics, University of the Aegean, G.R.832 00, Karlovassi, Samos, Greece, zimste@aegean.gr

Emmanouil Zoulias National Technical University of Athens, 9 Iroon Politechniou Str., 157 80 Athens, Greece, ezoulias@teemail.gr

Part I
Neural Networks and Applications

Neural networks represent a powerful data processing technique that has reached ripeness and broad application. When clearly understood and appropriately used, they are a compulsory component for the best use of the available data to build models, make predictions, process signals and images, recognize shapes, etc. This part is written by experts in the application of neural networks in different areas. The papers in this part cover the problems of the application of artificial neural networks such as handwriting knowledge based on parameterization for writer identification, identification surfaces family, neuro-fuzzy models and tobacco control, PNN for molecular-level selection detection, the automatic fine-tuning of artificial neural network parameters, a neuro-fuzzy network for supporting detection of diabetic symptoms, and empirical assessment of LR- and ANN-based fault prediction techniques

Part I
Neural Networks and Applications

Chapter 1
Handwriting Knowledge Based on Parameterization for Writer Identification

Carlos F. Romero, Carlos M. Travieso, Miguel A. Ferrer, and Jesús B. Alonso

Abstract This present paper has worked out and implemented a set of geometrical characteristics from observation of handwriting. In particular, this work has developed a proportionality index together with other parameters applied to handwritten words, and they have been used for writer identification. That set of characteristics has been tested with our offline handwriting database, which consists of 40 writers with 10 samples per writer. We have got a success rate of 97%, applying a neural network as classifier.

1.1 Introduction

Nowadays, advances in computer science and the proliferation of computers in modern society are unquestionable facts. But handwritten documents and personal writing styles continue to be of great importance.

Because of their wide use, many handwritten documents are subject to possible forgeries, deformations, or copies, generally for illicit use. Therefore, a high percentage of the routine work of experts and professionals in this field is to certify and to judge the authenticity or falsehood of handwritten documents (for example, testaments) in judicial procedures.

At present, two software tools are available for experts and professionals which are able to show certain characteristics of handwriting, but the experts have to use much time to reach their conclusions about a body of writing.

Therefore, these tools save neither time nor a meticulous analysis of the writing. They require the use of graph paper and templates for obtaining parameters (angles, dimensions of the line, directions, parallelisms, curvatures,

C.F. Romero (✉)
Centro Tecnológico para la Innovación en Comunicaciones (CETIC)
Dpto. de Señales y Comunicaciones, Universidad de Las Palmas de Gran Canaria,
Campus de Tafira, E35017, Las Palmas de Gran Canaria, Spain
e-mail: fabian_romero@ciudad.com.ar

N. Mastorakis et al. (eds.), *Proceedings of the European Computing Conference*, Lecture Notes in Electrical Engineering 27, DOI 10.1007/978-0-387-84814-3_1, © Springer Science+Business Media, LLC 2009

alignments, etc.). Too, they have to use a magnifying glass with graph paper for taking the measurements of angles and lines.

Writer identification is possible because the handwriting of each person is different, and everyone has personal characteristics. The scientific bases for this idea are from the human brain. If we try to do writing with the less skilful hand, there will be some parts or forms very similar to those used when writing with the skilful hand, because of the orders sent by the brain.

Generally, this effect is seen in handwriting because of two types of forces:

- Conscious or known, which can control one's own free will.
- Unconscious, which can escape the control of one's own free will. This type is divided into mechanical and emotional forces, the latter of which consist of feelings.

All people create their handwriting by means of their brains; and simultaneously the handwriting impulse, which manages the symbolism of space, adapts the dimensions of the writing into proportional forms, maintaining the size of the text—or a proportion of the natural size if the individual is forced to write in a reduced space.

Nowadays, handwriting identification is a great challenge because the research has not been as extensive as with identification based on fingerprints, hands, face, or iris (other biometric techniques); and this is mainly because the operation of the brain is very difficult to parameterize. On the other hand, the other techniques mentioned use widely researched biometric information.

Most of the techniques implemented until now offer information in the vertical and horizontal planes. In this present work, we have introduced a new parameter, the proportionality index, which projects in all directions, depending on the selected points.

The rest of the paper is organized in the following way. Section 1.2 contains a brief description of the structure of the system. In Sections 1.3 and 1.4 the creation of the database, and image preprocessing and segmentation of the words, respectively, are briefly described. In the following section the procedure for the extraction of the characteristics is explained. Section 1.6 contains the methods used for classification. And finally, in Section 1.7, the conclusion of this work is written up.

1.2 System Schedule

As the majority of the works up to now have proposed, the framework of the system of biometric recognition depends on the following basic phases (Fig. 1.1).

- Image preprocessing and segmentation: Preparation and modification of images, so that the module of segmentation produces the results desired. The segmentation separates the zones of interest (lines, words, or characters),

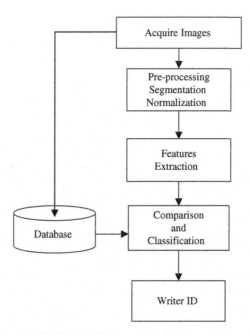

Fig. 1.1 System of writer identification

and it is key for the success or error of the following analysis (Feature
Extraction).

- Feature Extraction: These are qualitative and quantitative measures that
  permit obtaining a significant geometrical characterization of the style of
  writing, to differentiate writers from one another.
- Classification: A statistical analysis of the extracted characteristics is carried
  out, which permits comparison with the samples in our database, seeking the
  writer who possesses the most similarities.

1.3 Database

For the building of our database, we have used a paragraph of 15 lines from the
book *El Quijote de la Mancha*, in the Spanish language. With a sample this size,
writers can reveal their personal handwriting characteristics.

This database includes 40 writers, and each one has written this template
(the paragraph of 15 lines) 10 times. The size of the paper was DIN-A4
(297 × 210 mm). The sheet was written with a pen in black ink. Each writer in
our database had one week for doing the writing, which produced the effect of
multiple sessions.

The creation conditions of our database were normalized, with the same type
of paper, pen, and similar place of support (for doing the writing). In this way,

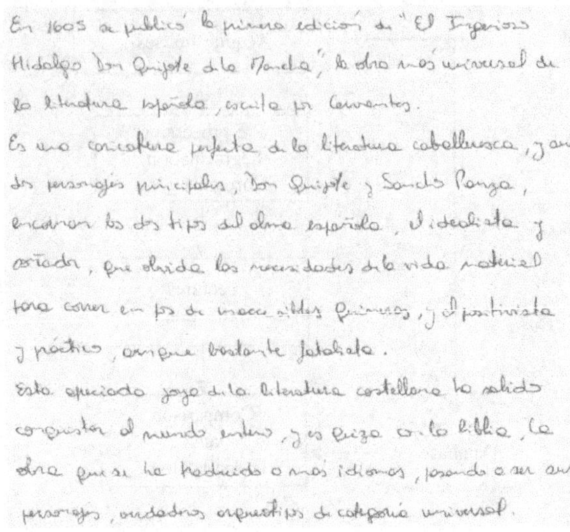

Fig. 1.2 A handwriting sample

our work was centred on the writing and the efficiency of the proposed parameters. In future work, we are going to change the rest of the variables.

The samples are scanned at 300 dpi, obtaining images in greyscale, with 8 bits of quantification (Fig. 1.2).

1.4 Image Preprocessing and Segmentation

The first step in image preprocessing consists of utilizing Otsu's method, which permits us to determine the necessary grey threshold value to carry out the binarization of the samples [1].

As a result of the binarization, in most cases, the line of writing remains with an irregular aspect. For that reason, another preprocessing is carried out, which permits us to smooth out the line, so that it remains well defined. It also eliminates the existing noise in the images after the scanning process.

As a previous step to the separation of connected words or components, the detection and elimination of punctuation marks (points, accents, and commas) is carried out.

Finally, we have segmented words, which compose the lines of writing (baselines), and they must establish the limits of each one of the words. For this estimation, the method of the "enclosed boxes" [2] is used, which provides us the coordinates that will permit segmentation of the words. The enclosed boxes are defined as the smallest rectangles that contain connections to the component.

1.5 Feature Extraction

In this step, a list is created containing the geometrical parameters of the different measures for analyzing documents. Afterwards, they will be measured with the samples obtained from the database.

So that the characteristics represent the style of writing, they should comply with the following requirement: the fluctuations in the writing of a person should be as small as possible, while the fluctuations among different writers should be as large as possible.

The extraction of the proportionality index is a new parameter in our system. The sample of the word is displayed in the window on the screen, and the operator, using the mouse, will mark the points of interest on the zoomed window. The selection of the points is random; that is to say, we have placed the most representative sites: they can be the ascending ones, the descending, the terminations, etc. The most representative points are shown in red, as in Fig. 1.3. On this same word, we have marked the same red points for each writer.

Next we unite the marked points (Fig. 1.3); each line formed between two points is considered a segment. Then we measure the length of each segment, obtaining a list of lengths (see Table 1.1).

Using this list, we calculate the average length and its variance, obtaining proportionality indices.

Unlike most of the characteristics, where the information is projected in the horizontal and vertical directions, the proportionality index offers information in all the directions.

This new characteristic is included in the list of the following characteristics already developed [3, 4]:

- length of the words,
- quantity of pixels in black,
- estimation of the widest one of the letters,
- height of the medium body of writing,
- heights of the ascending and descending portions,
- height relation between the ascending and medium body,
- height relation between the descending and medium body,

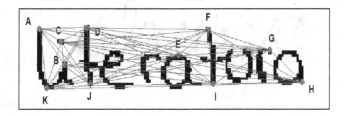

Fig. 1.3 Segments obtained when points are united

Table 1.1 List of distances of obtained segments, their average, and their variance

AB	14.95	CD	9.33	EJ	33.60		
AC	11.25	CE	45.37	EK	52.36		
AD	19.28	CF	57.13	FG	29.21		
AE	56.30	CG	84.93	FH	50.48		
AF	67.45	CH	104.20	FI	19.16		
AG	95.68	CI	62.05	FJ	48.02		
AH	115.23	CJ	20.73	FK	66.19		
AI	73.22	CK	17.16	GH	21.79		
AJ	31.08	DE	37.20	GI	26.46		
AK	20.64	DF	48.21	GJ	72.31		
BC	6.00	DG	76.43	GK	91.55		
BD	12.28	DH	96.16	HI	42.58		
BE	44.54	DI	54.52	HJ	89.46		
BF	57.31	DJ	19.56	HK	108.85		
BG	84.24	DK	24.27	IJ	46.96		
BH	102.82	EF	14.94	IK	66.35		
BI	60.35	EG	39.70	JK	19.39		
BJ	16.15	EH	58.96	**AVERAGE**	**48.63**		
BK	12.06	EI	18.08	**VARIANCE**	**30.09**		

- height relation between the descending and the ascending,
- height relation between the medium body and the width of the writing.

The quantity of black pixels and long words will give us an estimation of the dimension and thickness of the line, the width of the letters, and the height of the medium body. These are distinctive characteristics of the style of handwriting.

The estimation of the width of letters is carried out by seeking the row with the greatest amount of transition from black to white (0 to 1). The number of white pixels between each transition is counted, and this result is averaged.

In measuring the height of the medium body of the words, the goal is to determine the upper and lower baseline through maximum and minimum values, and to measure the distance among them (see Fig. 1.4).

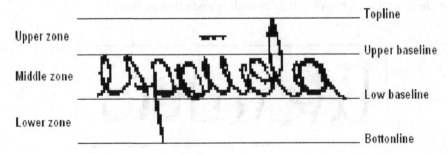

Fig. 1.4 Zones and baselines

To approach baselines of each word, it was decided to use the adjustment of minimum mean square error that is based on finding the equation (see expression 1.1) that can best be adjusted to a set of points "n" [5]. The equation is the following:

$$y = ax + b, \tag{1.1}$$

where the coefficients a and b determine the lineal polynomial regression by means of the following expressions:

$$a = \frac{n \sum_{i=1}^{n} x_i y_i - \left(\sum_{i=1}^{n} x_i \right) \left(\sum_{i=1}^{n} y_i \right)}{n \sum_{i=1}^{n} x_i^2 - \left(\sum_{i=1}^{n} x_i \right)^2} \tag{1.2}$$

$$b = \frac{\sum_{i=1}^{n} y_i - a \sum_{i=1}^{n} x_i}{n} \tag{1.3}$$

Those values of a and b, based on the coordinates of minimums or maximums detected in the contour of the word, are different baselines. Minimums are to approach the lower baseline and maximums the superior baseline.

1.6 Classification and Results

1.6.1 Neural Networks

In recent years several classification systems have been implemented using classifying techniques such as neural networks. These widely used neural networks techniques are well known in applications of pattern recognition. In particular, in this paper, they will be used for biometric systems (writer identification).

The perceptron of a simple layer establishes its correspondence with a rule of discrimination between classes based on lineal discrimination. However, it is possible to define discriminations for classes not lineally separable utilizing multilayer perceptrons, which are networks without the refreshing (feed-forward) of one or more layers of nodes between the input layer and the exit layer. These additional layers contain hidden neurons or nodes, which are directly connected to the input and output layer [6–8].

A neural network multilayer perceptron (NN-MLP) of three layers is shown in Fig. 1.5, with two layers of hidden neurons. Each neuron is associated with a weight and a bias; these weights and biases of the connections of the network will be trained to make their values suitable for distinguishing between the different classes.

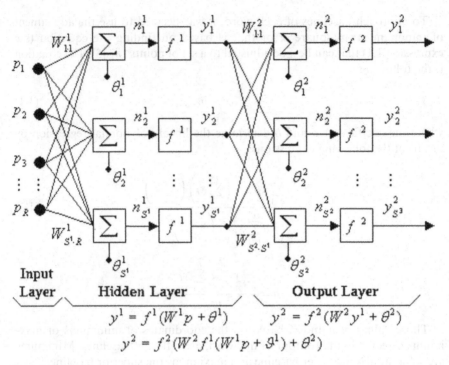

$$y^1 = f^1(W^1 p + \theta^1) \qquad y^2 = f^2(W^2 y^1 + \theta^2)$$
$$y^2 = f^2(W^2 f^1(W^1 p + \theta^1) + \theta^2)$$

Fig. 1.5 Multilayer perceptron

Particularly, the neural network used in the experiments is a Multilayer Perceptron (MLP) Feed-Forward with Back-Propagation training algorithm, and with only one hidden layer.

1.6.2 Experimental Methodology

Identification can be seen as a problem of the arrangement of N classes, in our case N writers. There are two variations of interest when the samples are compared: among the writings of a given writer, and between the writings of two different writers. The variation of writers among their own samples should be smaller than the variation among samples of two different writers.

For figuring a solution to this problem, the methodology of identification used was supervised classification. Therefore, we have a system with two modes, training and test mode.

For the training, we have used 50% of our database; the remainder is for carrying out the test mode. That is, words from five paragraphs have been chosen for training, and words from five other paragraphs for the test, since we have words from 10 paragraphs for each writer. Besides this, 34 words have been extracted from paragraphs, and there will be 34 words from the sample.

In order to calculate the proportionality index, we have worked on the word "Literatura" in each paragraph.

To calculate the rest of the characteristics, we have used 170 words (34 × 5 samples) in the process of training. The criterion for choosing the previous 34 words was their length, more than 5 letters—because with this length, they offer information more general than with a shorter size.

Experiments have been carried out five times, for which the results are shown by their average rates and their standard deviations. Each time, the training and test samples were chosen randomly.

As a classifier, we have used a feed-forward neural network (NN) with a back-propagation algorithm for training [6–8], in which the number of input units is given by the dimension of the vector of features, and the number of output units is given by the number of writers to identify. Also, we have experimented with different numbers of neurons in the hidden layer; and finally, 59 neurons were used because they presented the better results.

The average success rate of recognition is 90.15%, with a standard deviation of 2.38. But this result was improved using the method of the "more voted," in which we built a schedule with 15 neural networks (see Fig. 1.5), and we found a recognition rate of 96.5%, with a standard deviation of 0.

Finally, we increased the number of neuronal networks to 20, and we obtained a recognition rate of 97%, with a standard deviation of 0. This result has been compared with those of other authors, and our proposed work has obtained a good success rate (Fig. 1.6).

We have compared our results with our last work in [9], and with other works. In Table 1.2, it can be seen that, as with the proportionality index, the success rate has improved 2.34% with respect to [9], and with 10 writers more. Therefore, the proportionality index, along with the rest of the parameters, has a good grade of discrimination.

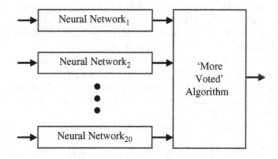

Fig. 1.6 Classification system with "more voted" algorithm, based on NN

Table 1.2 Comparison of results among different published methods vs. our work

Author	Number of writers	Success rates (%)
Said ([10])	40	95,00
Zois ([11])	50	92,50
Marti ([12])	20	90,70
Srihari ([13])	100	82,00
Hertel ([4])	50	90,70
Bensefia ([14])	150	86,00
Schomaker ([15])	100	95,00
Romero ([9])	30	94,66
This work	**40**	**97,00**

1.7 Conclusions

In this present work, we have proposed a set of characteristics for identifying a writer from offline handwritten text. We have used a back-propagation NN for the classification. And for improving results, we have implemented a "more voted" algorithm. The success rate is 97% for our database.

This result has been possible because of the use of a new parameter (proportionality index), together with other parameters (relations between heights and other geometrical measures). These parameters have introduced spatial information from words, and, in the case of the proportionality index, from all directions of words.

Acknowledgments The authors want to point out our acknowledgement to the members of the Signal and Communications Department from ULPGC-SPAIN, because of their help in building this database.

References

1. Otsu N (1979) A threshold selection method from gray-level histograms. IEEE Trans Syst Man Cybernet 9(1):62–66
2. Ha Jaekyu, Haralick RM, Phillips IT (1995) Document page decomposition by the bounding-box project. Proc IEEE Int Conf Doc Anal Recogn 2:1119
3. Romero CF, Travieso CM, Alonso JB, Ferrer MA (2007) Using off-line handwritten text for writer identification. WSEAS Trans Signal Process 3(1)
4. Hertel C, Bunke H (2003) A set of novel features for writer identification. In: Workshop on audio and video based biometric person authentication, pp 679–687
5. Chin W, Harvey M, Jennings A (1997) Skew detection in handwritten scripts. In: IEEE Region 10 Annual Conference, Speech and Image Technologies for Computing and Telecommunications 1:319–322
6. Bishop CM (1995) Neural networks for pattern recognition. Oxford University Press, Oxford
7. Hush DR, Horne BG (1993) Progress in supervised neural networks. IEEE Signal Process Mag 10(1):8–39

8. Juang BH, Rabiner LR (1992) Spectral representations for speech recognition by neural networks—a tutorial. In: Proceedings of the workshop neural networks for signal processing, pp 214–222
9. Romero CF, Travieso CM, Alonso JB, Ferrer MA (2006) Writer identification by handwritten text analysis. In: Proceedings of the 5th WSEAS international conference on system science and simulation in engineering, pp 204–208
10. Said HES, Peake GS, Tan TN, Baker KD (1998) Writer identification from non-uniformly skewed handwriting images. In: Proceedings of the 9th British machine vision conference, pp 478–487
11. Zois EN, Anastassopoulus V (2000) Morphological waveform coding for writer identification. Pat Recogn 33(3):385–398
12. Marti UV, Messerli R, Bunke H (2001) Writer identification using text line based features. In: Sixth international conference on document analysis and recognition, pp 101–105
13. Srihari S, Cha SH, Arora H, Lee S (2001) Individuality of handwriting: a validity study. In: Proceedings of the IEEE international conference on document analysis and recognition, Seattle (USA), pp 106–109
14. Bensefia A, Pasquet T, Heutte L (2005) Handwritten document analysis for automatic writer recognition. In: Electronic letters on computer vision and image analysis, pp 72–86
15. Schomaker L, Bulacu M (2004) Automatic writer identification using connected-component contours and edge-based features of uppercase western script. IEEE Trans Pat Anal Mach Intell 26(6):787–798

Chapter 2
Identification Surfaces Family

Virgil Teodor, Alexandru Epureanu, and Ciprian Cuzmin

Abstract In mechanical systems, the dimensional shape and the position deviations of the surfaces of the components represent very important attributes of
quality assessment. This is why technical specifications include a large number
of requirements regarding these attributes. At present the verification of these
requirements is based on the measurement of the coordinates of the points
belonging to the surface of the component. After the points of the coordinates
are obtained, the numerical model of the surface is fitted. Finally, the numerical
models are used to evaluate the actual dimensions of the features, to compare
these dimensions with the model dimensions, and to check for the tolerances.
Because of this there emerge some uncertainties regarding the dimensions, such
as the distance between two planes which are not actual parallel. This is why
there arises the need for grouping the component surfaces into families, for
obtaining cloud point coordinates for each surface, and for the coherent modeling of a family instead of the individual modeling of each surface. The quality
of the junction between two assemblies of components is given by the compatibility degree between surfaces belonging to one piece and the conjugate surfaces belonging to the other pieces which form the junction.

In this paper there are proposed two methods for geometric feature family
identification (using a genetic algorithm and using neural networks) for a better
evaluation of the deviations of the surfaces.

2.1 Introduction

In mechanical system manufacturing, the dimensional, shape, and position
deviations of the system components represent very important attributes in
quality assessment. For this reason technical specifications include a large
number of requirements regarding these attributes.

V. Teodor (✉)
Department of Manufacturing Engineering, "Dunărea de Jos" University of Galați,
Galați, 111 Domnească Street, Zip Code 800201, Romania
e-mail: virgil.teodor@ugal.ro

N. Mastorakis et al. (eds.), *Proceedings of the European*
Computing Conference, Lecture Notes in Electrical Engineering 27,
DOI 10.1007/978-0-387-84814-3_2, © Springer Science+Business Media, LLC 2009

At present these requirements are checked by measuring the coordinates of the points belonging to the component surface, using coordinates measuring machines or similar devices. A cloud of points for each of the explored surfaces is obtained. After this, the cloud of points is used to identify each surface individually, obtaining in this way a mathematical model of the surface with regard to a certain reference system.

Finally, the mathematical models of the surfaces are used for the actual evaluation of the feature dimensions, to compare these dimensions with the model dimensions, and to check the prescribed tolerances [1].

This method of inspecting the dimensions has the following shortcomings:

Firstly, the identification is done for each surface separately.

This is why a certain non-determination emerges, for example if the dimension is the distance between two plane surfaces. The mathematical models of the two planes may not be two parallel planes for which the distance is to be clearly defined, to give one of the simplest examples. There are many more complex cases.

Secondly, the target of geometry evaluation is the checking of the degree of compatibility between a group of surfaces which belongs to one piece and the conjugate group of surfaces which belongs to another piece that together form a junction, as, for example, in the case of bearing covers (see Fig. 2.1).

The tolerances assure a good enough superposition between the two surface groups (A-A', B-B', and C-C'). So it is more efficient to inspect the surface assembly ABC, fitting simultaneously three point clouds (gathered from A, B, and C) in order to obtain the numerical models of A, B, and C. The toleration of the minimum zone is made by limiting only certain form deviations of the assembly of surfaces. In this case the distance between A-C, the B surface diameter, the perpendicularity between A and B, and the parallelism between A and C are tolerated. These are the parameters of the surface form deviations.

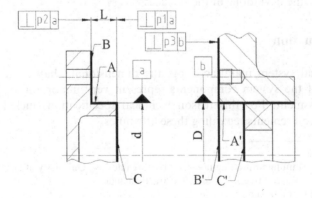

Fig. 2.1 Conjugate surfaces of junction

Thirdly, there may appear situations when the dimension of a surface refers to a reference system defined as regarding other surfaces of the piece. In this case, the modeling of the surface as the reference system modeling is important.

All of these reveal that the group surfaces should not be regarded separately.

The constituting of the family of geometric elements based on the functional relation between the components is needed. Moreover, the CAD models and also the numerical models should unitarily describe each family of geometric features.

In the literature there are presented methods which allow the evaluation of the deviations (statistical techniques [2], ants colony technique [3, 4], Grey theory [5], convex hull technique [6], the technique of finite differences [7], by Monte Carlo simulation, support vector regression [8], Chebyshev approximation [9], kinematic geometry [10], and a new approach to Newton's methods for nonlinear numerical minimization [11, 12]) only for the separate evaluation of various deviations which may appear at complex machining surfaces.

This paper intends to identify the family of geometric features by building a coherent numerical family model in order to better evaluate the geometrical compatibility of the conjugate surface families.

There are also presented two methods for the identification of the family of geometric features, based on the genetic algorithm approach and based on the neural network approach.

2.2 The Algorithm for the Identification of a Geometric Feature Family

As we have previously shown, in order to determine the machined surfaces' quality, these surfaces are inspected and the gathered points are processed.

The proposed algorithm presumes the followings steps:

1. Establishing that a surfaces group belongs to a piece, the surfaces of which will be in contact with another group of surfaces of the mechanical construction where the piece is mounted, forming a pair of surface assemblies. This pair of surface assemblies will be called a *topological structure*.
2. Gathering of the ordered points cloud from each of the surface pieces that form the topological structure.
3. Processing of these data so as to calculate the magnitude of the deviations which are tolerated in the technical specifications in order to determine the piece conformity with its CAD model.

For data processing there are proposed two methods: the genetic algorithm method and the neural network method. These methods ensure the in-cycle dimensional check with good precision and a minimum calculus effort.

The topological structure is established based on the criteria of form, dimension, and position restrictions for the structure elements.

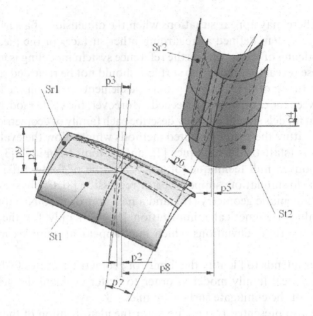

Fig. 2.2 Topological structure

We have to notice that a topological structure is not limited only on the surfaces machined in the current operation.

Each of the topological structure surfaces is characterized by its model and by the conformity parameters which describe the similitude between the actual piece surface and the model ($p_1, p_2, \ldots p_n$ parameters, see Fig. 2.2).

For each p_i parameter in the technical specification there is indicated a tolerance, representing the variation domain.

In Fig. 2.2 there is presented a topological structure selected by these criteria, which is composed of two theoretical surfaces S_{t1} and S_{t2}. Because of the machining errors, we can consider that the actual surfaces S_{r1} and S_{r2} were obtained, the position being determined by the p_1, p_2, p_3 and the p_4, p_5, p_6 parameters, respectively.

Assuming that in technical specifications the theoretical surface positions were restricted by the p_7, p_8, and p_9 parameters, it is interesting for us if these restrictions are within the tolerated field.

Each of the constraints imposed by form, position, and dimension, established when designing, will become a conformity parameter of the topological structure.

In the part-program there will be implemented a measuring phase. For this phase the trajectory of the stylus is established in order to gather the cloud of points. Because the measuring device stylus has a spherical end, the trajectory of the center of this sphere needs to be equidistant to the theoretical surface to be explored (see Fig. 2.3).

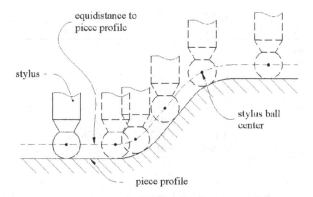

Fig. 2.3 Surface exploring trajectory

The points cloud will be processed using either the genetic algorithm or the neural network method.

2.3 The Identification of a Geometric Feature Family Using the Genetic Algorithm

We applied this method to identify a feature family composed of two parallel planes situated at a 10 mm distance.

In order to simulate the inspection of the actual surface, we used software which can generate a mesh with 25 points for each plane.

We used the plane equation:

$$n_1 \cdot x + n_2 \cdot y + n_3 \cdot z + p_1 = 0. \tag{2.1}$$

The software gives values on the x and y axis in domain $(-100 \ldots 100)$ with a step of 50 mm. For each couple (x,y) from (1), the value for z is calculated:

$$z = -\frac{n_1 \cdot x + n_2 \cdot y + p_1}{n_3}. \tag{2.2}$$

In this way we simulated the gathering of 50 points, 25 from each of the planes under discussion.

As an objective function, we considered the sum of the distances from each point to the theoretical planes P_1 and P_2:

$$P_1 : z_1 = p_1; P_2 : z_2 = p_2. \tag{2.3}$$

In order to minimize the objective function, we applied the least squares method.

Table 2.1 Numerical results

Parameter	Actual	Calculated	Relative error [%]
n_1	0.01	0.010208	−2.08
n_2	0.015	0.015312	−2.08
n_3	0.9	0.91881	−2.09
p_1	0.001	0.001022	−2.18
n_4	0.012	0.012023	−0.19167
n_5	0.014	0.014025	−0.17857
n_6	1.1	1.102	−0.18182
p_2	10.001	10.019	−0.17998

In this way the objective function becomes:

$$d = \sum_{i=1}^{25} (n_1 x_i + n_2 y_i + n_3 z_i + p_1)^2 + \sum_{i=26}^{50} (n_4 x_i + n_5 y_i + n_6 z_i + p_2)^2. \qquad (2.4)$$

For solving software we used the Genetic Algorithm Toolbox from Matlab, Version 7.

In Table 2.1 are shown the results obtained when applying the algorithm for two parallel planes.

The Genetic Algorithm Toolbox options were set to: Population Size = 20; Elite Count = 2; Crossover Fraction = 0.8; Migration Interval = 20; Migration Fraction = 0.2.

2.4 The Identification of a Geometric Feature Family Using the Neural Network

In order to model a topological structure by a neural network, one was generated by simulating a database filled in with coordinates of points belonging to the topological structure surfaces.

Each record of the database contains the x, y, z coordinate points from the cloud of points corresponding to a certain configuration $p_1, p_2, \ldots, p_n$ of the set of parameters.

This database is preprocessed by calculating the values of function:

$$f(x, y, z, p_1, \ldots p_n) = d \qquad (2.5)$$

in each of the generated points, where d is the distance from the point to the surface.

After this, each topological structure is neurally modeled with d from equation (2.5) as input data and with $p_1, p_2, \ldots p_n$ parameter values as output data.

2.4.1 Database Generation

On each surface belonging to the topological structure there are generated points corresponding to the established measuring trajectories. These measuring trajectories will be equidistant to the theoretical piece profile in order to simulate the positions of the stylus spherical surface center during the measuring process.

The coordinates of the generated points have a coordinate transformation from their own theoretical surface reference system to a reference system moved with the $p_1, p_2, \ldots p_6$ parameters set (see Fig. 2.4).

Fig. 2.4 shows:

$X_1 Y_1 Z_1$ is the theoretical surface of its own reference system;
XYZ—moved reference system;
$(p_1, p_2, p_3, p_4, p_5, p_6)$—conformity parameters set.

The coordinate transformation is given by the equation:

$$X_1 = \omega \cdot X + T, \tag{2.6}$$

where

$$T = \quad \| p_1 \quad p_2 \quad p_3 \|^T$$
$$\omega = \quad \omega_1^T(p_4) \cdot \omega_2^T(p_5) \cdot \omega_3^T(p_6) \tag{2.7}$$

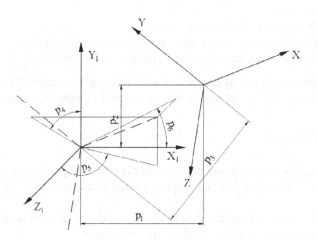

Fig. 2.4 Reference systems

or, after its development:

$$
\begin{Vmatrix} X_1 \\ Y_1 \\ Z_1 \end{Vmatrix} = \begin{Vmatrix} 1 & 0 & 0 \\ 0 & \cos p_4 & -\sin p_4 \\ 0 & \sin p_4 & \cos p_4 \end{Vmatrix} \cdot \begin{Vmatrix} \cos p_5 & 0 & \sin p_5 \\ 0 & 1 & 0 \\ -\sin p_5 & 0 & \cos p_5 \end{Vmatrix} \cdot \\
\cdot \begin{Vmatrix} \cos p_6 & \sin p_6 & 0 \\ \sin p_6 & \cos p_6 & 0 \\ 0 & 0 & 1 \end{Vmatrix} \cdot \begin{Vmatrix} X \\ Y \\ Z \end{Vmatrix} + \begin{Vmatrix} p_1 \\ p_2 \\ p_3 \end{Vmatrix}
\tag{2.8}
$$

If between the surface in discussion and the reference surface there are some constraints, these will generate an additional number of conformity parameters, and a new coordinate transformation is needed on type (2.6) between the surface of its own reference system and the reference system associated with the reference surface.

For each of the generated points there is calculated the value of $f(X_1, Y_1, Z_1, p_1, \ldots p_n)$ function, obtained by filling in the general equation of the theoretical surface with the point coordinates.

In this way there is obtained a database with M records and N fields for each record, M and N being given by:

$$
M = \prod_{i=1}^{m} v_i
$$
$$
N = \sum_{i=1}^{n} k_i,
\tag{2.9}
$$

where:

m is the number of the conformity parameters corresponding to the topological structure;

v_i—number of values for p_i parameter;

n—number of surfaces which make up the topological structure;

k_i—number of generated points on each theoretical surface.

The database obtained in this way is used as an input matrix for the neural network training. The output matrix will consist of the conformity parameters set.

2.4.2 Application

As an application we used this neural network modeling to identify a feature family composed of two parallel planes situated at a distance of 10 mm from each other.

In order to simulate the inspection of the actual surface we used software which generates a mesh with 25 points for each plane.

The theoretical planes are moved by coordinate transformation:

$$X_1 = \omega(\alpha_1) \cdot X \tag{2.10}$$

for the first plane, and

$$X_2 = \omega(\alpha_1 + \alpha_2) \cdot X + T \tag{2.11}$$

for the second plane.

In equations (2.10) and (2.11) ω and T are:

$$\omega(\alpha_1) = \begin{Vmatrix} 1 & 0 & 0 \\ 0 & \cos\alpha & -\sin\alpha \\ 0 & \sin\alpha & \cos\alpha \end{Vmatrix}$$

$$\omega(\alpha_1 + \alpha_2) = \begin{Vmatrix} 1 & 0 & 0 \\ 0 & \cos(\alpha_1 + \alpha_2) & -\sin(\alpha_1 + \alpha_2) \\ 0 & \sin(\alpha_1 + \alpha_2) & \cos(\alpha_1 + \alpha_2) \end{Vmatrix}. \tag{2.12}$$

$$T = \begin{Vmatrix} 0 & 0 & Z_0 \end{Vmatrix}^T$$

The software gives values on the X and Y axes in the domain $(-100\ldots100)$ with step 50 mm. In this way we simulate the gathering of 50 points, 25 from each plane taken into discussion.

For the neural modeling we used the NNMODEL software, Version 1.400.

In Table 2.2 are shown the results obtained by interrogating the network when the parameters belong to the set of training values.

In Table 2.3 are presented the results obtained when interrogating the network with parameters different from the set of training values.

The Neural Network options were set to: Hidden Neurons $= 50$; Maximum Training Count $= 10,000$; Fixed number of Hidden Neurons $=$ True; Training Method $=$ Standard BEP.

Table 2.2 Numerical results

Parameter	Actual	Calculated	Relative error [%]
α_1	−0.01047	−0.01036	1.069519
α_2	−0.01047	−0.0104	0.687548
Z_0	9.4	9.402	−0.02128

Table 2.3 Numerical results

Parameter	Actual	Calculated	Relative error [%]
α_1	−0.01396	−0.01388	0.594428
α_2	−0.01396	−0.0139	0.451192
Z_0	9.8	9.80307	−0.03133

2.5 Conclusion

For the sake of simplicity we have presented the simplest case, but the method was applied to more complex topological structures made up of plane and cylindrical surfaces and for various deviations (perpendicularity, parallelism, distances, etc.).

The results revealed that the differences between the deviations of exact values and the values obtained by applying the two methods are not larger than 5%.

The calculus time is longer for the genetic algorithm than for the neural network in the case of complicated topological structures (structures with more than 2 or 3 surfaces).

The numerical model built for a surface family as a whole is coherent and better describes the geometrical compatibility of conjugate surface families compared to the individual modeling of each surface.

References

1. Zhu LM, Xiong ZH, Ding H, Xiong YL (2004) A distance function based approach for localization and profile error evaluation of complex surface. J Manuf Sci Eng 126:542–554
2. Barcenas CC, Griffin PM (2001) Geometric tolerance verification using superquadrics. IIE Trans 33:1109–1120
3. Baskar N, Saravanan R, Asokan P, Prabhaharan G (2004) Ants colony algorithm approach for multi-objective optimization of surface grinding operations. Int J Adv Manuf Technol 23:311–317, DOI 10.1007 0/s 00170–002-1533–6
4. Li YY, Li LX, Wen QY, Yang YX (2006) Data fitting via chaotic ant swarm. Springer-Verlag Berlin Heidelberg, ICNC 2006, Part II, LNCS 4222, pp 180–183
5. Lin ZC, Lin WS (2001) The application of grey theory to the prediction of measurement points for circularity geometric tolerance. Int J Adv Manuf Technol, DOI 17:348–360
6. Orady E, Li S, Chen Y (2000) Evaluation of minimum zone straightness by a nonlinear optimization method. J Manuf Sci Eng 122:795–797
7. Gosavi A, Phatakwala S (2006) A finite-differences derivative-descent approach for estimating form error in precision-manufactured parts. J Manuf Sci Eng 128:355–359
8. Prakasvudhisarn C, Trafalis TB, Raman S (2003) Support vector regression for determination of minimum zone. Trans ASME 125:736–739
9. Zhu L, Ding Y, Ding H (2006) Algorithm for spatial straightness evaluation using theories of linear complex Chebyshev approximation and semi-infinite linear programming. J Manuf Sci Eng 128:167–174
10. Zhu LM, Ding H (2002) Application of kinematic geometry to computational metrology: distance function based hierarchical algorithms for cylindricity evaluation. Int J Mach Tools Manuf 43:203–215
11. Tucker T, Kurfess T (2002) Newton methods for parametric surface registration. Part I: Theory. Elsevier Science, Comput-Aid Des 35:107–114
12. Tucker T, Kurfess T (2002) Newton methods for parametric surface registration. Part II: Experimental validation. Elsevier Science, Comput-Aid Des 35:115–120

Chapter 3
Neuro-Fuzzy Models and Tobacco Control

Sonja Petrovic-Lazarevic and Jian Ying Zhang

Abstract This paper presents neuro-fuzzy models applications appropriate for tobacco control: the fuzzy control model, Adaptive Network-Based Fuzzy Inference System, Evolving Fuzzy Neural Network models, and EVOlving POLicies. We propose further the use of Fuzzy Causal Networks to help tobacco control decision makers develop policies and measure their impact on social regulation.

3.1 Introduction

Given that tobacco smoking is a largely preventable cause of death [13], the attempts to control tobacco usage have become part of contemporary government policy in many countries [16]. Tobacco control as social regulation is a public policy to reduce the incidence of smoking-related ill health and premature death. As such, tobacco control plays an important role in minimising the negative effects of tobacco use [5].

Governments, supported by medical evidence, antismoking social movements, and interest groups, are increasing their activities with regard to regulations concerning tobacco control. Development of any government policy, including smoking control policy, observed as a decision making process, is primarily supported by supplying data and information to decision makers. Decision support systems, of which neuro-fuzzy systems are one [7], are not often used in developing government regulations [8]. Very little academic research in this area, using either fuzzy logic or neuro-fuzzy system support, has been published. According to a literature review, the fuzzy logic approach has been implemented in energy policy planning [9], security policies [10], and health policies [13]. It seems, however, that there is no evidence in the literature

S. Petrovic-Lazarevic (✉)
Department of Management, Monash University, 26 Sir John Monash Drive,
Caulfield East, VIC 3145, Australia
e-mail: sonja.petrovic-lazarevic@buseco.monash.edu.au

N. Mastorakis et al. (eds.), *Proceedings of the European*
Computing Conference, Lecture Notes in Electrical Engineering 27,
DOI 10.1007/978-0-387-84814-3_3, © Springer Science+Business Media, LLC 2009

of using neuro-fuzzy systems in government social policy regulations, apart from [1, 5, 11, 12]. Highly commended preliminary research used both fuzzy control and neuro-fuzzy modelling to identify the factors influencing the effectiveness of measures to prevent tobacco smoking. That work provides a solid basis for evaluating the effectiveness of the reformed social regulatory measures affecting tobacco smoking.

With this paper we want to emphasize that the decision support systems based on neuro-fuzzy models help to improve the tobacco control social regulation process in order to either quit or minimise negative effects caused by cigarette smoking.

3.2 What Is Well Known

As a reasoning process, decision making leads to the selection of one course of action from among several considerations [18]. Governmental decisions' alternatives involve a large number of considerations. In the decision process they may not all be taken into consideration, with some of them being understood as more important than others. That understanding may be influenced by the different interest groups and therefore based on intuition, subjective judgment, and other nonobjective ways of selecting a decision alternative. By making the decision process as objective as possible, the use of decision support systems can significantly help in identifying the right solution to a given problem [2].

Because the decision making processes are applicable in different domains, the concept of decision support systems is differently interpreted [7, 18, 9, 8]. For the purpose of this paper we define decision support systems as any types of information systems that support decision making [14].

Both fuzzy-logic and neuro-fuzzy modelling contribute to the improvement of decision processes in tobacco control. Fuzzy logic is a family of methodologies used in the analysis of incomplete, imprecise, or unreliable information. The tobacco control social regulation field is full of imprecise, incomplete, and unreliable information that comes from two different interest groups. One comprises health improvement oriented groups, the other comes from the tobacco, hospitality, and entertainment industries [13]. Fuzzy logic enables rough reasoning, in which the rules of inference are approximate rather than exact [4]. Fuzzy (if...then) rules represent knowledge concerning the relationships between items. The accuracy of each rule is determined by membership functions [5].

The existing computational methods in tobacco control have been applied to investigate tobacco smoking policy regulations in Australia [5, 1, 12, 11]. The models, comprising two groups, neuro-fuzzy systems and evolutionary algorithms, were used to identify the factors influencing the effectiveness of measures to prevent tobacco smoking by people under 18 years of age.

The first group of methods, neuro-fuzzy systems, comprises both fuzzy logic and neural networks.

3.2.1 Fuzzy Control Model

The fuzzy control model implements "soft linguistic variables on a continuous range of truth values which allows intermediate values to be defined between conventional binary" [3] and is extensively used in artificial intelligence programs [5].

The fuzzy control model has been applied in estimating the type of social regulation of tobacco control in Australia. The model comprises the following variables: compliance rate by retailers' obedience, maximum enforcement according to protocol, and enforcement community education. Variables were presented in a membership form expressing explicit expert systems knowledge. If...then enforcement rules were introduced following the fuzzy control procedure. The accuracy of the model was tested with the data from local government areas in Melbourne, Australia. It indicated the influence of the different variables in different municipalities.

The model has limitations, because it only covers the explicit knowledge based on social policies and procedures. It does not reflect the tacit, indirect knowledge of the community based on local ethics and norms that can significantly reduce adolescent smoking rates. The model also does not provide government representatives with the answer to what extent to concentrate on available social regulation measures in anticipating enforcement efforts.

In spite of its drawbacks, the model demonstrates an estimate of the outcomes of social regulation, given its formal provision of the social regulation regime [5]. As such, it represents a first attempt to support tobacco control decision processes.

3.2.2 Neural Networks Models

The neuro-fuzzy modelling is a type of artificial intelligence program based on neural networks and fuzzy models [12]. "A neural network is an information processing concept that is inspired by the way biological nervous systems process information. The key element of this concept is the novel structure of the information processing system composed of a large number of highly interconnected processing elements (neurones) working in unison to solve specific problems" [15]. Neural networks combine "simple processing elements, high degree of interconnection, simple scalar messages, and adaptive interaction between variables" [10]. Neuro-fuzzy modelling enables handling imprecision and uncertainty in data and refining them by a learning algorithm [3]. It creates fuzzy rules in easy-to-comprehend linguistic terms [11].

With neuro-fuzzy modelling, the derivation of if...then rules and corresponding membership functions depends heavily on a priori knowledge about the system under consideration. The system can utilize human expertise by storing its essential components in rule bases and databases, and perform fuzzy reasoning to infer the overall output value. However, since there is no systematic way to transform the experiences or knowledge of human experts to the knowledge base, the adaptive network-based fuzzy inference system (ANFIS) and evolving fuzzy neural network (EFuNN) models were introduced to apply the neuro-fuzzy support of knowledge management in social regulation [12]. Thus, the explicit knowledge was based on social policies and procedures to reduce smoking among youngsters, and the tacit knowledge was expressed through the applied membership functions. Empirical results showed the dependability of the proposed techniques. Simulations were done with the data provided for the local government areas. Each data set was represented by three input variables and two output variables. The input variables considered were: compliance rate by retailers, enforcement according to protocol, and community education. The corresponding output variables were compliance rate by retailers, and compliance rate by retailers projected as the estimated rate of smoking uptake by minors. ANFIS performed better than EfuNN in terms of performance error. EfuNN performed approximately 12 times faster that ANFIS. Depending on governmental requests, it is possible to compromise between performance error and computational time.

Important disadvantages of ANFIS and EfuNN include the determination of the network parameters like number and type of membership functions for each input variable, membership functions for each output variable, and the optimal learning parameters.

Both ANFIS and EfuNN approaches represent a systematic way to transform experiences of prior tobacco control regulations into a knowledge base. As such, in spite of the limitations, they can provide useful information for helping tobacco control decision makers have sufficient information and minimise nonobjective ways of selecting a decision alternative.

3.2.3 EVOlving POLicies

With modelling comprising both tacit and explicit knowledge, a selection of optimal parameters may be formulated as an evolutionary search. This makes the neuro-fuzzy systems fully adaptable and optimal, according to government representatives' requests, by providing the answer to what extent to concentrate on available social regulation measures in anticipating smoking enforcement efforts.

Evolutionary algorithms transform a set of objects, each with an associate fitness value, into a new population using operations, based on Darwinian principles of reproduction and survival of the fittest, and naturally occurring genetic operations. The evolutionary algorithms learning technique can optimize human knowledge from the database [17]. In particular, the evolutionary algorithms technique may be helpful in the cases where expert knowledge is explained by a natural language or written words. Its usefulness is in encoding the fuzzy rules of the method of automatic database learning in the fuzzy control and neural networks learning models. It also minimizes the number of rules by including only the most significant ones [6]. The EvoPol (EVOlving POLicies), an evolutionary computation technique, was used to optimize the if…then rules to support governmental policy analysis in restricting recruitment of smokers. The proposed EvoPol technique is simple and efficient when compared to the neuro-fuzzy approach. It is useful in indicating to decision makers when to choose specific social regulation measures to control tobacco use. However, EvoPol attracts extra computational costs due to the population-based hierarchical search process [1].

All created and tested models are aimed at adolescents. Although each one of them has the potential to be further developed to include all tobacco users and therefore to minimise the negative effects of tobacco use, it seems that the knowledge-based models provide more sufficient information to decision makers. Consequently, we propose a new learning technique; the fuzzy causal networks (FCNs).

3.3 The Novelty of the Paper

FCN is a dynamic networked system with feedback [20]. It evolved from a cognitive map and has been used as a qualitative tool for representing causality and dynamic information modelling. In any FCN there are three kinds of elements, namely the concepts, the causal relationships between concepts, and the effects of one concept influencing the other. By convention, we use vertex to represent the concept, directed arc to represent the causal relationship between two concepts, and numerical values in $[-1, 1]$ associated with the directed arc to represent the effect of one concept on another. For each vertex, we assign a state value, which is quantified as a real number, to measure the degree of occurrence of a fuzzy event at a discrete time t. At any time, when a vertex receives a series of external stimuli, its state value is updated at the next time according to a state-transition function. Once constructed, the fuzzy causal network allows us to perform a qualitative simulation of the system and experiment with the model [19].

FCN is useful in tobacco control, since at the national/local government level within any country, all the factors involved in tobacco control, such as smokers, nonsmokers, researchers, doctors, advocates, tobacco industries, and national/local economies, form a discrete dynamic system. Also, these factors interact and influence the effectiveness of tobacco control policies. So in nature the complex tobacco control system can be regarded as a dynamic networked system with feedback. In the application of FCN in tobacco control, the vertex represents a fuzzy event, such as smokers' behaviour and the tobacco industry's response to tobacco control policies. According to scientific evidence, expert opinion, and the nature of the tobacco market, we identify and choose some suitable vertex state values at a specific time as an initial condition. Also, we assign a state transition function for each vertex. After constructing a discrete dynamic system for the purpose of tobacco use control, we perform a qualitative simulation for this system. As a result, we can provide knowledge discovery and decision support to decision makers for tobacco control policy planning and development.

3.4 Conclusion

Changing tobacco control policies based on adequate information no doubt can contribute to minimising negative effects of tobacco use in any society. Therefore, attempts to use the decision support in the form of neuro-fuzzy models in tobacco control should be welcomed by each government. In this paper we have pointed out the theoretical contributions to this important field of social regulation, highlighting the advantages and disadvantages of each tobacco control method created and tested. We have also emphasized the use of Fuzzy Causal Networks to further help improve the governmental decision making processes by including all factors involved in tobacco control.

References

1. Abraham A, Petrovic-Lazarevic S, Coghill K (2006) EvoPol: A framework for optimising social regulation policies. Kybernetes 35(6):814–824
2. Alter SL (1980) Decision support systems: current practice and continuing challenges. Addison-Wesley, Reading, MA, USA
3. Bezdek JC, Dubois D, Prade H (1999) Fuzzy sets in approximate reasoning and information systems. Kluwer Academic Publishers, Boston
4. Carlson C, Fedrizzi M, Fuller R (2004) Fuzzy logic in management. Kluwer Academic Publishers, Boston
5. Coghill K, Petrovic-Lazarevic S (2002) Self-organisation of the community: democratic republic of anarchic utopia. In: Dimitrov V and Korotkich V (eds) Fuzzy logic: a framework for the new millenium. Springer-Verlag, New York, pp 79–93
6. Cordon O, Herrera F (1997) Evolutionary design of TSK fuzzy rule based systems using (μ, λ) evolution strategies. In: Proceedings of the sixth IEEE international conference on fuzzy systems 1, Spain, pp 509–514
7. Finlay PN (1994) Introducing decision support systems. Blackwell, Oxford
8. Keen PGW (1980) Adaptive design for decision support systems. Association for Computing Machinery, electronic resource, Monash University Library
9. Keen PGW, Morton SS (1978) Decision support systems: an organizational perspective. Addison-Wesley, Reading, MA. USA
10. Nguyen HT, Walker EA (2000) A first course in fuzzy logic. Chapman & Hall/CRC, London
11. Petrovic-Lazarevic S, Abraham A, Coghill K (2002) Neuro-fuzzy support of knowledge management in social regulation. In: Dubois D (ed) Computing anticipatory systems: CSYS 2001 fifth international conference, American institute of physics, New York, pp 387–400

12. Petrovic-Lazarevic S, Abraham A, Coghill K (2004) Neuro-fuzzy modelling in support of knowledge management in social regulation of access to cigarettes by minors. Knowl-Based Syst 17(1):57–60
13. Petrovic-Lazarevic S, Coghill K (2006) Tobacco smoking policy processes in Australia. In: Ogunmokun, G, Gabbay R, Rose J (eds) Conference proceedings, 2nd biennial conference of the academy of world business, marketing and management development 2(1):535–550
14. Power J (1997) What is a DSS? The on-line executive. J Data-Intensive Decision Supp 1(3)
15. Stergiou C, Siganos D () Neural networks. http://www.doc.ic.ac.uk/~nd/surprise_96/journal/vol4/cs11/report.html (Accessed 7 March 2007)
16. Studlar DT (2002) Tobacco control. Broadview Press, Canada
17. Tran C, Lakhami J, Abraham A (2002) Adaptive database learning in decision support systems using evolutionary fuzzy systems: a generic framework. In: First international workshop on hybrid intelligent systems, Adelaide. Springer Verlag, Germany, pp 237–252
18. Turban E, Aronson JE (2001) Decision support systems and intelligent systems. Prentice Hall, Englewood Cliffs, NJ, USA
19. Zhang JY, Liu ZQ, Zhou S (2006) Dynamic domination in fuzzy causal networks. IEEE T Fuzzy Syst 14(1):42–57
20. Zhou S, Liu ZQ, Zhang JY (2006) Fuzzy causal networks: general model, inference and convergence. IEEE T Fuzzy Syst 14(3):412–420

Chapter 4
PNN for Molecular Level Selection Detection

Krzysztof A. Cyran

Abstract Contemporary population genetics has developed several statistical tests designed for the detection of natural selection at the molecular level. However, the appropriate interpretation of the test results is often hard. This is because such factors as population growth, migration, and recombination can produce similar values for some of these tests. To overcome these difficulties, the author has proposed a so-called multi-null methodology, and he has used it in search of natural selection in ATM, RECQL, WRN, and BLM, i.e., in four human familial cancer genes. However, this methodology is not appropriate for fast detection because of the long-lasting computer simulations required for estimating critical values under nonclassical null hypotheses. Here the author presents the results of another study based on the application of probabilistic neural networks for the detection of natural selection at the molecular level. The advantage of the proposed method is that it not so time-consuming and, because of the good recognition abilities of probabilistic neural networks, it gives low decision error levels in cross validation.

4.1 Introduction

Population genetics in the post-genomic age is armed with quite a number of statistical tests [7, 8, 11, 16] whose purpose is to detect signatures of natural selection operating at the molecular level. However, because of such factors as recombination, population growth, and/or population subdivision, the appropriate interpretation of the test results is very often troublesome [13]. The problem is that these departures from the selectively-neutral classical model (i.e., a model with a panmictic, constant size population with no recombination) can produce results for some of these tests similar to results produced by the existence of natural selection.

K.A. Cyran (✉)
Institute of Informatics, Silesian University of Technology, Gliwice, Poland

N. Mastorakis et al. (eds.), *Proceedings of the European Computing Conference*, Lecture Notes in Electrical Engineering 27, DOI 10.1007/978-0-387-84814-3_4, © Springer Science+Business Media, LLC 2009

Nevertheless, since the time of Kimura's famous book [12] until the present, geneticists have been searching for signatures of natural selection, treating the model of neutral evolution at the molecular level proposed by Kimura as a convenient null hypothesis, which is not fulfilled for particular loci under detectable selection. Without going into the details of the hot debate which took place when Kimura's book was published, let us only stress that after a short period it was widely accepted as a theory which does not contradict the existence of natural selection at the molecular level. It only states that selection cannot be the major determinant for genetic variability because of the too large reproductive cost required by it. By moving the emphasis from selective forces to random genetic drift and neutral mutations, the neutral theory of molecular evolution gave birth to the above mentioned non-neutrality tests [7, 8, 11, 13, 16], which treat this theory as a null model, and in which statistically significant departures from it, discovered in the loci under study, can be interpreted in favor of natural selection. The existence of a rare, positive selection has been confirmed, for example, in an ASPM locus that contributes to the size of the brain in primates [6, 17].

An interesting example of another, not so frequent type of selection, called balancing selection, has been detected by the author in ATM and RECQL loci. To overcome serious interpretation difficulties while searching for selection in ATM, RECQL, WRN, and BLM, i.e., in four human familial cancer genes, the author has proposed the idea of a so-called multi-null methodology (part of this methodology was published in an informal way in [3, 4]). However, this methodology is not appropriate for fast detection because of the long-lasting computer simulations required for estimating critical values under nonclassical null hypotheses.

Yet, armed with reliable conclusions about balancing the selection at ATM and RECQL and no evidence of such selection at WRN and BLM, after a time-consuming search with the use of computer simulations, the author has proposed the use of machine learning methodology, based only on the knowledge of critical values for classical null hypotheses. Fortunately, critical values for classical nulls are known for all proposed non-neutrality tests, and therefore the outcomes of such tests can be used as inputs for artificial intelligence classifiers without additional computer stochastic simulations of alternative models. The results of the application of rough set-based theory for knowledge acquisition and processing were published in [2]. In the current paper, the author presents the results of another study, based on the application of probabilistic neural networks (PNN) for the detection of natural selection at the molecular level. The advantage of the method proposed here is that it not so time–consuming, and because of good the recognition abilities of probabilistic neural networks it gives low decision error levels in cross validation (see results).

4.2 Problem Formulation and Methodology

4.2.1 Genetic Material

In this study the single nucleotide polymorphisms (SNP) data were used. Such SNPs form haplotypes with frequencies estimated by the EM algorithm [14], useful in the investigation of genetic diversity as well as complex genetic disease associations. Blood samples for this study were obtained from the residents of Houston, TX, belonging to four major ethnic groups: Caucasians (Europeans), Asians, Hispanics, and African-Americans.

The three out of four genes used in the analysis were human helicases RECQL, Bloom's syndrome (BLM) [10] and Werner's syndrome (WRN) [5], coding enzymes involved in various types of DNA repair. The fourth locus analyzed was ataxia telangiectasia mutated (ATM) [1, 9, 15]. The ATM gene product is a member of a family of large proteins implicated in the regulation of the cell cycle and the response to DNA damage.

For the detection of statistically meaningful departures from Kimura's neutral model of evolution, seven different tests were used: Kelly's (1997) Z_{nS} test, Fu and Li's (1993) F^* and D^* tests, Tajima's (1989) T test (in order to avoid the name conflicts we use here the nomenclature proposed by Fu [8] and Wall [16], but originally Tajima T was called D test), Strobeck's S test, and finally Wall's (1999) Q and B tests. The reader interested in definitions of these test statistics should refer to the original works of the inventors; or, if only a brief form is required, the Cyran et al. study [4] may by sufficient.

4.2.2 PNN-Based Methodology

In order to interpret the outcomes of the battery of seven tests mentioned, the machine learning technique was applied. As was mentioned in the introduction, the problem with interpretation is caused by various actual demographic factors which are not incorporated into the classical neutral model. Therefore a simple inference based on the test results is not possible, and more complex methods (like the multi-null methodology which the author is working on) are required.

Since in the pilot studies [3, 4] some of the ideas of complex multi-null methodology were applied to get answers about balancing selection in the four genes considered, the author also obtained considerably valid labels (balancing selection or no evidence of such selection) for given combinations of tests results computed with the assumption of the classical null hypothesis. The goal of the current paper is to show that the information preserved in these test results (even computed without taking into account factors like population growth, recombination, and population substructure) is valuable enough to lead to reliable inferences.

As a tool for this study, a probabilistic neural network was used. It is a specialized radial basis function (RBF) network, applicable almost exclusively for problems of classification in probabilistic uncertainty models. The network generates as its outputs likelihood functions $p(x|C_j)$ of input vectors x belonging to a given class C_j. One should notice that the likelihood functions also comprise random abstract classes defined in a probabilistic uncertainty model.

On the other hand, the likelihood function, after multiplying it by prior probabilities of classes (approximated by the frequencies of class representatives in a training set), and after dividing the result by the normalizing factor having the same value for all classes (and therefore negligible in a decision rule discriminating between classes), yields posterior probability $P(C_j|x)$ of the given class C_j, given the input vector x. However, this posterior probability is also the main criterion of a decision rule in a probabilistic uncertainty model implemented by Bayesian classifiers. The above-mentioned decision rule is very simple, assuming the same cost of any incorrect decision (i.e., in the case considered, treating false positive and false negative answers equally). It can be simply reduced to the choice of the class with maximum posterior probability $P(C_j|x)$.

Moreover, assuming the same frequencies of the representatives of all classes in a training set—which is the case in this study—the above rule is equivalent to the choice of the class with maximum likelihood $p(x,C_j)$. Since likelihood functions are generated by the output neurons of a probabilistic neural network, therefore to obtain a decision, one has to drive inputs of the PNN with the given vector x, and choose the class corresponding to the neuron with the highest level of response.

The training of the probabilistic neural network is a one-epoch process, given the value of the parameter s denoting the width of the kernel in the pattern layer. Since the results of the classification are strongly dependent on the proper value of this parameter, in reality the one-epoch training should be repeated many times in a framework used for optimization results with respect to s. Fortunately, the shape of the optimized criterion in one-dimensional space of parameter s in the majority of cases is not too complex, with one global extreme having a respectable basin of gravity. If the width parameter s is normalized by the dimensionality of the input data N in an argument of the kernel function, then the proper value of s is very often within a range from 10 to 10^{-1}. In this study, where there was applied the minimization of the decision error serving as a criterion, the optimal value of s proved to be 0.175.

4.3 Experimental Results and Conclusions

Table 4.1 presents the results of PNN classification during jackknife cross validation for s equal to 0.175.

In this paper the application of PNN to the problem of the interpretation of a battery of non-neutrality tests was presented. Such interpretation without machine learning methodology is troublesome because of the existence of factors which, for some of these tests, produce patterns similar to those caused by natural selection. However, the exact influence of such extra-selective factors is different for different tests, so the collection of such tests preserves enough information to discriminate between the mentioned non-selective factors and the selection. For this purpose three PNNs were trained, each with a different

Table 4.1 The results of jackknife cross validation procedure for the probabilistic neural network with parameter $s = 0.175$ (93.5% correct decisions)

Test number	Number of correct decisions	Percentage of correct decisions (%)	Decision error
1	2	100	0
2	1	50	0.5
3	1	100	0
4	2	100	0
5	2	100	0
6	2	100	0
7	2	100	0
8	2	100	0
Average	15/16	93.75	0.0625

width of the kernel function. In jackknife cross validation the PNN with $s = 0.175$ gave the best results. The decision error of this classifier in testing was equal to only 6.25%, with the estimated standard deviation of this error equal to 0.067.

Acknowledgments The author would like to acknowledge his financial support under the habilitation grant number BW/RGH-5/Rau-0/2007, under SUT statutory activities BK2007, and under MNiSW grant number 3T11F 010 29. Also the author would like to thank Prof. M. Kimmel from the Department of Statistics at Rice University in Houston TX, USA, for advice and long discussions concerning the statistical and biological aspects of the research using non-neutrality tests for the detection of natural selection operating at the molecular level.

References

1. Biton S, Gropp M, Itsykson P, Pereg Y, Mittelman L, Johe K, Reubinoff B, Shiloh Y (2007) ATM-mediated response to DNA double strand breaks in human neurons derived from stem cells. DNA Repair (Amst) 6:128–134
2. Cyran K (2007) Rough sets in the interpretation of statistical tests outcomes for genes under hypothetical balancing selection. In: Kryszkiewicz M, Peters J, Rybinski H, Skowron A (eds) Springer-Verlag, Lecture notes in artificial intelligence, pp 716–725
3. Cyran KA, Polańska J, Chakraborty R, Nelson D, Kimmel M (2004) Signatures of selection at molecular level in two genes implicated in human familial cancers. In: 12th international conference on intelligent systems for molecular biology and 3rd European conference on computational biology. Glasgow UK, pp 162–162
4. Cyran KA, Polańska J, Kimmel M (2004) Testing for signatures of natural selection at molecular genes level. J Med Info Technol 8:31–39
5. Dhillon KK, Sidorova J, Saintigny Y, Poot M, Gollahon K, Rabinovitch PS, Monnat RJ Jr (2007) Functional role of the Werner syndrome RecQ helicase in human fibroblasts. Aging Cell 6:53–61
6. Evans PD, Anderson JR, Vallender EJ, Gilbert SL, Malcom ChM, et al (2004) Adaptive evolution of ASPM, a major determinant of cerebral cortical size in humans. Hum Mol Genet 13:489–494

7. Fu YX (1997) Statistical tests of neutrality of mutations against population growth, hitchhiking and background selection. Genetics 147:915–925

8. Fu YX, Li WH (1993) Statistical tests of neutrality of mutations. Genetics 133: 693–709

9. Golding SE, Rosenberg E, Neill S, Dent P, Povirk LF, Valerie K (2007) Extracellular signal-related kinase positively regulates ataxia telangiectasia mutated, homologous recombination repair, and the DNA damage response. Cancer Res. 67:1046–1053

10. Karmakar P, Seki M, Kanamori M, Hashiguchi K, Ohtsuki M, Murata E, Inoue E, Tada S, Lan L, Yasui A, Enomoto T (2006) BLM is an early responder to DNA double-strand breaks. Biochem Biophys Res Commun 348:62–69

11. Kelly JK (1997) A test of neutrality based on interlocus associations. Genetics 146:1197–1206

12. Kimura M (1983) The neutral theory of molecular evolution. Cambridge University Press, Cambridge

13. Nielsen R (2001) Statistical tests of selective neutrality in the age of genomics. Heredity 86:641–647

14. Polańska J (2003) The EM algorithm and its implementation for the estimation of the frequencies of SNP-haplotypes. Int J Appl Math Comput Sci 13:419–429

15. Schneider J, Philipp M, Yamini P, Dork T, Woitowitz HJ (2007) ATM gene mutations in former uranium miners of SDAG Wismut: a pilot study. Oncol Rep 17:477–482

16. Wall JD (1999) Recombination and the power of statistical tests of neutrality. Genet Res 74:65–79

17. Zhang J (2003) Evolution of the human ASPM gene, a major determinant of brain size. Genetics 165:2063–2070

Chapter 5
Fine-Tune Artificial Neural Networks Automatically

Francisco Reinaldo, Rui Camacho, Luís P. Reis, and Demétrio Renó Magalhães

Abstract To get the most out of powerful tools, expert knowledge is often required. Experts are the ones with the suitable knowledge to tune the tools' parameters. In this paper we assess several techniques which can automatically fine-tune ANN parameters. Those techniques include the use of GA and stratified sampling. The fine-tuning includes the choice of the best ANN structure and the best network biases and their weights. Empirical results achieved in experiments performed using nine heterogeneous data sets show that the use of the proposed Stratified Sampling technique is advantageous.

5.1 Introduction

Artificial neural networks (ANN) are now a widely used technique to solve complex problems in a broad range of real applications [1]. Unfortunately, there is no cookbook which explains how to fine-tune the technical properties that define an ANN characteristic. We foresee that the main reason why this problem is largely unsolved is because of the substantial varying of the parameter values. Thus, in this study, we propose an automatic method for common users to develop and fine-tune ANNs by interlacing machine learning (ML) and probabilistic techniques.

Several studies concerning the automatic tuning of ANN parameters may be found in the literature. Most of them use a genetic algorithm (GA) as a stochastic search method to find solutions [2]. Others have studied suitable methods to initialise an ANN training by tuning the first weights [3, 4]. Unfortunately, these approaches work only with traditional parameters, instead of a complete set. Moreover, the process of weights extraction from a unique range of values is not effective because it will hardly avoid saturation areas. However,

F. Reinaldo (✉)
DEEC & LIACC, Universidade do Porto, R. Dr Roberto Frias, s/n, 4200-465 Porto, Portugal
e-mail: reifeup@fe.up.pt

N. Mastorakis et al. (eds.), *Proceedings of the European Computing Conference*, Lecture Notes in Electrical Engineering 27, DOI 10.1007/978-0-387-84814-3_5, © Springer Science+Business Media, LLC 2009

none of the proposed solutions consider a hybrid approach to automatically fine-tune an ANN structure and weights such as ML and statistical methods, or offer a harmonic solution to the problem.

To overcome these drawbacks, we propose a method to automatically achieve fine-tuning results and to obtain low error rates by the use of AFRANCI [5]. The method is developed in two main parts: a GA is first used as a wrapper to choose adequate values for the ANN parameters. Afterwards, we use stratified sampling (SS) in order to balance the first weights on different synaptic connections.

The rest of the paper is structured as follows: Section 5.2 describes how genetic algorithms were applied to tune the different ANN parameters and structure. Section 5.3 explains how we improve the ANN weights values by using stratified sampling. Experiments and results are presented in Section 5.4. We draw the conclusions in Section 5.5.

5.2 The Automatic Tuning of Artificial Neural Networks

Almost all machine learning systems have parameter values that must be tuned to achieve a good quality for the constructed models. This is most often a severe obstacle to the widespread use of such systems.

As proposed by John [6], one possible approach to overcome such a situation is by using a *wrapper*. In our tool the *wrapper* is a genetic algorithm that will improve the ANN construction procedure, obtaining low error rates. This automatic tuning of parameters completely hides the details of the learning algorithms from the users. The next paragraph shows all the ingredients used to fine-tune the ANN.

A **first ingredient** for using GAs is to encode the chromosomes as linear chains of binary digits, using features such as the learning rate; the momentum rate; the steepness rate; the bias for each hidden output layer; the transfer functions in every neuron of the hidden layers and the output layer; and the number of neurons in every hidden layer. **Another ingredient** concerns the evaluation of the solutions (chromosomes). Using linear scaling, the fitness function was implemented in GAlib. The error rate extracted from the current candidate is encoded to a proportional non-negative transformed rating fitness, or fitness score. A **third ingredient** involves the implementation of the roulette wheel selection method. This method selects the most evolved candidate for reproduction based on the highest fitness score relative to the remaining part of the population. A **fourth item** required to implement a GA is the combination of existing candidates and the use of a sexual crossover with one-cut point cross-over technique, where the selection point is random. After that, the parents change the right side, generating two offspring and preserving the same previous population size. In sequence, we have used the mutation operator that "disturbs" the chromosomes.

Finally, the last process sets the stop measure of the GA search. We chose the number of generations achieved because after several evolutional steps, the last generation brings the "best" candidates with the highest fitness scores. After that, the GA evolves.

5.3 Improving the Learning Process of Artificial Neural Networks

The performance of ANN learning algorithms can be strongly affected by the initial values of weights of the connection links. The weights can influence the global minimisation of an ANN training error function. Traditionally, the initial weight values are randomly generated within a specified range. However, arbitrary values selected from a unique range can be so sparse that the initial input signal falls in a saturate region, or so small that they cause extremely slow learning.

To improve the choice of the initial weights we propose the use of the stratified sampling (SS) technique to work with several weight regions.

In the SS technique, a population is split into s smaller proportional non-overlapping segments or strata. The SS technique guarantees that all regions have at least one representative sample.

Considering that each ANN has n synaptic connections $(w_1, w_2, w_3, \ldots, w_n)$ that need to be populated by samples from the subsets $(A_1, A_2, A_3, \ldots, A_s)$, we use arrangement with repetition to distribute the samples on the connection links. Additionally, we use the random sampling technique of the SS method to select the proportional number of samples from each s to populate every n. In other words, we will have s^n possible solutions. For example, assuming that six different weights $(w_1, w_2, \ldots, w_6)$ have two strata $A_1=[-0.5, 0]$ and $A_2=[0, 0.5]$, using the arrangement with repetition, the number of ANN being trained will be $T = s^n = (2^6) = 128$, taking into account that SS will be called $S = T * n = 768$ times both strata. Finally, the best set of first weights is discovered when the error rate of the test phase is closer to zero.

5.4 Experiments

5.4.1 Research Data and Experiment Design

The techniques presented in Sections 5.2 and 5.3 were evaluated using nine heterogeneous data sets from the UCI [7] repository. The classification data sets are Letter, RingNorm, Splice, Titanic, and TwoNorm; the regression data sets are Abalone, Addd10, Boston, and Hwang.

Three sets of experiments (tests) were devised in order to produce a fair comparison. In the **first experiment** the ANN was set by hand-tuning. ANN has been set to: three layers; back-propagation learning algorithm; random weights initialisation from $[-0.5, +0.5]$; five neurons set to sigmoid transfer function in the hidden layer, and bias set to value 1; one neuron set to sigmoid transfer function in the output layer, and bias set to value 1; learning rate, momentum rate, and steepness rate set to 0.8, 0.2, and 1, respectively; stop the training phase when the error rate gets below 0.1 or the training epochs reach 50. In the **second experiment** the ANN was set by using a GA-based *wrapper*. GA has been set to: 50 candidates; 30 generations; mutation probability set to 1%; crossover probability of 90%; population replacement percentage set to 25%. Consequently, the ANN has been set to: back-propagation learning algorithm; random weights initialisation from $[-0.5, +0.5]$; the limit of three times more neurons in the hidden layer than in the input layer; one out of seven transfer functions in each hidden and output neurons, and bias set to value 1; one neuron in the output layer, and bias set to value 1; learning rate, momentum rate, and steepness rate set to respective default internal range; stop the training phase when the error rate achieves 0.1 or the training epochs achieve 50. In the **third experiment** the ANN was set by using a stratified sampling approach to split the set of weights into two intervals $A_1=[-0.5, 0]$ and $A_2=[0, +0.5]$. Other ANN parameters have been set using the parameter values of the first experiment.

5.4.2 Experimental Results

The experimental results are reported in Table 5.1. Both the average error rate and the standard deviation of the training and test datasets are presented. The average represents the result obtained from the arithmetic sum of five cycles (K-fold technique) of the same dataset together, and then the total is divided by the number of cycles. The winning result percentage was obtained by the variation coefficient of the tests.

Table 5.1 Comparing among tests

Data set	ANN hand-tuning test1(T1)	ANN with GA test2(T2)	ANN with SS test3(T3)	Winner
Letter	123.6(13.8)	102.6(1.9)	109.2(1.6)	T3(0.5%)
RingNorm	981.7(123.0)	267.9(10.7)	238.0(7.0)	T3(1.1%)
Splice	16.8(2.2)	16.8(0.6)	16.5(2.3)	T2(10.1%)
Titanic	11.3(0.9)	9.5(0.5)	9.2(0.3)	T3(1.2%)
TwoNorm	475.2(257.2)	141.8(23.5)	120.2(3.3)	T3(13.9%)
Abalone	99.9(7.5)	70.1(5.1)	77.9(12.9)	T2(9.3%)
Addd10	31223(950)	30460(855)	30445(910)	T2(0.2%)
Boston	107982(27495)	84110(17309)	95979(95691)	T3(6.6%)
Hwang	1342.8(15.9)	1174.6(24.6)	1354(24.42)	T3(0.3%)

The test values in the **table** were all multiplied by $1E\text{-}10$.

From Table 5.1, the following conclusions can be drawn. First, it is worth noting that all tests listed in this study show good results, but the error rate decreased because of the use of a wrapper. Second, as can be observed, the use of GA turns out to be better in three cases out of nine due to the structural risk minimisation principle of ANN, such as local minima and overfitting. Finally, the use of stratified sampling is better in six cases out of nine. The SS technique was a surprise because of the reducing error rates.

5.5   Conclusion

In this paper two techniques to fine-tune ANNs using AFRANCI have been described. The techniques show that ANNs can be tuned using either a genetic algorithm (GA) or a stratified sampling (SS) to achieve good results. Under a wrapper, GA builds the best ANN structure by choosing the correct transfer functions, the right biases, and other essential ANN parameters. On the other hand, SS balances the connection weights of an ANN structure hand-tuned. Since it is easy to implement, GA is the most used fine-tuning technique to draw the best ANN structure. However, empirical evaluation on nine popular data sets has confirmed that the SS obtained better results with the lowest error rates. Thus, the evaluation of the proposed tuning suggests that better results can be achieved when the SS is used.

References

1. Reinaldo F, Certo J, Cordeiro N, Reis LP, Camacho R, Lau N (2005) Applying biological paradigms to emerge behaviour in robocup rescue team. In: Bento C, Cardoso A, Dias G (eds) EPIA. Springer, Lecture Notes in Computer Science 3808:422–434
2. Shamseldin AY, Nasr AE, O'Connor KM (2002) Comparison of different forms of the multi-layer feed-forward neural network method used for river flow forecasting. Hydrol Earth Syst Sci (HESS) 6(4):671–684
3. Erdogmus D, Fontenla-Romero O, Principe J, Alonso-Betanzos A, Castillo E, Jenssen R (2003) Accurate initialization of neural network weights by backpropagation of the desired response. In: Proceedings of the IEEE International Joint Conference on Neural Networks 3:2005–2010
4. Jamett M, Acuña G (2006) An interval approach for weights initialization of feed-forward neural networks. In: Gelbukh AF, Garća CAR (eds) MICAI. Springer, Lecture Notes in Computer Science, 4293:305–315
5. Reinaldo F, Siqueira M, Camacho R, Reis LP (2006) Multi-strategy learning made easy. WSEAS Transactions on Systems, Greece, 5(10):2378–2384
6. John HG (1994) Cross-validated c4.5: using error estimation for automatic parameter selection. Technical note stan-cs-tn-94-12, Computer Science Department, Stanford University, California
7. Newman CBDJ, Hettich S, Merz C (1998) UCI repository of machine learning databases

Chapter 6
A Neurofuzzy Network for Supporting Detection of Diabetic Symptoms

Leonarda Carnimeo

Abstract In this paper a neurofuzzy network able to enhance contrast of retinal images for the detection of suspect diabetic symptoms is synthesized. Required fuzzy parameters are determined by ad hoc neural networks. Contrast-enhanced images are then segmented to isolate suspect areas by an adequate thresholding, which minimizes classification errors. In output images suspect diabetic regions are isolated. Capabilities and performances of the suggested network are reported and compared to scientific results.

6.1 Introduction

In ophthalmology several researchers have been recently attempting to develop diagnostic tools for detecting symptoms related to diabetic retinopathies, because computational intelligence may help clinicians in managing large databases [1]. Diabetic retinopathies can be revealed by specific symptoms such as *exudates*, which appear as bright areas in digital retinal images. In this regard, some interesting contributions have already been proposed in [2–4]. In [2], exudates are found using their high grey level variation and contours are determined by means of morphological reconstruction techniques. In [3], diabetic symptoms are revealed by combining techniques of region growing and edge detection. Both these solutions give interesting results, but heuristic thresholds are needed. In [4], a three-step method based on the use of a median filter and a multilevel thresholding technique is proposed. Unfortunately, drawbacks arise because of the amount of filtering variables to be evaluated. Moreover, a heavy computational burden is involved in processing retinal images. In this regard, in [5], four textural features considered for training a fuzzy neural

L. Carnimeo (✉)
Dipartimento di Elettrotecnica ed Elettronica, Politecnico di Bari, Via Orabona,
4 70125 BARI – Italy
e-mail: carnimeo@deemail.poliba.it

N. Mastorakis et al. (eds.), *Proceedings of the European*
Computing Conference, Lecture Notes in Electrical Engineering 27,
DOI 10.1007/978-0-387-84814-3_6, © Springer Science+Business Media, LLC 2009

network, were shown to be successful for the detection of vague mass lesions in mammograms.

On the basis of these considerations, in [6], the authors developed a first attempt to detect suspect diabetic symptoms by synthesizing a fuzzy architecture for retinal image processing. Quality performances of the proposed system revealed improvements with respect to actual results. Nevertheless, the problem of the determination of fuzzy parameters in the synthesis procedure was not solved. In this paper, a neurofuzzy network is proposed, taking into account that vagueness is present also in retinal images. The indices defined in [5, 7] are considered in order to find optimal values for the network fuzzy parameters a and b. Adequate MLP neural networks are synthesized to determine an optimal threshold which can minimize classification errors if contrast-enhanced images are segmented. In output binary images, black regions identify suspect areas. The whole system is developed considering neural networks to reduce computational burdens in the detection of cited symptoms. Precise criteria for the evaluation of each parameter are given, and performances are compared to other scientific results.

6.2 Neurofuzzy System

In Fig. 6.1 the behavioural diagram of the proposed neurofuzzy system is shown.

Vague pale areas, suspected to be diabetic symptoms, have to be detected in image I. For this purpose, an adequate image segmentation has to be carried out in order to segment each *fundus* image into two suspect/not suspect sets, each one regarded as distinguishing a clinically significant area. Unfortunately, an automatic computation of a proper threshold is not feasible, both because of the vagueness of the pale regions and because of the strong nonlinearity of the histograms in *fundus* image I. Moreover, a thresholding can be computationally effective if the histograms of the analyzed images are revealed as bimodal, that is, if the two sets to be identified have the maximum distance in brightness space. This can be obtained via a neurofuzzy image contrast enhancement subsystem. For this purpose, a neurofuzzy contrast enhancement is performed. In detail, each image I can be processed to obtain a contrast-enhanced image I_f, whose histogram is bimodal. As shown in [6], right-angled triangular

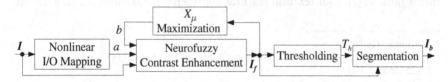

Fig. 6.1 Block diagram of the proposed neurofuzzy system behaviour

membership functions are adopted for the antecedents of the proposed system. Therefore, two parameters a and $b \in [0; 255]$ uniquely identify these functions. In order to provide the parameter a, a proper block, which maps p specific input features of each retinal image I into the corresponding value of a, is synthesized. The optimal value of the fuzzy parameter b is determined by maximizing the contrast of I. For this purpose, the index X_μ, defined as in [7], is adopted. A thresholding is successively performed, by evaluating an optimum threshold T_h that minimizes errors in classifying suspect regions. After evaluating each threshold T_h, each image I_f can be globally segmented, providing the output binary image I_b, in which black pixels identify suspect regions and optic disks in *fundus* image I. The proposed system is shown in Fig. 6.2 and successively described.

6.2.1 Nonlinear I/O Mapping

Let $f : p \in \Re^p \to a \in \Re$ be a function which maps a vector p of p input image features into the fuzzy parameter a. Due to the fact that the choice of $f(p)$ could reveal a difficult task, this function is determined by adopting a neural approach. In detail, four specific features are first extracted, which are not sensitive to shapes and areas of suspect symptoms. For this purpose, the co-occurrence matrix $\mathbf{CM} = \{c_{ij}\}$, $i,j = 0, \dots, 255$, is defined as in [5]. Then, matrix $\mathbf{CM}$ can be evaluated by considering an artificial image P_d computed as:

$$P_d = P_0 * I + P_{45} * I + P_{90} * I + P_{135} * I$$

being $c_{ij} = \text{cardinality}\{(k, l) \mid P_d(k, l) = 4(i+j)\}$ and

$$P_0 = \begin{pmatrix} 0 & 0 & 0 \\ 0 & 1 & 1 \\ 0 & 0 & 0 \end{pmatrix} \quad P_{45} = \begin{pmatrix} 0 & 0 & 0 \\ 0 & 1 & 0 \\ 0 & 0 & 1 \end{pmatrix} \quad P_{90} = \begin{pmatrix} 0 & 0 & 0 \\ 0 & 1 & 0 \\ 0 & 1 & 0 \end{pmatrix} \quad P_{135} = \begin{pmatrix} 0 & 0 & 0 \\ 0 & 1 & 0 \\ 1 & 0 & 0 \end{pmatrix}$$

Four features, defined in [5], are considered herein integrated with a neuro-fuzzy network, on the basis of the matrix $\mathbf{CM} = \{c_{ij}\}$:

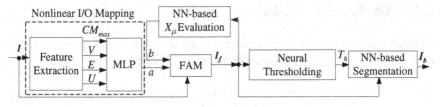

Fig. 6.2 Proposed neurofuzzy system

$$\text{maximum co} - \text{occurrence matrix element } CM_{max} = \max_{i,j} c_{ij} \qquad (6.1)$$

$$\text{contrast } V = \sum_i \sum_j (i-j)^2 c_{ij} \qquad (6.2)$$

$$\text{entropy } E = -\sum_i \sum_j c_{ij} \log c_{ij} \qquad (6.3)$$

$$\text{uniformity } U = \sum_i \sum_j c_{ij}^2 \qquad (6.4)$$

Selected features are not sensitive to shapes or areas of suspect regions, but depend on the histogram of each image. In particular, CM_{max} gives information about the most frequent pixel pairs in the image to be submitted, contrast V is a measure of the gray levels difference among neighboring pixels, entropy E is associated with the randomness of the information within the areas of interest, whereas uniformity U is a measure of their orderliness [5].

Let us consider a database of $(M x N)$-dimension images containing suspect symptoms caused by diabetic retinopathies. Let T be the $(N_T x 4)$-dimension matrix, whose generic row is given by the (1×4)-vector $v = [CM_i \ V_i \ E_i \ U_i]$, normalized in the range $[0; 1]$, with $0 \le i \le N_T$. Each row of the matrix T can be considered as the input of a multilayer perceptron network (MLP) with four input neurons. Each neuron has a logarithmic sigmoid transfer function. The network output is the optimal value a for the first membership function of the fuzzy associative memory (FAM). The network is trained by minimizing the mean square error (MSE), defined as:

$$MSE = \frac{1}{P} \sum_{i=1}^{P} (y-t)^2 \qquad (6.5)$$

y and t being the output and the target of the network, respectively. The minimization of this index allows us to determine the number of neurons in the hidden layer of the network. The synthesized network MLP is finally able to approximate the function $f(p)$.

6.2.2 NN-Based X_μ Evaluation

Optimal values of the fuzzy parameter b are evaluated by maximizing the contrast of I via the mean absolute contrast X_μ, defined as [7]:

$$X_\mu(b) = \frac{1}{mn} \sum_i \sum_j abs(C(i,j)) \qquad (6.6)$$

with $1 \leq i \leq m$, $1 \leq j \leq n$, being $C(i,j)$ the (i,j)-th pixel of the image C, obtained by processing image I_f as reported in [7]. The index X_μ can be considered as a function X_μ (b) of the variable b. The local contrast value $C(i, j)$ in (6.6) describes each pixel separately, but the index X_μ gives a measure of the overall contrast quality of image I_f. When pathologic details need to be detected, the contrast is required to be maximum. Nevertheless, as the contrast increases, the histogram of image I_f is emphasized toward extreme values of gray levels with respect to image I. Due to a saturation phenomenon, this implies that an information loss about details takes place. A compromise is to be found, that is, the optimal value of b should enhance image contrast, preserving its details. For this reason, given the value of parameter a, obtained by the MLP net, let $X_{\mu\ max}$ and $X_{\mu\ min}$ be the maximum and the minimum values of X_μ (b), respectively. Then, the optimal parameter b_{opt} can be found as:

$$X_\mu(b_{opt}) = 0.5(X_{\mu\max} + X_{\mu\min}) \tag{6.7}$$

6.2.3 Fuzzy Associative Memory

Given the parameters a and b, a Hopfield-like sparsely-connected neural network can now be synthesized to behave as the proposed fuzzy system by adopting the synthesis procedure developed in [6]. In this way, in the resulting image I_f, contrast is maximum, whereas the loss of information is minimum. This condition is optimal when suspect bright areas need to be detected. In fact, the histogram of I_f presents two significant peaks. The first peak concerns information about deep areas in image I, the last one concerns pale ones.

6.2.4 Neural Thresholding

Retinal suspect areas can be isolated in each contrast-enhanced image I_f by an adequate segmentation. An optimal thresholding is developed by requiring that errors in classifying suspect regions be minimized. For this purpose, a NN-based subsystem formed by two multi-layer perceptron networks MLP_D and MLP_P is designed for an optimal thresholding. Let $h_f(g)$, $g = 0, 1, \ldots, 255$ be the histogram of I_f. Let m denote the maximum gray level in $[1;254]$ such that $h(m)$ is a relative minimum of $h_f(g)$. Three vectors $g, h_D, h_P \in \mathbb{N}^{255 \times 1}$ can be defined:

$$g = [1, \ldots, 255]^T h_D = [h_1, ..h_m, 0.., 0]_T h_P = [0, \ldots, 0, h_{m+1}, \ldots h_{255}]^T$$

containing occurrences of deep/pale gray level values only, respectively, and the matrices $H_D = [\, g, h_D \,] \in \mathbb{N}^{255 \times 2}$ and $H_P = [\, g, h_P \,] \in \mathbb{N}^{255 \times 2}$, which contain information about deep areas and pale ones of contrast-enhanced images I_f.

Matrices H_D and H_P provide proper sets for training the networks MLP$_D$ and MLP$_P$ to recognize one mode of the bimodal histogram $h_f(g)$. The optimal threshold T_h, that minimizes errors in segmentation, is given by $h_D(T_h) = h_P(T_h)$.

The networks MLP$_D$ and MLP$_P$ are characterized by an input layer and an output layer each formed by one neuron with a logarithmic sigmoid transfer function. One hidden layer is considered, whose optimal number of neurons can be established by minimizing the mean square error index. These networks are trained using the Levenberg-Marquardt back-propagation (LMBP) training algorithm. Both networks approximate histograms $h_D(g)$ and $h_P(g)$ by fitting discrete values of both modes, which form histogram $h_f(g)$.

6.2.5 NN-Based Segmentation

After evaluating the threshold T_h, each image I_f can be globally segmented as

$$I_b(i,j) = 255 \quad \text{if} \quad I_f(i,j) < T_h \quad \text{and} \quad I_b(i,j) = 0 \quad \text{if} \quad I_f(i,j) > T_h$$

for $0 \le i \le M$ and $0 \le j \le N$, where in image I_b, black pixels identify suspect regions and the optic disk of the original *fundus* image I.

6.3 Numerical Results

The capabilities of the proposed system have been investigated by considering a database, constituted by sixty (450×530)-dimension retinal images. In Fig. 6.3 the green layer of a selected *fundus* image I and its histogram $h(g)$ are reported. Diabetic symptoms given by the vague pale regions can be noted.

The training set of the MLP network which evaluates the parameter a is constituted by $N_T = 48$ samples. By solving equation (6.7), values of $a = 25$ and $b = 193$ have been found.

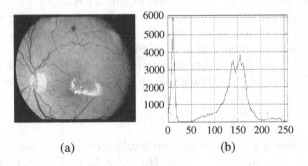

(a)   (b)

Fig. 6.3 (a) Input *fundus* image I; (b) its histogram $h(g)$

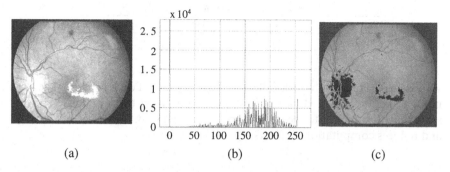

Fig. 6.4 (a) Contrast-enhanced image I_f; (b) its histogram $h_f(g)$; (c) superimposition of output image I_b to *fundus* image I

A (4×4) Hopfield-like neural network, behaving as a fuzzy associative memory, has been implemented. Fig. 6.4 shows the contrast-enhanced image I_f and its bimodal histogram obtained by processing image I with the designed neurofuzzy system. Then the MLP_D/MLP_P networks with an optimal number of 9 hidden neurons have been trained. An optimal global threshold computation is now feasible and gives $T_h = 237$ for image I_f.

In Fig. 6.4(c) the output image I_b is shown, where black pixels identify suspect diabetic areas and the optic disk in image I. Performances of the proposed architecture are evaluated by comparing output images to a *gold standard* one, provided by clinicians. Results are discussed by determining the values of true positives (TP), false negatives (FN), and false positives (FP), as defined in [2]. In detail, the quantity TP gives the number of pixels that the system correctly classifies as symptoms; the values FN and FP indicate wrongly classified pixels with respect to expected ones. As reported in [3], the values of the indices FN, FP, and TP have been computed in a circular window of radius equal to 156 pixels, centered on the fovea. This radius corresponds to a linear distance of about 6000 μm, computed as 2DD, DD being the diameter of the optic disc. Then performances are evaluated by computing the *correct recognition rate (CCR%)*, defined as [8]:

$$CCR\% = 100\,\frac{\text{Number of correctly classified cases}}{\text{Total Number of cases}}$$
$$= 100\left(1 - \frac{\text{FP} + \text{FN}}{\text{Total Number of pixels}}\right)$$

This quantity gives a percentage measure of correctly classified regions in comparison with known results. Table 6.1 shows results obtained with the proposed system compared with results obtained by applying methods in [2, 4] and [3]. It can be observed that the results are quite similar, but in [2] a preprocessing step is necessary in order to eliminate anatomic details like blood

Table 6.1 Comparison of CCR% values

Walter et al.	Kavitha et al.	Li et al.	Proposed system
99.92 %	98.55%	99.69%	99.79%

vessels; in [4] a method to seek valleys in the histogram of each *fundus* image also needs to be implemented; in [3] two image layers need to be processed at the same time. It can be noted that the proposed system simplifies image processing and reduces computational efforts.

6.4 Conclusions

A contribution to improve detection of suspect diabetic symptoms in retinal images by means of a neurofuzzy network has been proposed. After evaluating an optimal threshold to minimize pixel classification errors, enhanced contrast images have been segmented. In binary output images, suspect areas and the optic disk have been isolated. Quality performances of the proposed neurofuzzy system are revealed to be quite satisfactory when compared with other results.

References

1. Patton N, Aslam TM, MacGillivray T, Deary IJ, Dhillon B, Eikelboom RH, Yogesan K, Constable IJ (2006) Retinal image analysis: concepts, applications and potential. Prog Retinal Eye Res 25:99–127
2. Walter T, Klein J-C, Massin P, Erginay A (2002) A contribution of image processing to the diagnosis of diabetic retinopathy—detection of exudates in color fundus images of the human retina. IEEE T Med Imaging 21(10):1236–1243
3. Li H, Chutatape O (2004) Automated feature extraction in color retinal images by a model based approach. IEEE T Biomed Eng 51(2):246–254
4. Kavitha D, Shenbaga DS (2005) Automatic detection of optic disc and exudates in retinal images. IEEE International conference on intelligent sensing and information processing (ICISIP), pp 501–506
5. Cheng HD, Cui M (2004) Mass lesion detection with a fuzzy neural network. Patt Recogn 37(6):1189–1200
6. Carnimeo L, Giaquinto A (2007) A fuzzy architecture for detecting suspect diabetic symptoms in retinal images. In: 3rd WSEAS International conference on cellular and molecular biology, biophysics and bioengineering (BIO'07), pp 42–45
7. Brendel M, Roska T (2002) Adaptive image sensing and enhancement using the cellular neural network universal machine. Int J Circ Theory Appl 30(2–3):287–312
8. Ennett CM, Frize M, Charrette E (2004) Improvement and automation of artificial neural networks to estimate medical outcomes. Med Eng Phys 26:321–328

Chapter 7
Empirical Assessment of LR- and ANN-Based Fault Prediction Techniques

Bindu Goel and Yogesh Singh

Abstract At the present time, because of our reliance on software systems, there is a need for dynamic dependability assessment to ensure that these systems will perform as specified under various conditions. One approach to achieving this is to dynamically assess the modules for software fault predictions. Software fault prediction, as well as inspection and testing, are still the prevalent methods of assuring the quality of software. Software metrics-based approaches to build quality models can predict whether a software module will be fault-prone or not. The application of these models can assist to focus quality improvement efforts on modules that are likely to be faulty during operations, thereby cost-effectively utilizing the software quality testing and enhancement resources. In the present paper, the statistical model, such as logistic regression (LR), and the machine learning approaches, such as artificial neural networks (ANN), have been investigated for predicting fault proneness. We evaluate the two predictor models on three main components: a single data sample, a common evaluation parameter, and cross validations. The study shows that ANN techniques perform better than LR; but that LR, being a simpler technique, is also a good quality indicator technique.

7.1 Introduction

In a world interwoven economically the rapid development of hardware and software has led to an increase in competition among companies producing and delivering software products and services. In addition, there has been a growing need to produce low cost, high quality software in a short time. In real-time systems there is a need for assessing the dependability of the software under various normal and catastrophic conditions. All these requirements have forced

B. Goel (✉)
University School of Information Technology, Guru Gobind Singh Indraprastha University, Kashmere Gate, Delhi 06, India
e-mail: bindu_delus@yahoo.com

N. Mastorakis et al. (eds.), *Proceedings of the European Computing Conference*, Lecture Notes in Electrical Engineering 27, DOI 10.1007/978-0-387-84814-3_7, © Springer Science+Business Media, LLC 2009

researchers to focus more on planning and allocating resources for testing and analysis.

Early identification or prediction of faulty modules can help in regulating and executing quality improvement activities. The quality indicators are generally evaluated indirectly by collecting data from earlier projects. The underlying theory of software quality prediction is that a module currently under development is fault-prone if a module with similar products or process metrics in an earlier project (or release) developed in the same environment was fault-prone [1]. Measuring structural design properties of a software system, such as coupling, cohesion, or complexity, is a promising approach towards early quality assessments. Quality models are built to quantitatively describe how these internal structural properties relate to relevant external system qualities such as reliability or maintainability.

Fault-proneness models can be built using many different methods that mostly belong to a few main classes: machine learning principles, probabilistic approaches, statistical techniques, and mixed techniques. Machine learning techniques like decision trees have been investigated in [2, 3], and neural networks by Khoshgoftaar et al. [4]. Probabilistic approaches have been exploited by Fenton and Neil, who propose the use of Bayesian belief networks [5]. Statistical techniques have been investigated by Khoshgoftaar, who applied discriminant analysis with Munson [6] and logistic regression with Allen et al. [7]. Mixed techniques have been suggested by Briand et al., who applied optimized set reduction [8], and by Morasca and Ruhe, who worked by combining rough set analysis and logistic regression [9]. Many researchers have conducted studies predicting fault-proneness using different object-oriented (OO) metrics. They constructed highly accurate mathematical models to calculate fault-proneness [10–13]. Munson and Khoshgoftaar used software complexity metrics and logistic regression analysis to detect fault-prone modules [14]. Basili et al. also used logistic regression for detection of fault-proneness using object-oriented metrics [15].

In the present study we have focused on comparing the performance of models predicted using LR and ANN methods. The quality of the models can be further evaluated by comparing the efficiency of the models on subsets of the available data by means of leave-more-out cross validation. Cross validation measures the quality of models by referring to the data available for the same software product. The technique uses all data for both computing the models and evaluating their quality, thus allowing for evaluating the quality of prediction even if we have only a limited amount of data.

The paper is organized as follows: Section 7.2 gives the data set used and summarizes the metrics studied for the model. Section 7.3 presents the data analysis and research methodology followed in this paper. The results of the study are presented in Section 7.4. The models are evaluated on similar parameters and cross validated in Section 7.5. The paper is concluded in Section 7.6.

7.2 Descriptions of Data and Metric Suite

In this section we present the data set used for the study. A brief introduction to the metric suite is also given.

7.2.1 Data Set

The present study makes use of public domain data set KC1 posted online at the NASA Metrics Data Program web site [16]. The data in KC1 was collected from a storage management system for receiving/processing ground data, which was implemented in the C++ programming language. It consists of 145 classes that comprise 2107 methods and a total of 40 K lines of code. It consists of 60 faulty classes and the rest without any faults.

7.2.2 Metric Suite

Table 7.1 shows the different types of predictor software metrics (independent variables) used in our study. It is common to investigate large numbers of measures, as they tend to be exploratory for studies. The static OO measures [17] collected during the development process are taken for the study. We have taken all these measures as independent variables. They include various measures of class size, inheritance, coupling, and cohesion; they are all at class

Table 7.1 Metric suite at class level used for the study

Metric	Definition
Coupling_between_objects (CBO)	The number of distinct non-inheritance-related classes on which a class depends.
Depth_of_inheritance (DIT)	The level for a class. Depth indicates at what level a class is located within its class hierarchy.
Lack_of_cohesion (LCOM)	This counts the number of null pairs of methods that do not have common attributes.
Fan-in (FANIN)	This is a count of calls by higher modules.
Response_for_class (RFC)	A count of methods implemented within a class plus the number of methods accessible to an object class due to inheritance.
Weighted_methods_per_class (WMPC)	A count of methods implemented within a class (rather than all methods accessible within the class hierarchy).
Number_of_children (NOC)	This is the number of classes derived from a specified class.
Cyclomatic_complexity (CC)	This is a measure of the complexity of a modules decision structure. It is the number of linearly independent paths.
Source_lines_of_code (SLOC)	Source lines of code that contain only code and white space.
Num_of_methods_per_class (NMC)	This is the count of methods per class.

levels. The traditional static complexity and size metrics include well known metrics, such as McCabe's cyclomatic complexity and executable lines of code [18]. These metrics are posted at the site for method level, but we have converted them to class level and used the average values of them. Another traditional metric fan-in given by Henry and Kafura [19] is also used to measure interclass coupling by counting the number of calls passed to the class. Also, we have computed a metric called number of methods per class. It counts the number of modules in a class. The underlying rationale for this is that the higher the number of methods per class, the more it is fault-prone. The dependent variable used is the defect level metric, which refers to whether the module is fault-prone or fault-free. So our dependent variable, or the response variable, is the defect level whose response is true if the class contains one or more defects, and false if does not contain a fault. We have converted the variable into a numeric, giving it the value 1 if it contains a fault and 0 otherwise.

7.3 Data Analysis and Research Methodology

7.3.1 Descriptive Statistics

Within each case study, the distribution maximum, mean, and variance in the form of standard deviation of each independent variable is examined. Low variance measures do not differentiate classes very well and therefore are not likely to be useful. Presenting and analyzing the distribution of measures is important for the comparison of the different metrics used. All the measures with more than six non-zero data points are considered for further analysis.

7.3.2 Logistic Regression Model

Logistic regression (LR) [20] is widely applied to data where the dependent variable (DV) is dichotomous, i.e., it is either present or absent. It is used to predict the likelihood for an event to occur, e.g., fault detection. We present logistic regression in the essence of the relevance in the paper. LR, unlike linear regression (where the goal is to determine some form of functional dependency, e.g., linear, between the explanatory variables and the dependent variable), does not assume any strict functional form to link explanatory variables and the probability function. Instead, this functional correspondence has a flexible shape that can adjust itself to several different cases. LR is based on maximum likelihood and assumes that all observations are independent. Outlier analysis is done to find data points that are over–influential, and removing them is essential. To identify multivariate outliers, we calculate for each data point the Mahalanobis jackknife distance. Details on outlier analysis can be found in [21]. LR is of two types: (i) univariate logistic regression (ULR), and (ii) multivariate logistic regression (MLR).The MLR model is defined by the following equation (if it contains only one independent variable (IV), then we have a ULR model):

$$\frac{\pi(X_1, X_2, ..., X_N)}{1 - \pi} = \frac{\Pr ob(event)}{\Pr ob(noevent)} = e^{B_0 + B_1 X_1 + ... + B_N X_N} \qquad (7.1)$$

where X_is are the independent variables and π is the probability of occurrence of a fault. MLR is done in order to build an accurate prediction model for the dependent variable. It looks at the relationships between IVs and the DV, but considers the former in combination, as covariates in a multivariate model, in order to better explain the variance of the DV and ultimately obtain accurate predictions. To measure the prediction accuracy, different modeling techniques have specific measures of goodness-of-fit of the model. In the present study, the following statistics are used to illustrate and evaluate the experimental results obtained:

- B_is are the estimated regression coefficients of the LR equation. They show the extent of the impact of each explanatory variable on the estimated probability, and therefore the importance of each explanatory variable. The larger the absolute value of the coefficient, the stronger the impact (positive or negative, according to the sign of the coefficient) of the explanatory variable on the probability of a fault being detected in a class.
- p, the statistical significance of the LR coefficients, provides an insight into the accuracy of the coefficients estimates. A significance threshold of $\alpha = 0.05$ (i.e., 5% probability) has often been used to determine whether a variable is a significant predictor.
- R^2, called the r-square coefficient, is the goodness-of-fit, not to be confused with the least-square regression R^2. The two are built upon very different formulae, even though they both range between 0 and 1. The higher the R^2, the higher the effect of the model explanatory variables, the more accurate the model. However, this value is rarely high for LR. We have used Cox and Snell's R-square in the present study, which is an imitation of the interpretation of multiple R-square for binary LR.

It is known that the examined metrics are not totally independent and may capture redundant information. Also, the validation studies described here are exploratory in nature; that is, we do not have a strong theory that tells us which variables should be included in the prediction model and which not. In this state of affairs, a selection process can be used in which prediction models are built in a stepwise manner, each step consisting of one variable entering or leaving the model. The general backwards elimination procedure starts with a model that includes all independent variables. Variables are selected one at a time to be deleted from the model, until a stopping criterion is fulfilled. The MLR model is also tested for multicollinearity. If $X_1, \ldots, X_n$ are the covariates of the model, then a principal component analysis on these variables gives I_{max} to be the largest eigenvalue and I_{min} the smallest eigenvalue of the principal components. The conditional number is then defined as $\lambda = \sqrt{I_{max}/I_{min}}$. A large conditional number (i.e., discrepancy between minimum and maximum eigenvalues) indicates the presence of multicollinearity and should be under 30 for acceptable limits [22].

7.3.3 Artificial Neural Network Model

The network used in this work belongs to the architectural family of N—M—1 network. That is, to the network class with one hidden layer of M nodes and N inputs and one output node. In this paper, a ten-input and one-output network is developed. The input node(s) are connected to every node of the hidden layer but are not (directly) connected to the output node. All the ten nodes of the hidden layer are connected to the output node. Thus, the network does not have any lateral connection or any shortcut connection. The training goal was typically achieved within 1000 epochs of training. The training algorithm used in this paper is the Bayesian regularization algorithm as implemented in the Neural Network toolbox of Matlab 7.1. Trainbr is a network training function that updates the weight and bias values according to the Levenberg-Marquardt optimization. It minimizes a combination of squared errors and weights, and then determines the correct combination so as to produce a network that generalizes well and does not over-fit itself. The process is called Bayesian regularization. The input to the neural network is preprocessed by first normalizing the input vectors, so that they have zero mean and unity variance.

The output is evaluated using a classifier and ROC curve, as explained in next section.

7.3.4 Model Evaluation

To evaluate our model we have considered three metrics generally used for classifiers. The model predicts classes as either fault-prone (FP) or not fault-prone (NFP). The metrics used are derived from the model in Fig. 7.1. Accuracy shows the ratio of correctly predicted modules to entire modules. Low accuracy means that a high percentage of the classes being classified as either fault-prone or not fault-prone are classified wrongly. We want accuracy to be high so that classification is done rightly. Recall is the ratio of modules correctly predicted as FP to the number of entire modules actually FP. It is an indicator of fault-prone classes and it should also be high. This metric is of significance, because detecting not fault-prone classes is a waste of resources.

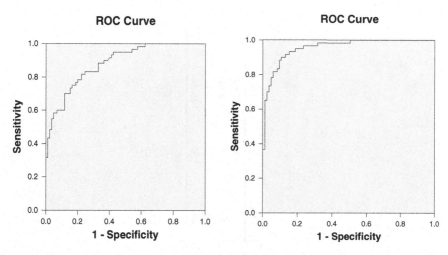

Fig. 7.1 ROC curve for LR and ANN models

Precision is the ratio of modules correctly predicted as FP to the number of entire modules predicted as FP.

Receiver operating characteristic (ROC) analysis: The model outputs are also evaluated for performance using ROC analysis. The ROC curve is a plot of sensitivity on the y coordinate versus its 1-specificity on the x coordinate. The sensitivity is the probability of correctly identifying a positive (TPR). 1-specifity is the probability of correctly identifying a negative (FPR). It is established as an effective method of evaluating the quality or performance of predicted models [23]. An ROC curve demonstrates the trade-off between sensitivity and specificity (any increase in sensitivity will be accompanied by a decrease in specificity). And the closer the curve follows the left-hand border and then the top border of the ROC space, the more accurate the test. The area under the curve is a measure of text accuracy and a combined measure of sensitivity and specificity. In order to compute the accuracy of the predicted models, we have stated the area under the ROC curve and the standard error of the ROC curve.

7.4 Analysis Results

7.4.1 Descriptive Statistics

Table 7.2 presents the descriptive statistics as shown below. There are 145 classes for which values of all the metrics are available.

Table 7.2 Descriptive statistics of the data used

	CBO	DIT	LCOM	FANIN	RFC	WMPC	NOC	CC	SLOC	NMC
Mean	8.32	2.00	68.72	.63	34.38	17.42	.21	2.46	11.83	14.53
Median	8.00	2.00	84.00	1.00	28.00	12.00	.00	1.86	8.67	10.00
Std. Dev.	6.38	1.26	36.89	.69	36.20	17.45	.69	1.73	12.12	15.26
Variance	40.66	1.58	1360.77	.48	1310.65	304.47	.49	2.99	146.88	232.99

Table 7.3 Multivariate model

Covariates	CBO	DIT	LCOM	RFC	WMPC	SLOC	NMC	Const.
Coefficient	.23	−1.3	−.02	.04	−.24	.07	.26	−.92
P(sig)	.001	.002	.028	.091	.079	.021	.061	.150
−2 Log likelihood	118.792							
R^2	.416							

7.4.2 Logistic Regression Results

The model is made using all 145 classes for the training. The model consists of various metrics as shown. NOC, CC, and FANIN are not there in the model. A classifier at threshold $= 0.5$ is used to classify the classes as fault-prone and fault-free. The predicted results are cross tabulated with the actual data. The results are presented in Table 7.3. The test for multicollinearity for the model is done by computing the conditional number: $\sqrt{(2.521/2.321)} = 1.04$, which is well below 30. Also, no influential outlier was found for the model. The accuracy of the model is 79.31%, the recall is 71.67%, and the precision is 76.79%.

7.4.3 Artificial Neural Network Results

A threshold of $P_o < = 0.5$ was chosen. Classes with predicted probability above 0.50 are classified as fault-prone and below this threshold classified as not fault-prone. The table gives only a partial picture, as other cut-off values are possible. The accuracy of the model is 88.28%, the recall is 85%, and the precision is 86.44%.

7.5 Model Evaluations

In this section we present a comparison of both methods and also assess predictive performances using leave-more-out cross validations.

7.5.1 Comparison of LR and ANN Using ROC Curve

Figure 7.1 presents the evaluation of the models using the ROC curve. The ROC curve for the LR and ANN models are shown in Fig. 7.1. The area under the curve (AUC) is .883 (SE 0.027), whereas the area under the ROC curve for the ANN model was 0.957 (SE 0.015). The results show that the ANN model was better, compared to cross validation results of the LR model.

7.5.2 Leave-More-Out Cross Validations

The predicted accuracy of the models is somewhat optimistic, since the models are applied to the same data set from which they are derived. To assess the accuracy of the model we applied leave-more-out cross validations. We randomly partitioned the data set into a training data set and a test data set. We used the training data set to build the estimation model and the test data set to evaluate the predictive performances of the model. We repeated this procedure for different partitions and averaged the prediction accuracy of the different models. The results are shown in Table 7.4. Each column is labeled L_X-T_{100-X} and shows the performances of the model obtained by randomly partitioning the data set into a learning data set composed of $X\%$ of the observations and a test data set composed of the remaining observations. As expected, the performances of the accuracy measures generally decrease as the size of the learning set decreases, although it should be remarked that the partitions are randomly selected. It is worth noting that in the leave-one-out cross validation, the test set is composed of only one observation and the number of test sets is equal to the number of observations.

Table 7.4 Leave-more-out cross validation with random partitions

$L_{training\%}$ -$T_{testing\%}$		L_{90}–T_{10}	L_{80}–T_{20}	L_{70}–T_{30}	L_{60}–T_{40}	L_{50}–T_{50}
LR	Accuracy	80	86.21	77.27	73.33	75
	Precision	100	80	53.33	61.90	72
	Recall	75	57.14	72.72	61.90	62.07
ANN	Accuracy	93.33	82.76	79.54	76.67	73.61
	Precision	100	75	57.14	66.67	72.73
	Recall	91.67	42.85	72.73	66.67	55.17

7.6 Conclusions

One of the main objectives of this paper is to assess which technique of the two (i.e., LR and ANN) is better. The ANN techniques perform better than LR because they are capable of modeling complex functions. But the accuracy levels of the LR models also establish that they can perform reasonably well with software quality models, and they are always simpler to implement than ANN techniques.

The paper also establishes the fact that OO metrics and static measures can make an accurate fault-proneness model. When used in this manner, the model can in fact be helpful by focusing verification effort on faulty classes. The predicted defect detection probabilities are realistic, based on actual fault data, i.e., the fault-proneness class is accurate. However, the model is tested on the same data sets. The model could further be generalized over the software systems from different environments. As with any empirical study, our conclusions are biased according to what data we used to generate them. For example, the sample used here comes from NASA and NASA works in a particularly unique market niche. Nevertheless, we contend that results from NASA are relevant to the general software engineering industry.

The application of neural networks and statistical techniques in predicting software fault-proneness are established. And they are a viable option.

References

1. Khoshgoftaar TM, Allen EB, Ross FD, Munikoti R, Goel N, Nandi A (1997) Predicting fault-prone modules with case-based reasoning. In: ISSRE 1997, IEEE computer society, the eighth international symposium on software engineering, pp 27–35
2. Porter AA, Selby RW (1990) Empirically guided software development using metric-based classification trees. IEEE Software 7(2):46–54
3. Selby RW, Porter AA (1988) Learning from examples: generation and evaluation of decision trees for software resource analysis. IEEE T Software Eng 14(12):1743–1757
4. Khoshgoftaar TM, Lanning DL, Pandya AS (1994) A comparative study of pattern-recognition techniques for quality evaluation of telecommunications software. IEEE J Sel Area Comm 12(2):279–291
5. Fenton NE, Neil M (1999) A critique of software defect prediction models. IEEE T Software Eng 25(5):675–689
6. Munson JC, Khoshgoftaar TM (1992) The detection of fault-prone programs. IEEE T Software Eng 18(5):423–33
7. Khoshgoftaar TM, Allen EB, Halstead R, Trio GP, Flass RM (1998) Using process history to predict software quality. Computer 31(4):66–72
8. Briand L, Basili V, Thomas W (1992) A pattern recognition approach for software engineering data analysis. IEEE T Software Eng 18(11):931–942
9. Morasca S, Ruhe G (2000) A hybrid approach to analyze empirical software engineering data and its application to predict module fault-proneness in maintenance. J Syst Software 53(3):225–237
10. Evanco W (1997) Poisson analyses of defects for small software components. J Syst Software 38:27–35
11. El-Emam K, Melo W, Machado J (1999) The prediction of faulty classes using object-oriented design metrics. J Syst Software
12. Thwin MMT, Quah TS (2002) Application of neural network for predicting software development faults using object-oriented design metrics. In: Proceedings of the 9th international conference on neural information processing, pp 2312–2316
13. Osamu M, Shiro I, Shuya N, Tohru K (2007) Spam filter based approach for finding fault-prone software modules. In: 29th international conference on software engineering workshops (ICSEW'07)
14. Munson JC, Khoshgoftaar TM (1992) The detection of fault-prone programs. IEEE T Software Eng 18(5):423–433
15. Basili VR, Briand LC, Melo WL (1996) A validation of object-oriented metrics as quality indicators. IEEE T Software Eng 22(10):751–761
16. Metrics Data Program, NASA IV&V Facility: http://mdp.ivv.nasa.gov/
17. Chidamber SR, Kemerer CF (1994) A metrics suite for object-oriented design. IEEE T Software Eng, 20(6):476–493
18. McCabe TJ (1976) A complexity measure. IEEE T Software Eng SE-2(4):308–320
19. Henry S, Kafura D (1981) Software structure metrics based on information flow. IEEE T Software Eng SE-7(5):510–518

20. Hosmer D, Lemeshow S (1989) Applied logistic regression. Wiley-Interscience, New York
21. Barnett V, Price T (1995) Outliers in statistical data. John Wiley & Sons, New York
22. Belsley D, Kuh E, Welsch R (1980) Regression diagnostics: identifying influential data and sources of collinearity. John Wiley & Sons, New York
23. Hanley J, McNeil BJ (1982) The meaning and use of the area under a receiver operating characteristic ROC curve. Radiology 143:29–36

Part II
Advances in Image Processing

The processing of gray-scale or color images has become an important research and investigation tool in many areas of science and engineering. Image processing is a key technology for the operational exploitation of images coming from different sources and systems. In this part we introduce advanced theoretical concepts and practical issues associated with image processing. The papers examine the practical applications of image processing for analyzing images from different systems. They treat the problems of image watermarking, diffusing the vector fields of gray-scale images for image segmentation, image processing via synchronization in self-organizing oscillatory networks, feature-based color stereo matching algorithms using restricted search, and edge-preserving lossy image coders. Emphasis is placed on gaining a practical understanding of the principles behind each technique and a consideration of their appropriateness in different applications. Emphasis also is placed on the importance of image interpretation.

Chapter 8
A New Scheme of Image Watermarking

Chokri Chemak, Jean Christophe Lapayre, and Mohamed Salim Bouhlel

Abstract This paper is an attempt to describe a new scheme of image water-marking based on 5/3 wavelet decomposition and turbo code. This new watermarking algorithm is based on embedding a mark (signature) in the multi-resolution field. For the purpose of increasing the image watermarking robustness against attacks of an image transmission, we encode with a turbo code an image-embedded mark. This new scheme of image watermarking is able to embed 2000 bits in medical images. Results of experiments carried out on a database of 30 256×256-pixel medical images show that watermarks are robust to noises, filter attacks, and JPEG compression. Results demonstrate that fidelity can be improved by incorporating a turbo code mark that is shaped into the embedding process. Other advantages of our embedding process are preserving intellectual features and securing the mark in the image. The image degradation is measured by the relative peak signal to noise ratio (RPSNR). Experimental results show that this unit of measurement is the best distortion metric which is correlated with the human visual system (HVS), and is therefore more suitable for digital watermarking.

8.1 Introduction

Image watermarking allows owners or providers to hide an invisible but robust signature inside images, often for security purposes—in particular for owner or content authentication [1, 2]. Medicine has benefited from watermarking research to preserve medical deontology [3, 4] and facilitate distant diagnosis [5]. Watermarking is a solution to conserve the intellectual properties of a diagnostic image, as well as to maintain the perceptual fidelity.

C. Chemak (✉)
Research Unit: Sciences and Technologies of Image and Telecommunications
(SETIT), Higher Institute of Biotechnology of Sfax, Sfax, Tunisia
e-mail: cchemak@univ-fcomte.fr

N. Mastorakis et al. (eds.), *Proceedings of the European*
Computing Conference, Lecture Notes in Electrical Engineering 27,
DOI 10.1007/978-0-387-84814-3_8, © Springer Science+Business Media, LLC 2009

There are three parameters in digital watermarking: data payload, fidelity, and robustness. The reader is directed to [6] for a detailed discussion of these concepts.

In this paper, we develop a new scheme of a robust image-watermark algorithm able to embed large data payloads. This scheme attempts to attain an optimal trade-off between estimates of perceptual fidelity, data payload, and robustness.

This new scheme is able to embed 2000 bits in medical images with 256×256-sized pixels. We propose to embed the mark in the multi-resolution domain.

The embedded mark is coded with an error correcting code (ECC): the turbo code. The ECC has been successfully used in digital communication systems and data storage applications in order to achieve reliable transmission on a noisy channel. Therefore, in the more recent literature, one can find water-marking techniques that use more powerful error correcting codes, such as convolutional codes [7, 8], BCH codes [9–11], or concatenated codes based on a convolutional code followed by a Reed-Solomon code [12], or even convolu-tional turbo codes [13–15]. In line with our goal, we chose the turbo code as an ECC in our watermarking scheme. Since it achieves a high error-correction capability with reasonable decoding complexity, the turbo code uses the soft output Viterbi algorithm (SOVA) [8]. Turbo code allows us to embed large data payloads in medical images while keeping their intellectual properties. This is achieved by increasing the correlation between the extracted marks and the embedded ones.

Our developed watermarking scheme will be tested on 30 256×256-pixel medical images against different attacks, such as noises, filtering and JPEG compression.

For the main goal of preserving image quality and perceptual fidelity, we present the RPSNR as a distortion metric of perceptual image fidelity to estimate image degradation; and in experimental results we demonstrate that it is the best distortion metric to measure image degradation that is correlated with the HVS.

The paper is organized as follows: Section 8.2 describes the new scheme of image watermarking and the main steps relevant to our new watermarking algorithm. We also show a short diagram to explain the choice of the multi-resolution field with 5/3 wavelet decomposition. We explain the turbo code used in our paper. Furthermore, we describe the utilities of using a powerful error correcting code, such as turbo code. Section 8.3 incorporates perceptual shap-ing, based on a distortion metric for perceptual image quality, the RPSNR. This metric is able to reduce the perceptual distance between attacked watermarked images and unmarked ones. We list some simulation results showing that RPSNR is the best metric correlated with HVS to evaluate image degradation after different attacks. Section 8.4 reports a set of selected simulation results which clearly demonstrate that, even after rather severe addition of noise, filtering, and JPEG compression, all 2000 bits are correctly detected in at least 90% of watermarked images, while preserving image fidelity after the watermarking scheme. Finally, Section 8.5 comprises some concluding remarks.

8.2 Our New Image Watermarking Algorithm

In this paper, we propose to embed the mark in the multi-resolution field by 5/3 wavelet decomposition. Our developed scheme is presented in Figs. 8.1 and 8.2.

8.2.1 The Watermarking Embedding Algorithm

- Decomposition of medical image in 5/3 wavelet.
- Choice of pixels embedded: if we embed marks in high frequency zones, they may be altered or lost with high-pass filtering attacks. Futhermore, if we embed marks in low frequency zones, the image can lose its perceptual quality because the low-frequency contains the

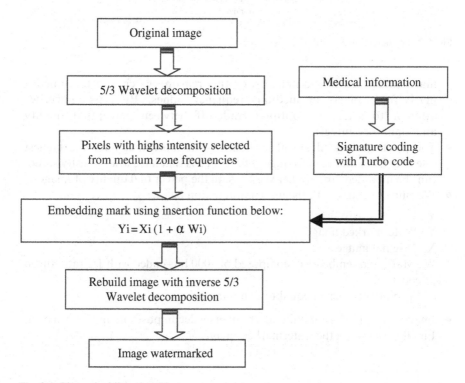

Fig. 8.1 The embedding algorithm

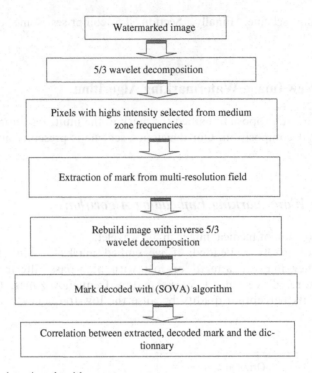

Fig. 8.2 The detection algorithm

mark embedded in the image. For this reason, we choose to embed a mark from pixels in medium-frequency zones. By this choice we attempt to attain an optimal trade-off between perceptual fidelity and signature integrity.

- The pixels embedded are also chosen from those that have the highest intensity from the mean frequency field. Pixels that have low intensity values can[1] be modified; and we here risk losing the perceptual quality of images.
- We embed a mark with the insertion function below:

$Y_i = X_i (1 + \alpha W_i)$:

Y_i: Watermarked image

X_i: Original image

W_i: Mark to be embedded, composed of 2000 bits coded with 1/2 ratio turbo coder.

α: Visibility coefficient equal to 2 in our work.

- Image embedded is rebuilt after inverse decomposition in 5/3 wavelet. Finally, we obtain the watermarked image.

---

[1] Soft output Viterbi algorithm: algorithm of soft turbo code decoding.

8.2.2 The Public Watermarking Detection Algorithm

- 5/3 wavelet medical image decomposition.
- In the first step, pixels with high intensity are selected from the medium-zone frequencies.
- After that, we extract a mark from the multi-resolution field with binary inverse transformation.
- The extracted mark is decoded by the use of the soft output Viterbi algorithm (SOVA).

We make a comparison between the extracted and the decoded mark and a dictionary of 800 marks with the same nature to the embedded mark, containing the reference mark (original mark). This comparison allows us to identify the similarity degree between the extracted marks and the original mark. If the reference mark presents a high value of correlation, we can say that the detection of the mark has succeeded. However, if the reference mark hasn't the maximum correlation, the detection mark is lost.

8.2.3 Use of Multi-Resolution Field: The 5/3 Wavelet Decomposition

Our choice of investigating the 5/3 wavelet decomposition in our double watermarking algorithm is motivated by many reasons:

The 5/3 wavelet decomposition is an integer-to-integer transform adapted for JPEG2000 compression and comes out ahead for its frequent use in the JPEG2000 norm. Consequently, the image watermarked is robust against JPEG compression attacks [16, 17].

In order to keep image fidelity after the watermarking process, we need a perceptual metric correlated with human visual system. Nevertheless, in multi-resolution fields the image decomposition in sample bands is near to the perception canal decomposition, so we can easily choose a psycho-visual model to measure image degradation [2].

The 5/3 wavelet equation decompositions are below:

$$d[n] = d_0[n] - \left\lfloor \frac{1}{2}(d[n] + d[n-1]) \right\rfloor \qquad (8.1)$$

$$s[n] = s_0[n] + \left\lfloor \frac{1}{4}(d[n] + d[n-1]) + \frac{1}{2} \right\rfloor \qquad (8.2)$$

To mark the selected coefficients, we use an embedding function. Therefore, the embedding function has an important role: to specify the applied field of the watermarking. The following function is used with our watermark image:

$$yi = xi(1 + wi).$$

Yi: Watermarked image

Xi: Original image

Wi: Mark to be embedded, composed of 2000 bits coded with 1/2 ratio turbo coder.

α: Visibility coefficient.

8.2.4 Presentation of the Turbo Coder

The system transmission of the turbo coder or parallel concatenation code consists in setting in parallel form the recursive systematic convolutionnel coders (RSC) C_1 and C_2 [18]. The structural diagram of the turbo codes is represented in Fig. 8.3.

d_k is the input bit information. X_k is systematic output bit forming code words.

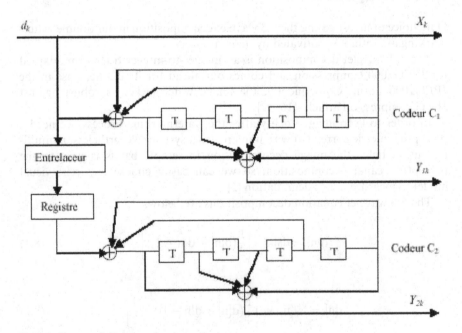

Fig. 8.3 Structural diagram of turbo coder

$Y_{1\,k}$ and $Y_{2\,K}$ are the parities bits coming respectively from the first and the second recursive coder after interleaving the systematic bits or input bi

Turbo codes have been successfully used in digital communication systems in order to achieve reliable transmission on a noisy channel [19–21].

The idea of using turbo code in our "double watermarking" algorithm scheme comes from the efficiency of this ECC to increase robustness against noises. Furthermore, the incorporation of turbo code into the formatting of the watermark increases the number of bits to hide in order to achieve higher payloads. Consequently, the number of repetitions of each bit of the watermark decreases in the same proportion [22]. Fortunately, turbo code using SOVA for the decoding technique is an attractive solution for this application because it achieves a powerful error-correction capability and higher payloads.

8.3 Perceptual Quality Metrics

Nowadays, the most popular distortion measures in the field of images are the signal-to-noise ratio (SNR), and the peak signal-to-noise ratio (PSNR). They are usually measured in decibels (dB): SNR (dB) = 10log10 (SNR).

Their popularity is very likely due to the simplicity of the metric. However, it is well known that these distortion metrics are not correlated with human vision [23]. This might be a problem for their application in digital watermarking, since sophisticated watermarking methods exploit the HVS in one way or another.

In addition, using the above metric to quantify the distortion caused by a watermarking process might result in misleading quantitative distortion measurements [24].

Furthermore, these metrics are usually applied to the luminance and chrominance channels of images [25], and they give a distortion value for all color channels.

In this paper, we introduce a metric of an image-quality evaluation entitled relative peak signal-to-noise ratio (RPSNR). This distortion metric, which has no relation to the content characteristics of the image, fits with the HVS and therefore is more suitable for digital watermarking.

In addition, this metric allows comparison even if the distortion is in a different color channel [26].

The estimation error for RPSNR is a function of packet loss rate and average loss burst length metric that represents path quality under different loss patterns.

The relative pick signal-to-noise ratio (RPSNR) is used to evaluate the image quality by calculating the relative mean square error (RMSE) between the images being compared.

The equation is as follows:

$$RMSE = \frac{1}{MN} \sum_{m=0}^{M-1} \sum_{n=0}^{N-1} \left[2 * \frac{|x(m,n) - y(m,n)|}{|x(m,n) + y(m,n)|} \right]^2 \tag{8.3}$$

$$RPSNR = 10_{Log\,10} \left(\frac{(\text{valeur max du signal})_2}{RMSE} \right) \tag{8.4}$$

M and N are the numbers of pixels lengthwise and width-wise per image; x and y are the grey scales of the images being compared.

In order to properly demonstrate the performance of RPSNR in watermarking schemes and allow a fair comparison between different perceptual quality metrics, the setup test conditions are of crucial importance. Table 8.1 lists different mean values of PSNR, WPSNR, and RPSNR after the most famous types of distortion and attacks on a database of 30 256×256-pixel medical images. Figure 8.4 presents some images from the above-mentioned database.

Table 8.1 Different mean values of PSNR, WPSNR, and RPSNR after different attacks in image banks

Distortion type	PSNR	WPSNR	RPSNR
Mean shift	24.6090	35.6873	64.3123
Contrast stretching	24.6003	35.7453	57.913
Impulsive Salt and paper noise	24.6499	35.4654	63.6996
Multiplicative speckle noise	24.6186	34.6206	66.9165
Additive Gaussian noise	24.5906	35.5855	62.4892
Floue	24.6054	45.0642	63.4734
JPEG compression	24.7849	38.7652	62.6179

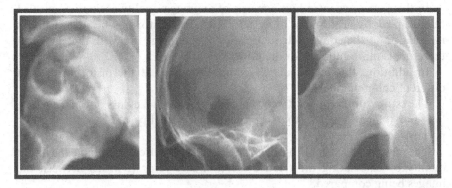

Fig. 8.4 Some figures from the database of 30 medical images

8.4 Preliminary Results

8.4.1 Robustness Against Attacks

This section displays first the experimental results carried out on a database of 800 marks in which the extracted and decoded marks are tested. The test technique is made by correlation between the extracted and decoded marks and the dictionary.

Figures 8.5, 8.6, and 8.7 demonstrate that the marks are correctly detected in our simulation results from watermarked images after noises attacks, JPEG compression attacks, and filtering attacks.

8.4.2 Fidelity of Watermarked Images After Attacks

We present in the following legend a summary of attacks against which the watermarked images are tested and evaluated, in order to validate the perceptual fidelity in our proposed approach, after the embedding of 2000 bits.

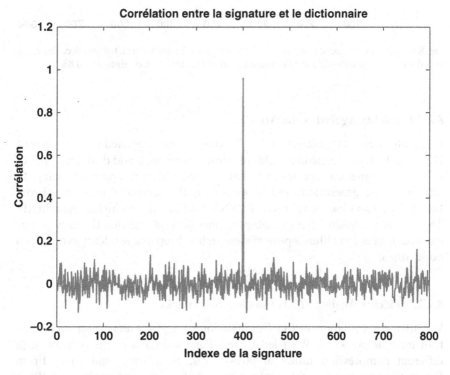

Fig. 8.5 Succeeded detection of a mark after correlation between extracted and decoded mark and the dictionary following Gaussian noise (0.03) attack. Correlation = 0.90

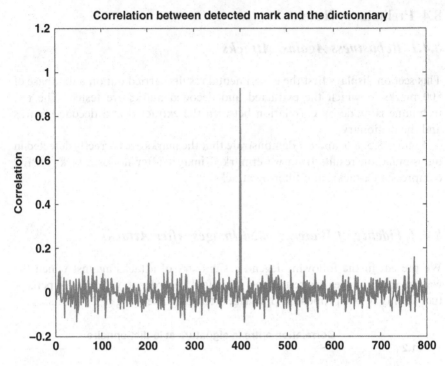

Fig. 8.6 Succeeded detection of a mark after correlation between extracted and decoded mark and the dictionary following image compression of Quality 60. Correlation = 0.83

8.4.2.1 Fidelity Against Noise Attacks

It is quite relevant to evaluate the robustness of the suggested method against Noise. In fact, in the medical field, the instruments used add different types of noise to the medical images. We have tested our new approach using 10 different noise generations and by modifying the variances each time. From Fig. 8.8, we can observe values of RPSNR that are always higher than 30 dB. This makes it obvious that the image quality is good and that this new watermarked images algorithm is powerful enough to keep image fidelity even after a noise attack.

8.4.2.2 Fidelity Against JPEG Compression Attacks

We often need to apply JPEG Compression to medical images for archive or transmission purposes. We tried to test the robustness of our scheme with different compression ratios of 90%, 70%, 50%, 30% and 10%. From Fig. 8.9, we can see that fidelity is improved after different qualities of JPEG compression attacks.

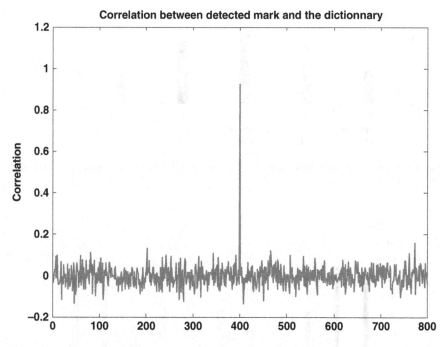

Fig. 8.7 Succeeded detection of a mark after correlation between extracted and decoded mark and the dictionary following Gaussian filter. Correlation = 0.92

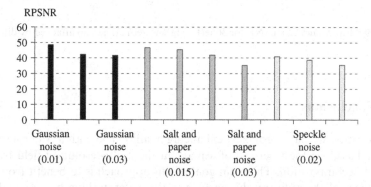

Fig. 8.8 Mean values of RPSNR for 30 test images watermarked and attacked by different types of noises

8.4.2.3 Fidelity Against Filtering Attacks

We have tested the robustness of our proposed method face against attacks of four filter types: Gaussian, unsharp, average, and motion. Figure 8.10 displays good values of RPSNR. It follows then that fidelity in images is improved after filtering attacks.

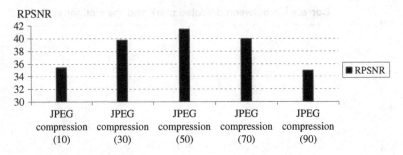

Fig. 8.9 Mean values of RPSNR for 30 test images watermarked and attacked by different ratios of JPEG compression

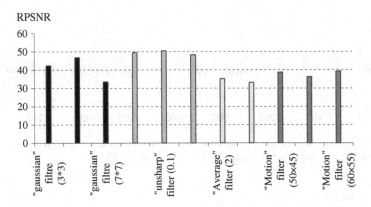

Fig. 8.10 Mean values of RPSNR for 30 test images watermarked and attacked by different filters

8.5 Conclusions

In this paper, we have proposed enhancing a still image with a watermarking scheme based on the insertion of a mark in the multi-resolution field by 5/3 wavelet decomposition. The main goal of this approach is to benefit from the advantages of the multi-resolution field in the watermarking process and the turbo code process. The watermarking scheme uses powerful error-correcting turbo codes to improve resistance against attacks. Simulation results show that for a given payload of 2000 bits, and after the severe addition of noises, JPEG compression, and filtering, all bits are detected in at least 90% of watermarked images from the multi-resolution field. Furthermore, for different tested attacks, the perceptual metric incorporating the RPSNR allows us to keep perfect values of images distorted above 30 dB. Consequently, we have good perceptual fidelity in the images after the watermarking process.

References

1. Cox IJ, Miller ML, Bloom JA (2001) Digital watermarking. Morgan Kaufmann, New York
2. Rey C, Dugelay JL (2000) Tech demo. Image watermarking for owner and content authentication. ACM Multimedia, Los Angeles
3. Chemak C, Bouhlel MS, Lapayre JC (2006) Algorithme de Tatouage Robuste et Aveugle pour la Déontologie et le Transfert des Informations Médicales: Le Tatouage Combinée ATRADTIM. Brevet déposés à l'INNORPI, SN06448
4. Kong X, Feng R (2001) Watermarking medical signals for telemedicine. IEEE T Inf Technol Biomed 5(3):195–201
5. Tricili H, Bouhlel MS, Derbel N, Kamoun L (2002) A new medical image watermarking scheme for a better telediagnosis. IEEE conference on systems, man and cybernetics. SMC, Hammamet, Tunisia
6. Miller ML, Doerr GJ, Cox IJ (2004) Applying informed coding and informed embedding to design a robust high capacity watermark. IEEE T Image Process 6(13):792–807
7. Miller ML, Doerr GJ, Cox IJ (2002) Dirty-paper trellis codes for watermarking. In: Proceedings of the International Conference on Image Processing. Rochester, NY, USA, 2:129–132
8. Hernández JR, Delaigle JF, Macq B (2000) Improving data hiding by using convolutional codes and soft-decision decoding. In: Proceedings of SPIE—Security and Watermarking of Multimedia Contents II. San Jose, CA, USA, pp 24–48
9. Darbon J, Sankur B, Maître H (2000) Error correcting code performance for watermark protection. In: Proceedings of SPIE—Security and Watermarking of Multimedia Contents III. San Jose, CA, USA
10. Baudry S, Delaigle JF, Sankur B, Macq B, Maître H (2001) Analyses of error correction strategies for typical communication channels in watermarking. Signal Process 81(6):1239–1250
11. Tefas A, Pitas I (2000) Multi-bit image watermarking robust to geometric distortions. IEEE international conference on image processing, Vancouver, Canada, September 2000
12. Inoue H, Miyazaki A, Yamamoto A, Katsura T (1999) A digital watermark based on the wavelet transform and its robustness on image compression and transformation. IEICE T Fund Electron Common Comput Sci E82-A: 2–10
13. Pereira S, Voloshynovskiy S, Pun T (2000) Effective channel coding for DCT watermarks. IEEE international conference on image processing. Vancouver, Canada
14. Brandão T, Queluz MP, Rodrigues A (2001) Improving spread spectrum-based image watermarking through non-binary channel coding. In: Proceedings of 3rd conference on telecommunications. Figueira da Foz, Portugal
15. Kesal M, Mýhçak MK, Koetter R, Moulin P (2000) Iteratively decodable codes for watermarking applications. Proceedings of 2nd international symposium on turbo codes and related topics. Brest, France
16. Chemak C, Bouhlel MS, Lapayre JC (2007) Un schéma de tatouage basé sur les ondlettes 5/3 et le turbo code pour la robustesse du tatouage contre les bruits gaussien et la compression JPEG 5èmes Journées Scientifiques de l'Ecole de l'aviation de Borj el Amri,Tunisie.
17. Adams MD, Kossentini F (2000) Reversible integer-to-integer wavelet transforms for image compression against attack of JPEG compression, evaluation and analysis. IEEE T Image Process 9(6):1010–1024
18. Berrou C, Glavieux A (1996) Near optimum error correcting coding and decoding: turbo codes. IEEE T Comm pp1261–1271

19. Chemak C, Bouhlel MS (2006) Near Shannon limit for turbo code with short frames. In: 2nd IEEE international conference on information and communication technologies from theory to applications (ICTTA'06), Damascus, Syria. (ISBN: 0–7803–9521–2)

20. Chan F (1999)Adaptive Viterbi decoding of turbo codes with short frames. In: Proceedings of the IEEE Commun. interleaver Good component codes have also been Theory Mini-Conference, in conjunction with ICC'99, pp 47–51

21. Chappelier V, Guillemot C, Marinkovic S (2003) Turbo trellis coded quantization. In: Proceedings of the International Symposium on Turbo Codes. Brest, France

22. Rey C, Amis K, Dugelay JL, Pyndiah R, Picart A (2003) Enhanced robustness in image watermarking using block turbo codes. In: Proceedings of SPIE—Security and Watermarking of Multimedia Contents V, 5020:330–336

23. Rey C, Dugelay JL (2002) A fair benchmark for image watermarking systems. Technical Report. Institut EURECOM, Sophia Antipolis, France

24. Winkler S (1999) A perceptual distortion metric for digital color video. In: SPIE proceedings of human vision and electronic imaging. San Jose, CA, Vol. 3644

25. Westen SJP, Lagendijk RL, Biemond J (1995) Perceptual image quality based on a multiple channel HVS model. In: Proceedings of ICASP, 4:2351–2354

26. van den Branden L, Christian J, Farrell JE (1996) Perceptual quality metric for digitally coded color images. In: Proceedings of EUSIPCO. Trieste, Italy, pp 1175–1178

Chapter 9
Diffusing Vector Field of Gray-Scale Images for Image Segmentation

X.D. Zhuang and N.E. Mastorakis

Abstract In this paper, a novel vector field transform is proposed for gray-scale images, which is based on the electrostatic analogy. By introducing the item of gray-scale difference into the repulsion vector between image points, the diffusing vector field is obtained by the transform. Then the primitive areas can be extracted based on the diffusing vector field with a proposed area-expanding method. A new image segmentation method is then presented by merging primitive areas. The experimental results prove the effectiveness of the proposed image segmentation method.

9.1 Introduction

Image transform is one of the fundamental techniques in image processing [1]. The image transform generates another space or field, where some characteristics of the generated space may be exploited for effective and efficient processing of the image. Classical image transform includes mathematical transform such as the Fourier transform, the Walsh transform, etc. A relatively new technique is the wavelet transform. In these techniques, the digital image is regarded as a discrete 2-D function and is transformed to the coefficient space. A more general view of image transform may include the transformation to the feature space. The gradient field can be a typical case, which is generated by the convolution of the image and the gradient template. In the gradient field, the edge feature of the digital image can be extracted.

Many image transform methods result in a space or field of scalar coefficients or scalar feature values. Some others can result in a vector field, such as the gradient field. The gradient templates can extract the components of the image gradient on the direction of the x-coordinate and y-coordinate

X.D. Zhuang (✉)
WSEAS Headquarters, Agiou Ioannou Theologou 17–23, 15773, Zografou, Athens, Greece

respectively. A general idea about image transform may include transformation to both scalar space and vector field.

Because the vector field possesses information of both intensity and direction, the vector field transform may give a detailed representation of image structure and features. Some physics-based approaches have been applied in image processing, which take an electrostatic analogy in the transformation from the image to the vector field [2, 3]. Such methods have got effective and promising results in skeletonization, shape representation, human ear recognition, etc. [4–7].

In this paper, a novel vector field is proposed to represent image structure of different areas in the image. The diffusing vector field is defined by extending the vector field of the electrostatic analogy to a more generalized form. Based on the diffusing vector field, the source points of diffusing vectors can be extracted and the image can be decomposed to primitive areas, based on which the image segmentation can be implemented by merging the primitive areas. The experimental results indicate the effectiveness of the proposed segmentation method.

9.2 Diffusing Vector Field of Gray-Scale Images

In physics, a charged area with a certain distribution of charge generates its electric field within and outside the area. In this section, a novel vector transform of a gray-scale image is proposed based on an electrostatic analogy, in which the image is regarded as a charged area. In the proposed transform, the form of the field force is extended by introducing the gray-scale difference between the related image points. With such a definition of the transform, in the generated field the vectors in a homogeneous area diffuse towards the outside of that area, and the generated field is named the diffusing vector field.

9.2.1 The Form of Electrostatic Field Force

The force of two charges q_1 and q_2 is given as the following [8]:

$$\vec{F}_{12} = \frac{1}{4\pi\varepsilon} \cdot \frac{q_1 q_2}{r_{12}^2} \cdot \frac{\vec{r}_{12}}{r_{12}} \tag{9.1}$$

where $\vec{F}_{12}$ is the force of q_1 on q_2, $\vec{r}_{12}$ is the vector from q_1 to q_2, r_{12} is the length of $\vec{r}_{12}$, and $4\pi\varepsilon$ is an item of constant.

The form of the electrostatic field force can be introduced into the vector field transform of images [2, 3]. If two image points are regarded as two charged particles, the force vector generated by one point on the other can be defined.

9.2.2 The Repulsion Vector Between Image Points

The form of electronic force formula has some characteristics as follows:

1. The formula has the power of distance r as one of the denominators. The larger the distance between two charged particles, the smaller the force. In images, this causes a kind of local feature extraction. One image point has a strong effect on the points nearby, but has little effect on distant points.
2. The force between two charged particles is related to the electric quantity of both charged particles. In images, the effect of one image point on the other point can also be defined with relation to the intensities (i.e., gray-scale values) of the two image points. Thus certain image features may be extracted by the vector field transform.

In this paper, the vector generated by one image point $g(i, j)$ on another position (x, y) is defined with direct relation to the reciprocal of the intensity difference of the two image points. The definition is proposed to generate a repulsion vector between neighboring points in homogeneous areas. The repulsion vector is defined as follows:

$$\vec{V} = \frac{A}{(|g(i,j) - g(x,y)| + \varepsilon) \cdot r^2_{(i,j)\rightarrow(x,y)}} \cdot \frac{\vec{r}_{(i,j)\rightarrow(x,y)}}{r_{(i,j)\rightarrow(x,y)}} \quad (9.2)$$

where $\vec{V}$ is the vector generated by image point (i, j) on position (x, y), g represents the intensity of image points, $\vec{r}_{(i,j)\rightarrow(x,y)}$ is the vector from (i, j) to (x, y), $r_{(i,j)\rightarrow(x,y)}$ is the length of $\vec{r}_{(i,j)\rightarrow(x,y)}$, ε is a pre-defined small positive value which guarantees that the above definition is still valid when $g(i, j)$ is equal to $g(x, y)$, and A is a pre-defined item of constant. According to the above definition, the two components of $\vec{V}$ are as follows:

$$V_x = \frac{A \cdot (x - i)}{(|g(i,j) - g(x,y)| + \varepsilon) \cdot ((x - i)^2 + (y - j)^2)^{3/2}} \quad (9.3)$$

$$V_y = \frac{A \cdot (y - j)}{(|g(i,j) - g(x,y)| + \varepsilon) \cdot ((x - i)^2 + (y - j)^2)^{3/2}} \quad (9.4)$$

where V_x and V_y are the components on the direction of the x-coordinate and the y-coordinate respectively.

9.2.3 The Diffusing Vector Field of Images

In Section 9.2.2, a definition of the repulsion vector is proposed for one image point on another. Based on the repulsion vector, the vector field transform can be defined for the whole image by summing up the vectors produced by all image points on any image points. The vector generated by the whole image on point (x, y) is defined as follows:

$$\vec{V}(x,y) = \sum_{\substack{i=1 \\ (i,j) \neq (x,y)}}^{W} \sum_{j=1}^{H} A \cdot \frac{\vec{r}_{(i,j) \to (x,y)}}{(|g(i,j) - g(x,y)| + \varepsilon) \cdot r^3_{(i,j) \to (x,y)}} \tag{9.5}$$

where $\vec{V}(x, y)$ is the vector produced by the transform on position (x, y), and W and H are the width and height of the image respectively. According to the above definition, the two components of $\vec{V}(x, y)$ are as follows:

$$V_y(x,y) = \sum_{\substack{i=1 \\ (i,j) \neq (x,y)}}^{W} \sum_{j=1}^{H} \frac{A \cdot (x - i)}{(|g(i,j) - g(x,y)| + \varepsilon) \cdot r^3_{(i,j) \to (x,y)}} \tag{9.6}$$

$$V_y(x,y) = \sum_{\substack{i=1 \\ (i,j) \neq (x,y)}}^{W} \sum_{j=1}^{H} \frac{A \cdot (y - j)}{(|g(i,j) - g(x,y)| + \varepsilon) \cdot r^3_{(i,j) \to (x,y)}} \tag{9.7}$$

where $V_x(x, y)$ and $V_y(x, y)$ are the components on the directions of the x-coordinate and y-coordinate respectively.

Because the effect of an image point on another decreases quickly with the increase of distance, the vector on any image point is determined by two major factors: the strong effect of a few neighboring points, and the accumulated effect of a large number of distant points. In the definition of the diffusing vector field, the smaller the gray-scale difference, the relatively larger the vector length. Therefore, a diffusing vector field will appear in each homogeneous area because of the strong "repulsion" between similar image points. On the other hand, at the boundary of two different areas, the vector fields at one side of the boundary will be in the opposite directions of those at the other side.

To investigate the property of the proposed transform, several simple test images are transformed to the diffusing vector field. The algorithm is implemented under the Visual C + + 6.0 development environment. Three of the test images are shown in Figs. 9.1, 9.4, and 9.7. These images are of size 32×32. For a clearer view, they are also shown 4 times the original size. Figures 9.2, 9.5, and 9.8 show the length of each vector in the transformed field respectively, where larger gray-scale values correspond to larger vector lengths. The results are also shown 4 times original size for a clearer view. The direction of each vector in the transformed field is digitalized into 8 discrete directions for further processing. Figures 9.3, 9.6, and 9.9 show the direction of the transformed field for each test image.

The image *test1* is an image of monotonous gray-scale, i.e., the whole image is a homogeneous area. In the transformed field of *test1*, it is obvious that the whole field is diffusing from the center of the image towards the outside.

Fig. 9.1 The first image *test1* (the original image on the left, and 4 times the original size on the right)

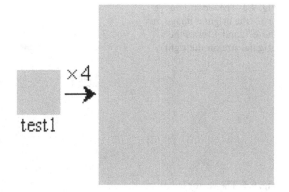

Fig. 9.2 The vector length in the transformed field of *test1* (the original image; 4 times original size on the right)

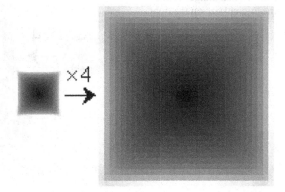

Fig. 9.3 The direction of each vector in the transformed field of *test1*

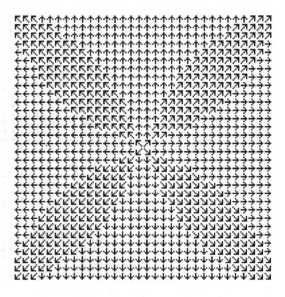

Fig. 9.4 The second image *test2* (the original image on the left, and 4 times the original size on the right)

Fig. 9.5 The vector length in the transformed field of *test2* (the original image; 4 times the original size on the right)

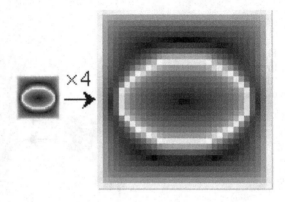

Fig. 9.6 The direction of each vector in the transformed field of *test2*

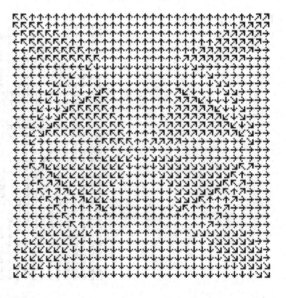

Fig. 9.7 The third image *test3* (the original image on the left, and 4 times the original size on the right)

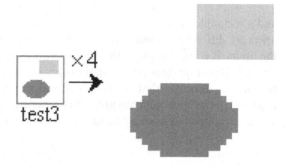

Fig. 9.8 The vector length in the transformed field of *test3* (the original image; 4 times the original size on the right)

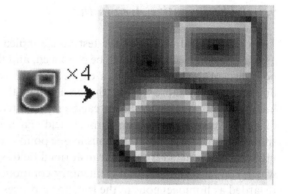

Fig. 9.9 The direction of each vector in the transformed field of *test3*

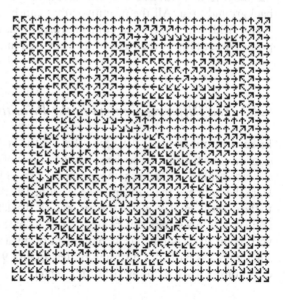

There is an ellipse area in image *test2*. In image *test3,* there are an ellipse area and a rectangle area. In their transformed fields, the fields in the homogeneous areas are diffusing outward from the center of each area. On the boundaries of the areas, it is obvious that the vectors at one side of the boundary line have opposite directions from those on the other side. The experimental results of the test images indicates that the proposed transform produces a diffusing vector field within the homogeneous areas, but generates vectors of opposite directions at the two opposite sides along the area boundary.

9.3 Image Segmentation by the Diffusing Vector Field

9.3.1 The Primitive Area in Images

The experimental results of the test image indicate that in the homogeneous area a diffusing vector field will be produced; and the diffusing field ends at the boundary of the homogeneous area, because the vectors outside have opposite directions from those within the area along the boundary. Therefore, the homogeneous areas in the image can be represented by the areas with consistent diffusing vectors in the transformed field. Each diffusing vector area corresponds to an area of homogeneous image points. The area of consistent diffusing vectors extracted from the transformed field is defined as a primitive area, which can be regarded as an elementary component of an image because it is regarded as homogeneous in the transform process.

According to the definition, the image *test1* is a whole primitive area, while the image *test3* has at least two primitive areas: the ellipse, the rectangle, and the background area. All the primitive areas can be extracted from the diffusing vector field, which can be exploited in further image analysis. In this paper, the primitive area forms the basis of the proposed image segmentation method.

9.3.2 The Diffusing Centers in the Primitive Area

In each primitive area, the vector field diffuses from the center towards the outside, thus the area center becomes the source of the diffusing field. Therefore, the area centers are the beginning points from which to extract primitive areas. Here the source of the diffusing field is defined as the diffusing center. According to the experimental results of the test images, the definition of the diffusing center is given as follows: for a square area consisting of four image points, if none of the vectors on these points has a component of inward direction into the area, the square area is part of a diffusing center. Figure 9.10 shows the allowed vector directions on each point in a diffusing center.

Fig. 9.10 The allowed
vector directions in a
diffusing center

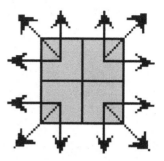

In Figs. 9.3, 9.6, and 9.9, according to the above definition, the diffusing centers can be found, which are shown in Figs. 9.11, 9.12, and 9.13. The source points in the diffusing centers are indicated in gray.

The image *test1* is a homogeneous area, therefore there is only one diffusing center found in Fig. 9.11. There is an area of ellipse in the image *test2,* and the diffusing center of the ellipse can be found in Fig. 9.12. Moreover, there are also four other diffusing centers found in the background area. The image *test3* has an ellipse and a rectangle. Correspondingly, in Fig. 9.13 there is one diffusing center for the ellipse, one for the rectangle, and five for the background area. It is indicated that in a large and irregular area there may be more than one diffusing center found, such as the background area.

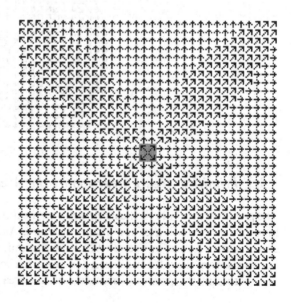

Fig. 9.11 The diffusing
centers in Fig. 9.3

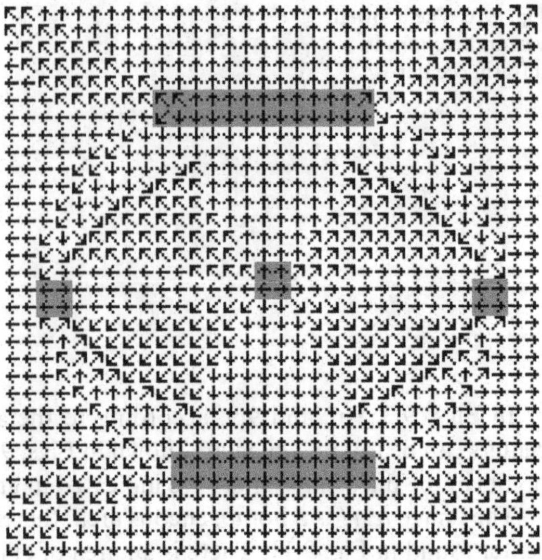

Fig. 9.12 The diffusing
centers in Fig. 9.6

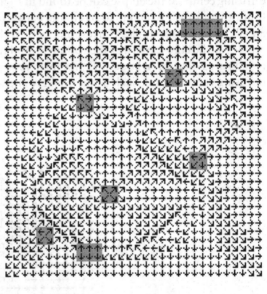

Fig. 9.13 The diffusing
centers in Fig. 9.9

9.3.3 Primitive Area Extraction by the Area-Expanding Method

The primitive areas are the basic elements in the diffusing vector field, which is a kind of representation of the image structure. As shown by the analysis and experimental results in Section 9.2.3, in a primitive area the vectors diffuse outwards from the diffusing center (i.e., the area center). Moreover, the diffusing vectors in the primitive area end at the area boundary, where opposite vectors on the outside are encountered. Therefore, the primitive area can be extracted by expanding outwards from the diffusing center along the directions of the diffusing vectors. The proposed area-expanding method to extract the primitive area is as follows:

Step 1: Get the diffusing vector field of the image by the transform proposed in Section 9.2.3; each image point now has a vector on it (the vector is discretized into 8 directions).

Step 2: Get the diffusing center points in the diffusing vector field according to the definition in Section 9.3.2.

Step 3: Assign each diffusing center a unique area label (here a unique area number is given to the points in each diffusing center, while the points not in the diffusing center are left unlabeled).

Step 4: Then a process of area expanding in the diffusing vector field is implemented to extract the primitive areas.

For each labeled point, select five of its neighboring points that are nearest to its vector's direction. For each of these points, if it is unlabeled and its vector is not opposite to the labeled point's vector (i.e., the area boundary is not reached), it is labeled the same area number as the labeled one. Otherwise, if the neighboring point has been labeled with another area number, a principle of least gray-scale difference is applied. The difference between its gray-scale and either area's average gray-scale is calculated. The point will belong to the area with less gray-scale difference. The primitive area can expand by iteration until the area boundary is reached. The above process is repeated until the areas all stop expanding.

Step 5: If there are still unlabeled points when the expanding of the areas stops, the principle of least gray-scale difference is applied. For each unlabeled point, calculate the difference between its gray-scale and the average gray-scale of its neighboring areas. Then this unlabeled point is merged into the neighboring area that is of the least difference.

The primitive areas are extracted for the three test images in Section 9.2.2 by area expanding in the diffusing vector fields. The experimental results are shown in Figs. 9.14, 9.15, and 9.16. In these three figures, the original images and the results of primitive area extraction are shown. The results are also shown 4 times the original size for a clearer view. In these figures, different primitive areas are distinguished from each other by different gray-scale values.

The image *test1* is a homogeneous area. Therefore the primitive area extracted in *test1* is only one complete area (i.e., the image itself).

The image *test2* contains an ellipse, and 3 primitive areas are obtained. The ellipse is extracted as 1 primitive area, and there are 2 other primitive areas extracted in the background area of *test2*.

Fig. 9.14 The result of
primitive area extraction
for *test1*

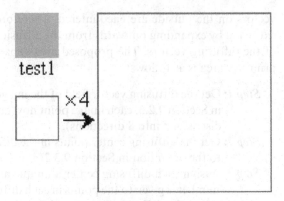

Fig. 9.15 The result of
primitive area extraction
for *test2*

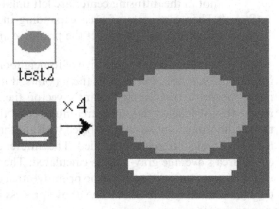

Fig. 9.16 The result of
primitive area extraction
for *test3*

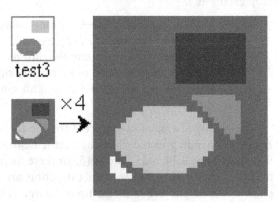

The image *test3* contains an ellipse and a rectangle, and 5 primitive areas are obtained. The ellipse and rectangle are extracted as two primitive areas, and there are 3 other primitive areas extracted in the background area of *test3*.

The experimental results for the test images show that the object areas can be extracted as primitive areas, such as the ellipse in *test2* and the ellipse and rectangle in *test3*. On the other hand, the number of primitive areas may be less than the number of diffusing centers extracted. This is because two or more diffusing centers may merge into one area in *Step 4* in the proposed area-expanding method.

9.3.4 Gray-Scale Image Segmentation Based on the Diffusing Vector Field

Compared with the test images, practical images obtained in the real world are more complex and contain many more objects. The boundaries between areas in these images are not as clear and distinguishable as in the test images. In the experiments, the primitive areas are also extracted for the pepper image, the cameraman image, and the house image. These images are of the size 128×128. The experimental results show that there are a large number of primitive areas extracted from the practical images. There are 341 primitive areas in the pepper image, 305 in the cameraman image, and 263 in the house image. This is because of the complexity of these images.

The primitive areas are a kind of representation of the image structure. To implement meaningful image segmentation, area merging must be done to get a more practically useful result. An area merging method is proposed to combine primitive areas based on the least gray-scale difference principle. First the expected number of remaining areas after merging is given. Then the following steps are carried out to merge areas until the expected area number is reached:

Step 1: For each area in the image, calculate its average gray scale.
Step 2: Find the pair of neighboring areas with the least average gray-scale difference, and merge them into one area.
Step 3: If the current area number is larger than the final area number, return to *Step 1*; otherwise, end the merging process.

Figures 9.17, 9.18, and 9.19 show the original images of the peppers, the cameraman, and the house, and also the results of merging the primitive areas respectively. In the results of primitive area merging, different areas are distinguished from each other by different gray-scale values. Figure 9.17 shows the result of merging 341 primitive areas into 20 areas for the peppers image.

Fig. 9.17 The image of peppers and the result of merging the primitive areas (20 areas remained)

Fig. 9.18 The image of a cameraman and the result of merging the primitive areas (12 areas remained)

Fig. 9.19 The image of a house and the result of merging the primitive areas (20 areas remained)

Figure 9.18 shows the result of merging 305 primitive areas into 12 areas for the cameraman image. Figure 9.19 shows the result of merging 263 primitive areas into 20 areas. The experimental results indicate that the primitive area-merging method can effectively implement image segmentation, and the main objects in the images can be successfully extracted by the proposed method.

As described in the previous sections, here a novel image segmentation method is proposed, based on the diffusing vector fields as follows:

Step 1: Get the diffusing vector fields of the image.
Step 2: Get the diffusing center points.
Step 3: Extract the primitive areas.
Step 4: Merge the primitive areas according to the requirement of the final area number.

The effectiveness of this method has been indicated by the above experimental results for the practical images from Figs. 9.17, 9.18, and 9.19.

9.4 Conclusion

In the study of image transforms, vector field transformation is a promising methodology, in which both vector length and vector direction can be exploited for feature extraction and analysis. Electrostatic analogy has become a useful way of designing vector field transform of images.

In this paper, the diffusing vector field transform is proposed by introducing the factor of gray-scale difference into the electrostatic analogy. In the diffusing vector field of images, homogeneous areas are expressed as the areas with a vector group diffusing outwards from the center.

Based on the proposed transform, an effective image segmentation method is presented. By finding the area center and the area-expanding method, primitive areas can be extracted. Then image segmentation is implemented by merging the primitive areas. The experimental results indicate the effectiveness of the segmentation method. Objects can be successfully extracted in practical images in the real world with the proposed method. Further research work will investigate more applications of the diffusing vector field transform in other tasks of image processing and analysis.

References

1. YuJin Z (1999) Image engineering: image processing and analysis. TUP Press, Beijing, China
2. Hurley DJ, Nixon MS, Carter JN (2005) Force field feature extraction for ear biometrics. Comput Vis Image Und 98(3):491–512
3. Hurley DJ, Nixon MS, Carter JN (1999) Force field energy functionals for image feature extraction. In: Proceedings of the British machine vision conference, pp 604–613
4. Luo B, Cross AD, Hancock ER (1999) Corner detection via topographic analysis of vector potential. Pattern Recogn Lett 20(6):635–650
5. Ahuja N, Chuang JH (1997) Shape representation using a generalized potential field model. IEEE T PAMI 19(2):169–176
6. Grogorishin T, Abdel-Hamid G, Yang YH (1996) Skeletonization: An electrostatic field-based approach. Pattern Anal Appl 1(3):163–177
7. Abdel-Hamid G, Yang YH (1994) Multiscale skeletonization: an electrostatic field-based approach. In: Proc. IEEE international conference on image processing. Austin, Texas, 1:949–953
8. Grant IS, Phillips WR (1990) Electromagnetism. 2nd ed. Wiley

Chapter 10
Image Processing via Synchronization in a Self-Organizing Oscillatory Network

Eugene Grichuk, Margarita Kuzmina, and Eduard Manykin

Abstract We develop a biologically inspired method of image processing based on synchronization-based performance of an oscillatory network with controllable self-organized coupling. The oscillatory network, obtained from a previously designed biologically motivated oscillatory neural network model of the brain's visual cortex, provides automatic, adaptive, and active image segmentation. Being tuned by an image to be processed, the network dynamics realizes network decomposition into a set of synchronized ensembles of oscillators, corresponding to image decomposition into the required set of image fragments. The current network model version provides: (a) full segmentation of real grey-level and colored images; and (b) selective image segmentation (extraction of a subset of image fragments with brightness values contained inside a given arbitrary brightness interval).

10.1 Introduction

Currently there is a significant interest in neuromorphic methods of image processing, based on the imitation of neurobiological processes in the brain's neuronal structures, despite a great variety of traditional methods developed in the field of computer vision. Such advantages as self-organized automatic performance and the capability of adaptive and active processing are usually inherent to biologically inspired methods. Specifically, interest in oscillatory methods of image processing was related to synchronized oscillations of neural activity that were experimentally discovered in the brain's visual cortex (VC) in 1988–1989 and were confirmed in later experiments [1–3]. The synchronized oscillations are believed to accompany visual information processing in the brain's visual structures. The attention to oscillatory aspects of visual information processing resulted in the creation of a series of oscillatory network

E. Grichuk (✉)
Moscow Engineering Physics Institute, Kashirskoe Schosse, 115409 Moscow, Russia
e-mail: es@t-25.ru

N. Mastorakis et al. (eds.), *Proceedings of the European Computing Conference*, Lecture Notes in Electrical Engineering 27, DOI 10.1007/978-0-387-84814-3_10, © Springer Science+Business Media, LLC 2009

models for image processing, demonstrating synchronization capabilities [4–18]. Oscillatory network models developed by D. Wang and Z. Li [4–12] are most closely related to our model, but are nevertheless different. The relation of our model to those by D. Wang and Z. Li was discussed in detail in [16]. Our oscillatory network, providing a dynamic method of image processing, was obtained by reduction from previously designed oscillatory network models of the primary visual cortex. The starting model simulated self-organized collective behavior of orientation selective cells of the primary visual cortex at a low (pre-attentive) level of visual information processing. The active network unit is the neural oscillator, formed by a pair of interconnected cortical neurons. It is a relaxational (limit cycle) oscillator with dynamics, controlled by image characteristics. The spatial architecture of the 3D starting model imitated the columnar structure of the VC. A network coupling principle was designed based on known neurobiological data about connections in the VC and also on general principles of connection formation in the brain's neural structures. The coupling principle of the working 2D oscillatory network model causes self-organized emergence of synchronization in the network and creates the simplest type of dynamical binding (on brightness). The current model version provides a workable dynamical method of image segmentation. It is capable of processing both grey-level and colored real multipixel images. In addition, it allows a natural method for selective image segmentation that can be regarded as a simple kind of active image processing.

10.2 Main Characteristics of the Oscillatory Network

The 2D oscillatory network is designed for brightness image segmentation tasks. We mean image segmentation as image decomposition into a set of image fragments—massive subregions of image pixels with a constant level of brightness. Oscillators of the network are located at the nodes of a two-dimensional square lattice, being in one-to-one correspondence with the pixel array of the segmented image. Image segmentation is carried out by the oscillatory network via synchronization of network assemblies, corresponding to image fragments of various brightness levels. If the image to be segmented is defined by $M \times N$–matrix $[I_{jm}]$ of pixel brightness values, the network state is defined by $M \times N$–matrix $\hat{u} = [u_{jm}]$ of complex-valued variables, defining states of all network oscillators. The system of ODE, governing oscillatory network dynamics, can be written as:

$$du_{jm}/dt = f(u_{jm}; I_{jm}) + \sum_{j',m'}^{N} W_{jmj'm'} \cdot (u_{f'm'} - u_{jm}), \qquad (10.1)$$

$$j = 1, \ldots, M; \ m = 1, \ldots, N$$

Here the function $f(u_{jm}; I_{jm})$ defines the internal dynamics of isolated network oscillators, whereas the second term defines the contribution into dynamics via oscillator coupling. The single network oscillator is a limit cycle oscillator defined by a pair of real-valued variables (u_1, u_2). A dynamical system, governing single oscillator dynamics, can be written in the form of an ODE for the complex-valued variable $u = u_1 + iu_2$:

$$du/dt = f(u, I) \tag{10.2}$$

where

$$f(u, I) = (\rho^2 + i\omega) - |u - \rho(1 + i)|^2)(u - \rho(1 + i) - \alpha T(\rho)[u - \rho(1 + i)] \tag{10.3}$$

$$\rho = \rho(I); \quad T(\rho) = 0.5[||th(\sigma(\rho - h_*))|| - th(\sigma(\rho - h_*))] \tag{10.4}$$

The limit cycle of dynamical system (2)–(4) is the circle of radius ρ, the circle center being located at the point with coordinates $u_{10} = u_{20} = \rho$ in phase plane (u_1, u_2). Dynamical system (2)–(4) contains the following parameters: ρ, defining limit cycle radius (free parameter which can be specified by arbitrary monotone continuous function of brightness, $\rho = \rho(I)$); ω is the frequency of free oscillations; h_* is the parameter defining brightness threshold value, below which Hopf bifurcation of converting the limit cycle into stable focus occurs; α is the parameter defining quickness of oscillation damping after limit cycle converting into focus; σ is a constant ($\sigma = 1$). The oscillator "response" to pixel brightness variation at $\rho(I) = \alpha I$ is depicted in Fig. 10.1, where time behaviors $u_1(t)$ and $u_2(t)$ and the corresponding phase trajectory of the oscillator dynamical system are shown.

The values $W_{jj'mm'}$, defining the coupling strength of network oscillators (j,m) and $(j'm')$, are designed in the form of nonlinear functions dependent on oscillation amplitudes (limit cycle radii) of the oscillator pair and spatial distance between oscillators in the network:

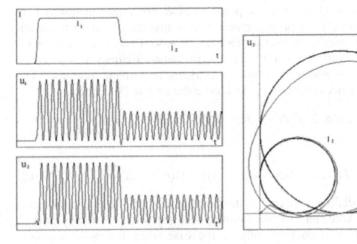

Fig. 10.1 Oscillator dynamics response to pixel brightness variation

$$W_{jj'mm'} = P_{jj'mm'}(\rho,\rho') \cdot D_{jj'mm'}(|r-r'|) \tag{10.5}$$

The cofactors $P_{jj'mm'}$, providing the dependence of network coupling on oscillation amplitudes, are specified as:

$$P_{jj'mm'}(\rho,\rho') = w_0 \cdot H(\rho_{jm}\rho_{j'm'} - h) \tag{10.6}$$

where $H(x)$ is a continuous step function and w_0 is a constant, defining total strength of network interaction. The cofactors $D_{jj'mm'}(|r-r'|)$, providing coupling spatial restriction, can be specified by any function, vanishing at some finite distance. As a result, any pair of network oscillators is proved to be coupled if they both possess sufficiently great oscillation amplitudes and are separated by a distance not exceeding the prescribed radius of spatial interaction. Otherwise the connection is absent.

10.3 Network Segmentation Capabilities

10.3.1 Grey-Level Image Segmentation

The oscillatory network performance consists of two steps: (1) preliminary tuning of oscillator dynamics by pixel brightness values of an image to be segmented (after the tuning operation's own limit cycle size has been specified for each network oscillator); and (2) network relaxation into the state of cluster synchronization, that is, to the state at which the oscillatory network is decomposed into the set of internally synchronized but mutually desynchronized oscillator ensembles (clusters), each ensemble being correspondent to the appropriate image fragment.

The gradual type of oscillator response to pixel brightness, guaranteed by oscillator dynamics (2)–(4), plays a crucial role in providing high segmentation accuracy. An improved coupling rule has been also used besides the initial biologically motivated coupling rule (5) to raise segmentation accuracy. It is based on prescribing to each oscillator some "mask," restricting its coupling "response." The modified coupling rule is defined by modified cofactor $\tilde{P}$ in (5), namely

$$\tilde{P}_{jj'mm'}(\rho,\Delta;\rho',\Delta') = T(\rho,\Delta)T(\rho',\Delta')P_{jj'mm'}(\rho;\rho') \tag{10.7}$$

where

$$T(\rho,\Delta) = 0.5[th(\sigma(\rho+\Delta)) - th(\sigma(\rho-\Delta))] \tag{10.8}$$

Here $T(\rho,\Delta)$ defines a "mask," restricting the size Δ of the oscillator interaction vicinity. According to the coupling rule (5) with $P = \tilde{P}$ any pair of network oscillators is coupled only in the case when the mask supports $[-\Delta, \Delta]$ and $[-\Delta', \Delta']$ of both the oscillators are intersected.

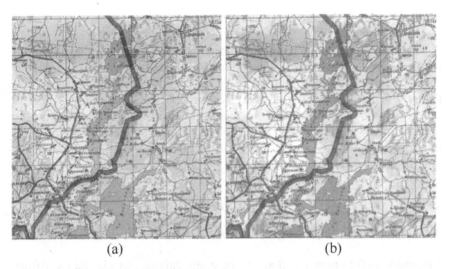

Fig. 10.2 Map fragment segmentation (492×475 pixels)

A flexible code ONN was created for computer experiments. An adaptive 5th-order Cash- Karp Runge-Kutta scheme has been incorporated for the ODE system integration. A series of computer experiments on real image segmentation have been performed. The example of map fragment segmentation at $\rho (I) = \alpha I$ and coupling rule (5) is presented in Fig. 10.2, where (a) is the original image and (b) is the segmentation result.

Fig. 10.3 Colored image segmentation (524×374 pixels)

10.3.2 Colored Image Segmentation

A new ONN code version has been created for colored image segmentation. In the first step, the pixel array of an original colored image is decomposed into three subarrays, corresponding to the red, blue, and green components of pixel colors. Further, these three subarrays are processed by the usual ONN code independently. Visualization of the segmentation result is performed via the conjunction of all three subarrays into a single array. An example of colored image segmentation is presented in Fig. 10.3, (where (a) is the original image and (b) is the final segmentation result).

10.4 Selective Image Segmentation

The most useful operation that can be easily carried out via the oscillatory network model is selective image segmentation. Selective segmentation can be viewed as the simplest type of active image processing, and consists in the extraction of a desirable subset of image fragments, the brightness values of which are contained inside some given interval. As is intuitively clear, the selective segmentation can be often more informative than the usual complete segmentation of the same image. Oscillator dynamics (10.2), (10.3), and (10.4) provides a very natural way of selective segmentation realization. It is sufficient to introduce a new function $\tilde{\rho}(I)$ instead of $\rho(I)$ in eq. (10.3), putting $\tilde{\rho}(I) = \rho(I)F(I)$, where $F(I)$ is a "filtering" function. If one desires to select only image fragments of brightness values $I \in [I^*, I^{**}]$, we choose $F(I)$ to be equal to 1 inside the interval $[I^*, I^{**}]$, and vanishing outside the interval.

For example, one can use:

$$F^{(1)}(I) = 0.5 \cdot \{th[\gamma(I - I^*)] - th[\gamma(I - I^{**})]\}, \quad \gamma \gg 1 \qquad (10.9)$$

In this case, only the oscillators, corresponding to image fragments with brightness values $I \in [I^*, I^{**}]$, will be "active," whereas the other oscillators will drop out of network interaction because of zero oscillation amplitudes. Similarly, the selection of an arbitrary collection of image fragments of given brightness levels is possible. Fig. 10.4 demonstrates the informative character of selective segmentation. Here one can compare complete segmentation (picture (b)) of the image (a human brain section) with three different cases of selective segmentation (pictures (c), (d), and (e)). In each case of selective segmentation, only several image fragments with brightness values inside a narrow interval have been selected.

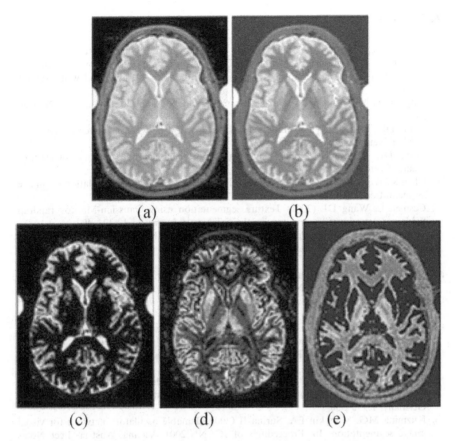

Fig. 10.4 Selective image segmentation. (**a**) original image; (**b**) complete image segmentation; (**c**) extraction of several of the brightest image fragments; (**d**) extraction of a set of fragments of middle brightness; (**e**) extraction of several of the least bright fragments

10.5 Conclusion

A synchronization-based approach to image processing via a tunable oscillatory network with self-organized coupling is presented. The following advantages are inherent in this approach:

(a) the parallel and automatic character of processing;
(b) an adaptive type of processing (in that both background level and noise reduction can be easily controlled);
(c) active image processing (due to the capability of selective segmentation).

The designed oscillatory network is actually closely related to multi-agent systems—distributed networks of active processing units with complicated controllable internal dynamics and a reorganizable structure of cooperative connections.

References

1. Kreiter AK, Singer W (1996) On the role of neural synchrony in primate visual cortex. In: Aertsen A, Braitenberg V (eds) Brain Theory, Amsterdam: Elsevier pp 201–226
2. Gray CM (1999) The temporal correlation hypothesis is still alive and well. Neuron 24:31–47
3. Singer W (1999) Neuronal synchrony: a versatile code for definition of relations? Neuron 24:49–65
4. Wang DL, Terman D (1995) Emergent synchrony in locally coupled neural oscillators. IEEE T Neural Networks 6:283–286
5. Wang DL, Terman D (1997) Image segmentation based on oscillatory correlations Neural Comput 9:805–836
6. Chen K, Wang D, Liu X (2000) Weight adaptation and oscillatory correlation for image segmentation. IEEE T Neural Networks 11:1106–1126
7. Cesmeli E, Wang DL (2001) Texture segmentation using Gaussian-Markov random fields and neural oscillator networks. IEEE T Neural Networks 12:394–409
8. Wang D (2005) The time dimension for scene analysis. IEEE T Neural Networks
9. Li Z (1998) A neural model of contour integration in the primary visual cortex. Neural Comput 10:903–940
10. Li Z (1999) Visual segmentation by contextual influences via intra-cortical interactions in the primary visual cortex. Networks 10:187–212
11. Li Z (2000) Pre-attentive segmentation in the primary visual cortex. Spatial Vision 13:25–50
12. Li Z (2001) Computational design and nonlinear dynamics of a recurrent network model of the primary visual cortex. Neural Comput 13(8):1749–1780
13. Kazanovich Y, Borisuyk R (2002) Object selection by an oscllatory neural network. Biosystems 67:103–111
14. Kuzmina MG, Manykin EA, Surina II (2000) Spatially distributed ocillatory networks related to modeling of the brain visual cortex. In: Proceedings of NOLTA'2000, Dresden, Germany. 1:335–338
15. Kuzmina MG, Manykin EA, Surina II (2001) Tunable oscillatory network for visual image segmentation. In: Proceedings of ICANN'2001. Vienna, Austria. Lect Notes Comput Sci 2130:1013–1019
16. Kuzmina MG, Manykin EA, Surina II (2004) Oscillatory network with self-organized dynamical connections for synchronization-based visual image segmentation. BioSystems 76:43–53
17. Kuzmina MG, Manykin EA, Surina II (2004) Biologically motivated oscillatory network model for dynamical image segmentation. In: Biologically inspired cognitive systems (BICS'04). Stirling, Scotland, Aug 29–Sep 1
18. Grichuk ES, Kuzmina MG, Manykin EA (2006) Oscillatory network for synchronization-based adaptive image segmentation. In: IEEE world congress on computational intelligence (WCCI 2006). July 16–21 2006, Vancouver, BC, Canada; WCCI 2006 CD, IEEE Catalog Number: 06CH37726D, ISSN: 0–7803–9489–5

Chapter 11
Feature Based Color Stereo Matching Algorithm Using Restricted Search

Hajar Sadeghi, Payman Moallem, and S. Amirhassan Monadjemi

Abstract The reconstruction of a dynamic complex 3D scene from multiple images is a fundamental problem in the field of computer vision. Given a set of images of a 3D scene, in order to recover the lost third dimension, depth, it is necessary to extract the relationship between images through their correspondence. Reduction of the search region in stereo correspondence can increase the performances of the matching process in both execution time and accuracy. In this study we employ edge-based stereo matching and hierarchical multiresolution techniques as fast and reliable methods, in which some matching constraints such as epipolar line, disparity limit, ordering, and limit of directional derivative of disparity are satisfied as well. The proposed algorithm has two stages: feature extraction and feature matching. We use color stereo images to increase the accuracy and link detected feature points into chains. Then the matching process is completed by comparing some of the feature points from different chains. We apply this new algorithm on some color stereo images and compare the results with those of gray level stereo images. The comparison suggests that the accuracy of our proposed method is increased around 20–55%.

11.1 Introduction

In stereo vision-based methods, as in the human eye, two images are taken from two different viewpoints, and then the information is inferred into 3D structures. Our stereo algorithm has two stages: feature extraction and feature matching. In the feature extraction stage, from the left and right images specific feature points such as edges, corners, centroids, and textured areas would be extracted. Next, in the feature matching stage, we extract correspondences. In this study, we focus on edges since they represent the high frequency and

H. Sadeghi (✉)
Faculty of Engineering, University of Isfahan, Isfahan, Iran
e-mail: hsadeghi@eng.ui.ac.ir

N. Mastorakis et al. (eds.), *Proceedings of the European Computing Conference*, Lecture Notes in Electrical Engineering 27, DOI 10.1007/978-0-387-84814-3_11, © Springer Science+Business Media, LLC 2009

structure of the shapes; we also focus on color images since color is an important feature too, such that its change is measured by the MSE[1] factor between left and right images. Different constraints such as epipolar constraint, disparity limit constraint, and limit of directional derivative of disparity are also applied in order to decrease the search space. We also use the chains of feature points to decrease both the computing time and matching error. It should be mentioned that, along with the chain of features, using color stereo images is effective in decreasing the matching error. The search space of successive connected features is reduced by the empirical limit of directional derivative of disparity. The color makes the match less sensitive to occlusion, considering the fact that occlusion most often causes color discontinuities.

11.2 Linking Feature Points into Chains

In this paper, we reduce the search space for the successive connected features from a predefined disparity within a small range. Therefore, the algorithm can run much faster than other corresponding algorithms [7]. To make the algorithm run efficiently, we test only the first two feature points from each chain using the MSE similarity measure. If both of the features have high correlation scores, the tested pair of chains from each image is defined as chains correspondence. The experimental results show that an average of 92% of the feature chains have a length less than or equal to 5. This indicates that about 40% of the feature point's correspondences are evaluated.

11.3 Color Stereo Vision

Color plays an important role in human perception. Obviously the red pixels cannot match the blue ones, even through their intensities are equal or similar. Consequently, the existing computational approaches to color stereo correspondence [1, 2, 3, 8] have shown that the matching results can be considerably improved when using color information. Drumheller and Poggio [3] presented one of the first stereo approaches using color. In using color images, we use the similarity measure MSE, that is defined in Eq. 11.1 below:

$$MSE_{color}(x, y, \Delta) = \frac{1}{n^2} \sum_{i=-k}^{k} \sum_{j=-k}^{k} dist_c(C_R(x+i, y+j),$$

$$C_L(x+i+\Delta, y+j)) \tag{11.1}$$

$$dist_c(c_1, c_2) = (R_1 - R_2)^2 + (G_1 - G_2)^2 + (B_1 - B_2)^2 \tag{11.2}$$

---

[1] Mean square error.

In Eq. 11.1, Δ is disparity, and c_1, c_2 in Eq. 11.2 are two points from color left and right images.

11.4 Proposed Algorithm

In this section, we are going to present a novel color feature-based stereo algorithm with chains. Our algorithm consists of two stages: feature extraction and feature matching. We can reduce the search space considerably by using Table 11.1 in the feature matching stage [5, 6, 4].

11.4.1 Feature Extraction

This stage is implemented for both of the left and right images, and non-horizontal thinned edge chains are extracted using a 3×3 Sobel filter in the horizontal direction. The thinned edge points are classified into two groups: positive and negative, depending on the intensity difference between the two sides of the feature points in the horizontal direction in any color band.

A non-horizontal thinned positive edge in the left image is localized to a pixel such that the filter response has to exceed a positive threshold $\rho^+{}_0$ and has to obtain a local maximum in the x direction; therefore:

$$\text{Threshold} \;\rightarrow\; \rho_l(x,y) > \rho_0^+$$
$$\text{Local Maximum} \begin{cases} \rho_l(x,y) > \rho_l(x-1,y) \\ \rho_l(x,y) > \rho_l(x+1,y) \end{cases} \tag{11.3}$$

Consider $\rho^+{}_0$ as the mean of the positive values of the filter response.

The extraction of non-horizontal negative thinned edge points is similar to the positive extraction. Now we should try to extract feature chains with a length greater than 3. Each 2 sequence feature points in the same chain should be in sequence scan lines, and the disparity in the x direction should be less than 2. Chains whose length is less than 3 are ignored.

Table 11.1 The relationship between Δx_l and the search region for CIL stereo dataset, when $\Delta y = 0$ and Δx_l is between 2 and 10 pixels

$x_l\,\Delta$	2	3	4	5	6	7	8	9	10
Min(Δx_l)	0	1	1	2	2	3	3	4	4
Max(Δx_l)	3	5	6	8	9	11	13	14	16

11.4.2 Feature Matching

Once the correspondence between the two images is known, the depth information of the objects in the scene can be obtained easily. The matching feature points, however, should be the same whether positive or negative. Thus, our algorithm includes two phases:

Phase I:

(a) Do a systematic scan from left to right.
(b) If the current point is not a feature point, go to (a).
(c) If the disparity was already computed for the current feature point, store its x value as $x0_l$ and go to (a).
(d) If the current point is not on the feature chain, go to (a), else call the x value of the current feature point xc_l. If there is not any $x0_l$ then go to (e), else compute $\Delta x_l = xc_l - x0_l$ and then compute the search space in the right image recording (Table 11.1) and then go to (f).
(e) Compute the search space based on the disparity range.
(f) Find the correspondence point of the current feature point in the right image. If there is not any correspondence point, go to (a), else go to phase II.

Phase II: If $L(x_{1\,l}, y)$ and $R(x_{1r}, y)$ are features correspondent:

(a) Choose the next feature points, $L(x_{2l}, y+1)$ and $R(x_{2r}, y+1)$, from the same feature chains in the left and right chains separately. Test the similarity between them; if the similarity score is higher than the threshold, record the feature chains correspondence and delete two correspondence chains from the left and right images. Note that the corresponding feature chains should be recorded with the same length, and then go to (a) in phase I, else go to (b) in phase II.
(b) Use the same method to test the third feature points, $L(x_{3l}, y+2)$ and $R(x_{3r}, y+2)$, from the feature chains. If they are features correspondent, record the feature chains as feature chain correspondence and delete two correspondence chains from the left and right images, and then go to (a).

The output of the algorithm is a disparity map in which each pixel in the map represents the disparity of the matching pixels between two images, or otherwise their depth in the scene.

11.5 Important Notes in the New Algorithm

In this paper, we presented the reduction of the search space in an edge-based stereo correspondence, using the context of the maximum disparity gradient. The continuous non-horizontal edges in chains are used to increase the computing speed, and we also employed color stereo images to increase the accuracy of matching. For reducing the search region, we used the conventional multiresolution technique with the Haar wavelet in three levels. The MSE with window sizes of 5×5, 3×3, and 3×3 was also applied for the coarse, medium, and fine level respectively, while the threshold was 1000.

For the first point in any chain, we should use only the epipolar line and the maximum disparity constraint. Therefore we apply a left-right consistency checking technique to reduce the invalid matching, since the matching result of these points is used for matching the next points on the chain; only two, or if needed, three points in each chain are matched.

11.6 Conclusions

In this study, we tested our algorithm on eight stereo scenes from the Middlebury database. All the scenes used are colored images in 380×432, and their disparity ranges are all considered ±30 and ±70.

The results of the proposed algorithm are shown in Table 11.2. The total of the correct matches, the error matches, and the not matched points in the left image in the matching stage are shown in the matched (M), error (E), and not matched (NOTM) columns respectively. The percentages of the error matched points with respect to the total matched points are shown in the ERR% column.

The implementation results on two stereo scenes from the Middlebury database are shown in Figs. 11.1 and 11.2. These images are depicted in different colors: the black color shows the not matching feature points, and the red color shows the error matched feature points; the yellow and the dark

Table 11. 2 The implementation results of the algorithm

Algorithm	Stereo scene	M	E	NOTM	ERR%
ColorChainStereo	Ball	13933	484	451	0.033
	Barn1	19278	597		0.030
	Barn2	14608	742	1069	0.048
	Cones	15799	872	504	0.052
	Poster	16843	712	2173	0.040
	Saw Tooth	14159	618	1356	0.041
	Teddy	12292	875	986	0.066
	Venus	12667	585	1907	0.044
				993	
GrayLevelChainStereo	Ball	13275	675	660	0.048
	Barn1	19086	1027	936	0.051
	Barn2	13941	1239	558	0.082
	Cones	15548	1947	1659	0.11
	Poster	17440	1238	880	0.066
	Saw Tooth	13841	1234	797	0.082
	Teddy	11929	1425	1809	0.10
	Venus	12373	982	788	0.073

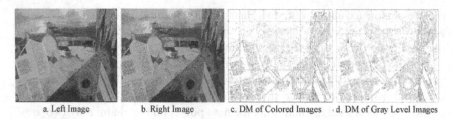

| a. Left Image | b. Right Image | c. DM of Colored Images | d. DM of Gray Level Images |

Fig. 11.1 The implementation result on Barn1. DM is disparity map

Fig. 11.2 The implementation result on poster

blue colors show the feature points with minimum and maximum disparities respectively.

Table 11.2 shows a comparison of the results on color stereo images and gray level stereo images. Compared to the percentage of failures in stereo matching, the error in the color matching is reduced by 20–55%; however, the time for the matching is increased by only about 10%. Therefore, the reduction of error is considerably high, and this makes our algorithm rational and advisable.

References

1. Cabani I, Toulminet G, Bensrhair A (2006) A fast and self-adaptive color stereo vision matching; a first step for road obstacle detection. In: IEEE intelligent vehicles symposium. Tokyo, Japan, pp 58–63
2. Hua X, Yokomichi M, Kono M (2005) Stereo correspondence using color based on competitive-cooperative neural networks. Parallel and Distributed Computing, Applications and Technologies, 10.1109/PDCAT.2005.227, pp 856–860
3. Koschan A (1993) What is new in computational stereo since 1989: a survey on current stereo papers. Technical Report 93-22, Technical University of Berlin
4. Moallem P, Ashorian M, Mirzaeian B, Ataei M (2006) A novel fast feature based stereo matching algorithm with low invalid matching. WSEAS T Comput 5(3):469–477
5. Moallem P, Faez K, Haddadnia J (2002) Fast edge-based stereo matching algorithms through search space reduction. IEICE T Inf Syst E85-D(11-20021101):1859–1871

6 Moallem P, Faez K (2005) Effective parameters in search space reduction used in a fast edge-based stereo matching. J Circuit Syst Comput 14(2):249–266

7 Tang B, Ait-Boudaoud D, Matuszewski BJ, Shark LK (2006) An efficient feature based matching algorithm for stereo images. In: Proceedings of the geometric modeling and imaging—new trends (GMAI'06), pp 195–202

8 Yang Q, Wang L, Yang R, Stewenius H, Nister D (2006) Stereo matching with color-weighted correlation, hierarchical belief propagation and occlusion handling. In: Computer Society conference on computer vision and pattern recognition—Volume 2 (CVPR'06), pp 2347–2354

Chapter 12
A Novel Edge-Preserving Lossy Image Coder

Osslan Osiris Vergara Villegas, Manuel de Jesús Nandayapa Alfaro, and Vianey Guadalupe Cruz Sánchez

Abstract Several lossy image coders are designed without considering the nature of the images. Images are composed at least by textures, edges, and details associated with edges. Sometimes important information in the images, such as edges that are used for image understanding, is lost in the coding quantization stage. In this paper we present a novel edge-preserving lossy image coder. The core of the proposed coder is a modification of the SPIHT algorithm and the definition of the edges in the wavelet and contourlet domain. Additionally, we show the results obtained in order to demonstrate the effectiveness of the proposed coder compared with two existing coders, even at very low bit rates.

12.1 Introduction

Nowadays, there has been a lot of research on designing image compression algorithms, mainly classified into lossy and lossless coders. The research is intended to design image coders yielding both big compression factors and good quality decompressed images [1]. The main reason to design image coders is the great growth in the use of the Internet and mobile communications devices in which information delivery needs to be efficient.

The efficient representation of visual information is one of the main goals for several image processing tasks, such as classification, compression, filtering, and feature extraction. Efficiency in the representation is the ability to capture significant information about an object in a small description [2].

When a lossy coding/decoding process is made, several important details of a better image understanding are eliminated; for this reason information preserving is very important. In addition, for areas such as medicine, the ways to

O.O. Vergara Villegas (✉)
Universidad Autónoma de Ciudad Juárez (*UACJ*), Avenida del Charro 450 Norte
Ciudad Juárez Chihuahua México
e-mail: overgara@uacj.mx

N. Mastorakis et al. (eds.), *Proceedings of the European Computing Conference*, Lecture Notes in Electrical Engineering 27, DOI 10.1007/978-0-387-84814-3_12, © Springer Science+Business Media, LLC 2009

handle image information are regulated by laws; therefore, image feature preserving becomes imperative [3, 4]. Feature preservation means that the *location, strength,* and *shape* of features are unchanged after the application of a general filter; of course, natural differences occur because of changes in resolution [5].

To solve the feature-preserving problem, we can use two different approaches: (a) send side information about the features and (b) use different algorithms to code different features [6]. The model presented in this paper differs with these approaches, because we do not send side information and we use only one coder to compress the complete image.

12.2 Proposed Model

An image contains several features which offer information about the objects presented in the image. In order to design a feature-preserving lossy coder, it is important to identify the image features to preserve, and then use more bits to represent important information and fewer bits in other image regions [3]. The stages of the proposed model are explained in the following sections.

12.2.1 Image Selection and Feature of Interest Definition

We need to select a variety of images with different frequency (edges) information, with the goal of ensuring good performance of the image coder in different environments (images). For this, we use the statistical properties offered by the wavelet coefficients, by measuring the "norm-2 energy" (equation 12.1) [7]:

$$E_1 = \frac{1}{N^2} \sum_{k=1}^{N} |C_k|^2 \tag{12.1}$$

The interval used for classification based in the *LL* subband is: (a) low frequency, greater than or equal to 99.5%; (b) medium frequency, from 99 to 99.49%; and (c) high frequency, less than 99% [5]. After image selection we define the features to preserve; for this paper the edges will be preserved.

12.2.2 Feature of Interest Map (FOIM) Extraction

To extract the feature of interest map (FOIM) we use an edge detector known as the smallest univalue segment assimilating nucleus (SUSAN). SUSAN is more robust and effective than Canny, in the sense that it provides much better edge localization and connectivity. SUSAN uses a predetermined window centered on each pixel of the image, applying a locally acting set of rules to give an edge response. This response is then processed to give as output a set of edges [8]. In brief, SUSAN involves three steps: (a) place a circular mask (37 pixels) around the pixel in question (the nucleus); (b) calculate the number of pixels within the circular mask which have similar brightness to the nucleus; and (c) subtract the USAN size from the geometric threshold to produce an edge strength image.

12.2.3 Domain Transform

We use two transformations: the discrete wavelet transform (DWT) and the novel wavelet-based contourlet transform (WBCT).

The DWT is obtained by convolving the image columns and rows with a low pass filter (scaling function Φ wavelet father) and a high pass filter (wavelet function wavelet mother). The images are decomposed log_2 *(image size)* -1 levels and the filter used is the biorthogonal 2.2. The WBCT is a novel transform presented in [9], the images are decomposed log_2 *(image size)*—2 levels and 5 directions at the finest wavelet subband. The filter used for the wavelet is the biorthogonal 2.2 and the directional filter is the PKVA. For the WBCT we ensure the perfect reconstruction for the best case; the challenge is the contourlet inside the wavelet.

12.2.4 Pixel Mapping to Domain Transform

The FOIM is used to map the edge points from the original to the transform domain. At the transform domain a coefficient of scale i has an area of *2i x 2i* positions of the original domain; there exists a hierarchic relation between the coefficients which allows defining a structure known as a spatial orientation tree. A pixel in the lower subband is father of three coefficients at the same position in the high frequency subbands at the same scale. Any other coefficient outside the lower band has four children.

The steps of the mapping process are: (a) compute the FOIM, (b) double the position of a pixel of the FOIM (2×2), (c) downsample the image, (d) repeat b and c an n number of levels, and (e) mark the descendent coefficients as a part of the FOIM until the original size is reached. In Fig. 12.1 an example of this mapping process is shown.

12.2.5 Image Coding with Modified SPIHT and Arithmetic Coding

In 1996 Said and Pearlman presented an improved version of the embedded zerotree wavelet (EZW) coder by proposing a different tree structure. The algorithm was called set partitioning in hierarchical trees (SPIHT) [10].

SPIHT uses the significance of a pixel coded under a defined threshold. A pixel is significant if its value is bigger than or equal to a given threshold. We

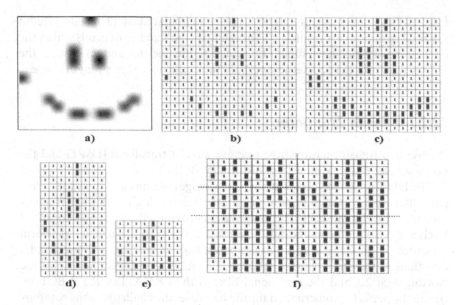

Fig. 12.1 Mapping process. (**a**) smiley face, (**b**) FOIM, (**c**) pixel duplicated image, (**d**) columns downsampling, (**e**) rows downsampling, and (**f**) final edge map

made a modification of the original SPIHT in which the pixel significance is defined not only by magnitude, but also by the corresponding pixel position in the transform domain.

Table 12.1 shows an example for a modified SPIHT and is compared to the original SPIHT shown in Table 12.2. The final bit stream obtained is entropy coded. In this process we do not have information loss; here we remove the redundancies of the bit stream. The algorithm used is the arithmetic coder. We add data about the color plane of the image, the levels of decomposition, and the image size.

12.2.6 Arithmetic and SPIHT Decoding

The first stage of the decompression consists of obtaining the information resulting from the compression process, such as: image color plane, image size, transform used, etc. We need to apply the arithmetic decoding algorithm to obtain the necessary information, and the bit stream to be used in the SPIHT decoding stage. Again the algorithm used is the arithmetic decoder.

After arithmetic decoding we use the original SPIHT decoder in order to obtain the decompressed images. After SPIHT decoding we obtain a transform domain image (wavelet or contourlet). In order to obtain the reconstructed image, we apply the inverse discrete wavelet transform (IDWT) or the inverse

Table 12.1 Modified SPIHT

Comment	Pixel or set tested	Output bit	Action	Control list
(1)				LIS = {(0,1)A, (1,0)A,(1,1)A}
				LIP = {(0,0), (0,1), (1,0), (1,1)}
				LSP = ϕ
(2)	(0,0)	0	none	
	(0,1)	1-	(0,1) to LSP	LIS = {(0,0), (1,0),(1,1)}
				LSP = {(0,1)}
	(1,0)	0	none	
	(1,1)	0	none	
(3)	D(0,1)	1	Test offsprings	LIS = {(0,1)A, (1,0)A,(1,1)A}
	(0,2)	1+	(0,2) to LSP	LSP = {(0,1), (0,2)}
	(0,3)	0	(0,3) to LIP	LIP = {(0,0), (1,0), (1,1), (0,3), (1,2)}
	(1,2)	0	(1,2) to LIP	LIP = {(0,0), (1,0), (1,1), (0,3), (1,2)}
	(1,3)	0	(1,3) to LIP	LIP = {(0,0), (1,0), (1,1), (0,3), (1,2), (1,3)}
				LIS = {(1,0)A, (1,1)A }
(4)	D(1,0)	0	none	
(5)	D(1,1)	0	none	

Table 12.2 Original SPIHT

Comment	Pixel or set tested	Output bit	Action	Control list
(1)				LIS = {(0,1)A, (1,0)A,(1,1)A}
				LIP = {(0,0), (0,1), (1,0), (1,1)}
				LSP = ϕ
(2)	(0,0)	1+	(0,0) to LSP	LIS = {(0,1), (1,0),(1,1)}
				LSP = {0,0}
	(0,1)	1-	(0,1) to LSP	LIS = { (1,0),(1,1)}
				LSP = {(0,0), (1,1)}
	(1,0)	0	none	
	(1,1)	0	none	
(3)	D(0,1)	1	Test offsprings	LIS = {(0,1)A, (1,0)A,(1,1)A}
	(0,2)	1+	(0,2) to LSP	LSP = {(0,1)A, (1,0)A, (1,1)A}
	(0,3)	0	(0,3) to LIP	LIP = { (1,0), (1,1), (0,3), (1,2)}
	(1,2)	0	(1,2) to LIP	LIP = {(0,0), (1,0), (1,1), (0,3), (1,2)}
	(1,3)	0	(1,3) to LIP	LIP = { (1,0), (1,1), (0,3), (1,2), (1,3)}
				LIS = {(1,0)A, (1,1)A }
(4)	D(1,0)	0	none	
(5)	D(1,1)	0	none	

wavelet-based contourlet transform (IWBCT)—as a result we obtain the final decompressed image.

12.3 Results, Discussion, and Comparisons

The goal of these tests is to compare the performance between the proposed coder and other existing coders. We select three 512×512 images called "Camman," "Lena," and "Peppers," classified into high, medium, and low frequency.

We apply the proposed model to Camman with a bit rate of 0.1 with wavelets and contourlets; Lena with a bit rate of 0.2; and Peppers with a bit rate of 0.4 with wavelets. The images are compressed under conditions similar to those used by Mertins [1] and Schilling [3]. In Fig. 12.2 we show the original images and their edge maps.

We perform a subjective comparison of the images by an observing process, and an objective comparison using the peak signal-to-noise ratio (PSNR) measure. The decompressed images for Camman and the respective PSNR obtained are shown in Fig. 12.3. In Fig. 12.4 we show the results obtained with Lena and Peppers. The objective measures cannot give information about the quality of edge preserving; even with these measures the proposed model visually performs better than the other methods, because more details are preserved.

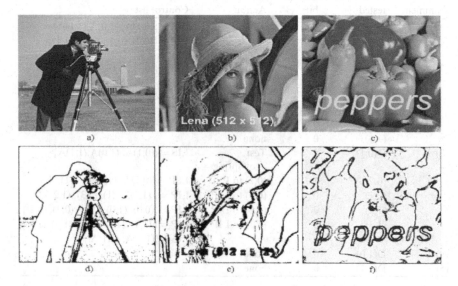

Fig. 12.2 Original images and maps with a 50 threshold. (**a**) Camman, (**b**) Lena, (**c**) Peppers, (**d**) Camman edge map, (**e**) Lena edge map, and (**f**) Peppers edge map

PSNR: 22.39 dB   PSNR: 22.39 dB   PSNR: 22.75 dB   PSNR: 22.60 dB

Fig. 12.3 Camman decompressed. (**a**) Mertins coder, (**b**) Schilling coder, (**c**) proposed coder and wavelets, and (**d**) proposed coder and contourlets

PSNR: 28.9 dB   PSNR: 23.42 dB   PSNR: 22.01 dB   PSNR: 25.37 dB

Fig. 12.4 Lena and Peppers decompressed. (**a**) Mertins coder, (**b**) proposed coder and wavelets, (**c**) Schilling coder, and (**d**) proposed model and wavelets

From the visual test, the poorest image is obtained with the proposed model and contourlets; but the edges are well preserved, even with the granularity introduced by the mapping. The better image is obtained with the proposed model and wavelets; we do not lose information about the tripod and the details of the face, and the background towers and the camera are very clear. With the Mertins and Schilling images, a great part of the camera tripod and details of the face and the background are lost.

For Lena the images are visually similar, even with the letters added to the image to augment the reconstruction complexity. For Peppers the image obtained with the proposed coder is better. The edges of the objects and letters are well reconstructed. The same occurs if we use the corresponding contourlet images.

12.4 Conclusions

We have presented a novel wavelet- and contourlet-based lossy image coder which allows edge preserving; this coder can be used if an early recognition of images is needed and in environments in which there are laws for the use of original images and feature preserving is imperative. The model and the tests presented show the ability of the coder to reconstruct images by spending more bits and giving more quality to those coordinates defined by an edge map. After the comparison we conclude that the coder presented is better than those proposed in the literature in the past.

In the future, we are going to work on finding a method to reduce the ringing that appears in some images, by handling the image-associated details and solving the problem of mapping in the contourlet domain, in order to avoid granularity in images.

References

1. Mertins A (1999) Image compression via edge-based wavelet transform. J Opt Eng 6:991–100
2. Do MN (2001) Directional multiresolution image representations. Ph.D. thesis, Lausanne Federal Polytechnic School (EPFL), Lausanne, Switzerland
3. Schilling D, Cosman P (2001) Feature-preserving image coding for very low bit rates. In: IEEE data compression conference (DCC). Snowbird, Utah, USA, pp 103–112
4. Namuduri KR, Ramaswamy VN (2003) Feature preserving image compression. J Patt Recogn Lett 15:2767–2776
5. Barnard HJ (1994) Image and video coding using a wavelet decomposition. Ph.D. thesis, Delft University of Technology, Department of Electrical Engineering, Information Theory Group, The Netherlands
6. Vergara Villegas OO, Pinto Elías R, Rayón Villela P, Magadán Salazar A (2006) Edge preserving lossy image compression with wavelets and contourlets. In: Electronics, robotics and automotive mechanics conference (CERMA), Cuernavaca, Morelos, Mexico, pp 3–8
7. Muneeswaran K, Ganesan L, Arumugam S, Ruba Soundar K (2005) Texture classification with combined rotation and scale invariant wavelet features. J Patt Recogn 10:1495–1506
8. Smith SM, Brady JM (1997) SUSAN—a new approach to low level image processing. Int J Comput Vision 1:45–78
9. Eslami R, Radha H (2004) Wavelet-based contourlet transform and its application to image coding. In: International conference on image processing (ICIP), Republic of Singapore, pp 24–27
10. Said A, Pearlman W (1996) A new fast and efficient image codec based on set partitioning in hierarchical trees. IEEE T Circ Syst Video Technol 3:243–250

Part III
Modeling and Simulation

Modeling and simulation are everywhere nowadays in life, science, and applications. In most science disciplines, simulation has replaced many experiments in labs and in the field. Historians and social scientists use simulation to explore and validate theories. Manufacturers use simulation in design to replace construction of expensive prototypes. The military uses simulation extensively, to train soldiers and to design and evaluate equipment. In this part we present some advanced concepts and the use of modeling and simulation in different areas. The papers consider equivalent circuit extraction for passive devices, sleep quality differences according to a statistical continuous sleep model, simulation modeling in support of a European airspace study, a 3D measurement system for car deformation analysis, adaptive MV ARMA identification under the presence of noise, Planck's law simulation using particle systems, simulation results using shock absorbers with primary intelligence—VZN, pattern generation via computer, 3D reconstructions estimating depth of holes from 2D camera perspectives, modeling diversity in recovery computer systems, cell automata models for a reaction-diffusion system, 3D reconstruction of solid models from 2D camera perspectives, VM-based distributed programmable network node architecture, a semantics of behavioral concepts for open virtual enterprises, a semantics of community related concepts in the ODP enterprise language, modeling of the speech process including anatomical structure of the vocal tract, carbon nanotube FET with asymmetrical contacts, and optimizing prosthesis design by using virtual environments. In the variety of applications presented, the authors put forward their original contributions in the area of modeling and simulation of different phenomena.

Chapter 13
Equivalent Circuit Extraction for Passive Devices

Yi-Ting Kuo, Hsueh-Yung (Robert) Chao, and Yiming Li

Abstract In this study, a minimal equivalent circuit extraction scheme is provided for passive elements such as interconnects and packages. Instead of common poles, the non-common poles are used as the dominant poles of each frequency response when approximating those frequencies into a rational form. An illustration is given to show the relation between the electrical lengths of the interconnects/packages and the corresponding dominant poles. Finally, a three-port interconnect is used to validate the proposed approach. The circuit extracted by the proposed scheme has fewer components than the circuit from the conventional scheme, and decreases the computational load of the circuit simulation.

13.1 Introduction

As the complexity and density of IC designs increase, high frequency effects modeling has played a crucial role in modern high-speed circuit design [1, 2]. To accurately describe those high frequency effects, the corresponding frequency responses are useful; however, active devices are mainly described in the time domain because of nonlinear effects. Hence the mixed frequency/time domain problem needs to be solved both accurately and efficiently [3–5]. A popular method of solving the problem is through the equivalent circuit extraction for passive elements, and there are many different schemes in circuit extraction [6–10]. Among all these circuit extraction schemes, the state-space-based equivalent circuit generation technique not only generates the equivalent circuit, but also can be incorporated into circuit equations directly [5, 11–13]. This method first approximates the admittance matrix into the rational form, and then generates the corresponding state-space equation from the resulting rational approximation. Finally, the equivalent circuit can be constructed

Y.-T. Kuo (✉)
Department of Communication Engineering, National Chiao Tung University,
1001 Ta-Hsueh Road, Hsinchu 300, Taiwan

N. Mastorakis et al. (eds.), *Proceedings of the European
Computing Conference*, Lecture Notes in Electrical Engineering 27,
DOI 10.1007/978-0-387-84814-3_13, © Springer Science+Business Media, LLC 2009

based on the state-space equation simultaneously. The state-space equation is also proven as a minimal realization in solving circuit equations [14]; however, the extracted circuit is not guaranteed to be minimal.

This study presents a minimal equivalent circuit extraction scheme based on the state-space equation. The traditional state-space-based equivalent circuit extraction has been shown to be inefficient for networks with a large number of ports. This work applies a separate set of poles (we name them non-common poles in the following sections) to construct the equivalent circuits from given frequency responses. An easy test case is used to explain the reason that the proposed method is useful to extract the minimal equivalent circuit. The remainder of this chapter is organized as follows. Section 13.2 discusses the derivation of the state-space equation using non-common poles. Section 13.3 presents a numerical example to explain the reason that the proposed method is helpful in circuit size reduction. Section 13.4 draws conclusions.

13.2 Formulation of the Proposed Method

The frequency responses for passive elements are generally used to accurately describe their behaviors in the frequency domain. For a multi-port element, the frequency responses (e.g., admittance matrix, etc.) are:

$$
\overline{\mathbf{Y}}(s) = \begin{bmatrix} y^{1,1}(s) & y^{1,2}(s) & \cdots & y^{1,N}(s) \\ y^{2,1}(s) & y^{2,2}(s) & \cdots & y^{2,N}(s) \\ \vdots & \vdots & \ddots & \vdots \\ y^{N,1}(s) & y^{N,2}(s) & \cdots & y^{N,N}(s) \end{bmatrix}
\tag{13.1}
$$

where $y^{m,n}$ represents the (m, n)th entry of the matrix, and N represents the number of ports of the network. To convert Eq. (13.1) into the equivalent circuit, each entry of the above matrix is first approximated into the rational form and expressed as:

$$
y^{m,n}(s) = d^{m,n} + \sum_{t=1}^{N_p^{m,n}} \frac{res_t^{m,n}}{s - pole_t^{m,n}},
\tag{13.2}
$$

where $d^{m,n}$ is the direct coupling term, and $pole_t^{m,n}$ and $res_t^{m,n}$ denote the tth pole and residue for the (m, n)th entry, respectively. $N_p^{m,n}$ is the number of dominant poles for the (m, n)th entry. The corresponding cell block diagram for the (m, n)th entry with a specific complex-conjugate pole-residue pair is constructed immediately [18, 19], and it is illustrated in Fig. 13.1.

Furthermore, the state-space equation is created based on the block diagram, which is depicted as follows:

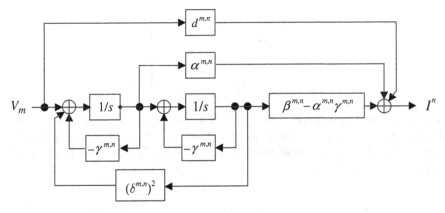

Fig. 13.1 Cell block diagram for the (m, n)th entry of the admittance matrix. The entry is described by a specific complex-conjugate pole-residue pair

$$\begin{cases} \dot{\mathbf{X}} = \overline{\mathbf{A}}\mathbf{X} + \overline{\mathbf{B}}\mathbf{V}, \\ \mathbf{I} = \overline{\mathbf{C}}\mathbf{X} + \overline{\mathbf{D}}\mathbf{V}, \end{cases} \tag{13.3}$$

where

$$\overline{\mathbf{A}} = \begin{bmatrix} \ddots & & 0 & & & \\ & & \vdots & & \cdots & 0 \\ 0 & \cdots & -\gamma^{m,n} & 1 & \cdots & 0 \\ & & (\delta^{m,n})^2 & -\gamma^{m,n} & & \\ 0 & \cdots & \vdots & & \ddots & \\ & & 0 & & & \end{bmatrix},$$

$$\overline{\mathbf{B}} = \begin{bmatrix} & \vdots & & & \\ 0 & \cdots & \cdots & \cdots & 0 \\ 0 & \cdots & 1 & \cdots & 0 \\ & \vdots & & & \end{bmatrix},$$

$$\overline{\mathbf{C}} = \begin{bmatrix} \ddots & \vdots & & \vdots & & \ddots \\ & 0 & & 0 & & \\ \cdots & \beta^{m,n}\alpha^{m,n} - \alpha^{m,n}\gamma^{m,n} & \alpha^{m,n} & \cdots \\ & 0 & & 0 & & \\ \ddots & \vdots & & \vdots & & \ddots \end{bmatrix}, \text{ and}$$

$$
\overline{\mathbf{D}} = \begin{bmatrix} d^{1,1} & d^{1,2} & \cdots & d^{1,N} \\ d^{2,1} & d^{2,2} & \cdots & d^{2,N} \\ \vdots & \vdots & \ddots & \vdots \\ d^{N,1} & d^{N,2} & \cdots & d^{N,N} \end{bmatrix},
$$

where the entry in position $(((m-1) \times N + (2n-1)), ((m-1) \times N + 2n))$ of matrix $\overline{\mathbf{A}}$ is 1, and the entry in $\overline{\mathbf{B}}$ valued 1 is located in the position $(((m-1) \times N + 2n), n)$. Additionally, the $(m, ((m-1) \times N + 2n))$th entry of $\overline{\mathbf{C}}$ is $\alpha^{m,n}$ (Fig. 13.2).

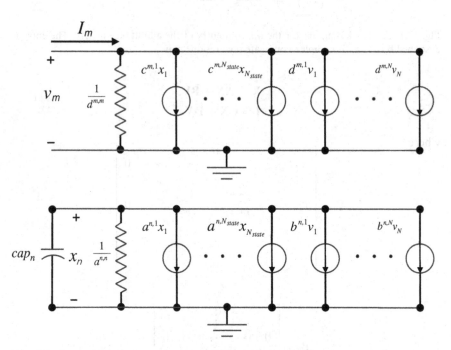

Fig. 13.2 The equivalent circuit based on the state-space equation. $a^{m,n}$, $b^{m,n}$, $c^{m,n}$, and $d^{m,n}$ denote the (m, n)th entry of the matrices $\overline{\mathbf{A}}$, $\overline{\mathbf{B}}$, $\overline{\mathbf{C}}$, and $\overline{\mathbf{D}}$

13.3 Computational Results

The given frequency responses shown in Fig. 13.3 are studied to explain the reason that the non-common pole technique can generate the minimal equivalent circuit for a network with a large number of ports. The corresponding poles are shown in Fig. 13.4. The resonant frequencies in each frequency response correspond to the imaginary part of the poles. In Fig. 13.3 the different

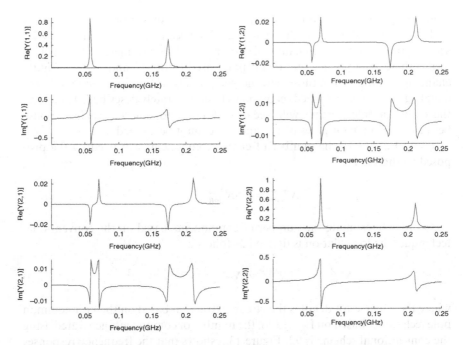

Fig. 13.3 The given frequency responses, $y^{1,2} = y^{2,1}$ in this case. The *solid line* represents the original responses, and the *dashed line* and *dotted line* denote the responses from the conventional and proposed methods, respectively

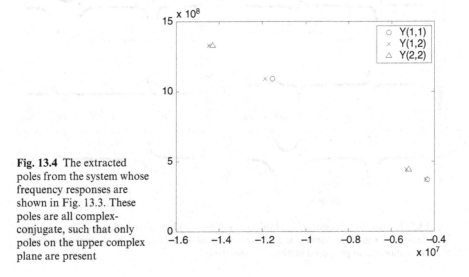

Fig. 13.4 The extracted poles from the system whose frequency responses are shown in Fig. 13.3. These poles are all complex-conjugate, such that only poles on the upper complex plane are present

dominant poles of $y^{1,1}$ and $y^{2,2}$ are expected based on these resonant frequencies of the responses. It is clearly shown in Fig. 13.4, because the circles and triangles are separated. If the dominant poles of $y^{1,1}$ and $y^{2,2}$ are merged together, it means there are four new dominant poles used to fit $y^{1,1}$ and $y^{2,2}$ by slightly changing residues. Even those new merged poles fit $y^{1,1}$ and $y^{2,2}$ quite well; the coupling effect, i.e., $y^{1,2}$, needs eight poles to be accurately described. Therefore, the common-pole scheme will need eight poles to describe each response because the dominant poles of $y^{1,1}$ and $y^{2,2}$ cannot be merged.

Using Eq. (13.4), the number of circuit components extracted by the proposed method is 112:

$$N_{comp}^{NCP} = 4.5 N_{state}^{NCP} + N^2 \qquad (13.4)$$

In order to estimate the number of components based on the conventional technique, the formulation is derived as follows:

$$N_{comp}^{CP} = 3.5 N_{pole}^{NCP} \times N + N^2 + N_{pole}^{NCP} \times N^2, \qquad (13.5)$$

where the superscript denotes the circuit extraction based on the non-common pole technique. Based on Eq. (13.5), the number of components generated using the conventional scheme is 92. Figure 13.5 shows that the frequency responses

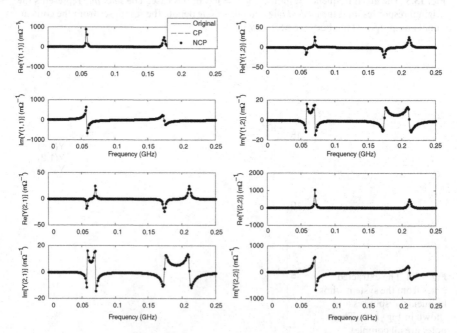

Fig. 13.5 The simulated results using both conventional and proposed schemes. The *solid line* represents the original responses, and the *dash line* and *dotted line* denote the responses from the conventional and proposed methods, respectively

from the extracted equivalent circuits by both conventional and proposed methods agree well with the original responses, and that the simulated results of the number of circuit components agree with the estimated ones from Eqs. (13.4) and (13.5). Although the result shows that the common-pole method performs better than the proposed technique for the two-port network, this example is used to explain why the proposed method will be superior to the conventional one for the network with a large number of ports.

From Eqs. (13.4) and (13.5), it is notable that if the dominant poles are all the same for each frequency response, the estimated results are equal to each other. But in most case, the common-pole method overapproximates some frequency responses by introducing extra dominant poles. This generally happens when the resonant frequencies of each frequency responses are different. From a geometrical point of view, the differences of the resonant frequencies mainly

Table 13.1 The information about extracted circuits by the proposed and conventional schemes

	Proposed method	Conventional method
Number of circuit components	2140	2190
Simulation time in HSPICE	1.72	2.67

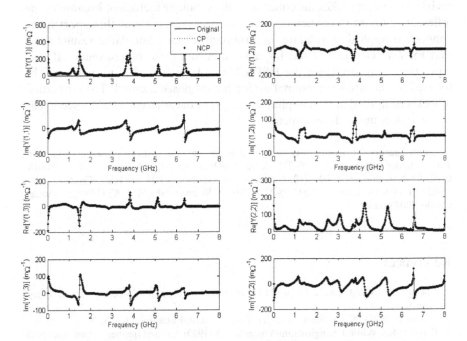

Fig. 13.6 The frequency responses for 3-port interconnects. The *solid line* represents the original responses, and the *dashed line* and *dotted line* denote the responses from the conventional and proposed methods, respectively

come from the different electrical lengths for each interconnect or signal trace. The above example has shown the difference of resonant frequencies. From Eq. (13.5), the overapproximated number of poles then causes the increasing number of poles with factor N. It implies that as the number of ports increases, the efficiency of the common-poles will decrease.

Finally, we use an example of interconnects to validate our technique. Here the three-port circuit topology refers to Example 3 in [20]. The simulated results are listed in Table 13.1. It shows that the proposed method generates the circuit with fewer components and decreases the load of the HSPICE simulator. The simulation results are partly illustrated in Fig. 13.6. The frequency responses from the extracted circuit agree well with the original ones and the results demonstrate the validity of the proposed method.

13.4 Conclusions

This study has developed the minimal equivalent circuit extraction based on the state-space-based equation subject to the non-common poles. An example is used to express why the proposed method is superior to the conventional one for the network with a large number of ports. The different electrical length of each interconnect makes the common-pole technique inefficient because of the different resonant frequencies. Furthermore, an example of three-port interconnects is applied to validate the proposed scheme. Simulated results show that the extracted circuit has fewer components than the conventional one. Also, the circuit decreases the HSPICE simulator time. Future work will be focused on the efficiency estimation for the proposed method. The estimation model is composed of the coupling effects from the adjacent interconnects. The model is now under development.

Acknowledgments This work was supported in part by the National Science Council of Taiwan under Contract NSC-96–2752-E-009-003-PAE, Contract NSC 96-2220-E-009-018, Contract NSC-96-2221-E-009-210, Contract NSC-95-2752-E-009-003-PAE, Contract NSC 95-2220-E-009-030, Contract NSC-95-2221-E-009-336; and by the MoE ATU Program under a 2006–2007 grant.

References

1. Cheng CK, Lillis J, Lin S, Chang N (1999) Interconnect analysis and synthesis. Wiley, New York
2. Young B (2001) Digital signal integrity. Prentice-Hall, Englewood Cliffs, NJ
3. Kundert KS, White J, Sangiovanni-Vincentelli A (1989) A mixed frequency-time approach for distortion analysis of switching filter circuits. IEEE J Solid-St Circ, 24:443–451
4. Griffith R, Nakhla MS (1992) Mixed frequency/time domain analysis of nonlinear circuits. IEEE T Comput Aid Design 11:1032–1043

5. Neumayer R, Stelzer A, Haslinger F, Weigel R (2002) On the synthesis of equivalent-circuit models for multiports characterized by frequency-dependent parameters. IEEE T Microw Theory Tech 50:2789–2795

6. Chiprout E, Nakhla M (1993) Addressing high-speed interconnect issues in asymptotic waveform evaluation. In: Proc. int. design automat.. Dallas, Texas, pp 732–736

7. Chiprout E, Nakhla M (1993) Transient waveform estimation of high-speed MCM networks using complex frequency hopping. In: Proceedings of the multi-chip module conference. Santa Cruz, CA, pp 134–139

8. Sanaie R, Chiprout E, Nakhla MS, Zhang QJ (1994) Integrating subnetworks characterized by measured data into moment-matching simulations. In: Proceedings of the multi-chip module conference. Santa Cruz, CA, pp 114–119

9. Na N, Choi J, Chun S, Swaminathan M, Srinivasan J (2000) Modeling and transient simulation of planes in electronics packages. IEEE T Adv Packaging 23:340–352

10. Morsey J, Cangellaris AC (2001) PRIME: Passive realization of interconnect models from measured data. In: Proceedings of the 10th topical meeting on elect. performance electron. packag, Cambridge, MA. pp 47–50

11. Achar R, Gunupudi PK, Nakhla M, Chiprout E (2000) Passive interconnect reduction algorithm for distributed/measured networks. IEEE T Circuits-II 47:287–301

12. Achar R, Nakhla MS (2001) Simulation of high-speed interconnects. Proc IEEE 89:693–728

13. Li EP, Liu EX, Li LW, Leong MS (2004) A coupled efficient and systematic full-wave time-domain macromodeling and circuit simulation method for signal integrity analysis of high-speed interconnects. IEEE T Adv Packaging 27:213–223

14. Achar R, Nakhla M (1998) Minimum realization of reduced-order models of high-speed interconnect macromodels. Kluwer, Boston, MA, pp 23–44

15. Grivet-Talocia S, Stievano IS, Canavero FG, Maio IA (2004) A systematic procedure for the macromodeling of complex interconnects and packages. In: Proceedings of the international symposium on electromagnetic compatibility (EMC Europe 2004), Eindhoven, The Netherlands, pp 414–419

16. Grivet-Talocia S, Brenner P, Canavero FG (2007) Fast macromodel-based signal integrity assessment for RF and mixed-signal modules. IEEE international symposium on EMC, Honolulu, Hawaii

17. Canavero F., Grivet-Talocia S, Maio IA, Stievano IS (2005) Linear and nonlinear macromodels for system-level signal integrity and EMC assessment. IEICE T Commun—Special Issue on EMC, E88-B:3121–3126

18. Chi-Tsong Chen (1999) Linear system theory and design. Oxford University Press, NJ

19. Kailath T (1980) Linear systems. Prentice-Hall, Englewood Cliffs, NJ

Chapter 14
Sleep Quality Differences According to a Statistical Continuous Sleep Model

A.G. Ravelo-García, F.D. Lorenzo-García, and J.L. Navarro-Mesa

Abstract This paper presents sleep quality differences between good and bad sleepers measured with a statistical continuous sleep model according to the Self-Rating Questionnaire for Sleep and Awakening Quality (SSA). Our main goal is to describe sleep continuous traces that take into account the sleep stage probability with a temporal resolution of 3 s, instead of the Rechtschaffen and Kales (R and K) resolution, which is 30 s. We adopt in our study the probability of being in stages W, S1, S2, S3, S4, and REM. The system uses only one electroencephalographic (EEG) channel. In order to achieve this goal we start by applying a hidden Markov model, in which the hidden states are associated with the sleep stages. These are probabilistic models that constitute the basis for the estimation of the sleep stage probabilities. The features that feed our model are based on the application of a discrete cosine transform to a vector of logarithmic energies at the output of a set of linearly spaced filters. In order to find differences between groups of sleepers, we define some measures based on the probabilistic traces. The experiments are performed over 24 recordings from the SIESTA database. The results show that our system performs well in finding differences in the presence of the Wake and S4 sleep stages for each group.

14.1 Introduction

Sleep is a basic necessity of life. A lack of sleep due to sleep loss or sleep disorders affects our health and quality of life.

One of the main tools used to study the quality of this process is the human sleep stage scoring. This representation is called a hypnogram; and the manual classification, scored by an expert, is a hard task.

A.G. Ravelo-García (✉)
Dpto. de Señales y Comunicaciones, Universidad de Las Palmas de Gran Canaria,
Campus de Tafira, E35017, Las Palmas de Gran Canaria, Spain
e-mail: aravelo@dsc.ulpgc.es

N. Mastorakis et al. (eds.), *Proceedings of the European Computing Conference*, Lecture Notes in Electrical Engineering 27, DOI 10.1007/978-0-387-84814-3_14, © Springer Science+Business Media, LLC 2009

Rechtschaffen and Kales (R and K) [1] is the set of rules which defines the sleep process as divided into six stages or epochs: W (wake), S1 (light sleep), S2, S3, and S4 (deep sleep), and REM activity (rapid eye movement).

R and K rules are subjective, resulting in low interscorer agreement. For example, two independent manual hypnograms scored by two experts have an agreement rate between 51% and 87% [2]. The reason for this result is that the rules are not useful in older people or in cases where sleep disorders exist. Moreover, R and K has a poor time resolution (30 s).

The work presented in this paper proposes other objectives, which are innovations based on previous research [3, 4]. We have three main goals in this work:

Firstly, to find a feature extraction technique with high definition capacity for the sleep process, taking only single-lead EEG recordings. Secondly, to define a set of representations that takes into account the sleep stage probability with a higher time resolution. And finally, to define a set of quality measures that could be useful in sleep quality screenings.

14.2 Database

The database consists of 24 recordings of subjects with ages between 20 and 69 years. Recordings belong to the SIESTA database [5] of 16 EEG signals (C3–M2) with a sampling frequency of 100 Hz. According to SSA criteria (Self-Rating Questionnaire for Sleep and Awakening Quality), we have randomly separated our database into 16 recordings with good sleep quality and 8 recordings with poor sleep quality. Eight recordings with good sleep quality were selected to train the model (group TR). On the other hand, 8 recordings with good sleep quality (group TG) and another 8 recordings with poor quality (group TP) were chosen as test recordings to validate the model.

The database also contains 130 manual hypnograms whose scorings have been done according to the R and K rules that will be used to propose a hypothesis about the possible transition among stages.

14.3 Feature Extraction

In the feature extraction process, it is necessary to segment the EEG signal. This segmentation takes into consideration the stationary criteria of the signal. Traditionally 1–30 s length EEG segments have been used. In our work we have proposed 3 s length, since it presents a good compromise between frequency and time resolution to track the changes in the stationary state of the signals.

The considered features were studied in previous work [3], with good results compared with other parameters techniques. The features are applied in two successive steps:

Filter bank log-energies (Fbank). There exists a clear correspondence between sleep stages and spectral power [2]. This evidence allows us to suggest the usefulness of the spectral (log) energies to discriminate among stages.

The analysis covers the whole frequency band. A bank of equally spaced filters is applied to the periodogram of each signal segment one second long. The periodogram is estimated by means of an FFT. In order to avoid dependencies with signal dynamics, we normalize each filter output with the total signal power. A logarithm operation is applied to the power values in order to reduce the dynamic range and therefore keep the whole frequency information.

Discrete cosine transform (DCT). A matrix of DCT coefficients is applied to the vectors of log-energies. This operation has the effect of decorrelating the log-energies, thus facilitating an increase in the discrimination capacity. This matrix transformation also results in a reduction of the dimensionality of the feature vectors that alleviates the computational burden. This fact is determinant in achieving good classification scores with HMM classifiers. Thus, a matrix transformation is applied to the Fbank vectors, where matrix coefficients are obtained with the DCT.

14.4 HMM Sleep Modelling

One has to notice that we are observers outside the brain, but at the output of an AD converter. A hidden Markov model (HMM) has been proposed as a modelling method of the hidden behaviour of the sleep stages, which at the same time allows us to model the time evolution of the characteristics.

The model represents the stages as "hidden" states and their dynamic evolutions in sleep periods.

Therefore, in what follows, the concepts of state and stage are equivalent for us. This model has two main components: the transition probability matrix among states, and the state-dependent probability density functions of the features.

The transition probability matrix defines the probabilistic nature of the dynamic state transitions (Fig. 14.1). A mixture of probability density functions (pdf) characterises the probabilistic nature of the features in each state.

In our work, we propose an association between states and sleep stages. This strategy makes our approach easier from two points of view. For one thing, we give sense to the concept of "state;" and on the other hand, it makes the training process easier.

To design an HMM, we must estimate the parameters (A, B, π) which optimize the probability of the training observation vectors set.

$$\lambda = (A, B, \pi) \tag{14.1}$$

where A, B and π are, respectively, the state transition probability matrix, the mixtures of probability density functions in each state, and the probabilities to be in a initial state in the initial observation period.

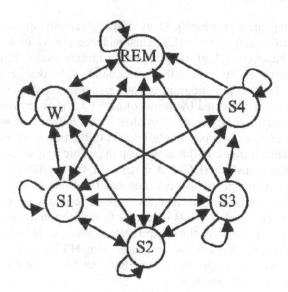

Fig. 14.1 Example of transition diagram among sleep stages

We consider a model with a topology as shown in Fig. 14.1, in which all the transitions are possible except W→S3, W→S4, REM→S3, and REM→S4. We get to this conclusion after analyzing the transition matrix which has been studied in more than 130 hypnograms, corresponding to the same number of different recordings.

In the initial instant all the patients are in W state. Thus the probabilities vector π is

$$\pi = [1, 0, \ldots 0] \tag{14.2}$$

The state-dependent pdf's are modelled by means of gaussian mixtures. These mixtures have been trained separately with the feature vectors assigned to the corresponding stages. For this purpose, we use a hypnogram for each one of the 8 training recordings which have been classified by the consensus of 2 human experts according to the R and K rules. The recordings are divided into segments of 3 s each and then parameterized to extract the features and grouped in a stage based on the hypnogram.

Now our goal is centred in estimating the parameters of the Gaussian mixtures of our HMM. In particular, for each state, we have a mixture with the following definition:

$$b_j(o_t) = \sum_{l=1}^{M} c_{jl} b_{jl}(o_t) = \sum_{l=1}^{M} c_{jl} N\left(o_t / \mu_{jl}, \sum_{jl}\right), 1 \leq j \leq N \tag{14.3}$$

where given a feature vector ot at time instant "t," the probability of being observed in the state p is given by the sum of M Gaussians, each one defined by its particular measure vector μ_{jl}, its covariance matrix Σ_{jl}, and the weighting coefficients c_{jl} of each Gaussian of the mixture.

For the estimation of $N(\mu_{jl}, \Sigma_{jl})$, we proceed with the application of the expectation maximization (EM) algorithm. In our training phase, the Gaussian mixtures showed good performance in B with EM isolated from the A matrix. This matrix was designed based on the same recordings used for B. In the test phase, using the Viterbi algorithm, the most probable state sequence is estimated. A detailed documentation can be studied in [6].

14.5 Experiments and Results

The experiments are in line with the goals detailed in Section 14.1. The time resolution of our system is three seconds, instead of the thirty seconds in the R and K rules. Although manual R and K–based scores have been taken into account to train the system, a final three seconds probability representation is obtained after the test process. This is in an attempt to achieve new insights in the short time structure of the sleep stages.

In Section 14.1 we described R and K as a set of rules that could be subjective to interpret. There may be present similar time-frequency components in different R and K stages, thus causing confusion. With the probabilities that we obtain from the HMM, a continuous sleep trace is represented for each stage, and it is possible to establish the weight of each stage in each 3-second segment, thus obtaining a better temporal resolution.

The probability traces in each stage for a patient of group TG is represented in Fig. 14.2. At the bottom there is the manual hypnogram. It is possible to find correspondences between manual hypnograms and the outputs in the form of continuous probability traces. In instants where the Wake probability trace is near one, the manual hypnogram represents the Wake stage. Similar considerations can be observed for S1, S2, S3, S4, and REM. In the case of patient G1 (good sleep) there is clearly a greater presence of S4 and S3 sleep stage probability and a lower presence of Wake.

On the other hand, a similar representation can be seen in Fig. 14.3 for a patient of the group with poor sleep quality (patient TP1). In this case, there exists a greater presence of Wake and a lower S3 and S4 representation.

One of the main goals of the present work is to find numerical differences between groups of patients. Table 14.1 summarizes the presence of S4 and Wake sleep stage probabilities for all the groups of patients and for each patient separately. It is possible to define a measure of the S4 presence as the ratio between the sum of probabilities to be in S4 all the night and the sum of probabilities to be in any other stage. An analogous consideration is taken into account for the Wake state. Final results in percentages are shown in Table 14.1.

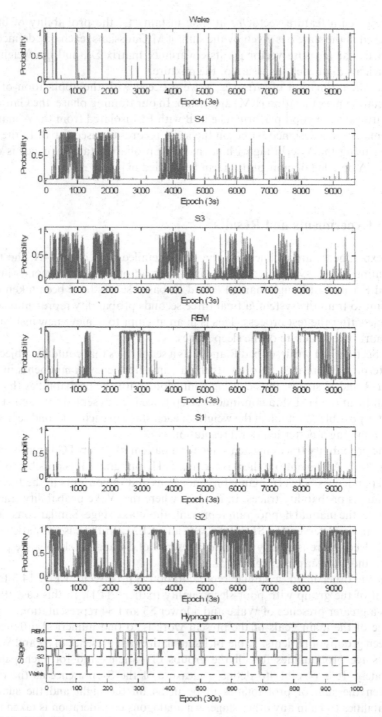

Fig. 14.2 Probability traces for each stage and manual hypnogram (*bottom*) for patient TGl (group TG)

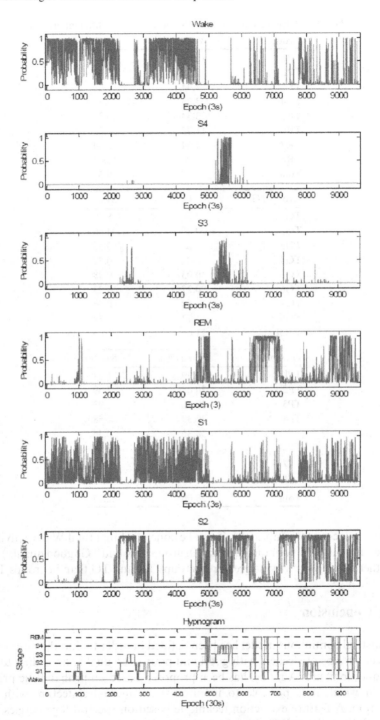

Fig. 14.3 Probability traces for each stage and manual hypnogram (*bottom*) for patient TP1 (group TP)

Table 14.1 Percentage of wake and S4 sleep stage for different groups

Group TR	Wake (%)	S4 (%)
TR1	12.37	0.54
TR2	7.78	13.97
TR3	3.44	17.93
TR4	3.29	0.32
TR5	3.34	10.27
TR6	11.72	6.38
TR7	14.34	13.34
TR8	4.26	13.28
Total	7.57	9.51
Group TG	Wake (%)	S4 (%)
TG1	2.89	8.89
TG2	10.47	4.56
TG3	7.85	3.53
TG4	3.29	0.32
TG5	10.03	0.98
TG6	20.27	15.23
TG7	21.60	0.34
TG8	11.72	6.38
Total	11.01	5.03
Group TP	Wake (%)	S4 (%)
TP1	31.99	1.37
TP2	20.08	3.59
TP3	20.82	2.92
TP4	30.39	4.58
TP5	38.36	0.36
TP6	47.45	1.59
TP7	10.30	0.16
TP8	25.08	2.76
Total	28.06	2.17

After inspection of Table 14.1, it can be concluded that total Wake activity is greater in group TP (poor quality) than in groups TR and TG (good quality). On the other hand, S4 activity is greater in groups TR and TG than in groups TP.

14.6 Conclusion

A statistical continuous model has been defined in order to find the differences between groups of sleepers with good and poor sleep quality according to the SSA questionnaire. A model has been proposed based on probabilistic principles, in which it is possible to monitor sleep micro-architecture with 3 s resolution. A feature extraction technique based on spectral log-energies and decorrelation with a DCT-based matrix has demonstrated a high capacity of definition of the sleep process. It is possible to find numerical differences between groups of patients mostly in the Wake and S4 sleep stages.

Acknowledgments We would like to thank Dr. Alpo Varri at Tampere University of Technology for providing us the database used in this research.

References

1. Rechtschaffen A, Kales A (1968) A manual of standardized terminology techniques and scoring system for sleep stages of human subjects. Brain Research Institute, UCLA, Los Angeles, USA. A threshold selection method from gray-level histograms. IEEE T Syst Man Cybernetics 9(1):62–66
2. Nielsen KD (1993) Computer assisted sleep analysis. Doctoral Thesis, Aalborg University, Denmark
3. Navarro JL, Ravelo AG, Lorenzo FD, Martín SI, Hernández E, Quintana P (2006) On the determination of differences between good and bad sleepers by means of a hidden Markov model. WSEAS T Comput Res 1:321–324
4. Flexer A, Dorffner G, Sykacek P, Rezek I (2002) An automatic, continuous and probabilistic sleep stager based on a hidden Markov model. Appl Artif Intell 16(3):199–207
5. http://www.ai.univie.ac.at/oefai/nn/siesta/
6. Rabiner LR, Juang BH (1993) Fundamentals of speech recognition. Prentice-Hall, Englewood Cliffs, NJ

Chapter 15
Simulation Modeling in Support of a European Airspace Study

John Morris, John James, James DeArmon, Kelly Connolly,
and Paula Mahoney

Abstract To address some questions about usage of European airspace by United States Air Force aircraft, two simulation models are employed. The first model offers a wide array of functions to represent aircraft movement and management of aggregates or "flows" of aircraft. The second model uses a Petri net approach to represent the complexity of a flight planning/replanning operation, in order to estimate staffing requirements.

15.1 Introduction

A recent study by The MITRE Corporation used two corporate owned and developed simulation models, which we describe hereunder. The MITRE Corporation is a not-for-profit research and systems engineering firm, chartered to perform work only in the public interest, i.e., working for governments, either domestic or foreign. Two major sponsors are the United States Air Force (USAF) and the Federal Aviation Administration (FAA), which supported the development of the simulation models.

Our team was tasked with answering questions regarding airspace usage by USAF military flights over Europe. The USAF must make investment decisions regarding avionic capabilities (i.e., electronic equipment on aircraft—radios, altimeters, navigation equipment, etc.), and our research looked into the advantages of full avionic equipage of the fleet (about a dozen types of aircraft). With full avionic equipage, civilian air traffic control (ATC) provides the best, expedited service. The problem was made more complex via the requirement that differing types of aircraft, departing from different airfields, rendezvous at a given time and place.

The two simulation models to be described in this paper are the Collaborative Routing and Coordination Tools (CRCT, pronounced "circuit") and

J. Morris (✉)
The MITRE Corporation, 202 Burlington Road, Bedford, MA 01730-1420, USA
e-mail: jemorris@mitre.org, www.mitre.org

N. Mastorakis et al. (eds.), *Proceedings of the European*
Computing Conference, Lecture Notes in Electrical Engineering 27,
DOI 10.1007/978-0-387-84814-3_15, © Springer Science+Business Media, LLC 2009

MSim. CRCT was used to model civilian air traffic and airspace, as well as USAF military flights' intersections with airspace sectors. MSim was used to model arrival times of aircraft, as well as the staffing requirements for the flight planning/replanning function.

15.2 Background on Airspace Study

The success of the USAF in the international arena hinges on the ability to access civilian airspace while dealing with a wide array of sovereign policies. Onboard aircraft equipage determines the level of ATC services and access to airspace—in general, the better the equipage, the better the service and access. The term Communication Navigation Surveillance/Air Traffic Management (CNS/ATM) is used to refer to these avionic capabilities. It is not cost effective for the USAF simply to equip every aircraft with the best-available avionics. Rather, a trade-off analysis emerges: what level of spending for upgrades coincides with a given level of performance? Lacking access to the best routes and altitudes, military aircraft are subject to increased mixing with the civil fleet, and may suffer congestion delay. We modeled this phenomenon for two future years by using two different simulation models.

Several hypothetical scenarios were considered. For each scenario, cases where military aircraft were "CNS capable" (enabled to meet CNS requirements to the extent possible) and "CNS not-capable" (lacking on one or more CNS capabilities) were examined. A hypothetical flight sortie was considered, whereby various aircraft are to rendezvous at a given time and location. Full results have been presented in other papers [1, 2], and are not repeated here. Rather, this paper discusses the underlying simulation models.

In general terms, CRCT was used to model civilian air traffic over Europe, and using some published equations for sector loading, the times and locations of congestion were computed. Next, the subject USAF military flights were likewise "flown" in the simulation model, and the times of sector-by-sector intersections were recorded. This data on congestion and military flight paths was next presented to the MSim simulation for processing. MSim modeled the dynamic interactions—military flights being delayed via reroute around congested sectors, or queuing for tankers to periodically refuel. Estimates of the lateness of flights was used as input into a second MSim model, to be used to model flight planning/replanning staffing at an operations center. This second MSim model is the one described below.

15.3 CRCT

In the field of air traffic management, two major functions exist: ATC provides separation services between pairs of aircraft, or between an aircraft and airspace; traffic flow management (TFM), by contrast, manages system resources such as airports, routes, and airspace. It is the job of TFM to ensure that the demand for resources does not exceed the available capacity. Generally, in light of a predicted demand-over-capacity situation, aircraft can be moved in either time (via delay assignment) or space (via reroute, in the case of airspace congestion).

CRCT is a set of functions designed to assist TFM personnel in their quest to balance air traffic demand with the capacity of airspace resources to accommodate that demand. Using a graphical user interface, flow managers may visualize airspace and forecast air traffic, and assess potential airspace congestion. To ameliorate congestion, a traffic flow manager typically works with "flows" or aggregates, as opposed to individual flights. CRCT supplies a "what-if" capability, allowing a traffic flow manager to try various strategies, or variants on pre-stored initiatives, in a virtual mode. The traffic flow manager would consider not only resource capacities, but also the impact on air carrier preferences. For example, some air carriers prefer a route deviation over a ground delay. Once satisfied that a candidate initiative would likely succeed, the traffic flow manager would disseminate the solution to airspace managers and users, expecting compliance from users, such as pilots and commercial air carriers. Although CRCT was developed for real-time, real-world application, it also includes an offline mode which will support simulation studies.

Developed by the Center for Advanced Aviation System Development (CAASD) at The MITRE Corporation, CRCT currently exists on a research platform in the CAASD computer laboratories, and as a concept-evaluation platform at several federal ATC facilities. For this project, we accessed the offline and playbook features of the software system—all of the real-time, real-world traffic management functions in CRCT can be used for simulation modeling. In the parlance of simulation modeling (see [3]), we have a dynamic, deterministic simulation model. It is dynamic in that the situation (state of flights, sector demand, etc.) change over time. It is deterministic in that there are no random variables—a set of flight plan inputs creates the same outputs each time the simulation model is run. Specifically, we used the following functions in our simulation modeling.

15.3.1 Traffic Flow and Demand Analysis

The CRCT adjunct supporting infrastructure performed flight plan and trajectory processing for the subject military flights. (A trajectory is the estimated future path of a flight, consisting of geographic position and altitude per time).

In addition to the subject military flight paths, multiple complete days of civilian traffic were modeled. Fig. 15.1 gives a snapshot—a "freeze" in time—of this busy environment. In the lower right of the figure, the label "FUTURE" is displayed, indicating that the user is in forecast mode. Note the very busy nexus on the continent—these are busy civilian airports in cities such as Paris and

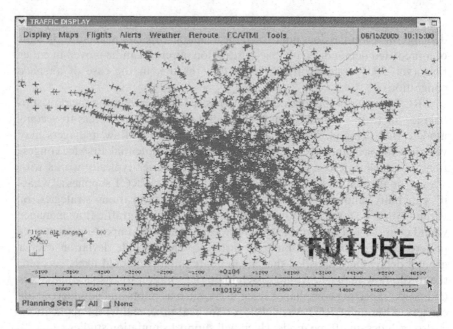

Fig. 15.1 CAPER traffic display for Europe

Frankfurt. Note also the flows on the far left, headed west. This is the morning (10 a.m. local time) push of outbound flights to North America.

To assess sector congestion and the potential impact on military flights, an array of flight geometry information per sector was captured and analyzed. Per published equations and thresholds, sector loading was assessed using traffic level, number of altitude transitions, number and type of pairwise aircraft proximity events, etc.

15.3.2 Aircraft Reroute Definition

It was decided, for our model representation, that military flights would avoid congested sectors by re-routing around them. It was hence necessary to determine the additional flight time for congestion-avoiding flights. The re-routes would not be represented dynamically in the CRCT modeling—rather, a probability distribution of typical re-route times was developed: several analysts played the role of traffic managers. Using the CRCT graphical user interface (GUI), the analysts selected 30 sectors at random, and worked-out re-route paths around these sectors. These 30 delay times were then fitted to a log-normal distribution, and used in MSim, as described below. Fig. 15.2 shows military routes overlaid on civilian traffic, at a zoom-in magnification greater than that of Fig. 15.1.

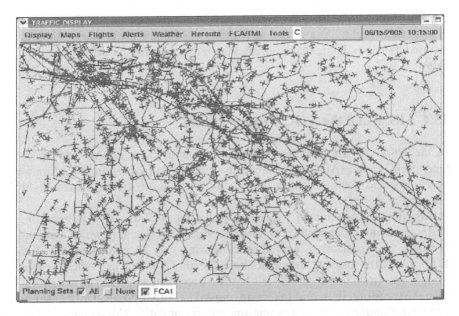

Fig. 15.2 Military routes overlaid on civilian traffic

15.3.3 Capture of Dynamic Data

CRCT allows capture and recording of dynamic data. In the simulation mode, the dynamism is a function of events scheduled and executed per the simulation system clock (in the real-time mode, the dynamics are obviously a function of real-world actions and events unfolding). Dynamic data on sector-specific flight geometries were captured and evaluated to determine times and locations of congested sectors.

15.3.4 Data Transfer from CRCT to MSim

In summary, three sets of information are transferred from the CRCT simulation to the MSim simulation runs:

- Times and locations of congested sectors;
- Parameters for the probability distribution representing congested sector delays;
- Military flight trajectories with sequences of sector entry and exit times.

MSim uses these datasets, in concert with information about tanker capacities and their en route locations, to model the dynamical time ordering of events, in light of a complex network of interacting entities. The goal is to assess the probability of rendezvous of specific sets of airborne assets, i.e., the number of late arrival flights. As a second application of MSim, these lateness estimates are used to help estimate the requirements for flight planning staff, as described below.

15.4 MSim and the AOC Process Model

In the USAF, flight planning/replanning takes place at an Air and Space Operations Center (AOC). An elaborate sequence and synchrony of information flows and decisions must precede any execution of coordinated flight planning activities.

An AOC process model was developed using MSim, which is based on Petri net methodology. Petri nets were developed for systems in which communication, synchronization, and resource sharing play an important role. See Fig. 15.3 for the simplest atomic example. Starting from the left, when the two input places (note that Input Place 1 has two precedent activities) have a token, then the transition fires, and tokens are moved to output places. The transition can be configured to consume some amount of simulation clock time.

MSim is a prototype simulation tool that was developed at The MITRE Corporation initially to model the performance of distributed computer systems, but later used to analyze the performance of business processes [4]. MSim may be categorized as a dynamic, stochastic simulation model. It is dynamic in that the situation (rendezvous success rate, requirement for flight planning staff, etc.) changes over time. It is stochastic in that random variables, e.g., time to refuel or time to replan a coordinated flight package, create "chance effects" and a probabilistic outcome.

MSim also utilizes a high level definition of system "threads" to specify the routing of tokens in the Petri net. A thread is a path through the model of the system for a given system stimulus. (For readers familiar with the Unified Modeling Language (UML), this corresponds to the term "scenario," as used by the Object Management Group [5].) It may be open or closed and is drawn in the model diagrams by associating a thread of a specific color with a set of edges, as illustrated in Fig. 15.4. Threads also have a name and a priority, and

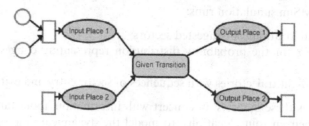

Fig. 15.3 A simple two-input, two-output Petri net

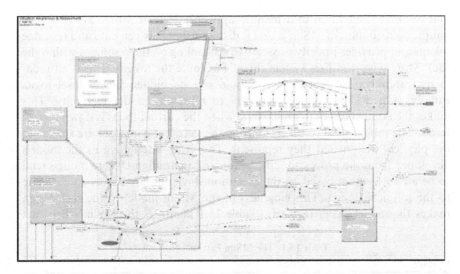

Fig. 15.4 Portion of AOC Petri net model

model diagrams can show multiple threads at the same time, which is a valuable feature, both for debugging and for appreciation of complex interrelationships.

For a given scenario, MSim produces performance metrics such as resource utilization, component throughput, and thread response time. These metrics can be used to: (1) determine if the process modeled meets its operational performance requirements, (2) find the performance bottlenecks, and (3) evaluate the performance effectiveness of proposed process changes. The metrics may be exported directly to Excel for plotting and subsequent use in other automation products. MSim models can participate as a member of a simulation federation [6] through integration using High Level Architecture (HLA), such that federation time and MSim time are synchronized.

Figure 15.4 shows part of the AOC model that is implemented in MSim. Petri net transitions (rectangular box) model the activities that do work and produce outputs. Places (circles or ellipses) represent the type of data that the transition needs for input and the type of data that the transition produces as output. Tokens represent the instances of data created and consumed by the transitions. Tokens are also associated with thread types and their movement is restricted along arcs that are associated with the same types.

The execution behavior of an ordinary Petri net follows two simple rules: (1) Once all the input places to a given transition have a token, then the transition can fire (occur) and is said to be enabled. (2) When a transition fires, it takes a token from each input place and puts a token in each output place. A timed Petri net allows discrete event simulation to be modeled, and in this case, transitions usually have timing functions that introduce time delays in processing.

The timed Petri net methodology is useful for modeling the performance of real-time distributed systems partly because of its ability to analyze concurrency and resource contention in a manner appropriate to real-time systems. The unique contribution of MSim is that it distinguishes between data and resource tokens and provides priority preemptive scheduling on the resources within the tool. Moreover, it utilizes system threads to route the tokens. This is a fundamental architectural feature that should not be hidden in low-level logic. Finally, it provides in-place hierarchy to represent model hierarchy. This makes the context easier to understand when the current focus is down several levels. The models that result from using the Petri net paradigm are similar to the physical design that they represent. This is an alternative to the models developed from the process-oriented simulation paradigm, where models tend to be an abstraction based on the execution flow, and are harder to correlate to the physical design. This closeness of the MSim model to the real design makes the model easier to verify. Table 15.1 identifies the MSim's Petri net

Table 15.1 The MSim Petri net features

Feature	Description
High level Petri net	A place is marked by a multi-set of structured tokens. The tokens have a thread type and can carry a data structure as well as a synchronization value (i.e., control fork and join operations).
Timed Petri net	The transition firing takes a user-defined amount (i.e., distribution) of time.
Arcs	Associate thread type with the arc of the net and only allow movement of tokens along edges having the same type as the token.
Transitions	Every transition has a code expression which can have a Boolean guard function that must evaluate to true for the transition to fire. The code expression can change the token's type and set values in the data structure within the token.
Specify the Petri net	Allows fused places and transitions so that a Petri net can be made up of a set of pages with common features.
	Hierarchy is used as a shorthand way to specify a set of places connected to a given transition or a place connected to a set of transitions.
	Binding is a set of input data tokens—one from each input place. It represents the necessary data and resources required to do the work in a transition. It has an inherent thread type associated with its tokens and an associated priority as defined by the thread type. The inherent thread type of a ready-to-run transition controls the type of resource tokens used by the firing of the transition. Thus, a different resource could be used depending on the type of data flowing through the transition.
Resource allocation algorithm	A built in, optional resource allocation algorithm determines the highest priority bindings that should be running with the required resources. The algorithm is a variation of a standard combinatorial problem called the provisioning problem [7]. Each binding corresponds to the items being provisioned, and the resources are the provisions. Their cost is the binding's priority. This algorithm is run whenever there are transitions that could fire to select the transitions that will be running next.

features, which are more general than ordinary Petri nets. In summary, MSim was successfully used to model military flight arrivals under conditions of a rendezvous requirement, as well as the staffing requirements for the flight planning/replanning function.

15.5 Conclusion

Simulation modeling is one of the most prominent analytical solutions available for modern, complex applications. It has been said: "When all else fails, simulate," suggesting that closed-form and other algorithmic solutions, though desirable, may be inadequate to represent the vagaries of the real world.

Two quite different simulation models have been described. CRCT was developed as a real-time tool for air traffic managers, but offers a wide array of offline facilities that can support simulation modeling. These facilities were used for our European airspace study—civilian traffic for two future years, and associated sector congestion; and flight path trajectory of military flights pursuing a rendezvous point with minimal delay.

MSim, by contrast, was used to represent time sequencing and dynamic processes such as queuing for refueling resources, or demand for flight planning/replanning staff. Using a Petri net technology, MSim supports classical discrete event simulation, and allows transparency with respect to "threads" or transaction sequence paths through multiple heterogeneous processes.

References

1. Wigfield E, Connolly K, Alshtein A, DeArmon J, Flournoy R, Hershey W, James J, Mahoney P, Mathieu J, Maurer J, Ostwald P (2006) Mission effectiveness and European airspace: U.S. Air Force CNS/ATM planning for future years. In: Integrated command and control research and technology symposium (ICCRTS), http://www.dodccrp.org/events/11th_ICCRTS/html/papers/139.pdf
2. Wigfield E, Connolly K, Alshtein A, DeArmon J, Flournoy R, Hershey W, James J, Mahoney P, Mathieu J, Maurer J, Ostwald P (2006) Mission effectiveness and European airspace: U.S. Air Force CNS/ATM planning for future years. Accepted for publication in the Journal of Defense Modeling and Simulation, San Diego, CA
3. Law A, Kelton WD (2000) Simulation modeling and analysis. McGraw-Hill, New York
4. James JH, Schaffner SC (1994) Visualization of the dynamics of performance simulations. In: Proceedings of the summer computer simulation conference, LaJolla, CA
5. Object Management Group (2007) Website on unified modeling language. http://www.uml.org/
6. Kuhl F, Weatherly R, Dahmann J (1999) Creating computer simulation systems: an introduction to the high level architecture. Prentice, Jersey City, NJ
7. Lawler E (1976) Combinatorial optimization: networks and matroids. University of California at Berkeley, Holt, Rinehart, and Winston

Chapter 16
3D Measurement System for Car Deformation Analysis

András Rövid, Takeshi Hashimoto, and Péter Várlaki

Abstract Nowadays many applications require for their processing knowledge about the shapes of objects. In a lot of cases, the shapes to be acquired can have the kind of segments for which measurement is difficult, because of the occlusion problem. Shapes with such properties can be observed on crashed car bodies for example, where the deformation can be of complex shapes. In this paper a new measurement system is introduced, the aim of which is to accurately acquire the deformed parts of a car body and obtain useful information supporting car deformation analysis.

16.1 Introduction

There are many applications which require precise information about the shape of objects, e.g., in the field of robotics, gathering information about the environment of a robot; in the field of archaeology, 3D documentation of cultural heritage artifacts and statues; in the field of car crash analysis, the modeling of the deformation [1, 5, 8], etc. The objects to be measured can be of arbitrary shape, so in many cases their acquisition is problematic because of their complexity.

For acquisition of the objects, 3D measurement devices are used, based on different principles. The purpose of a 3D measurement device is to analyze a real-world object or environment to retrieve information about its shape. From such collected data three-dimensional models can be constructed, which have a wide variety of applications.

Different kinds of 3D shape measurement methods have been developed to obtain information about the shape of the objects, but only a few from them are able to acquire complex shapes as well.

A. Rövid (✉)
Department of Chassis and Lightweight Structures, Budapest University of
Technology and Economics, Bertalan Lajos u. 2. 7. em., 1111 Budapest, Hungary

N. Mastorakis et al. (eds.), *Proceedings of the European*
Computing Conference, Lecture Notes in Electrical Engineering 27,
DOI 10.1007/978-0-387-84814-3_16, © Springer Science+Business Media, LLC 2009

Here we will focus on measuring the shape of the deformation on crashed cars, to support a car-body deformation analysis.

Through the analysis of traffic accidents we can obtain information concerning the vehicle, which can be of help in modifying its structure/parameters to improve its future safety. The energy absorbed by the deformed car body is one of the most important factors affecting accidents, thus it plays a very important role in car crash tests and accident analysis. There is an ever-increasing need for more correct techniques, which need less computational time and can be more widely used. Thus, new modeling and calculating methods are highly welcome in deformation analysis. One of the important parameters of after-crash deformation analysis is the shape of the resulting deformation.

In car crash analyses there are many situations when information about the deformation of the crashed car body is needed, i.e., knowledge about the shape of the deformation is necessary. It is known that crashed cars can be of different—in most cases very complicated—shapes. They can contain many overlapping parts, acquisition of which is among the most common problems of many 3D measurement methods.

As mentioned before, there are several concepts of 3D shape measurement [10, 11]. Each concept has its advantages and disadvantages; the applicability of them depends on the criteria which need to be met by the concrete application. For example, laser range scanners using triangulation can be very advantageously applied in high accuracy 3D measurement. With the help of the projected laser stripe, the point correspondence problem is effectively solved, which is the general source of the measuring error in many 3D reconstruction methods. As another well known example, the "time of flight" concept can be mentioned. In this concept, because the speed of light is precisely known, a distance can be estimated by measuring the time that a light pulse takes to travel from an observed target to a reference point [7]. 3D scanners based on structured or coded light project a pattern—in most cases a grid or a line stripe pattern—on the object [2, 3, 4].

Shapes containing many overlapping parts cannot be extracted by scanning devices, because the position and orientation of the device cannot be changed to get the occluded parts within sight of the cameras. Devices that are based on pattern projection also have limitations because of the reconstructed object's reflectance properties. The method introduced in this paper eliminates these limitations and enables the measurement of complicated surfaces by maintaining accuracy. The algorithm is optimized for real-time processing and provides a way to measure the shape of the deformed car body with high accuracy.

The paper is organized as follows: Section 16.2 introduces the basic concept of the proposed algorithm. Section 16.3 describes the main tasks of the measurement principle more detail. In Section 16.4 the noise reduction and structural arrangement of 3D points are described. Section 16.5 presents examples. And finally, Section 16.6 reports conclusions.

16.2 The Basic Concept of the Method

The proposed method is based on the combination of a stereo camera and a camera-laser-projector system. The task of the stereo camera system is to estimate the position and orientation of the camera-laser-projector device in the world coordinate system using four markers of different colors (see Fig. 16.1). Using different colors is advantageous when searching for the corresponding markers in the stereo images, which were taken by the stereo camera system. Searching for the corresponding markers in such a way can be performed much more quickly than when using markers of the same color.

Performing the search in the corresponding camera images is very important. When searching for the markers is performed using non-corresponding images, the calculated 3D coordinates may not be valid. For this reason an important task of this method is to synchronize the cameras in order to obtain corresponding images.

The markers are mounted onto the scanning device, which consists of a camera and a laser line projector. Using the information involved in the stereo images taken by the stereo camera system, the position and orientation of the scanning device can be estimated. In the calibration of the cameras the direct linear transformation method (DLT) was used. The laser line projector projects a line onto the object, which can be easily recognized. These points are located on the plane formed by the laser strip. The 3D coordinates of such a point can be calculated as the crossing point of the laser plane and the line of sight corresponding to that point (see Fig. 16.2).

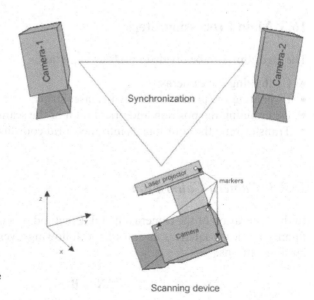

Fig. 16.1 Illustration of the proposed system

Fig. 16.2 Illustrating the
measuring principle of the
scanning device (the line of
sight intersects the plane
formed by the laser stripe,
which forms the solution
point)

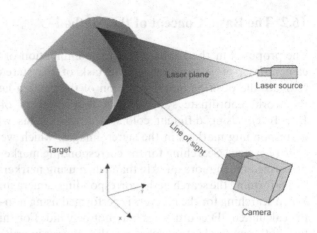

The obtained coordinates are in the local coordinate system of the scanning device. These coordinates should be transformed into the world coordinate system and, depending on the estimated position and orientation of the scanning device, they should be multiplied by the corresponding rotation and translation matrix. In this way we can obtain 3D models of complex shapes by moving the scanning device freely by hand and measuring the position of each point of the target.

The implementation of this concept was realized in C + +. The software code was optimized to enable real-time processing on a single personal computer.

16.3 Main Processing Steps

The algorithm can be divided into four main tasks:

- Calibrating the cameras;
- Estimating the plane formed by the laser stripe;
- Determining the position and orientation of the scanning device;
- Transforming the coordinates into the world coordinate system.

16.3.1 Camera Calibration

In the calibration of the cameras the DLT method was used, which is based on finding the least square solution of the following overdetermined system of linear equations:

$$AX = B \qquad (16.1)$$

$$A = \begin{bmatrix} \mathbf{M}_i & \mathbf{0} & -u_iX_i & -u_iY_i & -u_iZ_i \\ \mathbf{0} & \mathbf{M}_i & -v_iX_i & -v_iY_i & -v_iZ_i \\ & & \vdots & & \end{bmatrix}, \tag{16.2}$$

$$\mathbf{X} = [\mathbf{P}_1 \quad \mathbf{P}_2 \quad \mathbf{P}_3]^T, \tag{16.3}$$

$$\mathbf{P}_k = [p_{k1} \quad p_{k2} \quad p_{k3} \quad p_{k4}], \; k = 1..3, \tag{16.4}$$

$$\mathbf{B} = [u_i \quad v_i \quad \cdots]^T, \tag{16.5}$$

where $\mathbf{M}_i = [X_i; \, Y_i; \, Z_i]$ and u_i, v_i stand for the 3D coordinates of the calibration points and their projections onto the camera image plane respectively. The values p_{ij} represent the parameters of the camera, which should be estimated. At least 6 points are needed for the calibration, but to achieve higher accuracy we propose to increase the number of calibration points.

16.3.2 Estimating the Plane Formed by the Laser Stripe

For the estimation of the plane formed by the laser stripe, at least three spatial points are needed. When using more then three points, higher accuracy can be achieved. These points can be obtained by setting two planes of known equations in front of the camera and detecting their intersections with the laser stripe. Among the intersection points at least three points have to be chosen. Figure 16.3 illustrates the situation, i.e., the points and the vectors $\mathbf{p}$, $\mathbf{q}$ formed by the selected three points. The vector $\mathbf{n} = \mathbf{p} \times \mathbf{q}$ represents the normal vector of the laser plane.

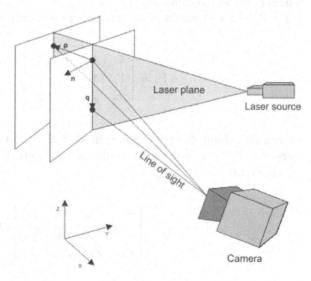

Fig. 16.3 Illustrating the principle of the laser plane estimation (in the image the vector $\mathbf{n} = \mathbf{p} \times \mathbf{q}$ represents the normal vector of the laser plane)

16.3.3 Determining the Position and Orientation of the Scanning Device

This part deals with the question of how to follow the motion of the scanning device accurately; i.e., we will focus on the estimation of two important parameters, the position and the orientation of the device. To determine these parameters, four markers are attached to the scanning device. The arrangement of the markers influences the sensitivity of the motion registration.

The detection of the markers is performed by setting the gamma value low and searching for the pixels whose RGB values are higher then a predefined threshold. As a result, a set of points is obtained. These points should be clustered, and for each cluster the center of gravity should be calculated.

After the marker's image coordinates in both camera images have been estimated, the 3D position of each marker should be determined. These coordinates can be calculated as the intersection point of the lines of sights corresponding to the stereo cameras. Because the markers are represented by light-emitting diodes of different colors, the correspondence problem can be easily and quickly solved. For the determination of the intersection point of the lines of sight, the least square solution was used.

After the coordinates of the markers are estimated, the relative rotation and translation matrix should be determined according to a reference position and the orientation of the scanning device. Such reference information can be obtained, for example, by saving or remembering the position and orientation of the scanning device when estimating the transformation matrixes between the world and local coordinate systems. The scanning device has its own local coordinate system, and the points calculated by the device should be transformed into the world coordinate system, which corresponds to the coordinate system of the stereo cameras.

Let $\mathbf{P}_r$ be a 3×4 matrix of the reference points defined as follows:

$$\mathbf{P}_r = \begin{bmatrix} X_1 & X_2 & X_3 & X_4 \\ Y_1 & Y_2 & Y_3 & Y_4 \\ Z_1 & Z_2 & Z_3 & Z_4 \end{bmatrix}, \tag{16.6}$$

where each column corresponds to a marker point and contains its 3D coordinates. Let $\mathbf{P}_a$ be a 3×4 matrix containing the actual 3D coordinates of the marker points:

$$\mathbf{P}_a = \begin{bmatrix} X_{1a} & X_{2a} & X_{3a} & X_{4a} \\ Y_{1a} & Y_{2a} & Y_{3a} & Y_{4a} \\ Z_{1a} & Z_{2a} & Z_{3a} & Z_{4a} \end{bmatrix}, \tag{16.7}$$

where the columns represent the actual coordinates of the markers. A least squares fit should be made in order to get the constraints exactly satisfied. Using the marker points in $\mathbf{P}_r$ and $\mathbf{P}_a$ the following matrices can be defined:

$$\mathbf{V} = \begin{bmatrix} v_{11} & v_{12} & v_{13} \\ v_{21} & v_{22} & v_{23} \\ v_{31} & v_{32} & v_{33} \end{bmatrix}, \tag{16.8}$$

$$\mathbf{V}' = \begin{bmatrix} v'_{11} & v'_{12} & v'_{13} \\ v'_{21} & v'_{22} & v'_{23} \\ v'_{31} & v'_{32} & v'_{33} \end{bmatrix}, \tag{16.9}$$

where the elements v_{ij} and v'_{ij} are defined as follows:

$$v_{1j} = \frac{X_{1a} - X_{(j+1)a}}{|\mathbf{V}_j|}, \quad v'_{1j} = \frac{X_1 - X_{(j+1)}}{|\mathbf{V}'_j|} \tag{16.10}$$

$$v_{2j} = \frac{Y_{1a} - Y_{(j+1)a}}{|\mathbf{V}_j|}, \quad v'_{2j} = \frac{Y_1 - Y_{(j+1)}}{|\mathbf{V}'_j|} \tag{16.11}$$

$$v_{3j} = \frac{Z_{1a} - Z_{(j+1)a}}{|\mathbf{V}_j|}, \quad v'_{3j} = \frac{Z_1 - Z_{(j+1)}}{|\mathbf{V}'_j|}. \tag{16.12}$$

Here $\mathbf{V}_j$ and $\mathbf{V}'_j$ stand for the vector in the jth column of $\mathbf{V}$ and $\mathbf{V}'$ respectively. We can add additional columns (vectors) to the matrices $\mathbf{V}$ and $\mathbf{V}'$ to improve the accuracy. Using the above matrices, the rotation matrix can be determined as the least square solution of the following system of linear equations:

$$\mathbf{V} = \mathbf{R}\mathbf{V}', \tag{16.13}$$

where $\mathbf{R}$ stands for the 3×4 rotation matrix and has the form of:

$$\mathbf{R} = \begin{bmatrix} r_{11} & r_{12} & r_{13} & r_{14} \\ r_{21} & r_{22} & r_{23} & r_{24} \\ r_{31} & r_{32} & r_{33} & r_{34} \end{bmatrix}. \tag{16.14}$$

The translation vector can be determined as follows:

$$\mathbf{T} = \sum_{j=1}^{n_{col}} \left(\mathbf{P}_{aj} - \mathbf{RP}_{rj} \right) / n_{col}, \qquad (16.15)$$

where n_{col} stands for the number of columns of matrices $\mathbf{P}_r$ and $\mathbf{P}_a$, and $\mathbf{P}_{rj}$ and $\mathbf{P}_{aj}$ represent the jth column (vector) of matrices $\mathbf{P}_r$ and $\mathbf{P}_a$. The reason why the average is taken is that noise can appear in the detection of the markers.

Next, the transformation between the local coordinate system of the scanning device and the world coordinate system should be accomplished. This procedure can be performed in the same way as the above determination of the motion parameters of the scanning device (rotation and translation), i.e., the coordinates of three arbitrary vectors should be given for both coordinate systems.

For each selected pixel of the projected laser stripe, the following steps should be performed to obtain the 3D coordinates in the world coordinate system:

- Estimation of the 3D coordinates in the local coordinate system of the scanning device, i.e., the 3D coordinates of the point which is the intersection of the plane formed by the laser stripe and the line of sight corresponding to the actual pixel analyzed.
- Transformation of the obtained point into the world coordinate system.
- Rotation and translation of the point according to the motion of the scanning device, which is estimated with the help of LED markers.

From the motion of the scanning device the whole object can be acquired.

16.4 Structuring and Noise Reduction

From the scanning, the measured data can contain noise, which has to be eliminated. Many points of the object to be acquired are measured more than one time from different positions and orientations of the scanning device. The accuracy of the measurement depends also on the accuracy of the motion estimation of the scanning device. This means that more measured data can correspond to a point of the object, with some dispersion. To eliminate this effect, the points falling into an elementary subspace are replaced by their average. This requires organizing the data effectively to enable their access within a desirable time limit. For this purpose the measurement space is divided into elementary subspaces. These subspaces are organized into an oct-tree structure [6, 9]. Such organizing fulfills the criteria mentioned, thus the accessing can be performed effectively. This structure is illustrated in

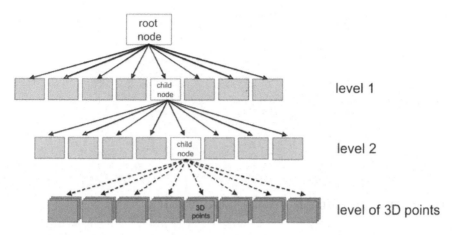

Fig. 16.4 Illustration of the oct-tree structure used

Fig. 16.4. Each node of the level of 3D points contains 3D data from those points which fall into the corresponding subspace. Searching in such a tree is of the order O(log n).

16.5 Examples

Figure 16.5 represents the picture of the object acquired by the proposed method. Its surface is similar to a deformed car-body part. In Figs. 16.6, 16.7, 16.8, and 16.9 the point cloud corresponding to the object in Figure 16.5 can be followed from different camera positions.

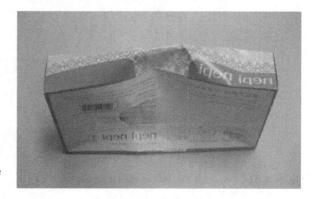

Fig. 16.5 The picture of the target object

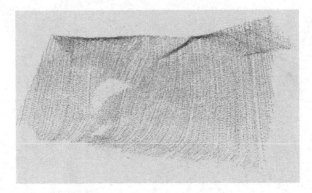

Fig. 16.6 The measured 3D data of the target object (View-1)

Fig. 16.7 The measured 3D data of the target object (View-2)

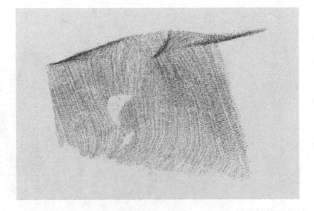

Fig. 16.8 The measured 3D data of the target object (View-3)

Fig. 16.9 The measured 3D data of the target object (View-4)

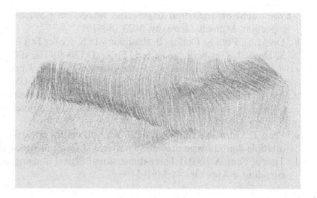

16.6 Conclusions

In this paper a new 3D scanning technique was introduced, which enables the high accuracy acquisition of high complexity objects. The method can advantageously be used for acquiring objects containing many hidden areas. The object to be scanned can be of various materials. The complexity of the method was reduced, so it works in real time, using a single personal computer. With regard to the applications of the scanning system introduced, it can advantageously be used to create 3D digital archives, to analyze car bodies, measuring their deformation, etc.

Acknowledgment This work was supported by the Japan Suzuki Foundation and by the Hungarian Scientific Research Fund (OTKA T042896, OTKA T048756).

References

1. Adolfsson B (1997) Documentation and reconstruction of architectural elements at the fortress of Fredriksborg and at the Swedish National Art Museum. In: CIPA International Archives of Photogrammetry and Remote Sensing XXXI, no. 5C1B, pp 161–167
2. Batlle J, Mouaddib E, Salvi J (1998) Recent progress in coded structured light as a technique to solve the correspondence problem. J Patt Recogn 31(7): 963–982
3. Bergmann D (1995) New approach for automatic surface reconstruction with coded light. In: Proceedings of Remote Sensing and Reconstruction for Three Dimensional Objects and Scenes, 2572:2–9
4. Coggrave CR, Huntley JM (2000) Optimization of a shape measurement system based on spatial light modulators. J Opt Eng 39(1):91–98
5. Conroy R, Dorrington A, Cree M, Künnemeyer R, Gabbitas B (2006) Shape and deformation measurement using heterodyne range imaging technology. In: Proceedings of 12th Asia-Pacific Conference on NDT, Auckland, New Zealand
6. Jackins CL Tanimoto SL (1980) Oct-trees and their use in representing three-dimensional objects. J Comput Graphics Image Process 14(3):249–270
7. Langea R, Seitza P, Bibera A, Schwarteb R (1999) Time-of-flight range imaging with a custom solid-state image sensor. In: Proceedings of the EOS/SPIE International

Symposium on Industrial Lasers and Inspection Conference on Laser Metrology and Inspection, Munich, Germany, 3823:180–191

8. Levoy M, Pulli K, Curless B, Rusinkiewicz S, Koller D, Pereira L, Ginzton M, Anderson S, Davis J, Ginsberg J, Shade J, Fulke D (2000) The digital Michelangelo project: 3d scanning of large statues. In: Proceedings of the 27th Annual Conference on Computer Graphics and Interactive Techniques, New York, NY, USA, pp 131–144

9. Mazumder P (1988) A new strategy for oct-tree representation of three-dimensional objects. In: Proceedings of IEEE Conference on Computer Vision and Pattern Recognition, pp 270–275

10. Peng T, Gupta SK, Lau K (2005) Algorithms for constructing 3-D point clouds using multiple digital fringe projection patterns. J Comput Aided Design Appl 2:737–746

11. Tian J, Peng X (2005) Three-dimensional digital imaging based on shifted point-array encoding. J Appl Opt 44:5491–5496

Chapter 17
Adaptive MV ARMA Identification Under the Presence of Noise

Stylianos Sp. Pappas, Vassilios C. Moussas, and Sokratis K. Katsikas

Abstract An adaptive method for simultaneous order estimation and parameter identification of multivariate (MV) ARMA models under the presence of noise is addressed. The proposed method is based on the well known multi-model partitioning (MMP) theory. Computer simulations indicate that the method is 100% successful in selecting the correct model order in very few steps. The results are compared with two other established order selection criteria, namely, Akaike's information criterion (AIC) and Schwarz's Bayesian information criterion (BIC).

17.1 Introduction

The problem of fitting a multivariate ARMA model to a given time series is an essential one in speech analysis, biomedical applications, hydrology, electric power systems, and many other applications [1–4].

In this paper, a new method for multivariate ARMA model order selection and parameter identification is presented, as an extension to the one proposed in [5] for MV AR models. The method is based on the well known adaptive multimodel partitioning theory [6, 7]. It is not restricted to the Gaussian case, it is applicable to online/adaptive operation, and it is computationally efficient. Furthermore, it identifies the correct model order very quickly.

An m-variate ARMA model of order (p, q) [ARMA (p, q)] for a stationary time series of vectors **y** observed at equally spaced instants $k = 1, 2, \ldots, n$ is defined as:

S.Sp. Pappas (✉)
Department of Information and Communication Systems Engineering, University of the Aegean, 83200 Karlovassi Samos, Greece
e-mail: spappas@aegean.gr

N. Mastorakis et al. (eds.), *Proceedings of the European Computing Conference*, Lecture Notes in Electrical Engineering 27, DOI 10.1007/978-0-387-84814-3_17, © Springer Science+Business Media, LLC 2009

$$\mathbf{y}_k = \sum_{i=1}^{p} \mathbf{A}_i \mathbf{y}_{k-i} + \sum_{j=1}^{q} \mathbf{B}_j \mathbf{v}_{k-j} + \mathbf{v}_k, \; E[\mathbf{v}_k \mathbf{v}_k^T] = \mathbf{R} \qquad (17.1)$$

where the m-dimensional vector $\mathbf{v}_k$ is uncorrelated random noise, not necessarily Gaussian, with zero mean and covariance matrix $\mathbf{R}$, $\theta = (p, q)$ is the order of the predictor, and $\mathbf{A}_1, \ldots, \mathbf{A}_p, \mathbf{B}_1, \ldots, \mathbf{B}_q$ are the $m \times m$ coefficient matrices of the MV ARMA model.

It is obvious that the problem is twofold. The first task, which is the most important for the problem under consideration, is the successful determination of the predictor's order $\theta = (p, q)$. Once the model order selection task is completed, one proceeds with the second task, i.e., the computation of the predictor's matrix coefficients $\{\mathbf{A}_i, \mathbf{B}_j\}$.

Determining the order of the ARMA process is usually the most delicate and crucial part of the problem. Over the past years substantial literature has been produced on this problem and various different criteria, such as Akaike's [8], Rissanen's [9, 10], Schwarz's [11], and Wax's [12], have been proposed to implement the order selection process.

The above mentioned criteria are not optimal and are also known to suffer from deficiencies; for example, Akaike's information criterion suffers from overfit [13]. Also, their performance depends on the assumption that the data are Gaussian and upon asymptotic results. In addition to this, their applicability is justified only for large samples; furthermore, they are two-pass methods, so they cannot be used in an online or adaptive fashion.

The paper is organized as follows. In Section 17.2, the MV ARMA model order selection problem is reformulated so that it can be fitted into the state space under the uncertainty estimation problem framework. In the same section the multi-model partitioning filter (MMPF) is briefly described and its application to the specific problem is discussed. In Section 17.3, simulation examples are presented which demonstrate the performance of our method in comparison to previously reported ones. Finally, Section 17.4 summarizes the conclusions.

17.2 Problem Reformulation

If we assume that the model order fitting the data is known and is equal to $\theta = (p, q)$, we can rewrite equation (17.1) in standard state-space form as:

$$\mathbf{x}(k + 1) = \mathbf{x}(k) \qquad (17.2)$$

$$\mathbf{y}(k) = \mathbf{H}(k)\mathbf{x}(k) + \mathbf{v}(k) \qquad (17.3)$$

where $\mathbf{x}(k)$ is an $m^2(p+q) \times 1$ vector made up from the coefficients of the matrices $\{\mathbf{A}_1, \ldots, \mathbf{A}_p, \mathbf{B}_1, \ldots, \mathbf{B}_q\}$, and $\mathbf{H}(k)$ is an $m \times m^2(p+q)$ observation history matrix of the process $\{\mathbf{y}(k)\}$ up to time $k-(p+q)$.

Assume that the general form of the matrix is:

$$\mathbf{A}_p \text{ is} \begin{bmatrix} a^p_{11} & \cdots & a^p_{1m} \\ \vdots & \ddots & \vdots \\ a^p_{m1} & \cdots & a^p_{mm} \end{bmatrix} \text{ and}$$

$$\mathbf{B}_q \text{ is} \begin{bmatrix} b^q_{11} & \cdots & b^q_{1m} \\ \vdots & \ddots & \vdots \\ b^q_{m1} & \cdots & b^q_{mm} \end{bmatrix} \text{ than}$$

$$\mathbf{x}(k) = [\alpha^1_{11}\,\alpha^1_{21}\cdots\alpha^1_{m1} \vdots \alpha^1_{12}\,\alpha^1_{22}\cdots\alpha^1_{m2} \vdots \cdots \alpha^1_{mm} \vdots \cdots \alpha^p_{mm} \vdots$$
$$b^1_{11}\,b^1_{21}\cdots b^1_{m1} \vdots b^1_{12}\,b^1_{22}\cdots b^1_{m2} \vdots \cdots b^1_{mm} \vdots \cdots b^q_{mm}]^T$$

$$\mathbf{H}(k) = [y_1(k-1)I \cdots y_m(k-1)I \vdots \cdots \vdots y_1(k-p)I \cdots y_m(k-p)I \vdots$$
$$v_1(k-1)I \cdots v_m(k-1)I \vdots \cdots \vdots v_1(k-q)I \cdots v_m(k-q)I]$$

where I is the $m \times m$ identity matrix and $\theta = (p, q)$ is the model order.

If the system model and its statistics were completely known, the Kalman filter (KF) in its various forms would be the optimal estimator in the minimum variance sense.

In the case where the prediction coefficients are subject to random perturbations, (17.2) becomes

$$\mathbf{x}(k+1) = \mathbf{x}(k) + \mathbf{w}(k) \tag{17.4}$$

$v(k)$, $w(k)$ are independent, zero-mean, white processes, not necessarily Gaussian.

$$\mathbf{w}(k) = [w^1_{11}\,w^1_{21}\cdots w^1_{m1} \vdots w^1_{12}\,w^1_{22}\cdots w^1_{m2} \vdots \cdots w^1_{mm} \vdots \cdots w^p_{mm} \vdots$$
$$w^1_{11}\,w^1_{21}\cdots w^1_{m1} \vdots w^1_{12}\,w^1_{22}\cdots w^1_{m2} \vdots \cdots w^1_{mm} \vdots \cdots w^q_{mm}]^T$$

A complete system description requires the value assignments of the variances of the random processes $w(k)$ and $v(k)$. We adopt the usual assumption that $w(k)$ and $v(k)$ at least are wide sense stationary processes, hence their variances; and $\mathbf{Q}$ and $\mathbf{R}$ respectively are time invariant. Obtaining these values is not always trivial. If $\mathbf{Q}$ and $\mathbf{R}$ are not known, they can be estimated by using a method such as the one described in [14]. In the case of coefficients constant in time, or slowly varying, Q is assumed to be zero (just as in equation (17.4)).

It is also necessary to assume an a priori mean and variance for each $\{A_i, B_i\}$. The a priori mean of the $A_i(0)$'s and $B_i(0)$'s can be set to zero if no knowledge about their values is available before any measurements are taken (the most likely case). On the other hand, the usual choice of the initial variance of the A_i's and B_i's, denoted by P_0, is $P_0 = nI$, where n is a large integer.

Let us now consider the case where the system model is not completely known. The adaptive multi-model partitioning filter (MMPF) is one of the most widely used approaches for similar problems. This approach was introduced by Lainiotis [6, 7] and summarizes the parametric model uncertainty into an unknown, finite-dimensional parameter vector whose values are assumed to lie within a known set of finite cardinality. A non-exhaustive list of the reformulation, extension, and application of the MMPF approach, as well as its application to a variety of problems by many authors, can be found in [15] and [16–19]. In our problem, assume that the model uncertainty is the lack of knowledge of the model order θ. Let us further assume that the model order θ lies within a known set of finite cardinality: $1 \leq \theta \leq M$, where $\theta = (p, q)$, is the model order.

The MMPF operates on the following discrete model:

$$\mathbf{x}(k + 1) = \mathbf{F}(k + 1, k / \theta)\mathbf{x}(k) + \mathbf{w}(k) \tag{17.5}$$

$$\mathbf{y}(k) = \mathbf{H}(k / \theta)\mathbf{x}(k) + \mathbf{v}(k) \tag{17.6}$$

where $\theta = (p, q)$ is the unknown parameter, the model order in this case. A block diagram of the MMPF is presented in Fig. 17.1.

In the Gaussian case, the optimal MMSE estimate of $\mathbf{x}(k)$ is given by

$$\hat{\mathbf{x}}(k/k) = \sum_{j=1}^{M} \hat{x}(k/k; \theta_j)\, p(\theta_j/k). \tag{17.7}$$

A finite set of models is designed, each matching one value of the parameter vector. If the prior probabilities $p(\theta_j / k)$ for each model are already known, these are assigned to each model. In the absence of any prior knowledge, these are set to $p(\theta_j/k) = 1/M$ where M is the cardinality of the model set.

A bank of conventional elemental filters (non-adaptive, e.g., Kalman) is then applied, one for each model, which can be run in parallel. At each iteration the MMPF selects the model which corresponds to the maximum posteriori probability as the correct one. This probability tends to one, while the others tend to zero. The overall optimal estimate can be taken to be either the individual estimate of the elemental filter exhibiting the highest posterior probability, called the maximum a posteriori (MAP) estimate,in [20], which is the case used in this paper; or the weighted average of the estimates produced by the elemental filters, as described in equation (17.7). The weights are determined by the posterior probability that each model in the model set is in fact the true model.

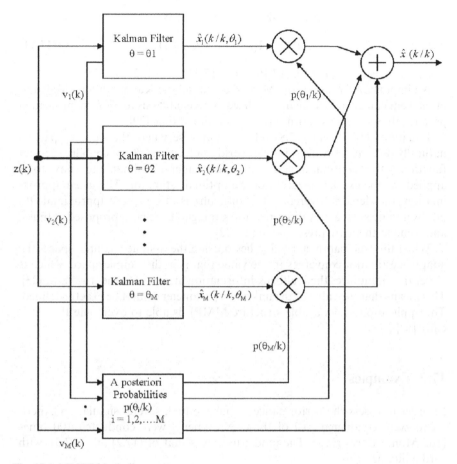

Fig. 17.1 MMPF block diagram

The posterior probabilities are calculated online in a recursive manner as follows:

$$p(\theta_j/k) = \frac{L(k/k; \theta_j)}{\sum\limits_{j=1}^{M} L(k/k; \theta_j)\, p(\theta_j/k-1)}\, p(\theta_j/k-1) \qquad (17.8)$$

$$L(k/k; \theta_j) = \left|\mathbf{P}_{\tilde{\mathbf{y}}}\ (k/k-1; \theta_j)\right|^{-1/2} \cdot$$
$$\exp[-\frac{1}{2}\tilde{\mathbf{y}}^{\mathrm{T}}(k/k-1; \theta_j)\,\mathbf{P}_{\tilde{\mathbf{y}}}^{-1}(k/k-1; \theta_j)\,\tilde{\mathbf{y}}\ (k/k-1; \theta_j)] \qquad (17.9)$$

where the innovation process

$$\tilde{\mathbf{y}}(k/k-1; \theta_j) = \mathbf{y}(k) - \mathbf{H}(k; \theta_j)\hat{\mathbf{x}}(k/k-1; \theta_j) \qquad (17.10)$$

is a zero mean white process with covariance matrix

$$\mathbf{P}_{\tilde{y}}\ (k/k-\ 1; \theta_j) = \mathbf{H}(k; \theta_j\ )\mathbf{P}(k/k; \theta_j)\mathbf{H}^{\mathrm{T}}(k; \theta_j\ ) + \mathbf{R} \qquad (17.11)$$

For equations (17.8), (17.9), (17.10), and (17.11), j = 1, 2, . . . , M.

An important feature of the MMPF is that all the Kalman filters needed to implement can be independently realized. This enables us to implement them in parallel, thus saving us enormous computational time [20].

Equations (17.7) and (17.8) refer to our case where the sample space is naturally discrete. However, in real world applications, θ's probability density function (pdf) is continuous and an infinite number of Kalman filters have to be applied for the exact realization of the optimal estimator. The usual approximation considered to overcome this difficulty is to somehow approximate θ's pdf by a finite sum. Many discretization strategies have been proposed at times, and some of them are presented in [21–22].

When the true parameter value lies outside the assumed sample space, the adaptive estimator converges to the value that is in the sample space, which is closer (i.e., minimizes the Kullback information measure) to the true value [23]. This means that the value of the unknown parameter cannot be exactly defined. The application of a variable structure MMPF is able to overcome this difficulty [17].

17.3 Examples

In order to assess the performance of our method, several simulation experiments were conducted. All of these experiments were conducted 100 times (100 Monte Carlo runs). The model used was that of (17.2) and (17.3), with cardinality $M = 10$.

17.3.1 Example 1

ARMA $(1, 1)$. $\theta = (1, 1) = 2$.

$$A = \begin{bmatrix} -0.85 & 0.75 \\ 0.65 & -0.55 \end{bmatrix} B = \begin{bmatrix} -1.9833 & 1.889 \\ 1.7 & 1.9833 \end{bmatrix} R = \begin{bmatrix} 1.5625 & 1.5 \\ 1.5 & 1.5625 \end{bmatrix}$$

Figure 17.2 depicts the posterior probabilities associated with each value of θ. Figure 17.3 shows the criteria comparison for two data sets, one relatively small (50 samples) and one larger (100 samples), and Table 17.1 shows the estimated ARMA parameter coefficients.

From Fig. 17.2, it is obvious that the MMPF identifies the correct $\theta = (1, 1) = 2$ very quickly, in just 17 steps. Convergence is taken to occur when the posterior probability of the model exceeds 0.9.

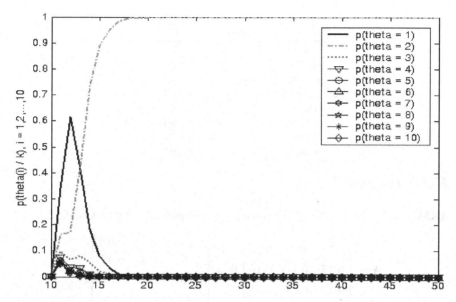

Fig. 17.2 Example 1, posterior probabilities

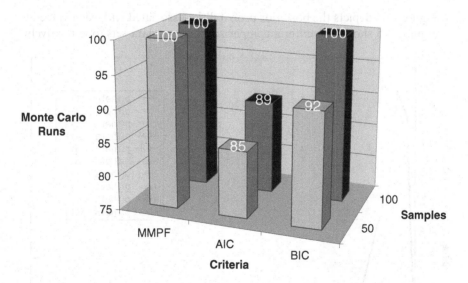

Fig. 17.3 Example 1, criteria comparison—correct model order identification

From Fig. 17.3, we deduce that MMPF is 100% successful in selecting the correct model order for both data sets, while only BIC matches its performance for the larger data set.

Also, Table 17.1 shows that the parameter coefficient estimation is very accurate. (RMSE—root mean square error—is very small).

Table 17.1 Example 1, estimated ARMA coefficient parameters

Estimated parameters	RMS error
−0.8499	0.0033
0.6508	0.0036
0.7501	0.0040
0.5501	0.0032
−1.9823	0.0092
1.7011	0.0064
1.8891	0.0074
1.9831	0.0057

17.3.2 Example 2

ARMA *(1, 1)*. $\theta = (1, 1) = 2$. This is a more complex MV ARMA since $m = 3$

$$\mathbf{A} = \begin{bmatrix} 1 & 0.2 & 0.23 \\ 0.15 & 0.18 & 0.16 \\ 0.17 & 0.24 & 0.21 \end{bmatrix} \mathbf{B} = \begin{bmatrix} 1 & 0.15 & 0.09 \\ 0.1 & -0.1 & 0.05 \\ 0.05 & 0.13 & 0.075 \end{bmatrix}$$

$$\mathbf{R} = \mathbf{diag}[(0.42, 0.01, 0.16)].$$

Figure 17.4 depicts the posterior probabilities associated with each value of θ. Figure 17.5 shows the criteria comparison for two data sets, one relatively

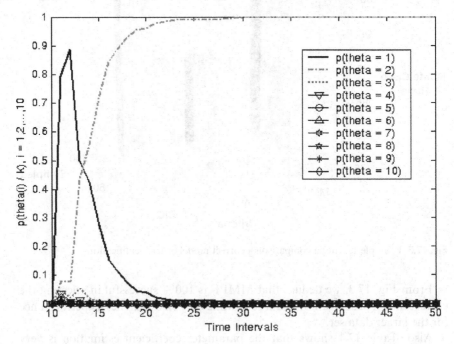

Fig. 17.4 Example 2, posterior probabilities

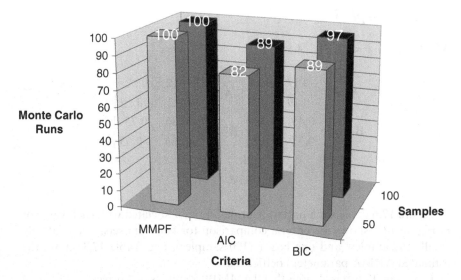

Fig. 17.5 Example 2, criteria comparison—correct model order identification

small (50 samples) and one larger (100 samples), and Table 17.2 shows the estimated ARMA parameter coefficients.

From Figure 17.4, it is obvious that the MMPF identifies the correct $\theta = (1, 1) = 2$ very quickly, in just 18 steps. Convergence is taken to occur when the posterior probability of the model exceeds 0.9.

From Figure 17.5, we deduce that MMPF is 100% successful in selecting the correct model order for both data sets, while none of the two other criteria achieve a similar performance for either data set.

As Table 17.2 clearly shows, the parameter estimation is again accurate because the root mean square error (RMSE) is very small.

Table 17. 2 Example 2, estimated ARMA coefficient parameters

Estimated parameters	RMS error		
0.9932	1.0217	0.0152	0.0026
0.2023	0.1516	0.0035	0.0016
0.2310	0.0894	0.0021	0.0059
0.1519	0.1013	0.0143	0.0023
0.1829	−0.1027	0.0091	−0.0127
0.1612	0.0059	0.0044	0.0019
0.1702	−0.0048	0.0030	0.0017
0.2408	0.1351	0.0371	0.0046
0.2143	0.0742	0.0045	0.0028

17.3.3 Example 3

ARMA $(2, 2)$. $\theta = (2, 2) = 4$.

$$\mathbf{A}_1 = \begin{bmatrix} -0.17 & 0.14 \\ -0.19 & -0.1 \end{bmatrix}, \mathbf{A}_2 = \begin{bmatrix} -0.2 & 0.12 \\ 0.22 & -0.25 \end{bmatrix}$$

$$\mathbf{B}_1 = \begin{bmatrix} -0.45 & 0.52 \\ -0.32 & -0.7 \end{bmatrix}, \mathbf{B}_2 = \begin{bmatrix} -0.85 & 0.75 \\ -0.65 & -0.55 \end{bmatrix},$$

$$\mathbf{R} = \begin{bmatrix} 1 & -0.08 \\ -0.08 & 1 \end{bmatrix}$$

Figure 17.6 depicts the posterior probabilities associated with each value of θ. Figure 17.7 shows the criteria comparison for two data sets, one relatively small (50 samples) and one larger (100 samples), and Table 17.3 shows the estimated ARMA parameter coefficients.

From Fig. 17.6, it is obvious that the MMPF identifies the correct $\theta = (1, 1) = 2$ very quickly, in just 24 steps. Convergence is taken to occur when the posterior probability of the model exceeds 0.9.

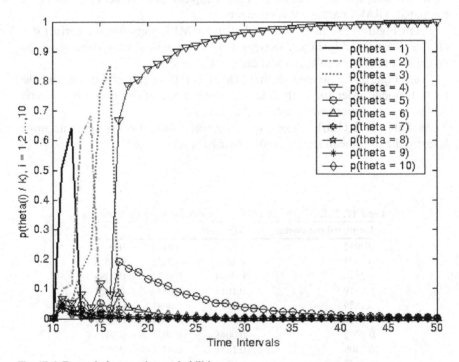

Fig. 17.6 Example 3, posterior probabilities

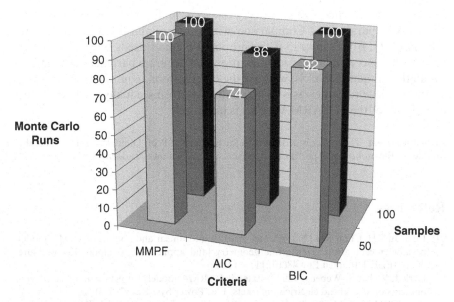

Fig. 17.7 Example 2, criteria comparison—correct model order identification

Table 17.3   Example 3, estimated ARMA coefficient parameters

Estimated parameters		RMS error	
−0.1691	−0.4535	0.0047	0.0094
−0.1896	−0.3260	0.0054	0.0109
0.1458	0.5244	0.0137	0.0090
−0.0899	−0.6931	0.0161	0.0108
−0.1982	−0.8441	0.0064	0.0122
0.2234	−0.6407	0.0073	0.0147
0.1154	0.7508	0.0094	0.0086
−0.2573	−0.5471	0.0114	0.0094

From Fig. 17.7, we deduce that MMPF is 100% successful in selecting the correct model order for both data sets, while only BIC matches its performance for the larger data set.

As Table 17.3 clearly shows, the parameter estimation is again accurate because the root mean square error (RMSE) is very small.

17.4 Conclusions

A new method for simultaneously selecting the order and estimating the parameters of a MV ARMA model has been developed, as an extension of the method proposed for the MV AR case. The proposed method successfully selects the correct model order in very few steps and identifies very accurately the ARMA parameters. Comparison with other established order selection criteria (AIC, BIC) shows that the proposed method needs only the shortest data set for successful order identification and accurate parameter estimation for all the simulated models; whereas the other criteria require longer data sets as the model order increases. The method performs equally well when the complexity of the MV ARMA model is increased.

Acknowledgment This paper is dedicated to the memory of Prof. Dimitrios G. Lainiotis, the founder of the multi-model partitioning theory, who passed away suddenly in 2006.

References

1. Lu S, Ju KH, Chon KH (2001) A new algorithm for linear and nonlinear ARMA model parameter estimation using affine geometry [and application to blood flow/pressure data]. IEEE T Biomed Eng 48(10):1116–1124
2. Derk J, Weber S, Weber C (2007) Extended ARMA models for estimating price developments on day-ahead electricity markets. Elec Power Syst Res 77(5–6):583–593
3. Chen CP, Bilmes JA (2007) MVA processing of speech features. IEEE T Audio Speech 15(1):257–270
4. Kourosh M, Eslami HR, Kahawita R (2006) Parameter estimation of an ARMA model for river flow forecasting using goal programming. J Hydrol 331(1–2):293–299
5. Pappas Sp S, Leros AK, Katsikas SK (2006) Joint order and parameter estimation of multivariate autoregressive models using multi-model partitioning theory. Dig Signal Process 16(6):782–795
6. Lainiotis DG (1976) Partitioning: a unifying framework for adaptive systems I: Estimation. Proc IEEE 64:1126–1143
7. Lainiotis DG (1976) Partitioning: a unifying framework for adaptive systems I: Control. Proc IEEE 64:1182–1198
8. Akaike H (1969) Fitting autoregressive models for prediction. Ann Inst Stat Math 21:243–247
9. Rissanen J (1978) Modelling by shortest data description. Automatica 14:465–471
10. Rissanen J (1986) A predictive least squares principle. IMA J Math Contr Inform 3:211–222
11. Schwarz G (1978) Estimation of the dimension of the model. Ann Statist 6:461–464
12. Wax M (1988) Order selection for AR models by predictive least squares. IEEE T Acoust Speech Signal Process 36:581–588
13. Lutkepohl H (1985) Comparison of criteria for estimating the order of a vector AR process J Time Ser Anal 6:35–52
14. Sage AP, Husa GW (1969) Adaptive filtering with unknown prior statistics. In: Proceedings of the Joint Automatic Control Conference, Boulder, Colorado, pp 760–769
15. Watanabe K (1992) Adaptive estimation and control: partitioning approach.:Prentice Hall, Englewood Cliffs, NJ
16. Katsikas SK, Leros AK, Lainiotis DG (1994) Passive tracking of a maneuvering target: an adaptive approach. IEEE T Signal Process 42(7):1820–1825
17. Katsikas SK, Likothanassis SD, Beligiannis GN, Berkeris KG, Fotakis DA (2001) Genetically determined variable structure multiple model estimation. IEEE T Signal Process 49(10):2253–2261
18. Lainiotis DG, Papaparaskeva PA (1998) Partitioned adaptive approach to nonlinear channel equalization. IEEE T Commun 46(10):1325–1336

19. Moussas VC, Likothanassis SD, Katsikas SK, Leros AK (2005) Adaptive online multiple source detection. In: IEEE International Conference on Acoustics Speech and Signal Processing Proceedings (ICASSP '05,), 4:1029–1032
20. Lainiotis DG, Katsikas SK, Likothanassis SD (1988) Adaptive deconvolution of seismic signals: performance, computational analysis parallelism. IEEE T Acoust Speech Signal Process 36(11):1715–1734
21. Sengbush RL, Lainiotis DG (1969) Simplified parameter quantization procedure for adaptive estimation. IEEE T Automat Contr AC-14:424–425
22. Anderson BDO, Brinsmead TS, Bruyne FDe, Hespanha J, Liberzon D, Morse AS (2000) Multiple model adaptive control part 1: finite controller coverings. Int J Robust Nonlinear Contr 10:909–929
23. Hawks RM, Moore JB (1976) Performance of Bayesian parameter estimators for linear signal models. IEEE T Automat Contr AC–21:523–527

Chapter 18
Planck's Law Simulation Using Particle Systems

Douglas Farias Cordeiro, Marcos Aurélio Batista, and Celia A. Zorzo Barcelos

Abstract The nature phenomena simulation has always appeared among the greatest Computer Graphics challenges. One of the mainly developed solutions for the representation of such phenomena types was the Particle System model. Several effects were proposed, starting with the application of this method. However, some problems, such as the black body radiation, still await simulation solutions. This chapter responds to such a need by presenting a simulation proposal of black body radiation through an analysis of mathematical, physics and computational concepts applied on the Particle Systems model.

18.1 Introduction

Until the beginning of the 1980s (1983), Computer Graphics engineers had known that although they could represent different objects, they could not provide a complex phenomena simulation. With this in mind, a theory was developed which was able to provide enough abstract power and description to represent the characteristic aspects of this phenomenon known as Particle Systems.

Despite the fact that this theory presents a standardized structure for most representations of natural phenomena, an elaboration which combines mathematical, physical and computational characteristics into a single analysis, in order to reach results that are as close to reality as possible – as is the case for black body radiation simulation – is presented in this work.

D.F. Cordeiro (✉)
Federal University of Uberlândia, 38.400-902, Uberlândia, MG, Brasil
e-mail: douglasf@pos.facom.ufu.br

N. Mastorakis et al. (eds.), *Proceedings of the European*
Computing Conference, Lecture Notes in Electrical Engineering 27,
DOI 10.1007/978-0-387-84814-3_18, © Springer Science+Business Media, LLC 2009

18.2 Black Body Radiation

Until the end of the 19th Century, physics remained in its classic state, although some scientists of the time had already foreseen the appearance of some changes.

During this period most of the intellectual community did not believe in the conception of theories that could revolutionize this new knowledge step. However, this new evolution allowed for the discovery of phenomena that could not be explained by concepts of acquaintance, and new fields of exploration appeared.

Basically, a black body can be defined as an ideal radiation absorber and emitter [5], that is, a body that's able to absorb and emit radiation in all wavelengths. Starting from this definition as well as other concepts of physics, several theories were developed in order to explain such phenomenon; however, none of them describe the problem behind the theory of black bodies in a valid and complete way, principally because most of the formalizations presented are only coherent with empirical data for values of high frequency, which became known as "ultraviolet catastrophe."

In 1900, the scientist Max Planck, during his search for a solution to this problem, proposed a new theory about energy propagation. In agreement with his theory, energy does not just spread in an undulatory or corpuscular form, but in a dual way by emitting wavelength dependents and indivisible energy packages called quanta.

Using this theory as a base, as well as other concepts developed by the study of black bodies – for example, the Wien Displacement Law, which shows that the product between the maximum wavelength $\lambda_{\max}$ and the absolute temperature T of a black body is always constant – Planck developed a coherent analysis with this empirical data, describing the spectral distribution of black body radiation in the following way:

$$\rho(\lambda, T) = \frac{8\pi hc}{\lambda^5 (e^{\frac{hc}{\lambda \kappa T}} - 1)} \tag{18.1}$$

where h is the Planck constant, c is the light velocity, e is the Euler number and κ is the Boltzmann constant.

Planck's discovery became known worldwide as the birth point of the Quantum Physics [3]. In spite of this, the empirical visualization of black body radiation still remains something extremely complex, due to the difficulties of building a mechanical simulator, which refreshes the need for methods of graphic simulation that demonstrate with fidelity the physical properties of such phenomenon.

18.3 Particle Systems

Until the early 1980s, computer graphics were still quite restricted. For example, their methods were not capable of generating complex scenes such as clouds, fires, dust, etc. This inflexibility implied the need for a robust technique capable of providing the necessary realism to render certain images.

In 1983, as a special effects project for the film "Star Trek II: The Wrath of Khan" was carried out, William T. Reeves added physics concepts to the computer graphics theory and reached results that depicted real environment satisfactorily, using the Particle Systems technique. Through his theory it was possible to reach the modeling of objects; not of the ones that presented discreet limits, but also forms of complex and continuous design [1].

In agreement with Reeves' proposal, each particle can be defined as an individual entity, that although possesses properties similar to the other particles of the system, it can bear differentiated values, obeying the modeling pre-defined for determined representations [6].

When rendering the scenes, the individual properties of such particles, such as velocity, color, size, among others, are modified in order to obtain the desired result during a stochastic life interval. After becoming extinct, each particle has new determined values added to its attributes, presenting a different visual behavior from that previous, which produces the illusion of being a new entity in the system.

Using this theory it is possible to obtain simulations of various natural phenomena, even those which in spite of being deterministic as a complete entity, presents unpredictableness when analyzed as isolated actions. This is due to the fact that one of the most important Particle System's characteristics is the stochastic characteristic of the isolated actions.

18.4 The Proposed System

Although discovered during the 17th Century, the study of black body radiation is still quite limited as relevant theoretical concepts are concerned, mainly because the equipment used for the experimental exploration presents a high acquisition cost and great maintenance complexity [2]. Based on this, this work presents a form of computational graphic simulation of the black body radiation phenomenon, using graphic computer concepts which refer to the Particle System's technique as a starting point, in order to arrive at the most faithful possible Planck spectral distribution.

When analyzing the black body radiation problem, particularly in relation to its visual disposition, one notices that its graphic simulation will basically consist of an analysis of the spatial emitted radiation, in agreement with the inherent physical relationships, which are described by the Planck theory. Analyzing such concepts, a basic structure of abstraction of the reference properties of each particle of the system was defined and presented under the following forms: Emissive Point, Linear displacement, Undulatory displacement, Velocity, Density, Transparency, Color and Life time.

Starting from this structure building, the geometric and computational elaboration of the black body radiation behavior was necessary through the use of mathematical and graphic computer concepts.

The first step considered in the systematization process of the simulation environment was defining the basic form of the emissive radiation body.

In the solution proposed, a spherical surface or ray r was determined as a base point, from which radiation would be emitted. Each emitted radiation bundle refers to the individual behavior of each one of the particles present in the system. In this case, such particles present the primary need of being organized in points upon the spherical surface.

To define the emissive point E_0 for each particle a point of black body superficial space was included, two stochastic angular values α and β associated to each particle were used, where $0 \leq \alpha, \beta \leq 2\pi$.

Starting from this point, the simulation environment will find itself in an initial state of particle disposition, thus allowing one to define the properties of the radiation emitted by the body in question.

In agreement with the theory proposed by Max Ludwig Planck for the problem of the black body spectral distribution, the behavior observed by the emitted radiation bundles comes under a dual nature; that is, the radiation can be explained as a corpuscular as well as undulatory entity [4]. In the present proposal the representation of the radiation's corpuscular nature was interpreted through its particle disposition in the graphic system used. However, due to the need for the radiation displacement representation in its undulatory form, the space transition of the particles was defined in agreement with the theory of British physicist James Clerk Maxwell on electromagnetic waves.

In agreement with the concepts postulated by Maxwell, the electromagnetic behavior of radiation is represented geometrically by two perpendicular planes, each one corresponding to the electric and magnetic oscillations. However, the mathematical exploration of wave displacement work individually on each plane causes a bi-dimensional analysis of the radiation transferance. As the solution for the simulation proposal presents three-dimensional character, it was necessary to employ geometric concepts for the projection of the undulatory movement of the particles in space.

To solve such a problem, the emissive point of each particle was considered as having vectorial character, with its origin in $O = (0, 0, 0)$. From this point, the velocity $\vec{V}$ of a determined particle was obtained as a fraction of the module of this vector. With the valor $\vec{V}$ calculated, a vectorial sum was attributed between the emissive point vector and the speed vector to the linear displacement E, computed in the successive renders. The module value of such displacement was then considered as an attribute for the wave equation used providing, in this way, the undulatory displacement S of each particle, that is, its special position in each unit of time.

Another important factor in black body graphic representation is the relationship between colors and radiation. In the present proposal an algorithm of simple conversion between wavelength and standard RGB was developed so that the parameters used for the simulation of the Planck Law could be used satisfactorily in the graphic representation.

Finally, it was necessary to make our radiation simulation agree with the Planck theory. In agreement with the Planck postulate, the black body radiation spectral density is a physical property dependent on the wavelength and absolute temperature of a specific body. However, these two attributes are intrinsically related; that can be easily observed by the Wien Displacement Law.

With this in mind, during the elaboration of the present proposal of the black body radiation graphic simulation, it was considered that the representation will just be tied to the Planck Law only. The others will be computationally calculated.

With the wavelength and the absolute temperature values of the body determined on a given instant, it is necessary to calculate the spectral density in agreement with Planck's proposed equation. The quantity of particles used in the system remains constant during all successive renderings; however, its characteristic attributes were randomly altered between birth and death. The spectral density, that refers to the amount of radiation per volume unit was associated to the area occupied by each particle in space, supplying a visual approach to the behavior of black body radiation.

18.5 Results

During the completion of this work proposal, a series of necessary requirements for the elaboration of a black body radiation simulation system were observed, which were able to present consistency with the empirically verified results by the Planck Law. To construct a concise base for such a need, it is necessary to unite inherent knowledge from three branches of human knowledge: Physics, supplying the basic rules for the study object; Mathematics, providing abstract forms of the phenomenon, and, finally, Computer Science, mainly through the computer graphics, supplying means of developing mechanisms capable of visually presenting the analyzed problem.

From this point, the scientific exploration of our analysis showed itself to be mature enough to provide the elaboration of a relationship among the necessary theories for the proposed simulation. This allowed for the development of a system which was capable of making a practical demonstration of the black body radiation.

The developed system allows the user to observe the radiation emitted in agreement with a certain wavelength. After defining the data input the system calculates the referent absolute temperature and, later on, the spectral density, supplying the visual result in agreement with Planck's Law. It is important to remember that the color presented by the radiation in a determined instant is directly related to the wavelength in agreement with the visible radiation spectrum. The obtained result for wavelengths equals to 420, 520, 570 and 650 nm is shown in Fig. 18.1.

Fig. 18.1 Simulations with
420, 520, 570 and 650 *nm*

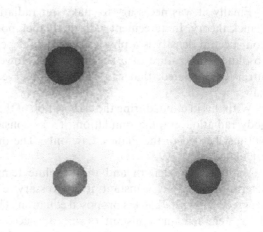

18.6 Conclusions

With the execution of the project one notices that the mathematical and computational analysis used to simulate the physical concepts of black body radiation provided satisfactory results capable of representing the visual behavior of this mysterious phenomenon of physics with clarity. This made it possible to elaborate coherent Particle Systems with the problem of the black body radiation, providing a satisfactory visual analysis of the phenomenon.

In future works, we intend to extend this idea through a study of the relative concepts of particle theory, providing a base which not only produces the simulation of the black body radiation problem, but also of various phenomena observed in nature.

References

1. Azevedo E, Conci A (2003) Computação Gráfica – Teoria e Prática. Editora Campus, Rio de Janeiro.
2. Cavalcante R (2005) Marisa Almeida anda Haag. Corpo negro e determinação experimental da constante de Planck. Revista da Sociedade Brasileira de Física 27(3):343–348.
3. Einstein A, Infeld L (1980) A Evolução da Física. Zahar Editora, Rio de Janeiro.
4. Lagemann RT (1968) Ciencia Fisica: Origenes y Principios. Editorial Hispano-Americana, México.
5. Planck M (1901) Über das gesetz der energieverteilung im normalspektrum. Annalen der Physik 4:553–563.
6. Reeves WT (1983) Particle systems: A technique for modeling a class of fuzzy objects. Comput Graph (SIGGRAPH 83 Proceedings) 17(3):359–376.

Chapter 19
Simulation Results Using Shock Absorbers with Primary Intelligence—VZN

Adrian-Ioan Niculescu, Tudor Sireteanu, and Daniel Dumitriu

Abstract This paper presents the new concept of a self-adjustable shock absorber called VZN, some simulation results at some specific excitation regimes, and a theoretical roll study. The acronym VZN stands for "VARIABLE ZETA necessary for well NAVIGATION", where ZETA represents the relative damping, which is changed stepwise automatically, according to the piston position. Both on rebound and compression, the damping coefficients have low values at the beginning of the stroke, becoming medium at the middle of the stroke, and high and very high at its end. Thus a better wheel/road holding, comfort, and protection at the end of the stroke are assured, compared to standard shock absorbers.

The VZN shock absorber, having primary intelligence, is much better than standard ones because this self-adjustable shock absorber confers the possibility of stepwise adjustment of the damping force as a function of the instantaneous piston position, correlated with the load and road conditions.

The simulation tests, using Matlab/Simulink software, conducted on a quarter car model excited by harmonic and multi-harmonic inputs, show better body stability, skyhook behaviour, lower body acceleration, and increased protection of car axles and body, as compared with those provided by conventional shock absorbers.

The theoretical study denotes progressive anti-roll torque, having an anti-gyration effect and favouring the redressing movement.

With other advantages like decreasing lift/squat at acceleration, decreasing dive/lift at braking, and decreasing pick/roll, the automotive self-adjustable shock absorber confers high performances, almost like semi-active suspensions, at low costs, comparable to those of standard shock absorbers.

A.-I. Niculescu (✉)
Institute of Solid Mechanics—Romanian Academy, Bucharest, Romania
e-mail: adrian_ioan_niculescu@yahoo.com

N. Mastorakis et al. (eds.), *Proceedings of the European Computing Conference*, Lecture Notes in Electrical Engineering 27, DOI 10.1007/978-0-387-84814-3_19, © Springer Science+Business Media, LLC 2009

19.1 Introduction

The acronym VZN stands for "VARIABLE ZETA necessary for well NAVIGATION," where ZETA represents the relative damping, which is changed stepwise automatically, according to the piston position (Fig. 19.1).

Both on rebound and compression, the damping coefficients have low values at the beginning of the stroke, becoming medium at the middle of the stroke, and high and very high at its end. Thus a better wheel/road adherence, comfort, and protection at the end of the stroke are assured, compared to standard shock absorbers.

The damping effect can be made with metering holes, or with damping valves. For simplicity and cost reasons, the solution with metering holes was studied.

19.2 The Matlab/Simulink Simulation

As shown in Fig. 19.2, $F_{e,\,stop\,bumper}$ increases linearly from 0 to -500 daN beginning at the touch point of the rebound bumper, up to the d_1 distance (stroke of the rebound bumper stop, under -500 daN), respectively decreases linearly from 0 to 1000 daN beginning at the touch point of the compression bumper, up to d_2 distance (stroke of the compression bumper stop). Otherwise, $F_{e,\,stop\,bumper}$ is null.

In Fig. 19.2, l represents the full stroke; l-$(d_1 + d_2)$ represents the free stroke; d represents the distance between the static middle piston position and the static equilibrium position of the piston for the current value of the sprung mass m_1, is given by:

$$d = \left(\frac{m_{1-empty} + m_{1-full}}{2} - m_1\right)\frac{g}{k_1} \tag{19.1}$$

The quarter car model, used to study the vertical interaction between car vehicle and road [2], is shown in Fig. 19.3. The numerical simulation in this section concerns only the vertical interaction for a rear wheel, neglecting the rolling and the pitch motion of the car. The model has two degrees of freedom, i.e., the vertical displacement x_1 of the car body (bounce), and the vertical displacement x_2 of the wheel center (wheel hop). At time t, the vertical profile of the road is denoted by $x_0(t)$. The model contains two levels of elastic and damping elements: one level between the wheel and the road, characterized by the stiffness coefficient k_2 of the tire and its damping coefficient c_2; the second level between the wheel and the body (vehicle suspension), including a spring with stiffness coefficient k_1 and a VZN shock absorber (or a standard shock absorber, as comparison variant) with damping coefficient c_1.

Figure 19.3 presents the car/road vertical interaction model, where:

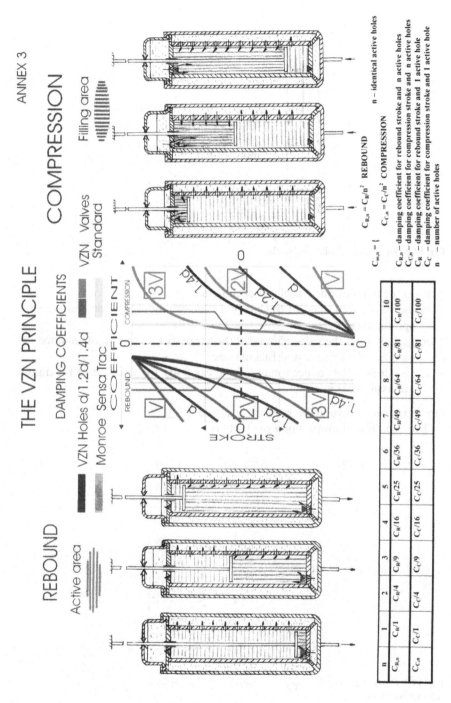

Fig. 19.1 The VZN principle, relative to Monroe Sensa Trac and standard ones

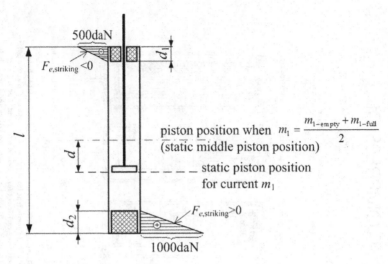

Fig. 19.2 Scheme of the shock absorber

m_2 is the mass of one wheel—unsprung mass;

m_1 is the reduced mass of the car body corresponding to one rear wheel—sprung mass;

$m_{1\,\text{empty}}$ denotes the reduced mass of the car body for the case of an unloaded car, including driver and fuel masses;

$m_{1\,\text{full}}$ denotes the reduced mass of the car body for the case of maximum admissible car loading.

Obviously, m_1 is a value between $m_{1\,\text{empty}}$ and $m_{1\,\text{full}}$.

The equation of motion of a car body is:

$$m_1\ddot{x}_1 + c_1(\dot{x}_1 - \dot{x}_2)^2 + k_1(x_1 - x_2) = F_{e,\text{stop bumper}} \qquad (19.2)$$

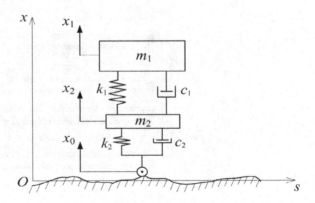

Fig. 19.3 Car/road vertical interaction model

where $F_{e,\text{ stop bumper}}$ represents the elastic striking force, being < 0 at rebound and > 0 on compression bumper.

The equation of motion of the wheel center is:

$$m_2\ddot{x}_2 - c_1(\dot{x}_1 - \dot{x}_2)^2 + c_2(\dot{x}_2 - \dot{x}_0) - k_1(x_1 - x_2)$$
$$+ k_2(x_2 - x_0) = -F_{e,\text{stop bumper}} \tag{19.3}$$

Denoting by v_1 the vertical velocity $\dot{x}_1$ of the car body and by v_2 the vertical velocity $\dot{x}_2$ of the wheel center, the second order differential equations (19.2) and (19.3) can be transformed into the following system of four first-order differential equations, ready to be numerically integrated by the usual methods, e.g., the Runge-Kutta method:

$$\begin{cases} \dot{x}_1 = v_1 \\ \dot{x}_2 = v_2 \\ \dot{v}_1 = \frac{F_{e,\text{stop bumper}}}{m_1} - \frac{1}{m_1}[c_1(v_1 - v_2)^2 + k_1(x_1 - x_2)] \\ \dot{v}_2 = -\frac{F_{e,\text{stop bumper}}}{m_2} + \frac{1}{m_2}[c_1(v_1 - v_2)^2 - c_2(v_2 - \dot{x}_0) + k_1(x_1 - x_2) - k_2(x_2 - x_0)] \end{cases} \tag{19.4}$$

The road/wheel adherence force is given by:

$$F_{\text{adh}} = \begin{cases} -[k_2(x_2 - x_0) + c_2(\dot{x}_2 - \dot{x}_0)], & \text{if contact} \\ 0 & , \text{ if contact lost} \end{cases} \tag{19.5}$$

The road-car vertical interaction has been simulated using Matlab/Simulink. The case of using a VZN shock absorber has been compared with the case of using a standard shock absorber. The considered car has the following characteristics:

- $m_{1-\text{empty}} = 240[\text{kg}]$, $m_{1-\text{full}} = 360\,[\text{kg}]$;
- $l = 0.236[\text{m}]$, $d_1 = 0.014[\text{m}]$, $d_2 = 0.040[\text{m}]$;
- $k_1 = 14.085[\text{kN/m}]$, $k_2 = 21.8[\text{kN/m}]$;
- $c_2 = \frac{k_2}{2\pi f} = \frac{21.8}{2\pi f}[\text{kN} \cdot \text{s/m}]$- the tire-damping coefficient, where f is the frequency of the sine wave test cycle under consideration;
- For a standard shock absorber of the car under consideration, the damping coefficient c_1 is given in 5 piston velocity steps [2] having values from 60.8 [N] up to 51716 [N].
- For the VZN shock absorber, the damping coefficient value increases more than 3000 times between minimal and maximal values, depending on the instantaneous piston position.

For a complex behavior evaluation a sum of three harmonic functions and a pure harmonic function were taken into consideration, conforming to Table 19.1.

Table 19.1 Amplitude and frequency of the harmonic functions considered as a road profile

Case	Excitation
1	$x_0 = 0.1 \sin 2\pi + 0.03 \sin(10\pi + 5.0815) + 0.01 \sin(20\pi + 1.2146)$
2	$x_0 = 0.03 \sin 24\pi$

The behavior of the VZN shock absorber has been compared with the behavior of a standard shock absorber.

In Figs. 19.4 and 19.5 are shown the following relevant time evolutions: the car body vertical accelerations and their root mean squares; the forces in stop bumpers and their root mean squares; and the car body vertical displacements.

For each situation, the comparison between the behaviors of the VZN shock absorber relative to the standard shock absorber has been realized, showing:

- The car body vertical accelerations and their root mean squares ($\ddot{x}_{1,RMS,VZN}$ and $\ddot{x}_{1,RMS,standard}$);
- The forces in stop bumpers and their root mean squares (RMS$_{VZN}$, RMS$_{standard}$);
- The car body vertical displacements ($x_{1,VZN}$; $x_{1,standard}$) and their dispersion ($\sigma_{x_{1,VZN}}$; $\sigma_{x_{1,standard}}$).

The results show the VZN shock absorbers compared to the standard shock absorbers:

- Root mean squares car body vertical accelerations, 32% decreased;
- No forces in stop bumpers at VZN, relative to standard sample with 2.67 [kN] root mean squares;
- Displacements dispersion 38% decreased;
- Car body squat 4 cm reduced;
- SCKYHOOK behavior.

19.3 On Roll Theoretical Consideration

The suspension reaction forces equilibrate the lateral forces, us the wind forces acting in the side pressure center, or the centrifugal forces acting in the gravity center.

Considering shock absorbers having 10 identical metering holes, so damping coefficients have variation of 100 times, along the stroke, at system unbalanced/ redressing, the shock absorbers act with anti-gyration/anti-redressing torques, the active metering holes according to the roll angle being indicated in Fig. 19.6 for both states.

In Fig. 19.6, the number of ordinate lines represents the number of active metering holes. The anti-gyration/anti-redressing damping coefficients and the relative torques have the values indicated in Table 19.2.

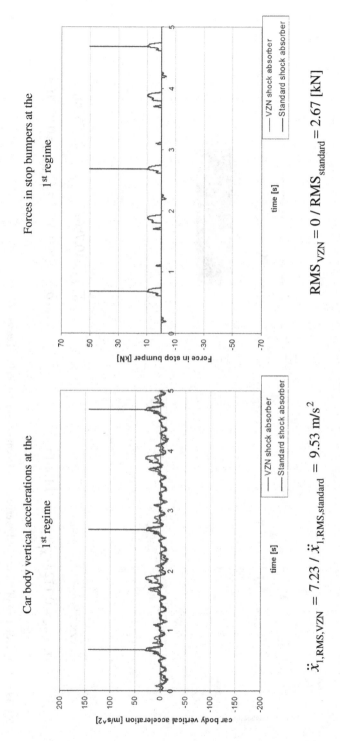

Fig. 19.4 Time evolution of the car body vertical accelerations and forces in stop bumpers at the 1st regime

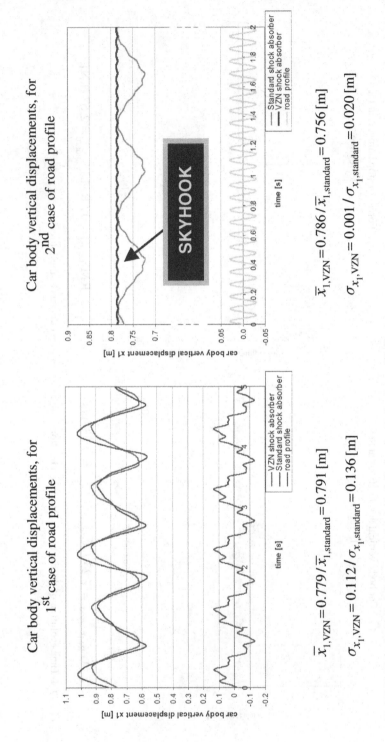

Fig. 19.5 Time evolution of the car body vertical displacements for 1st and 2nd cases

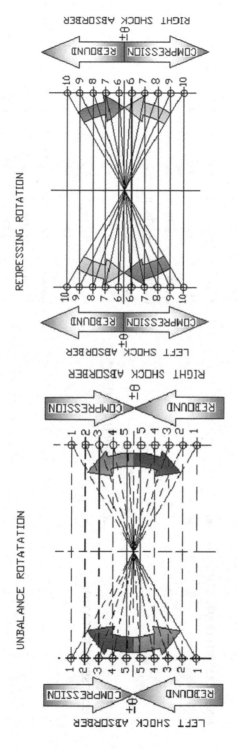

Fig. 19.6 The correlation between VZN active holes number and roll angle at unbalance/redressing rotations

Table 19.2 The VZN and standard roll damping coefficients, their ratio and relative roll angles at unbalancing rotation

ROTATION SENSE	The holes position $[\pm i]$	± 1	± 2	± 3	± 4	± 5
	Relative roll angle $\theta = f(u)$ $u = f(i,\delta)$	$\theta = \arcsin\left(\frac{2u}{b}\right) = \arcsin\left(\frac{\mp\delta\pm i\delta}{b}\right)$				
		u				
		$\pm\frac{\delta}{2}$	$\pm\frac{3\delta}{2}$	$\pm\frac{5\delta}{2}$	$\pm\frac{7\delta}{2}$	$\pm\frac{9\delta}{2}$
UNBALANCEING ROTATION	n_U	5	4	3	2	1
	n_U^2	25	16	9	4	1
	$c_{\theta U}^{VZN}$	$c_{\theta 0}/25$	$c_{\theta 0}/16$	$c_{\theta 0}/9$	$c_{\theta 0}/4$	$c_{\theta 0}$
	$c_{\theta U}^{S}$	$c_{\theta 0}/36$	$c_{\theta 0}/36$	$c_{\theta 0}/36$	$c_{\theta 0}/36$	$c_{\theta 0}/36$
	$c_{\theta U}^{VZN}\big/c_{\theta U}^{S}$	1.44	2.25	4	9	36
	$RAE_U = T_{cU}^{VZN}\big/T_{cU}^{S}$	1.44	2.25	4	9	36
REDRESSING ROTATION	n_R	6	7	8	9	10
	n_R^2	36	49	64	81	100
	$c_{\theta R}^{VZN}$	$c_{\theta 0}/36$	$c_{\theta 0}/49$	$c_{\theta 0}/64$	$c_{\theta 0}/81$	$c_{\theta 0}/100$
	$c_{\theta R}^{S}$	$c_{\theta 0}/36$	$c_{\theta 0}/36$	$c_{\theta 0}/36$	$c_{\theta 0}/36$	$c_{\theta 0}/36$
	$c_{\theta R}^{VZN}\big/c_{\theta R}^{S}$	1	0.73	0.56	0.44	0.36
	$RAE_R = T_{cR}^{VZN}\big/T_{cR}^{S}$	1	0.73	0.56	0.44	0.36

$$c_R = \frac{c_{0q}}{n^2} \quad q = \text{Rebound/Compression}; \; n = \text{number of actives holes} \quad (19.6)$$

$$c_\theta = \frac{T_c}{\varpi} = \frac{r(D_{Lq} + D_{Rq})}{\frac{V}{r}} = \frac{r^2(D_{Lq} + D_{Rq})}{V} = r^2(c_C + c_R)$$

$$= \frac{b^2}{4}(c_C + c_R) = \frac{b^2}{4}\frac{(c_{0C} + c_{0R})}{n^2} = \frac{c_{\theta 0}}{n^2} \qquad (19.7)$$

$D = cV^j$; for this case $D_q = c_q V$

c_{0q} – compression/rebound damping coefficient for one active hole,

D_L – the damping force of the left shock absorber,

D_R – the damping force of the right shock absorber,

T_c – the damping torque of the shock absorbers.

$$c_{\theta 0} = \frac{b^2}{4}(c_{0C} + c_{0R}) \qquad (19.8)$$

Based on the values from Table 19.2, Fig. 19.7 shows the VZN and standard relative anti-roll and anti-redressing effect.

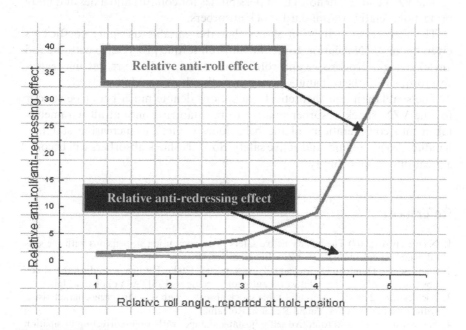

Fig. 19.7 VZN and standard relative anti-roll and anti-redressing effect

The theoretical analyses show the VZN shock absorbers confer better behavior at the roll relative to the standard ones, because for identical metering
holes, the anti-roll torque decreases with the square of the number of active
metering holes, varying usually more than 30 times.

The progressive damping coefficient given by the VZN solution confers:

- An anti-roll torque progressive with the roll angle, giving anti-roll relative
  effect;
- An anti-redressing torque regressive with the roll angle, giving anti-redressing
  relative effect.

19.4 Conclusions

The simulation results show a 32% decrease of body acceleration, no strikes on
the stop bumpers, 38% lower body vertical displacement, body squat 25% less,
and SKYHOOK behaviour at the VZN shock absorber, relative to the standard
one.

But more relevant are the maximal values, which can damage the passengers,
merchandise, axles, and car body.

So the maximal values are, at the VZN sample, five times less than at the
body acceleration, without force in stopper bumpers, more than 1.4 times less
body displacement than the standard one.

The VZN concept denotes high possibilities for comfort and axles and body
protection, relative to standard shock absorbers.

The theoretical evaluation shows that the variable damping coefficients
given by the VZN solution confer an anti-roll torque progressive with the roll
angle, giving a progressive anti-roll relative effect, and anti-redressing regressive torque with the roll angle, giving an anti-redressing relative effect.

The simulation and theoretical evaluations denote much better behaviour
for the VZN shock absorber relative to the standard ones at all parameters
taken into consideration: all the behaviours conferring increasing stability,
handling, road holding, and active safety. So VZN shock absorbers will become
a great future opportunity.

References

1. Niculescu A (2007) The vehicle active safety increasing, using suspensions with the self-
   adjustable shock absorbers—VZN. Beograd, Serbia
2. Niculescu A, Dumitriu D, Sireteanu T, Alexandru C (2006) The new self-adjustable shock
   absorber—VZN. In: The 31st world automotive congress FISITA, Yokohama, Japan
3. Niculescu A (2006) The self-adjustable shock absorber with self-correcting characteristic,
   with curve filling valves. Patent Request, Romania
4. Niculescu A (2005) Automotive self-adjustable damper with a self-correcting dissipation
   characteristic. European Patent EP 1190184

5. Niculescu A (2005) Some concepts for the self-adjustable shock absorbers tuning. Annual Symposium of the Institute of Solid Mechanics, Bucharest, Romania
6. Niculescu A, Sireteanu T, Stancioiu D (2004) Automotive self-adjustable shock absorber (VZN). In: The 30th World Automotive Congress FISITA, Barcelona, Spain

Chapter 20
Pattern Generation via Computer

Tatsuo Suzuki and Yasumichi Hasegawa

Abstract The method of computer-aided geometric pattern generation is discussed. The feature of our approach is to regard pattern as the behavior of some kind of dynamical system. The concrete procedure to construct the mathematical model of a given pattern is presented, which has been proposed recently by the authors. Some Japanese traditional pattern generations are treated as one of the typical applications of our approach.

20.1 Introduction

Geometrical patterns often appear in various scenes in our daily life. It is also well known that there are many traditional geometrical patterns all over the world. They frequently make our life more graceful and elegant.

This paper is concerned with the description and design of such geometrical patterns. The mathematical model which is represented by some kind of dynamical system is constructed from the pattern data such as pixel, hue, and so forth. Then a new system which describes any pattern faithfully is proposed. Using this system thus allows a description problem to be treated as a problem of realization and design.

In ordinary image processing today, images are often transformed into one-dimensional signals, which are then analyzed by means of various established methods in signal processing theory [1, 2]. Another common processing procedure employs tree structures [3].

Our approach is quite different from the above mentioned ones. In our approach, patterns are viewed as input/output relations with special features, which are special cases of the behavior of linear representation systems [4]. The processing method presented in this paper is built on this new insight and is very well adapted to input/output relations. It will illustrate how to design

T. Suzuki (✉)
Gifu University, 1-1 Yanagido, Gifu 501-1193, Japan
e-mail: tasuzuki@gifu-u.ac.jp

N. Mastorakis et al. (eds.), *Proceedings of the European Computing Conference*, Lecture Notes in Electrical Engineering 27, DOI 10.1007/978-0-387-84814-3_20, © Springer Science+Business Media, LLC 2009

geometrical patterns by computer through taking up the generation of some Japanese traditional patterns as examples.

Notation *N is the set of non-negative integers. $N^2 := N \times N$ denotes the product set in two sets of N. N/pN is a finite field of residue class, where p is a prime number. K is a field. $K[z_a, z_\beta]$ denotes the commutative K-algebra of polynomials in two variables. $K[[z_a^{-1}, z_\beta^{-1}]]$ is the K-linear space of formal power series in two variables. $F(X, Y)$ and $L(X, Y)$ denote the set of all functions and all linear maps from X to Y , respectively. $L(X)$ implies $L(X, X)$. $\ll S \gg$ denotes the smallest linear space which contains a set S. I_n denotes the n-dimensional identity matrix. $\oplus$ and $\otimes$ denote direct sum and tensor product, respectively.*

20.2 Pattern Description

20.2.1 Pattern as Input/Output Relation

A geometrical pattern is represented as the following array:

$$
\begin{vmatrix}
a(0,0) & a(0,1) & a(0,2) & \cdots \\
a(1,0) & a(1,1) & a(1,2) & \cdots \\
a(2,0) & a(2,1) & a(2,2) & \cdots \\
\vdots & \vdots & \vdots & \ddots
\end{vmatrix}
$$

where $a(i,j) \in Y = K^p$ for any $i,j \in N$. A pattern a is expressed by $a \in F(N^2, Y)$. It is also represented as the following formal power series in two variables, z_a and z_β:

$$
a = \sum_{i=0}^{\infty} \sum_{j=0}^{\infty} a(i,j) z_a^{-1} z_\beta^{-1} \in K^p[[z_a^{-1}, z_\beta^{-1}]]
$$

where i and j denote the position markers in two-directional axes for the position (i,j).

Consider a connection between this pattern and input/output relations. Let U be a set of input values (alphabet: a and β). Let U^* be a set of words generated by the alphabet set U. Then any input/output map which satisfies causality conditions and takes a value in Y can be expressed by an input response map $\underline{a} \in F(U^*, Y)$ which satisfies $\gamma(|\omega|) = \underline{a}(\omega)$, where $\gamma(|\omega|)$ denotes the value of the output when an input ω has been fed into the observed input/output map and $|\omega|$ denotes the length of input ω. Let us introduce a set $F_c(U^*, Y)$ of an input response map $\underline{a}$ which satisfies $\underline{a}(\omega_1|\omega_2) = \underline{a}(\omega_2|\omega_1) =$ for any $\omega_1, \omega_2 \in U^*$, where | denotes the concatenation operator. Then an isomorphic relation between the set $F(N^2, Y)$ and the set $F_c(U^*, Y)$ is obtained as follows:

$$F(N^2, Y) \rightarrow F_c(U^*, Y) : a \mapsto \underline{a} \text{ by setting}$$

$$a(i,j) = \underline{a}(\overbrace{a \cdots a}^{i} \mid \overbrace{\beta \cdots \beta}^{j}) \text{ for } i,j \in N \text{ and } a, \beta \in U.$$

Therefore, a pattern $a \in F(N^2, Y)$ is equivalent to an input response map $\underline{a} \in F_c(U^*, Y)$. This equivalence relation between $F(N^2, Y)$ and $F_c(U^*, Y)$ leads to the realization problem for a geometrical pattern. The realization problem is roughly stated as follows:

For given data which are typically considered as input/output data, find a mathematical model which realizes them, that is, which faithfully describes them.

It is well understood in the image processing field that description is the essential problem. Similarly, in input/output relation matters, realization is recognized as the most important problem. In order to solve the problem, we have introduced a new special class of the mathematical models known as linear representation systems [4], whose idea came from [5]. These new mathematical models for digital images are called 2-commutative linear representation systems [6], or 2-CLRS for short.

20.2.2 2-Commutative Linear Representation System

Definition 1
1. *A system represented by the following equations is written as a collection $\sigma = ((X, F_a, F_\beta), x^0, h)$ and is called a 2-CLRS.*

$$\begin{cases} x(i+1,j) = F_a x(i,j) \\ x(i,j+1) = F_\beta x(i,j) \\ \quad x(0,0) = x^0 \\ \quad \gamma(i,j) = h x(i,j) \end{cases}$$

for any $i,j \in N$, $x(i,j) \in X$, $\gamma(i,j) \in Y$, where X is a linear space over the field K, F_a and F_β are linear operators on X which satisfy $F_a F_\beta = F_\beta F_a$, $x^0 \in X$ is an initial state, and $h : X \rightarrow Y$ is a linear operator.
2. *The pattern $a_\sigma : N^2 \rightarrow Y; (i,j) \mapsto h F_a^i F_\beta^j x^0$ is called the behavior of σ.*
3. *σ which satisfies $a^\sigma = a \in F(N^2, Y)$ is called a realization of a.*
4. *A 2-CLRS σ is said to be quasi-reachable if the linear hull of the reachable set $\left\{ F_a^i F_\beta^j x^0 ; i,j \in N \right\}$ equals X.*
5. *A 2-CLRS σ is said to be distinguishable if $h F_a^i F_\beta^j x_1 = h F_a^i F_\beta^j x_2$ for any $i,j \in N$ implies $x_1 = x_2$.*

6. *A 2-CLRS σ is called canonical if σ is quasi-reachable and distinguishable.*

The $x(i,j)$ in the system equation of σ is the state that produces output value of a_σ at the place (i,j), while the linear operator $h : X \rightarrow Y$ generates the output value $a_\sigma(i,j)$ at the place (i,j). σ realizes a pattern a implies that σ is a faithful model for a.

Note that the behavior of the 2-CLRS is the same as the input/output map of the separable two-dimensional system [7].

Theorem 1 *The following 2-CLRSs are canonical realizations of any pattern* $a \in F(N^2, Y)$.

1. $\left(\left(K[z_a, Z_\beta]/_{=a}, \dot{z}_a, \dot{z}_\beta \right), [1], \dot{a} \right)$,

   where $K[z_a, z_\beta]/_{=a}$ is a quotient space obtained by the equivalence relation:

$$\sum_{i,j} \lambda_1(i,j) z_a^i z_\beta^j = \sum_{i,j} \lambda_2(i,j) z_a^i z_\beta^j \Leftrightarrow \sum_{i,j} \lambda_1(i,j) a(i,j) = \sum_{i,j} \lambda_2(i,j) a(i,j)$$

   *$\dot{z}_a, \dot{z}_\beta$ and $\dot{a}$ are given by a map $\dot{z}_a : K[z_a, z_\beta]/_{=a} \rightarrow K[z_a, z_\beta]/_{=a}; [\lambda] \mapsto [z_a\lambda]$, $\dot{z}_\beta : K[z_a, z_\beta]/_{=a} \rightarrow K[z_a, z_\beta]/_{=a}; [\lambda] \mapsto [z_\beta\lambda]$ and $\dot{a} : K[z_a, z_\beta]/_{=a} \rightarrow Y; [\lambda] \mapsto \dot{a}([\lambda]) = \sum_{i,j} \lambda(i,j) a(i,j)$, respectively, where $\lambda = \sum_{i,j} \lambda(i,j) z_a^i z_\beta^j \in K[z_a, z_\beta]$.
   1 is a unit element of multiplication.*

2. $\left(\left(\ll S_a^N S_\beta^N a \gg, S_a, S_\beta \right), a, (0,0) \right)$,

   where $S_a^N S_\beta^N a := \left\{ S_a^i S_\beta^j a; (i,j) \in N^2 \right\}$, $S_a a := N^2 \rightarrow Y; (i,j) \mapsto a(i+1, \ j)$, $S_\beta a := N^2 \rightarrow Y; (i,j) \mapsto a(i,j+1)$ and $(0,0) : F(N^2, Y) \rightarrow Y; a \rightarrow a(0,0)$.

Definition 2 *Let $\sigma_1 = ((X_1, F_{a_1}, F_{\beta_1}), x_1^0, h_1)$ and $\sigma_2 = ((X_2, F_{a_2}, F_{\beta_2}), x_2^0, h_2)$ be 2-CLRSs. Then a linear operator $T : X_1 \rightarrow X_2$ is said to be a 2-CLRS morphism $T : \sigma_1 \rightarrow \sigma_2$ if T satisfies $T F_{a_1} = F_{a_2} T$, $T F_{\beta_1} = F_{\beta_2} T$, $T x_1^0 = x_2^0$ and $h_1 = h_2 T$. If $T : X_1 \rightarrow X_2$ is bijective, then $T : \sigma_1 \rightarrow \sigma_2$ is said to be an isomorphism.*

Theorem 2 *Realization Theorem of 2-CLRSs*
1. *Existence: For any pattern $a \in F(N^2, Y)$, there exist at least two canonical 2-CLRSs which realize a.*
2. *Uniqueness: Let σ_1 and σ_2 be any two canonical 2-CLRSs that realize $a \in F(N^2, Y)$. Then there exists an isomorphism $T : \sigma_1 \rightarrow \sigma_2$.*
   2-CLRS $\sigma = ((X, F_a, F_\beta), x^0, h)$ is called a finite-dimensional if the state space X is a finite-dimensional linear space.

Lemma 1 *The following conditions are mutually equivalent.*
1. *$a \in F(N^2, Y)$ has the behavior of a finite-dimensional canonical 2-CLRS.*
2. *The quotient space $K[z_a, z_\beta]/_{=a}$ is finite-dimensional.*
3. *The linear space generated by $\left\{ S_a^i S_\beta^j a : i,j \in N \right\}$ is finite-dimensional, where $K[z_a, z_\beta]/_{=a}$ is a quotient space given by the equivalence relations: $a_1 = a_2 \Leftrightarrow a_1(i,j) = a_2(i,j)$ for any $i,j \in N$.*

Since finite-dimensional linear space, that is, n-dimensional linear space over the field K, is isomorphic to K^n and $L(K^n, K^m)$ is isomorphic to $K^{m \times n}$, a finite-dimensional 2-CLRS can be regarded as $\sigma = ((K^n, F_a, F_\beta), x^0, h)$, where $F_a, F_\beta \in K^{m \times n}$, $x^0 \in K^n$ and $h \in K^{p \times n}$ without loss of generality.

Definition 3 *A canonical 2-CLRS $\sigma_s = ((K^n, F_{a_s}, F_{\beta_s}), e_1, h_s)$ is called a quasi-reachable standard system with a vector index $v = (v_1, v_2, \cdots, v_k)$ if the following conditions hold:*

1. *An integer $v_j (1 \leq j \leq k)$ satisfies $n = \sum_{j=1}^{k} v_j$ and $0 \leq v_k \leq v_{k-1} \leq \cdots \leq v_2 \leq v_1$.*
2. *For any $i, j (1 \leq j \leq k, 1 \leq i \leq v_j)$,*

$$F_{\beta_s}^{j-1} F_{a_s}^{i-1} e_1 = e_{v_1 + v_2 + \cdots + v_{j-1} + i},$$

$$F_{\beta_s}^{j-1} F_{a_s}^{v_j} e_1 = \sum_{m=1}^{j} \sum_{l=1}^{v_m} c_{ml}^j F_{\beta_s}^{m-1} F_{a_s}^{l-1} e_1, \quad c_{ml}^j \in K,$$

$$F_{\beta_s}^{k} F_{a_s}^{v_j} e_1 = \sum_{m=1}^{j} \sum_{l=1}^{v_m} c_{ml}^{k+1} F_{\beta_s}^{m-1} F_{a_s}^{l-1} e_1, \quad c_{ml}^{k+1} \in K,$$

$$e_i = \left[0, \cdots, 0, \overset{i}{1}, 0, \cdots, 0 \right]^{\mathrm{T}} \in K^n.$$

3. $h_s = [a(0,0), a(1,0), \cdots, a(v_1 - 1, 0), a(0,1), \cdots, a(v_2 - 1, 1), \cdots, a(0, k-1), \cdots, a(v_k - 1, k-1)]$

Theorem 3 *Representation theorem for equivalence classes: For any finite-dimensional canonical 2-CLRS, there exists a uniquely determined isomorphic quasi-reachable standard system.*

For any pattern $a \in F(N^2, Y)$, the corresponding linear input/output map $A : (K[z_a, z_\beta], z_a, z_\beta) \to (F(N^2, Y), S_a, S_\beta)$ satisfies $A\left(z_a^i z_\beta^j\right) = S_a^i S_\beta^j$ for $i, j \in N$. Hence, A is represented by the following infinite Hankel matrix:

$$H_a = {}_{(\tilde{i}\tilde{j})} \left(\begin{array}{c} \overset{(i,j)}{\vdots} \\ \cdots \quad a(\tilde{i} + i, \tilde{j} + j) \end{array} \right),$$

where $a(\tilde{i} + i, \tilde{j} + j) = (0, 0) S_a^{\tilde{i}} S_\beta^{\tilde{j}} S_a^i S_\beta^j a$ and the column vectors of H_a are represented by $S_a^i S_\beta^j a$ for $i, j \in N$.

Theorem 4 *Theorem for existence criteria*
The following conditions are mutually equivalent:

1. *The pattern $a \in F(N^2, Y)$ is the behavior of the finite-dimensional canonical 2-CLRS.*

2. *There exist n-linearly independent vectors and no more than n-linearly independent vectors in a set* $\left\{ S_a^i S_\beta^j a; \ i+j \leq n-1 \ \text{for} \ i,j \in N \right\}$.
3. *rank* $H_a = n$.

20.2.3 Realization Procedure

Let a pattern $a \in F(N^2, Y)$ satisfy the condition of Theorem 8. Then the quasi-reachable standard system $\sigma_s = \left((K^n, F_{a_s}, F_{\beta_s}), e_1, h_s \right)$ which realizes a is obtained by the following procedure:

1. Find an integer v_1 and coefficients $\left\{ c_{1\,l}^1; \ 1 \leq l \leq v_1 \right\}$ such that the vectors $\left\{ S_a^i a; \ 1 \leq i \leq v_1 - 1 \right\}$ of the set $\left\{ S_a^i a; \ i \leq n-1, \ i \in N \right\}$ are linearly independent and $S_a^{v_1} a = \sum_{l=1}^{v_1} c_{1\,l}^1 S_a^{l-1} a$.
2. Find an integer v_2 and coefficients $\left\{ c_{ml}^2; \ 1 \leq l \leq v_m, \ 1 \leq m \leq 2 \right\}$ such that the vectors $\left\{ S_\beta^{j-1} S_a^{i-1} a; \ 1 \leq i \leq v_j - 1, \ 1 \leq j \leq 2 \right\}$ of the set $\left\{ S_\beta^j S_a^{ia}; \ i \leq n-1, \ j \leq n-2 \in N \right\}$ are linearly independent and

$$S_\beta S_a^{v_2} a = \sum_{m=1}^{2} \sum_{l=1}^{v_m} c_{ml}^2 S_\beta^{m-1} S_a^{l-1} a.$$

$\vdots$

k. Find an integer v_k and coefficients $\left\{ c_{ml}^k; \ 1 \leq l \leq v_l, \ 1 \leq m \leq k \right\}$ such that the vectors $\left\{ S_\beta^{j-1} S_a^{i-1} a; \ 1 \leq i \leq v_j, \ 1 \leq j \leq k \right\}$ of the set $\left\{ S_\beta^j S_a^i a; \ i \leq n-1, \ j \leq k-1 \in N \right\}$ are linearly independent, and

$$S_\beta^{k-1} S_a^{v_k} a = \sum_{m=1}^{k} \sum_{l=1}^{v_m} c_{ml}^k S_\beta^{m-1} S_a^{l-1} a,$$

$$S_\beta^k a = \sum_{m=1}^{k} \sum_{l=1}^{v_m} c_{ml}^{k+1} S_\beta^{m-1} S_a^{l-1} a.$$

k + 1. Let the state space be K^n, where $n = \sum_{i=1}^{k} v_i$. Let the initial state be e_1. Set matrices F_{a_s}, F_{β_s}, and h_s be those given in Definition 3.

20.2.4 Partial Realization

A partial realization problem of 2-CLRS is stated as follows:

For any given $\underline{a} \in F(\mathbf{L} \times \mathbf{M}, Y)$, find a partial realization σ of $\underline{a}$ such that the dimension of state space X of σ is minimum. This σ is said to be a minimal partial realization of $\underline{a}$.

Let $\underline{a} \in F(\mathbf{L} \times \mathbf{M}, Y)$ be a $(L+1) \times (M+1)$ finite-sized pattern, where $L, M \in N$, $\mathbf{L} := \{0, 1, \cdots, L-1, L\}$, $\mathbf{M} := \{0, 1, \cdots, M-1, M\}$. Then a finite-dimensional 2-CLRS $\sigma = ((X, F_a, F_\beta), x^0, h)$ is called a partial realization of $\underline{a}$ if $hF_a^i F_\beta^j x^0 = \underline{a}(i,j)$ holds for any $(i,j) \in \mathbf{L} \times \mathbf{M}$. Set $\underline{a}(i,j) = 0$ for any $(i,j) \notin \mathbf{L} \times \mathbf{M}$. Then $\underline{a} \in F(\mathbf{L} \times \mathbf{M}, Y)$. By virtue of Theorem 4, there exists a finite-dimensional partial realization of $\underline{a}$. Thus, there always exists a minimal partial realization of $\underline{a}$. Minimal partial realizations are not unique up to isomorphism in general. Thus, the concept of natural partial realization is introduced as follows;

If $X = \ll \{F_a^i F_\beta^j x^0; i \le l_1, j \le m_1\} \gg$ for a 2-CLRS $\sigma = ((X, F_a, F_\beta), x^0, h)$ and some $l_1, m_1 \in N$, then σ is said to be (l_1, m_1)-quasi-reachable. Let $l2$ and $m2$ be some integer. If $h F_a^i F_\beta^j x = 0$ implies $x = 0$ for any $i \le l_2$ and $j \le m_2$, then σ is said to be (l_2, m_2)-distinguishable. For a given $\underline{a} \in F(\mathbf{L} \times \mathbf{M}, Y)$, if there exist l_1, m_1 and $l_2, m_2 \in N$ such that $l_1 + l_2 < L$, $m_1 + m_2 < M$ and σ of its partial realization is (l_1, m_1)-quasi-reachable and (l_2, m_2)-distinguishable, then σ is said to be a natural partial realization of $\underline{a}$.

The following matrix, $H_{\underline{a}\,(l_1, m_1, L-l_1, M-m_1)}$, is called a finite-sized Hankel matrix of a partial finite-sized pattern $\underline{a} \in F(\mathbf{L} \times \mathbf{M}, Y)$.

$$H_{\underline{a}(l_1,m_1,L-l_1,M-m_1)} =_{(l,m)} \begin{pmatrix} & & \overset{(i,j)}{\vdots} & \\ \cdots & \underline{a}(i+l, j+m) & \end{pmatrix},$$

where $0 \le i \le l_1, 0 \le j \le m_1, 0 \le l \le L - l_1, 0 \le m \le M - m_1$.

Theorem 5 *There exists a natural partial realization of $\underline{a} \in F(\mathbf{L} \times \mathbf{M}, Y)$ if and only if the following condition holds for some $l_1 \in \mathbf{L}, m_1 \in \mathbf{M}$:*

rank $H_{\underline{a}\,(l_1, m_1, L-l_1, M-m_1)}$

$\quad = \text{rank}\, H_{\underline{a}\,(l_1+1, m_1, L-l_1-1, M-m_1-1)}$

$\quad = \text{rank}\, H_{\underline{a}\,(l_1, m_1+1, L-l_1, M-m_1-1)}$

$\quad = \text{rank}\, H_{\underline{a}\,(l_1, m_1+1, L-l_1-1, M-m_1-1)}$

$\quad = \text{rank}\, H_{\underline{a}\,(l_1, m_1, L-l_1-1, M-m_1-1)}$

Theorem 6 *There exists a natural partial realization of a given partial finite-sized pattern $\underline{a} \in F(\mathbf{L} \times \mathbf{M}, Y)$ if and only if the minimal partial realizations of $\underline{a}$ are unique up to isomorphism.*

20.3 Pattern Generation

Any finite-sized pattern can be categorized whether it is periodic or not. First, 2-CLRS σ which realize non-periodic patterns will be presented. Next, the periodic case is investigated. In this section, $a(i,j)$ is written by $a_{i,j}$.

Lemma 2 *There exists a 2-CLRS* $\sigma = ((K^{(L+1)\times(M+1)}, F_a, F_\beta), x^0, h)$ *which realizes any* $(L+1) \times (M+1)$*-sized non-periodic pattern* $\underline{a} \in F(\mathbf{L} \times \mathbf{M}, Y)$, *where*

$$F_a = I_{M+1} \otimes F_{L+1}, \quad F_\beta = F_{M+1} \otimes I_{L+1},$$

$$h = \underbrace{\left[a_{0,0} \cdots a_{L,0}, a_{0,1} \cdots a_{L,1} \cdots a_{0,M} \cdots a_{L,M} \right]}_{(L+1)\times(M+1)},$$

$$x^0 = \left.\begin{bmatrix} 1 \\ 0 \\ \vdots \\ 0 \end{bmatrix}\right\}(L+1)\times(M+1), F_q = \underbrace{\begin{bmatrix} 0 & \cdots & & 0 & 0 \\ 1 & \ddots & & \vdots & \vdots \\ 0 & \ddots & 0 & & \vdots \\ 0 & \cdots & 0 & 1 & 0 \end{bmatrix}}_{q} \text{ for } q = L+1 \text{ or }$$

$$M+1.$$

A pattern $a \in F(N^2, Y)$ with a period of ℓ length in the vertical and m length in the horizontal directions is called $\ell \times m$-periodic pattern.

Proposition 1 *Any* $\ell \times m$*-periodic pattern is realized by a 2-CLRS* $\sigma_p = ((K^{\ell\times m}, F_{ap}, F_{\beta p}), x_p^0, h_p)$, *where*

$$F_{ap} = I_m \otimes F_\ell, \quad F_{\beta p} = F_m \otimes I_\ell,$$

$$h_p = \underbrace{\left[a_{0,0} \cdots a_{\ell-1,0}, a_{0,1} \cdots a_{\ell-1,1} \cdots \cdots a_{\ell-1,m-1} \right]}_{\ell\times m},$$

$$x^0 = \left.\begin{bmatrix} 1 \\ 0 \\ \vdots \\ 0 \end{bmatrix}\right\}\ell \times m, F_q = \underbrace{\begin{bmatrix} 0 & \cdots & & 0 & 1 \\ 1 & \ddots & & \vdots & 0 \\ 0 & \ddots & 0 & & \vdots \\ 0 & \cdots & 0 & 1 & 0 \end{bmatrix}}_{q} \text{ for } q = \ell \text{ or } m.$$

Let us introduce a direct sum $\sigma_1 \oplus \sigma_2 = ((K^{n_1 \times n_2}, F_a, F_\beta), x^0, h)$ for the 2-CLRS $\sigma_1 = ((K^{n_1}, F_{a1}, F_{\beta 1}), x_1^0, h_1)$ and $\sigma_2 = ((K^{n_2}, F_{a2}, F_{\beta 2}), x_2^0, h_2)$. The behavior of $\sigma_1 \oplus \sigma_2$ is given by $a_{\sigma_1} + a_{\sigma_2}$, where

$$F_a = \begin{bmatrix} F_{a1} & 0 \\ 0 & F_{a2} \end{bmatrix}, \quad F_\beta = \begin{bmatrix} F_{\beta1} & 0 \\ 0 & F_{\beta2} \end{bmatrix}, \quad x^0 = \begin{bmatrix} x_1^0 & x_2^0 \end{bmatrix}^\mathrm{T}, \quad h = \begin{bmatrix} h_1 & h_2 \end{bmatrix}.$$

Here, the direct sum of 2-CLRS and number theory are introduced for rapid generation of a periodic pattern. The following lemma is a special case of Dirichlet's theorem [8].

Lemma 3 *There are infinite primes p which satisfy $p = L \times M + 1$ for a positive integer L and for a fixed integer M.*

Lemma 4 *Fermat's Lemma* [8]
If p is prime and x is not divisible by p, then $x^{p-1} \equiv 1 \pmod{p}$ holds.
   By virtue of Fermat's Lemma 4, if an integer L satisfies $L < p$, then $x^L - 1 \equiv 0 \pmod{p}$ has L different solutions.

Lemma 5 *Let p be a prime number such that $p = L \times M + 1$ for a positive integer L and a fixed integer M. Then $x^L - 1 \equiv (x - x_1)(x - x_2)(x - x_3) \cdots (x - x_L)$ $\pmod{p}$ holds.*

Definition 4 $\sigma_e = ((K^{\ell \times m}, F_{ae}, F_{\beta e}), x_e^0, h_e)$ *is called an Eigen Standard System, where*

$$F_{ae} = I_m \otimes F_{\ell e}, F_{\beta e} = \text{Block diag}[F_1, F_2, \cdots, F_m],$$

$$F_{\ell e} = \text{diag}[a_1, a_2, \cdots a_\ell], F_i = \text{diag}\underbrace{[\beta_i, \beta_i, \cdots \beta_i]}_{\ell} \text{ for } 1 \leq i \leq m,$$

$$x^\ell - 1 \equiv (x - a_1)(x - a_2) \cdots (x - a_\ell) \pmod{p}$$

$$y^m - 1 \equiv (y - \beta_1)(y - \beta_2) \cdots (y - \beta_m) \pmod{p}$$

$$x_e^0 = \underbrace{[1, 1, \cdots, 1]}_{\ell \times m}{}^{\mathrm{T}}.$$

Take the $(\ell \times m) \times (\ell \times m)$ matrix T_e:

$$T_e = \Big[x_e^0, F_{ae} x_e^0, \ldots, F_{ae}^{\ell-1} x_e^0, F_{\beta e} x_e^0, F_{ae} F_{\beta e} x_e^0, \ldots, F_{ae}^{\ell-1} F_{\beta e} x_e^0, \ldots$$
$$\ldots, F_{\beta e}^{m-1} x_e^0, \ldots, F_{ae}^{\ell-1} F_{\beta e}^{m-1} x_e^0 \Big]$$

Let $h_p := h_e T_e$. Then T_e is a 2-CLRS morphism: $T_e : \sigma_p \to \sigma_e$. Hence the behavior of σ_p is the same as that of σ_e.

Theorem 7 *Let σ_p be the 2-CLRS which realizes any pattern with a $\ell \times m$ period. Then σ_p is isomorphic to the Eigen Standard System σ_e.*

When we design any periodic pattern, σ_e can generate patterns more rapidly than using σ_p. To construct σ_e, it is necessary to find a prime number p for given integers l_1 and l_2 such that $p - 1 = m_1 \times l_1 = m_2 \times l_2$ for some integers m_1 and m_2. If such a prime number p is found for given integers l_1 and l_2, the polynomials $x^{l_1} - 1$ and $x^{l_2} - 1$ can be factorized simultaneously via modulo p. To find such a prime number, first calculate the greatest common divisor g and the least common multiple l of l_1 and l_2 by using the Euclidean algorithm and the well-known relation $l \times g = l_1 \times l_2$, and then find the minimum prime number which satisfies $p = m_1 \times l_1 + 1 = m_2 \times l_2 + 1$ and $p \geq l_1 \times l_2$. Upon finding the apparent minimum prime number, one only has to judge whether the given number is truly a prime or not.

Example 1 Consider the 2×2-periodic pattern shown in Fig. 20.1.

Let the set Y of output values be K^3, that is, the set of pixels, hues, and rotations. Let K be $N/N5\,N$. The map h_p is represented as follows:

$$h_p = \begin{bmatrix} 0 & 0 & 1 & 1 \\ 0 & 0 & 2 & 1 \\ 0 & 2 & 0 & 1 \end{bmatrix}.$$

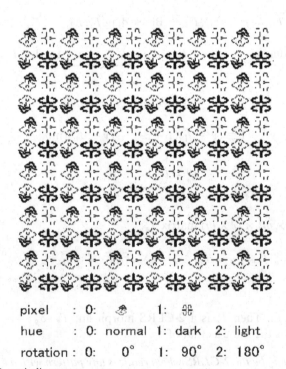

pixel   : 0:   1:

hue   : 0: normal 1: dark 2: light

rotation : 0:   0°   1: 90° 2: 180°

Fig. 20.1 2×2 -periodic pattern

Since $l = m = 2$, F_2, F_{ap}, $F_{\beta p}$ and x^0 are written as follows:

$$F_2 = \begin{bmatrix} 0 & 1 \\ 1 & 0 \end{bmatrix}, x^0 = [1 \quad 0 \quad 0 \quad 0]^T, F_{ap} = I_2 \otimes F_2, F_{\beta p} = F_2 \otimes I_2.$$

Let us construct the Eigen Standard System for rapid design. First, find the prime number p which can be factorized $x^l - 1$ and $x^m - 1$ simultaneously. Since $l = m = 2$, we find that p = 5. Since $x^2 - 1 \equiv (x - 1)(x - 4) \pmod 5$, we can set $a_1 = 1, a_2 = 4, \beta_1 = 1, \beta_2 = 4$. This leads to $F_{le}, F_1, F_2, x_e^0, F_{ae}, F_{\beta e}$ being as follows: $F_{le} = \mathrm{diag}[1, 4]$, $F_1 = I_2$, $F_2 = 4 \cdot I_2$, $x_e^0 = [1\ 1\ 1]^T$, $F_{ae} = \mathrm{diag}[1, 4, 1, 4]$, $F_{\beta e} = \mathrm{diag}[1, 1, 4, 4]$.

Since 2-CLRS morphism T_e is constructed as $T_e = [x_e^0, F_{ae}x_e^0, F_{\beta e}x_e^0, F_{ae}F_{\beta e}x_e^0]$ the map h_e will be as follows:

$$h_e = \begin{bmatrix} 3 & 0 & 2 & 0 \\ 2 & 4 & 3 & 1 \\ 2 & 3 & 4 & 1 \end{bmatrix}.$$

Example 2 Let us take up the 3 × 2-periodic pattern shown in Fig. 20.2.
Let K be $N/7N$. The map h_p is represented as $h_p = [1 \quad 3 \quad 5 \quad 0 \quad 2 \quad 4]$.

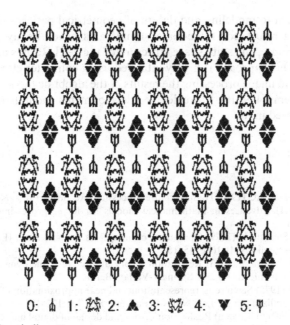

0: 🎐 1: 🎏 2: ▲ 3: 🎇 4: ▼ 5: 🎐

Fig. 20.2 3 × 2-periodic pattern

For this pattern, since $l = 3$, $m = 2$, F_2, F_3, F_{ap}, $F_{\beta p}$ and x^0 are written as follows:

$$F_2 = \begin{bmatrix} 0 & 1 \\ 1 & 0 \end{bmatrix}, F_3 = \begin{bmatrix} 0 & 0 & 1 \\ 1 & 0 & 0 \\ 0 & 1 & 0 \end{bmatrix}, x^0 = \begin{bmatrix} 1 & 0 & 0 & 0 & 0 & 0 \end{bmatrix}^T,$$

$$F_{ap} = I_2 \otimes F_3, F_{\beta p} = F_2 \otimes I_3.$$

Since $l = 3$, $m = 2$, we found p = 7 such that $x^l - 1$ and $x^m - 1$ can be factorized simultaneously. Since $x^3 - 1 \equiv (x-1)(x-2)(x-4)$ (mod7) and $x^2 - 1 \equiv (x-1)(x-6)$ (mod7), we can set $a_1 = 1$, $a_2 = 2$, $a_3 = 4$, $\beta_1 = 1$, $\beta_2 = 6$. This results in F_{le}, F_1, F_2, F_3, x_e^0, F_{ae} and $F_{\beta e}$ as follows: $F_{le} = \text{diag}[1,2,4]$, $F_1 = I_3$, $F_2 = 6 \cdot I_3$, $x_e^0 = [1\ 1\ 1\ 1\ 1\ 1]^T$, $F_{ae} = I_2 \otimes F_{le}$, $F_{\beta e} = \text{Block diag}[F_1, F_2]$.

The 2-CLRS morphism T_e will be constructed as:

$$T_e = \begin{bmatrix} x_e^0, F_{ae}x_e^0, F_{ae}^2x_e^0, F_{\beta e}x_e^0, F_{ae}F_{\beta e}x_e^0, F_{ae}^2F_{\beta e}x_e^0 \end{bmatrix}.$$

Then the map h_e will be obtained as $h_e = \begin{bmatrix} 6 & 3 & 2 & 4 & 0 & 0 \end{bmatrix}$.

20.4 Conclusion

A new approach to modeling and design of geometrical patterns is presented. Corresponding results for three-dimensional cases have already been developed. Furthermore, the state space structure of the 2-CLRS has already been investigated. Our approach is highly intended using computer effectively. Our results extend to image processing the results of the algebraic theory of linear or nonlinear systems [4, 9, 10]. More detailed developments and illustrative examples are presented in [6].

References

1. Rosenfeld A, Kak AC (1976) Digital picture processing. Academic Press, New York
2. Jain AK (1981) Advances in mathematical models for image processing. Proc IEEE 69(5):502–528
3. Hunter GM, Steiglitz K (1979) Operations on images using quad trees. IEEE T Pattern Anal Machine Intell PAMI-1(2):145–153
4. Matsuo T, Hasegawa Y (2003) Realization theory of discrete -Time dynamical systems. Lect Notes Contr Inform Sci 296, Springer-Verlag, Berlin
5. Sussmann HJ (1975) Semigroup representations, bilinear approximation of input/output maps and generalized inputs. Lect Notes Econ Math Syst 131:172–192
6. Hasegawa Y, Suzuki T (2006) Realization theory and design of digital images. Lect Notes Contr Inform Sci 342, Springer-Verlag, Berlin

7. Roesser RT (1975) A discrete state-space model for linear image processing. IEEE T Automat Contr AC-20(1):1–10
8. Hardy GH, Wright EM (1979) An introduction to the theory of numbers. 5th ed, Oxford University Press, Oxford
9. Matsuo T (1981) Realization theory of continuous–time dynamical systems. Lect Notes Contr Inform Sci 32, Springer-Verlag, Berlin
10. Kalman RE, Falb PL, Arbib MA (1969) Topics in mathematical system theory. McGraw-Hill, New York

Chapter 21
3D Reconstruction: Estimating Depth of Hole from 2D Camera Perspectives

Muhammad Abuzar Fahiem, Shaiq A. Haq, Farhat Saleemi, and Huma Tauseef

Abstract In this paper we have implemented a novel approach called perception-based vision (PBV) to retrieve depth information of a hole from a single camera perspective. Three dimensional modeling of real world objects is always of great concern for scientists and engineers. Different approaches are used for this purpose, e.g., 3D scanners, CAD modeling, and contour tracing by coordinate measuring machines (CMMs). This paper does not deal with 3D modeling as a whole but specifically addresses the issue of depth information retrieval of a hole. This is a cost effective, efficient, and accurate solution and requires just a single 2D camera perspective under ambient conditions.

21.1 Introduction

Three-dimensional modeling of real world objects is always of great concern for scientists and engineers. There is a wide range of applications which require this modeling. These applications may vary from CAD/CAM systems to aesthetic designing of costumes; from reconstruction of ancient sculptures to ergonomically suitable products; from astronomical predictions to archeological expeditions; from underwater exploration to on-ground welling. Different techniques and principles have been proposed by different researches for this 3D modeling. 3D scanners and cameras are being very widely used for this purpose. The major difference between cameras and scanners is that cameras collect color information, while the scanners collect distance information in their fields of view.

These 3D scanners employ different principles, such as the time of flight principle and the laser triangulation principle, with different types of light

M.A. Fahiem (✉)
University of Engineering and Technology, Lahore, Pakistan; Lahore College
for Women University, Lahore, Pakistan
e-mail: abuzar@uet.edu.pk

N. Mastorakis et al. (eds.), *Proceedings of the European
Computing Conference*, Lecture Notes in Electrical Engineering 27,
DOI 10.1007/978-0-387-84814-3_21, © Springer Science+Business Media, LLC 2009

sources, such as laser light, structured light, or colored structured light. There are some other scanners too which employ a stereoscopic approach and use ambient light in a controlled environment.

Before proceeding with 3D scanning, 3D reconstruction, and the retrieval of depth information, it is very crucial to set the view points and the processing pipeline. Some very good surveys are presented by Scott et al. [1] on view planning and Bernardini & Rushmeier [2] on the 3D model acquisition pipeline.

Another set of 3D modeling techniques involves different CAD drawings with orthographic projections.

Coordinate measuring machines (CMMs) and contact-based probing scanners are also used but have certain limitations and issues.

In this paper we are not concentrating on 3D modeling or 3D reconstruction in general; instead we are dealing with the retrieval of depth information of a hole: the lost third dimension in an image. A major goal of this research was to develop a cost effective system, yet one efficient and accurate enough to be acceptable. We have used just one camera, ambient light, perspective properties, and geometric calculations to achieve our goal.

Section 21.2 deals with contact-based scanners, while noncontact scanners are discussed in Section 21.3. An overview of CAD modeling is presented in Section 21.4, followed by a detailed discussion of our own approach, perception-based vision (PBV), in Section 21.5.

21.2 Contact/Mechanical/Electromechanical Scanners

These scanners have some sort of stylus that moves on the surface of an object to trace its contours. Different types of CMMs are an example of such scanners and are used for the retrieval of depth information and dimensioning 3D objects [3]. Major problems with these scanners are:

1. They are very slow.
2. They can cause deformation/destruction of sensitive surfaces.
3. The size of the stylus dictates the depth and diameter of the holes that can be traced from such scanners.

Another problem is that these scanners are very sensitive to the structure and texture of a surface, so a little roughness may distort the results considerably.

21.3 Non-Contact Scanners

Almost all the scanners that do not need any physical contact with the surface of an object use some kind of light source for probing onto the surface. Different scanners use different sources of light and/or different principles for retrieving depth information.

21.3.1 Time of Flight Scanners

Time of flight scanners [4] use a laser beam for probing onto the surface of an object, and the core technology is a laser range finder. The principle is very simple: a laser beam is produced; it strikes the surface of the object and reflects back to a detector. The time is noted for this whole trip and thus the distance is calculated. For example:

If the time taken by a laser beam at different locations on the object (Fig. 21.1) is as shown in Table 21.1, then the distance from the detector to the object's surface at each point can be calculated by the equation (21.1), where c is the velocity of light. The results are calculated with time in nanoseconds and distance in meters.

$$z = (c * t)/2 \qquad\qquad (21.1)$$

Fig. 21.1 Time of flight scanners: laser beam projections on the surface of an object and the position of detectors

Table 21.1 Distance calculated according to time of flight

Detector (n)	Time (t)	Distance (z)
0	4.0	0.600
1	6.5	0.975
2	6.5	0.975
3	6.5	0.975
4	6.5	0.975
5	6.5	0.975
6	6.5	0.975
7	6.5	0.975
8	6.5	0.975
9	4.0	0.600

These scanners are very efficient, but the major problem is the measurement of time for such a fast ray, traveling at the speed of light. The more accurate the results we require, the more extensive is the timing system to be used.

Another problem with these scanners is that the laser beam should be reflected at a set angle to the detector from the surface of the object. Any deflection will cause either no measure or the wrong measure of time. Such scanners are good at modeling distant objects like buildings, etc.

21.3.2 Laser Triangulation Scanners

These scanners also use laser beams for probing, but the principle is different from that of time of flight scanners. In these scanners a laser light is projected onto the surface of an object and a camera is used to capture the position of the laser dot produced on the surface. Thus, the laser source, the laser dot on the surface of the object, and the lens of the camera form three vertices [5] of a triangle as shown in Fig. 21.2.

The triangle can be solved and the depth information can be retrieved by equation (21.2).

$$z = \frac{d \tan \theta_1}{1 + \tan \theta_1 / \tan \theta_2} \tag{21.2}$$

Here, the distance between the lens and the laser source (d), the angle of the laser beam (θ_1), and the angle of the reflected beam (θ_2) all are known.

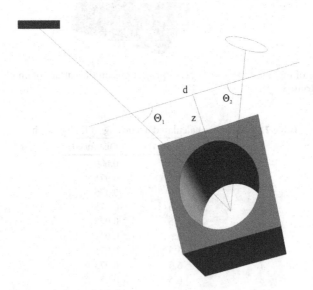

Fig. 21.2 Laser triangulation principle

The scanners operating on the laser triangulation principle are very accurate but extensive ones. The major limitation of such scanners is the distance between the scanner and the object; it should be only a few meters at the most.

These scanners suffer from a severe drawback in some cases when deep narrow holes are present in the object, because the projected and/or reflected beam may be distorted by the edges of the hole.

21.3.3 Structured Light

The basic principle behind the scanners using structured light [6] is that whenever structured light intersects an object, it produces a bright line on the surface of that object as shown in Fig. 21.3. This strip of bright light can be detected from a known angle through a camera, and the depth information may be retrieved. There are some scanners which use colored structured light [7, 8].

This technique is very good for retrieving height information but has limited use for depth retrieval in the case of holes and cavities. The strip of bright light generated by the structured light may break or become invisible when holes and cavities are encountered, as can be seen in Fig. 21.3 (B).

21.3.4 Stereoscopic Scanners

Stereoscopic scanners [9, 10] borrow a theme from the human vision system and use two cameras located at some distance. The pictures taken from both the cameras at different angles are then related to each other for the retrieval of

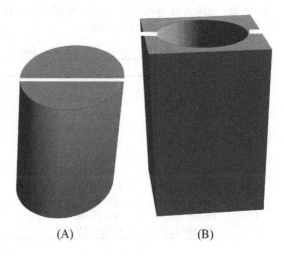

Fig. 21.3 Scanners using structured light: (**A**) *Bright strip* for heights, (**B**) *Bright strip* disappears in the case of holes

(A)   (B)

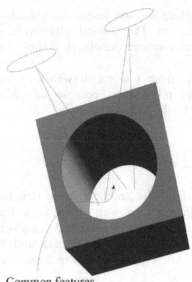

Common features

Fig. 21.4 Common features to be matched in stereoscopic vision

depth information. The correspondence of these images is based upon the matching of certain common features overlapping in both. This correspondence itself is a major problem for this kind of scanner; it requires heavy computation time and is not guaranteed, as far as accuracy is concerned.

A typical system using the stereoscopic technique is shown in Fig. 21.4.

21.3.5 Overview of Scanners

There are different types of commercially available scanners on the market. Table 21.2 is an overview of them.

Table 21.2 Some commercially available 3D scanners

Manufacturer	Make	Technology
3rd Tech	DeltaSphere	Time of flight
Callidus	Callidus CP 3200	Time of flight
Callidus	Callidus CT 900	Laser triangulation
Callidus	Callidus CT 180	Laser triangulation
Cyberware	Model 3030RGB/PS	Laser triangulation
Cyberware	Model 15	Laser triangulation
Brueckmann	Opto TOP-HE	Structured light
Brueckmann	FaceSCAN-III	Structured light
GOM	ATOS SO 4 M	Structured light
GOM	ATOS III	Structured light
Steinbichler	Comet IV 4 M	Structured light

21.4 CAD Modeling

There are some other computer-aided modeling techniques that are used to determine depth information from a view. These techniques can mainly be categorized as multiview and single view approaches. The multiview approach is based upon boundary representation or constructive solid geometry. Single view approaches are: the labeling approach, the gradient space approach, the linear programming approach, the perceptual approach, and the primitive identification approach. None of these single view approaches handle the depth information for holes, while multiview approaches may determine depth information of a hole in some cases. The major limitation of these techniques is that they require a wireframe model of an object represented in CAD drawings. Interested readers may refer to Wang and Grinstein [11] and Company et al. [12] for a detailed discussion of these techniques.

21.5 Our Approach—Perception Based Vision (PBV)

In our approach we have used the principle that humans have adopted: that is, retrieving depth information by perceiving. The main concept behind PBV is that the size of objects seems smaller as the distance increases. Thus when an object containing a hole is viewed from a camera, the near end of the hole is larger as compared to that of the far end. The deeper the hole, the smaller is the far end. This perception is employed in our technique, and we found promising results for cylindrical holes.

All we need is an orthographic view and symmetrical illumination to incorporate our technique. Figure 21.5 depicts our PBV execution model.

An image is input from a camera, a grayscale conversion is performed, and then different image preprocessing techniques are employed to enhance the image. We are not concentrating on these preprocessing techniques, because this is a very rich area and has evolved a lot already; instead, we are using already well developed techniques.

The next steps are binarization of the image, boundary extraction, and the geometric calculations, after which the depth information is output.

We experimented with our technique on different images, and a sample execution is shown in Fig. 21.6.

Once the boundary is extracted, geometric calculations are performed. We calibrated the camera at a set distance, and the difference between the circumference of the circles corresponding to the near edge and the far edge of a hole is used to calculate the depth of the hole. Equation (21.3) is the mathematical representation of this statement, where r_1 and r_2 are the radii of the far and near circles corresponding to the edges of the hole and d is the perpendicular distance between the camera and the object. The

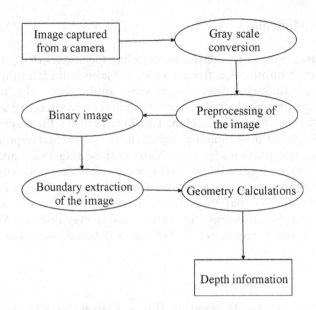

Fig. 21.5 PBV execution model

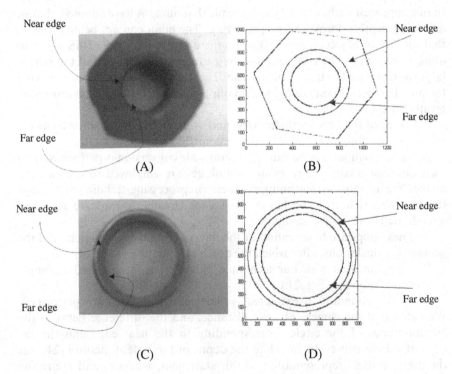

Fig. 21.6 PBV execution: (**A**) First sample image in grayscale, (**B**) Boundary extraction of (**A**), (**C**) Second image in grayscale, (**D**) Boundary extraction of (**C**)

depth of the hole is calculated as z, and f is the focal length of the lens of the camera.

$$z = \frac{(r_2 - r_1)d}{f} \tag{21.3}$$

PBV is a very low-cost yet efficient and accurate solution for determining the depth of a hole, which is of utmost importance in different applications like engineering, exploration, welling, and ergonomic designs.

21.6 Future Work

Our PBV approach can be extended to determine depth information for other types of holes, e.g., elliptical, conical, etc.

References

1. Scott WR, Roth G, Rivest JF (2003) View planning for automated three-dimensional object reconstruction and inspection. ACM Comput Surv 35(1):64–96
2. Bernardini F, Rushmeier H (2002) The 3D model acquisition pipeline. Comput Graph Forum 21(2):149–172
3. Carbone V, Carocci M, Savio E, Sansoni G, De-Chiffre L (2001) Combination of a vision system and a coordinate measuring machine for the reverse engineering of freeform surfaces. Int J Adv Manuf Technol 17:263–271
4. Blais F, Beraldin JA, El-Hakim S, Godin G (2003) New developments in 3D laser scanners: from static to dynamic multi-modal systems. In: 6th Conference on Optical 3-D Measurement Techniques
5. Xu Y, Xu C, Tian Y, Ma S, Luo M (1998) 3-D face image acquisition and reconstruction system. IEEE Instrumentation and Measurement Technology Conference
6. Dipanda A, Woo S (2004) Structured light system configuration determination for efficient 3D surface reconstruction. International Conference on Image Processing
7. Skydan OA, Lalor MJ, Burton DR (2002) Technique for phase measurement and surface reconstruction by use of colored structured light. Appl Opt 41(29):6104–6117
8. Li H, Straub R, Prautzsch H (2004) Fast subpixel accurate reconstruction using color structured light. In: Proceedings of the Fourth IASTED International Conference on Visualization, Imaging, and Image Processing
9. Gledhill D, Tian GY, Taylor D, Clarke D (2004) 3D reconstruction of a region of interest using structured light and stereo panoramic images. In: Proceedings of the Eighth International Conference on Information Visualisation
10. Fofi D, Salvi J, Mouaddib EM (2003) Uncalibrated reconstruction: an adaptation to structured light vision. Patt Recogn 36:1631–1644
11. Wang W, Grinstein GG (1993) A survey of 3D solid reconstruction from 2D projection line drawings. Comput Graph Forum 12(2):137–158
12. Company P, Piquer A, Contero M, Naya F (2005) A survey on geometrical reconstruction as a core technology to sketch-based modeling. Comput Graph 29:892–904

Chapter 22
Modeling Diversity in Recovery Computer Systems

George Popov and Valeri Mladenov

Abstract The goal of this paper is to present quantitative criteria for the measurement of diversity in recoverable computer systems. For this purpose a model of a diversity system with two failure types, detectable and undetectable, is presented, and a formula to calculate them is given.

22.1 Introduction

Diversity in multichannel systems (systems with dynamic program redundancy) is a powerful tool for failure detection and increasing safety. This is a method of solving a problem in two different ways (A and B) with identical input data, by virtue of which a criterion of the solution being perfect is the correspondence (in this particular case, identity) of the obtained output results [1]. The assumption is that there exist at least two ways of solving the problem.

For example, the input data (Fig. 22.1) are processed in two ways (A and B) and are compared in terms of their correspondence. When the system performs perfectly well, the comparison of the obtained results shows a positive output (OK). That is a condition for the normal work of the system which continues until there is a failure.

Error consequences are activated with certain input data and the flow of the algorithm. If, during one of the two ways of solving the problem (in one of the programs, e.g., processing, A), an error or defect is activated, there will not be a result, or the result will be incorrect at the respective exit. But because at the other channel the result is correct, the output results will not correspond, and the agreement (OK) is removed from the exit. The system passes on to a mode of detected failure. This event is visualized through the diagnostic information.

G. Popov (✉)
Department of Computer Science, Technical University–Sofia, 8 Kliment Ohridski St., Sofia 1000, Bulgaria
e-mail: popovg@tu-sofia.bg

N. Mastorakis et al. (eds.), *Proceedings of the European Computing Conference*, Lecture Notes in Electrical Engineering 27, DOI 10.1007/978-0-387-84814-3_22, © Springer Science+Business Media, LLC 2009

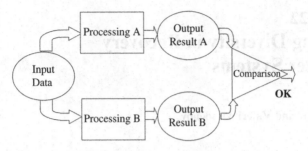

Fig. 22.1 A principle of diversity processing

An analogous result is obtained when a fault is activated on the other program. When errors or defects are activated on both programs, we get different output vectors because the causes and the processing channels are different. The probability of getting one and the same wrong result is quite negligible.

The difference between the output vectors under comparison is an indication that the processing is incorrect, and the work of the system is terminated. We search for the cause, and it is removed. In this way:

- we can identify errors and defects in an offline mode (in the testing period);
- we can terminate producing an incorrect guiding or control signal in the online mode of the real-time control systems and create a compulsion for the removal of the causes of failure.

The principle of diversity processing has long been known not only in the scientific literature, but also in the practice of the real-time systems used to control technological processes of great importance. In our country, such systems are used in rail transport.

The aim of this paper is to model diversity in such a way as to make it possible to determine the factors which influence the identification of failures. Then we will proceed by exploring the influence of the different factors and will suggest methods of enhancing the identifiably of failures.

22.2 Problem Formulation

Microcomputer systems are a basic element of modern technologies in such important areas as industry, military science, nuclear power engineering, transport, communications, medicine, etc. In these cases we speak about safety-critical control systems. The problems in the development of these systems arise from the tight schedules for their development and the requirements for their reliability [2, 3]. At the level of both hardware [1, 4] and software [5–13] surplus and diversity are the most common methods of enhancing reliability.

Avizienis and Laprie [10, 14] define a few aspects of reliability: readiness, good working order, safety, confidentiality, integrity, and the capability to be maintained and upgraded. A basic approach for enhancing reliability, according to them, is to include diversity in the architectural system.

Strunk [5] draws attention to the fact that the different variants of one and the same algorithm (program) must be developed independently, and a test must be run at different stages of the algorithm, so that the results can be compared. The benefit from such an approach is that the errors appear in the different versions at different times in the running of the programs. Avizienis [14], Horning and Sha [15, 16], and Rajkumar and Gagliardi [17] are of the same opinion regarding software diversity.

From what we have said so far it becomes clear that diversity is one of the main approaches of enhancing the reliability of embedded microprocessor systems, including alarm systems. In spite of their indisputable potential, there are scanty ideas in the literature about quantitative evaluation and the modeling of diversity [13, 17]; hence, few conclusions about their effectiveness (economic, technical) in their particular realizations. There is a difference between the two versions, yes—but how big is it, from what point of view, what metrics are used to measure it? And how does this difference, identified using the adopted metrics, affect the ability to discern (to detect) the causes of failure?

An attempt at solving this scientific problem has been made in [1, 4, 16]. It comes down to the following.

Two types of failures reflecting these facts have been introduced:

- α *failures:* detectable by comparing the output results;
- η *failures:* undetectable through comparison because they bring about the same mistakes in the compared results.

The division of the failures into these two classes is based on the presumption that despite all attempts to make it complete, diversity in practice is not absolute. An absolute diversity would mean that the failures in the two versions are absolutely independent (uncorrelated) and that there are no common causes for failure along the two paths which will lead to the same wrong result. In practice these causes are present in the common components of the systems: when entering information from a common source, in the only comparator for comparing the results, in synchronizing the work along the two channels, in the common power supply, and others—i.e., where incorrectness after failure or error is introduced in one and the same way in the two channels.

On this basis in [1, 4] a measure of diversity is introduced:

$$\Omega = \frac{\lambda_\alpha}{\lambda_\alpha + \lambda_\eta} \tag{22.1}$$

where λ_α, λ_η is the intensity of the two types of failures. If all failures are due to the same cause—for example, if the two programming versions A and B are the same and the errors in the two copies of the single program are duplicated—then the results will be wrong, but corresponding, and the detectability at comparison Ω is brought to zero. The deeper the diversity is, the closer Ω is to one, and the bigger the detectability of the errors and defects. Undetected, although detectable, will remain only those failures which by accident cause one and the same output result (vector- i) from the two processes.

There is a piece of research [4] concerning the assessment of this probability, according to which after failure you can get all possible vectors, including the anticipated functional vector. The probability $q(i)$ of occurrence of any i-vector ($i = \overline{1 - 2^w}$) or some sequence of vectors depends on the nature of the failure (errors, defects, accidental disturbances), the input data, and the algorithm of work. Because in each particular case, the distribution of probabilities is accidental and difficult to predict, we can accept the equally probable averaging out:

$$q(i) = 2^{-w} \tag{22.2}$$

which also refers to the probability, as a result of failure, of unpredictable occurrence of the functional (correct) vector. With such averaging out, the probability of wrong and, with the worst-case approach, potentially dangerous, failure will be:

$$Q_d(t) = q_d Q(t) = \frac{2^w - 1}{2^w} Q(t) \tag{22.3}$$

where w is the bit of the vectors under comparison which bear the results of the two processings, and $Q(t)$—the probability of any failure in the system.

These scientific studies and their results lie at the base of the new model of diversity, which the present paper claims to develop.

22.3 Modeling Diversity

In the context of probability logic we can assume that there will be an inability to identify the failure in two cases:

1. if α_a and α_b accidentally cause one and the same wrong results;
2. if a η failure has happened.

We introduce the Boolean function F_{ni} (non-identification), which is given in equation (22.4) and illustrated in Fig. 22.2.

The Boolean function of the inability to identify failures has the form:

$$F_{ni} = z^1_{\alpha_a} z^1_{\alpha_b} \vee z^1_{\eta} \tag{22.4}$$

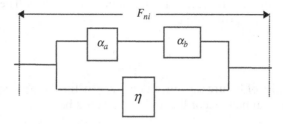

Fig. 22.2 Boolean function F_{ni} (non-identification)

where z_i^0 is the logic variable of the statement that «something» has not happened, and z_i^1 - that the «something» has happened.

In order to model, in terms of probability, the effect of diversity on the ability to identify failures, a logic-probability transition has to be carried out [4]:

$$F_{ni} = z_{\alpha_a}^1 z_{\alpha_b}^1 \vee z_\eta^1 = \overline{z_{\alpha_a}^1 \, z_{\alpha_b}^1 \, z_\eta^0} \qquad (22.5)$$

After applying the theorem of De Morgan we have arrived at a nonrecurrent Boolean function in a basic "conjunction-negation." When we make replacements in this function using the rules of the logic-probability transitions, we arrive at the probability that the failures will not be identified:

$$K_{ni} = 1 - \{1 - [K_{\alpha_a} K_{\alpha_b}]\}[1 - K_\eta] \qquad (22.6)$$

where:

- K_{ni} is the coefficient for non-identification;
- $K_{\alpha a}$ and $K_{\alpha b}$ are coefficients for unavailability for recognition of recognizable failures in both channels, which will result in accidentally equal but wrong output signals in the two processes;
- K_η is the coefficient for unavailability for recognition of failure because of a common cause, which generates as unidentifiable through comparison of output results from both channels.

If we assume that the two channels are equivalent,

$$K_{\alpha a} = K_{\alpha b} = K_{\alpha k} \qquad (22.7)$$

the formula is simplified as:

$$K_{ni} = 1 - \{1 - [K_{\alpha k}]^2\}[1 - K_\eta] \qquad (22.8)$$

As can be seen from (22.2), the probability for an output functional vector accidentally to appear after failure is $q_i = 2^{-w}$. And the probability for this vector to be wrong, but one and the same in both channels A and B, will be the result of multiplying these two probabilities:

$$q_{i2} = (2^{-w})^2 = 2^{-2w} \tag{22.9}$$

The probability of failure causing a previously defined output signal in both equally reliable channels, according to (22.2), would be:

$$K_{\alpha k} = \frac{K}{2^w} \tag{22.10}$$

When we replace (22.10) in (22.8) we get:

$$K_{ni} = 1 - \left\{ 1 - \left[\frac{K_\alpha}{2^w} \right]^2 \right\} [1 - K_\eta] \tag{22.11}$$

where:
- K_{ni} is the coefficient of getting an unidentified failure;
- K_α is the coefficient of getting an identifiable α-failure in one of the two channels of processing;
- K_η is the coefficient of getting an unidentifiable η-failure.

From [1] we know that $K_{a\alpha}$ (unavailability for recognition of recognizable failure) is:

$$K_{a\alpha} = \frac{\lambda_\alpha}{\lambda_\alpha + \mu} \tag{22.12}$$

where λ_α is intensity for recognizable failures and μ is intensity of recovery.
$K_{a\eta}$ (unavailability for recognition) is:

$$K_{a\eta} = \frac{\lambda_\eta}{\lambda_\eta + \mu} \tag{22.13}$$

After long system exploitation coefficient for non-identification K_{ni} has view (22.15):

$$K_{ni} = 1 - \left\{ 1 - \left[\frac{\lambda_\alpha}{2^w (\lambda_\alpha + \mu)} \right]^{2(2^w - 1)} \right\} \frac{\mu}{\lambda_\eta + \mu} \tag{22.14}$$

If we go to presentation of intensities λ_α and λ_η through the depth of diversity Ω and the total intensity of failures λ, as has been done for nonrecoverable systems, we will obtain:

$$K_{ni} = 1 - \left\{ 1 - \left[\frac{\Omega\lambda}{2^w(\Omega\lambda + \mu)} \right]^{2(2^w-1)} \right\} \frac{\mu}{(1 - \Omega)\lambda + \mu} \qquad (22.15)$$

It is clear that, if $\Omega = 1$, i.e., in the case of maximal possible diversity, the coefficient of nonrecognition of failure is minimal:

$$K_{ni_{min}} = \left[\frac{\lambda}{2^w(\lambda + \mu)} \right]^{2(2^w-1)} = \left[\frac{1}{2^w K_\alpha} \right]^{2(2^w-1)} \qquad (22.16)$$

and if we have no diversity, $\Omega = 0$.

$$K_{ni_{max}} = \frac{\lambda}{\lambda + \mu} = 1 - K_\alpha = K_{na} \qquad (22.17)$$

i.e., we become equal to the coefficient of unavailability. Hence, the probability of wrong identification as a result of failures increases up to the level for a single-channel system without protection.

To evaluate the influence of diversity on the coefficient K_{ni} for wrong identification of failure, we can take into account the relationship ξ_B for the marginal values in the case of zero diversity (22.17) and in the case of full diversity (22.16) (effectiveness of diversity):

$$\xi_B = \frac{K_{nimax}}{K_{nimin}} = \frac{\frac{\lambda}{\lambda + \mu}}{\left[\frac{\lambda}{2^w(\lambda+\mu)} \right]^{2(2^w-1)}} \qquad (22.18)$$

and after a transformation we have:

$$\xi_B = \frac{2^{2w(2^w-1)}}{(1 - K_\alpha)^{2^{w+1}-3}} \qquad (22.19)$$

In the upper formula, the intensity of failure and the intensity of recovery are not independent. They define the coefficient of availability: the main characteristic of recoverable systems through which the effect of diversity has been investigated.

The relation (22.19) is presented graphically in Fig. 22.3. It is clear that by increasing the coefficient of availability, the effectiveness of diversity increases rapidly and strongly depends of the length w of the vectors under comparison. The has the lowest value, as could be expected, when $w = 1$.

Obviously, the effectiveness of diversity in recoverable systems depends on the availability K_α. As this coefficient becomes higher, the diversity is more effective. The coefficient has its lowest value for diversity-based systems, where both channels have one binary output ($w = 1$):

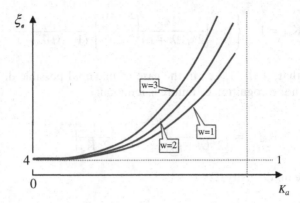

Fig. 22.3 Investigation of diversity effectiveness ξ_B depending on bit number w

$$\xi_B = \frac{2^{2w(2^w-1)}}{(1-K_\alpha)^{2^{w+1}-3}} = \frac{4}{1-K_\alpha} \tag{22.20}$$

When the availability is too low, as it is also for the nonrecoverable systems, the effectiveness is low, too, and is close to 4.

Compared to nonrecoverable systems [4], the effectiveness of diversity for recoverable systems becomes many times higher, as the availability coefficient is ever close to 1. For example, $K'_\alpha = 0{,}999$ and $\xi_B = 4000$.

Obviously, the coefficient of unidentifiable (dangerous) failure K_{ni} in diversity systems depends on:

- the depth of diversity Ω;
- the number of bits w, with which the output result from the processing in the diversity channels is represented;
- the total intensity of failures in the system λ;
- the total intensity of recovery in the system μ.

22.4 A Study of the Effect of Diversity on the Identifiability of Failures in the Systems

In formula (22.15) some calculations have been made for a fixed $\lambda = \mu = 0{,}00001$ and $w=1$ or $w=16$. In Fig. 22.4 the obtained results have been interpreted graphically.

We can see that with the increase in the depth Ω of the diversity, the coefficient of not identifying the failure decreases significantly the longer the input vector w, and the bigger the intensity of the failures of the microcomputer system. With $\Omega = 1$, which is practically unattainable, we can achieve a hundred times greater identifiability.

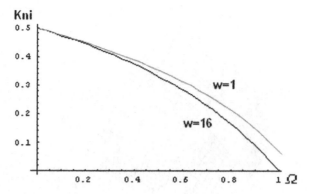

Fig. 22.4 Investigation of $K_{ni} = f(\Omega)$

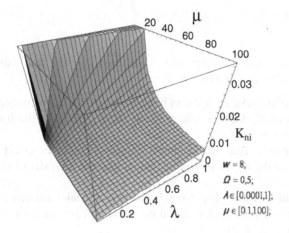

Fig. 22.5 Investigation for $K_{ni} = f(\lambda, \mu)$

In Fig. 22.5 shows dependency $K_{ni} = f(\lambda, \mu)$ at $\Omega = 0,5$ and $w = 8$.

In Fig. 22.6 shows dependency $K_{ni} = f(\Omega, \lambda)$. Once again it is seen that K_{ni} is dependent on the depth of diversity Ω.

22.5 Conclusion

This paper states and solves the issue of modeling diversity in computer systems. Adopting as a metric the depth of diversity, variable Ω as it is known in the literature, it has been proven that it increases, on the one hand, with the independence of the information processing channels of the hardware and the software and, on the other hand, with the decrease of the common diversity

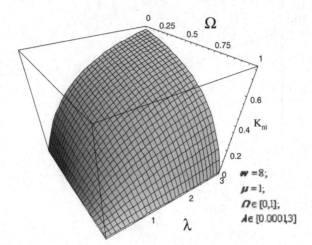

Fig. 22.6 Investigation for $K_{ni} = f(\Omega, \lambda)$

systems programming and apparatus components. On that basis we have obtained the following new results:

1. We have found a new correlation between the effect of diversity depth on the ability to identify failures which result from defects in the hardware and errors in the software.
2. We have done a study of the quantitative values of the effect of diversity depending on the bits of the input vectors and the intensity of the failures of the system.
3. We have put forward ideas for schematic/technical solutions of redundant systems with increased identifiability evolving from the conclusions drawn from the present study.

References

1. Hristov H, Trifonov V (2005) Reliability and safety for communications. Novi Znania, Sofia
2. Karakehayov Z, Kristensen KS, Winther O (1995) Embedded systems. Technical University of Denmark, Department of Applied Electronics
3. Isaksen U, Bowen JP, Nissanke N (1996) System and software safety in critical systems. The University of Reading, Department of Computer Science Whiteknights, PO Box 225, Reading, Berks RG6 6AY, UK
4. Popov G (2005) Modeling diversity as a method of detecting failures in nonrecovery computer systems. Information Technologies and Control, No 2
5. Strunk E (2003) Survivability in embedded systems. Ph.D. Dissertation
6. Törngren M, Torin J (1998) Conceptual design of dependable embedded control systems.
7. Burns A, Wellings AJ (1994) HRT-HOOD: A structured design method for hard real-time systems. J Real-Time Syst 6(1)

8. Rivera J, Danylyszyn A, Winstock CB, Sha L, Gagliardi MJ (1996) An architectural description of the simplex architecture. Technical report CMU/SEI-96-TR-006 ESC-TR-96–006, Carnegie Mellon University, Software Engineering Institute
9. Törngren M, Wikander J (1996) A decentralization methodology for real-time control applications. Control Eng Pract 4(2):219–228
10. Avizienis A (1985), The N-version approach to fault tolerant software. IEEE T Software Eng 11(12):1491–1501
11. Robyn R Lutz (2000) Software engineering for safety: a roadmap. The future of software engineering. ACM Press, New York
12. Isaksen U, Bowen JP, Nissanke N (1996) System and software safety in critical systems. The University of Reading, Department of Computer Science Whiteknights, PO Box 225, Reading, Berks RG6 6AY, UK
13. Leveson NG (1986) Software safety: Why, what, and how. Comput Surv 18(2):125–163
14. Avizienis A, Laprie J, Randel B (2001) Fundamental concepts of computer system dependability. IARP/IEEE-RAS Workshop on robot dependability: technological challenge of dependable robots in human environments. Seoul, Korea
15. Horning JJ, Lauer HC, Melliar-Smith PM, Randell B (1974) A program structure for error detection and recovery. Symposium on Operating Systems, pp 171–187
16. Sha L (2001) Using simplicity to control complexity. IEEE Soft 18(4):20–28
17. Sha L, Rajkumar R, Gagliardi M (1995) A software architecture for dependable and evolvable industrial computing systems. Technical Report CMU/SEI-95-TR-005, Carnegie Mellon University
18. Knight JC, Strunk EA, Sullivan KJ (2003) Towards a rigorous definition of information system survivability. DISCEX 2003, Washington, DC
19. Butler RW, Finelli GB (1991) The infeasibility of experimental quantification of life-critical software reliability. ACM SIGSOFT '91 Conference on Software for Critical Systems, New Orleans, LA
20. Sandoval M (2004) "Smart" sensors for civil infrastructure systems. A Dissertation Submitted to the Graduate School of the University of Notre Dame
21. Wilikens M, Masera M, Vallero D (1997) Integration of safety requirements in the initial phases of the project lifecycle of hardware/software systems. In: Proceedings of SAFE-COMP97, Springer-Verlag, ISBN 3–540–76191–8
22. Redell O (1998) Modelling of distributed real-time control systems: an appraoch for design and early analysis. Licentiate thesis, Department of Machine Design, Royal Institute of Technology, Stockholm
23. Littlewood B, Popov P, Strigini L (2001) Modeling software design diversity. ACM Comput Surv (CSUR) 33(2):177–208
24. Forrest S, Somayaji A, Ackley D (1997) Building diverse computer systems. IEEE Computer Society Press, Los Alamitos, CA, pp 67–72

Chapter 23
Cell Automata Models for a Reaction-Diffusion System

Costin-Radu Boldea and Afrodita Balasa Boldea

Abstract This paper presents two ultra-discrete versions (discrete time, space, and values) of an important reaction-diffusion system arising from the bio-mathematical domain, the so-called predator-prey interaction system.

23.1 Introduction

Cellular automata (CA) have been widely adopted in the sciences as simple but powerful models of the real world, because the complex patterns produced by their long-time behaviours can mimic observations with tremendous accuracy ([10]). However, the lack of mathematical tools makes prediction difficult in CA models. That there are integrable, predictable CAs, possessing solitons, was confirmed by the work of Tokihiro et al. [9]. The path they took was through ultra-discrete equations.

The first integrable ultra-discrete equations called *soliton cellular automata* were obtained by Takahashi et al. (see [8]). A method to ultra-discretize integrable discrete systems was developed in [9], followed by the study of different ultra-discrete versions of known integrable equations, including the Painlevé equations [7]. One open problem was the lack of an algorithmic method for finding new ultra-discrete equations possessing nonlinear properties such as soliton-like phenomena or pattern formation.

The reaction-diffusion modeling and simulations, particularly in the sense of chemical computations or in the domain of biophysics, have become a hot topic of computer science, physics, and chemistry. Our paper presents a direct ultra-discretisation procedure applied to a simple reaction-diffusion equation, the naïve predator-prey system. The method employed can be easily

C.-R. Boldea (✉)
Department of Computer Science, West University of Timisoara, 4,Vasile Parvan, Timisoara 1900, Romania
e-mail: cboldea@info.uvt.ro

N. Mastorakis et al. (eds.), *Proceedings of the European Computing Conference*, Lecture Notes in Electrical Engineering 27, DOI 10.1007/978-0-387-84814-3_23, © Springer Science+Business Media, LLC 2009

applied to some other reaction-diffusion systems, such as the Brusselator equation or the reaction-diffusion special system (RDS), and permits the generation of CA models simulating some nonlinear reaction diffusion phenomena.

23.2 The Reaction-Diffusion Systems

Many pattern formations and wave propagation phenomena that appear in nature can be described by systems of coupled nonlinear differential equations, generally known as reaction-diffusion equations [5].

The general form of RD systems is:

$$
\begin{cases}
\dfrac{\partial u}{\partial t} = k_1 \nabla^2 u + f(u, v) \\
\dfrac{\partial v}{\partial t} = k_2 \nabla^2 v + g(u, v)
\end{cases}
\tag{23.1}
$$

where k_i, $i = 0, 1$ are diffusion constants; x, y are the spatial coordinates; and u, v are functions of x, y, and t.

The equations of type (23.1) are the most popular models of reaction-diffusion. It is practically impossible to enumerate all the applications of such equations. We restrict ourselves to a few examples only:

- The primitive predator-prey system (PPS), as defined in [6], can be described by an extended second order PDE version of the Lotka-Voltera equation:

$$
\begin{cases}
\dfrac{\partial u}{\partial t} = \nabla^2 u - uv \\
\dfrac{\partial v}{\partial t} = \nabla^2 v + uv
\end{cases}
\tag{23.2}
$$

- The Brusselator model (as presented in [4]), where the functions u and v depend only on one space variable:

$$
\begin{cases}
\dfrac{\partial u}{\partial t} = k_1 \dfrac{\partial^2}{\partial x^2} u + A - (B + 1)u + u^2 v \\
\dfrac{\partial v}{\partial t} = k_2 \dfrac{\partial^2}{\partial x^2} v + Bv - u^2 v
\end{cases}
\tag{23.3}
$$

- The Jackiw-Teitelboim model of two-dimensional gravity with non-relativistic gauge [5], which appears as a particular case of the reaction-diffusion special system (the RDS system is part of a large hierarchy of completely integrable systems [1]):

$$\begin{cases} \dfrac{\partial u}{\partial t} = \dfrac{\partial^2}{\partial x^2}u + 2ku - 2u^2v \\ \dfrac{\partial v}{\partial t} = -\dfrac{\partial^2}{\partial x^2}v + 2kv + 2v^2u \end{cases} \tag{23.4}$$

23.3 Discrete and Ultra-Discrete Reaction Diffusion Systems Derived from the PPS Equation

In this paper we are interested in some ultra-discrete versions of the first of the equations enumerated above. The technique used can be easily applied to the two other models, or any other equation of the type (23.1).

23.3.1 Discretization Procedure

The first step in order to obtain a discrete valued, discrete time, discrete space variables system (an ultra-discrete system) from eq. (23.2), considered here only in $1+1$ dimensions, is to bypass a classical discrete version of this equation.

The discrete versions of the above RD systems are obtained by replacing the time derivative:

$$\frac{\partial u(x,t)}{\partial t} \rightarrow \frac{u(x,t+\Delta t) - u(x,t)}{\Delta t} \rightarrow u(x,t+1) - u(x,t) \tag{23.5}$$

and the space derivatives with:

$$\frac{\partial u(x,t)}{\partial x} \rightarrow u(x+1,t) - u(x,t) \tag{23.6}$$

$$\frac{\partial^2 u(x,t)}{\partial x^2} \rightarrow u(x+1,t) - 2u(x,t) + u(x-1,t) \tag{23.7}$$

By plugging these discretisations into eq. (23.2), one obtains

$$\begin{cases} u(x,t+1) = [u(x+1,t) + u(x-1,t)] - u(x,t)[1 + v(x,t)] \\ v(x,t+1) = [v(x+1,t) + v(x-1,t)] + v(x,t)[u(x,t) - 3] \end{cases} \tag{23.8}$$

respectively. Note that eq. (23.8) is similar to the quasilinear schemes (23.10) from [2].

23.3.2 Ultra-Discretization Procedures

Given a rational function in u and v, the ultra-discretisation method requires that we introduce new variables U and V defined by $u = \exp(U/\varepsilon)$, $v = \exp(V/\varepsilon)$. Then we take the limit $\varepsilon \cdot 0^+$ of the equations using the identities:

$$\lim_{\varepsilon \to 0+} \varepsilon \log\left[e^{A/\varepsilon} + e^{B/\varepsilon}\right] = \max(A,B), \lim_{\varepsilon \to 0+} \varepsilon \log\left[e^{A/\varepsilon} - e^{B/\varepsilon}\right]$$

$$= Alt \max(A,B)$$

$$(23.9)$$

where

$$Alt \max(A,B) = \begin{cases} A, & if\ A > B \\ 0, & if\ A = B \\ -B, & if\ A < B \end{cases} \qquad (23.10)$$

is the alternate maximum function. Ultra-discrete equations are naturally posed on the max-plus semi-ring (defined in [3]). The direct ultra-discretisation procedure transforms eq. (23.8) into:

$$\begin{cases} \max(u_x^{t+1}, u_x^t + \max(v_x^t, 0)) = \max[u_{x+1}^t, u_{x-1}^t] \\ \max(v_x^{t+1}, v_x^t + 1) = \max[v_{x+1}^t, v_{x-1}^t, v_x^t + u_x^t] \end{cases} \qquad (23.11)$$

modulo a scaling factor. (We used the operations $uv \cdot U + V$, $kv \cdot V + c$, and $u + v \cdot max(U,V)$ in the ultra-discrete limit, with $c = round(\ln(k))$ for any positive constant k.)

In order to obtain a cellular automaton model, the system (23.11) must be rewritten as:

$$\begin{cases} u_x^{t+1} = Alt \max\{\max(u_{x+1}^t, u_{x-1}^t), u_x^t + \max(v_x^t, 0)\} \\ v_x^{t+1} = Alt \max\{\max(v_{x+1}^t, v_{x-1}^t, v_x^t + u_x^t), v_x^t + 1\} \end{cases} \qquad (23.12)$$

The system (23.12) can be used to define a nonstandard cellular bivaluated automaton, granting to each cell a pair of values $C_n^t = (u_n^t, v_n^t)$, where $n, t \in$ IN, and u_n^t, v_n^t are entire functions.

Note that the model described by (23.12) does not characterize a predator-prey evolution. A supplementary reset condition must be introduced in order to simulate the physical phenomena, e.g., $u_n^{t+1} \leftarrow 0$, if the calculated value (from eq. 23.12) is negative, corresponding to a negative population.

An alternative version proposes to separate the diffusion part of the eq. (23.1) and the reaction part, and to apply the ultra-discretisation method to each one, the effects being cumulated at the end. The diffusion part generates a simple cell automaton:

$$\begin{cases} \dfrac{\partial u}{\partial t} = \dfrac{\partial^2}{\partial x^2} u \\ \dfrac{\partial v}{\partial t} = \dfrac{\partial^2}{\partial x^2} v \end{cases} \rightarrow \begin{cases} u_x^{t+1} = Alt \max\{\max(u_{x+1}^t, u_{x-1}^t), u_x^t\} \\ v_x^{t+1} = Alt \max\{\max(v_{x+1}^t, v_{x-1}^t,), v_x^t\} \end{cases} \quad (23.13)$$

The reaction part becomes $uv \rightarrow (u_x^t + v_x^t)$, so the ultra-discrete system will be:

$$\begin{cases} u_x^{t+1} = Alt \max\{\max(u_{x+1}^t, u_{x-1}^t), u_x^t\} - (u_x^t + v_x^t) \\ v_x^{t+1} = Alt \max\{\max(v_{x+1}^t, v_{x-1}^t), v_x^t\} + (u_x^t + v_x^t) \end{cases} \quad (23.14)$$

This model presents a very fast incrementation of the numerical values, so a reset condition $u_n^{t+1} \leftarrow 0$ applies if the calculated value (from (23.14)) is greater in absolute value than a fixed maximal limit M.

23.4 Numerical Simulations and Final Remarks

The more than 400 numerical experiments realized by the authors permit us to remark that the CA model described by eq. (23.12) is always superior, limited by a maximum value, and becomes periodical after a certain time. The effect of the reset condition is to reduce the period of the v-values and to annihilate the u-values after a limited time. The first CA model described above clearly produces nonlinear phenomena: a soliton-like interaction can be observed by choosing the initial condition $u_n^0 = v_n^0 = 0$, with the exception $u_2^0 = u_3^0 = u_7^0 = v_3^0 = v_4^0 = 2$, $v_7^0 = 1$. The values of u resulting from simulation are fragmentarily presented below:

```
t=3   2  -6   2 -3 -3  2 -3   2 -3  2 -1  ...
t=4  -6   2  -6   2  2 -3   2 -3   2 -3  2  ...
t=5   2 -3   2  -6 -3   2 -3  2  -3   2 -3  ...
t=6  -3 -3 -3   2   2 -3  2 -3   2  -3  2  ...
t=7  -3  2 -3 -3  -8   2 -3  2 -3  2  -3  ...
t=8   2  -5   2 -3  2  -8   2 -3  2 -3   2  ...
t=9  -5   2 -3  2 -3  2  -8  2  -3  2 -3
```

One observes clearly the phase shift between $t = 5$ and $t = 7$. We strongly expect that the system (23.12) is integrable.

The second CA model, considered with the reset limitation condition, is characterized by local pattern formation and strong chaotic behaviors. A large number of nonlinear interactions were empirically observed, and their classification will become the object of a future paper.

References

1. Boldea CR A hierarchy for the completely integrable RDS system. Tensor N S 66(3):260–267
2. Garvie MR (2006) Finite-difference schemes for reaction-diffusion equations modeling predator-prey interactions in MATLAB. Bull Math Biol 69(3):931–956
3. Joshi N, Ormerod C (2007) The general theory of linear difference equations over the max-plus semi-ring. Stud Appl Math 118(1):85–97(13)
4. Kang H, Pesin Y (2005) Dynamics of a discrete Brusselator model: escape to infinity and Julia set. Milan J Math 73:1–17
5. Martina I, Pashaev OK, Soliani G (1998) Phys Rev D 58 084025
6. Murray JD (1989) Mathematical biology. Springer-Verlag, New York
7. Ramani A, Takahashi D, Grammaticos B, Ohta Y (1998) The ultimate discretisation of the Painlevé equations. Physica D 114:185–196
8. Takahashi D, Satsuma J (1990) A soliton cellular automaton. J Phys Soc Japan 59:3514–3519
9. Tokihiro T, Takahashi D, Matsukidaira J, Satsuma J (1996) From soliton equations to integrable cellular automata through a limiting procedure. Phys Rev Lett 76:3247–3250
10. Wolfram S (2003) New constructions in cellular automata. In: Proceedings of the Conference, Santa Fe, NM, 1998. Oxford University Press, New York

Chapter 24
3D Reconstruction of Solid Models from 2D Camera Perspectives

Muhammad Abuzar Fahiem

Abstract Engineering objects are represented in the form of manual and computerized line drawings in the production and manufacturing industry. Digitization of manual drawings in some computerized vector format is one of the very crucial areas in the domain of geometric modeling and imaging. There are conventionally three views of an object; front, top, and side. Registration of these 2D views to form a 3D view is also a very important and critical research area. In this paper we have developed a technique for the reconstruction of 3D models of engineering objects from their 2D views. Our technique is versatile in the sense that 3D representation is in both the drawing exchange file (DXF) format recognized by various CAD tools as well as the scalable vector graphics (SVG) format recognized by various web-based tools. We have also described different comparison metrics and have compared our technique with existing techniques.

24.1 Introduction

Three-dimensional objects always suffer from a loss of depth information whenever captured from camera perspectives. Retrieval of this lost information has always been of great concern and is a problem faced by different researchers. There are many situations in which 2D projections are to be registered to form the 3D shape of an object. The problem gets intense when the views are angle independent; however, it is a bit simplified with orthographic projections, which is the case for most drawings of engineering objects. Traditionally, engineering objects are represented through three orthographic views: front, top, and side views. Most of the CAD tools provide this facility, but it is a

M.A. Fahiem (✉)
University of Engineering and Technology, Lahore, Pakistan; Lahore College for Women University, Lahore, Pakistan
e-mail: abuzar@uet.edu.pk

N. Mastorakis et al. (eds.), *Proceedings of the European Computing Conference*, Lecture Notes in Electrical Engineering 27, DOI 10.1007/978-0-387-84814-3_24, © Springer Science+Business Media, LLC 2009

challenging task to transform manual drawings into CAD or any other vector representation. The task gets even more complicated when 3D CAD drawings are to be generated from these 2D orthographic projections.

A lot of work has been done to solve this problem. In this paper we have developed a technique to reconstruct 3D models from these 2D perspectives. Our technique is versatile, because the 3D representation is also in scalable vector graphics (SVG) format [1], which is a new vector format for information exchange, especially over the Internet and mobile devices [2].

Different 3D reconstruction techniques mainly involve two kinds of approaches: single and multiple view approaches. Another type of classification is based upon bottom-up and top-down approaches. The bottom-up approach is also known as wireframe or boundary representation (B-Rep), while the top-down approach is also known as volume-based or constructive solid geometry (CSG). We will use these terms intermittently in the following text. The very first studies were done by Idesaws [3] and Aldefeld [4] for bottom-up and top-down approaches respectively. Different reviews and surveys were compiled on the research regarding these techniques by Wang et al. [5] in 1993, Company et al. [6] in 2004, and Fahiem et al. [7] in 2007.

We have discussed our technique in Section 24.2, followed by the description of different comparison metrics in Section 24.3, on the basis of which we have compared our technique with other ones. Section 24.4 is dedicated to the critical comparison of existing techniques. The comparison is summarized in Table 24.1 for multiview approaches and in Table 24.2 for single view approaches. The discussion is concluded with a summary in Section 24.5.

Table 24.1 Comparison of multiview techniques for 3D reconstruction

Parameters		References					
		[8]	[9]	[10]	[14]	[15]	Our approach
Technique	Bottom-up	x				x	x
	Top-down		x	x	x		
Drawing	Perfect	x	x	x	x	x	x
	Imperfect						
Object	Straight	x	x	x	x	x	x
	Curve	x	x	x	x	x	x
X-sectional view	Yes	x	x	x			
	No				x	x	x
Vectorization	DXF	x	x	x	x	x	x
	SVG						x
User interaction	Yes			x			
	No	x	x		x	x	x
Dead-end hole	Handled	x	x	x			
	Not Handled				x	x	x
Hidden line	Yes	x	x	x			
	No				x	x	x

Table 24.2 Comparison of single view techniques for 3D reconstruction

Parameters		References		
		[16]	[17]	[18]
Technique	Bottom-up	x		
	Top-down		x	x
Drawing	Perfect	x	x	x
	Imperfect			
Object	Straight	x	x	x
	Curve	x		x
X-sectional view	Yes			
	No	x	x	x
Vectorization	DXF	x	x	x
	SVG			
User interaction	Yes			x
	No	x	x	
Dead-end hole	Handled			
	Not handled	x	x	x
Hidden line	Yes			
	No	x	x	x

24.2 Our Approach

In all the bottom-up approaches, the following four steps [3] are executed:

1. Conversion of 2D to 3D vertices.
2. Extraction of 3D lines from 3D vertices.
3. Creation of 3D faces from 3D lines.
4. Production of 3D objects from 3D faces.

We have proposed a versatile technique based upon translation and rotation to eliminate the first three steps. In our technique we took advantage of camera perspectives to form 3D objects from 3D faces. In camera perspectives we already have 3D faces, so the problem is the registration of these faces into a 3D model. The input to the system is 2D orthographic projections from camera perspectives, and the output is 3D representations as files in DXF or SVG format.

The three conventional views for an engineering drawing (front, top, and side) are registered into a 3D hypothetical cuboid of dimensions such as will accommodate all three views. The three views are normalized and the rotations and/or translations are performed in the following manner:

For the front view: translation only, to adjust the view on the front face of the hypothetical cuboid.

For the top view: rotation of $-90°$ about the x-axis; translation to adjust the view on the top face of the hypothetical cuboid.

For the side view: rotation of $-90°$ about the y-axis; translation to adjust the view on the side face of the hypothetical cuboid.

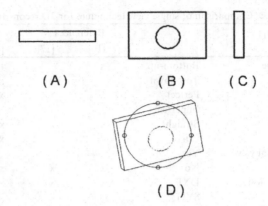

Fig. 24.1 Views: (**A**) top view (**B**) front view (**C**) side view (**D**) 3D view

The matrices for the rotation of the top and side views are:

$$\begin{bmatrix} 1 & 0 & 0 & 0 \\ 0 & 0 & -1 & 0 \\ 0 & 1 & 0 & 0 \\ 0 & 0 & 0 & 1 \end{bmatrix} \quad and \quad \begin{bmatrix} 0 & 0 & 1 & 0 \\ 0 & 1 & 0 & 0 \\ -1 & 0 & 0 & 0 \\ 0 & 0 & 0 & 1 \end{bmatrix}$$

respectively.

To normalize the views, a hypothetical rectangle is drawn around each view by dropping tangents on all four sides of the view. A sample interaction is shown in Fig. 24.1.

24.3 Comparison Metrics

The success of an approach can be determined on the basis of different metrics. The metrics on the basis of which our comparison is performed are as follows:

24.3.1 Number of Views

First of all, we categorize different approaches on the basis of the number of views used by a technique.

Multiview approaches mainly need three views: front, top, and side. However, for complete reconstruction of a solid model, these views may not be adequate. For example, the front and rear, the left and right, and the top and bottom views of an object may not be the same. In that case, six orthographic views may be needed, especially if hidden lines are not supplied in the three conventional views.

There is another aspect of dealing with multiple views to reconstruct a 3D model: there are situations when different geometric objects, like cylinders, spheres, circles, etc., may have the same orthographic projections. Yet another aspect is determining whether the line drawing of an object corresponds to its height or its depth. For example, a hole and a circular protrusion may have the same line drawing from one particular view. Such information is retrieved from the other views of that object when performing 3D reconstruction.

Single view approaches are truly speaking not reconstruction approaches; instead, these deal with the correspondence and meaning of a line in the drawing. These approaches need only a single projection to interpret the 3D modeling of an object.

Multiview approaches have evolved through the past four decades and are getting more and more practical.

24.3.2 Technique

Another categorization of an approach may be on the basis of the technique used, i.e., bottom-up or top-down. A bottom-up approach corresponds to the modeling of an object through 2D entities, while a top-down approach corresponds to the modeling of an object through 3D entities. For example, the object in Fig. 24.2 will be represented by two circles (with a center point and the diameter) and two lines (with starting and ending coordinates) in the bottom-up approach. On the other hand, it will be represented as a cylinder (with center point, diameter, and height) in the top-down approach.

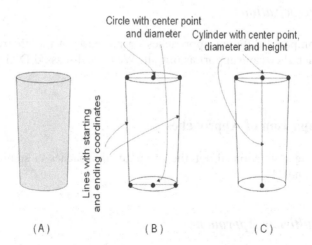

Fig. 24.2 Different approaches: (**A**) 3D object, (**B**) bottom-up, (**C**) top-down

24.3.3 Drawing

It is also very important whether only perfect line drawings are needed, or imperfect drawings having some noise or distortions may also be used. There is also another aspect of drawing: whether camera perspectives or scanned manual drawings are used.

24.3.4 Object

The strength of an approach is also determined on the basis of the type of objects that it can handle. Some approaches can handle only straight lines in a drawing, while others can handle both straight lines as well as curves.

24.3.5 Cross-Sectional Views, Hidden Lines, and Dead-End Holes

Different approaches are also compared on the basis of their ability to handle cross-sectional views and dead-end holes. Incorporation of hidden lines is also one of the comparison metrics.

24.3.6 User Interaction

It is also very important to distinguish between fully automated and nonautomated approaches. Fully automated approaches, of course, are more robust.

24.3.7 Vectorization

We have compared different approaches on the basis of the different vector formats which they can support as output. We have discussed DXF and SVG formats only.

24.4 Comparison of Approaches

The following discussion about the different approaches is summarized in Tables 24.1 and 24.2.

24.4.1 Multiview Approaches

In this section we will focus on the different approaches dealing with multiviews.

24.4.1.1 Liu et al. [8]

This technique uses a wireframe approach and requires three orthographic projections. Different objects such as planar, cylindrical, and conical faces can be handled by this technique. The beauty of the technique is the use of matrices to represent conic faces for the reconstruction process. The authors have also proved that the minimum number of views required to represent conics are three.

24.4.1.2 Cicek and Gulesin [9]

This approach is based upon CSG and requires three orthographic views. Extrusion, revolution, and Boolean operations (subtraction, intersection) are used to construct feature volumes from 2D projections. This technique handles different objects, such as lines, center lines, hidden lines, arcs, circles, etc. The technique is powerful in handling blind pockets, through pockets, circular pockets, through holes, blind holes, counter-bored through holes, counter-bored blind holes, stepped countersunk through holes, and stepped counter-sunk blind holes.

24.4.1.3 Dimri and Gurumoorthy [10]

This is a volume-based approach requiring three views. The novelty of this technique is the reconstruction from cross-sectional views. Different types of cross-sections (full sections, removed sections, revolved sections, half sections, broken out sections, offset sections) are discussed. Handling of sectional views is also discussed by Wesley and Markowsky [11], but their approach is limited to full sectional views only. Aldefeld and Richter [12] and Bin [13] have also considered sectional views but require user interaction. The technique of Dimri and Gurumoorthy [10] takes into account full sectional, half sectional, offset sectional, and broken out sectional views, but does not provide removed sectional and revolved sectional views. This technique can handle straight and circular edges using sweep and Boolean operations (union, difference), as well as objects with protrusion, depression, and seek through holes. The technique requires the type of sectional view to be entered by the user.

24.4.1.4 Lee and Han [14]

They have proposed a CSG-based approach to handling the solids of revolution from orthographic views. To recognize solids of revolution from orthographic views, a hint-based method is used. An interesting feature of this technique is the handling of intersecting as well as isolated objects. Extrusion and Boolean operations (subtraction, intersection) are used for the construction of objects, including the spherical ones too, but is limited to axis-aligned solids only. Their approach uses existing CSG techniques to construct solids other than solids of revolution.

24.4.1.5 Gong et al. [15]

This technique handles natural quadrics (sphere, cylinder, cone, plane) and interactive as well as isolated objects by hint-based pattern matching using B-Rep. In this paper a new hybrid wireframe consisting of geometry, topology, vertices, and edges is proposed. This requires three views with perfect line drawings for the handling of different objects such as line, circle, arc, ellipse, circular arc, elliptical arc, and higher order curves. Higher order curves are approximated in polyline.

24.4.2 Single View Approach

This section covers the approaches using a single view.

24.4.2.1 Cooper [16]

Cooper's [16] approach deals with the construction of a wireframe from a single view and uses a labeling technique. He has also given the necessary and sufficient conditions for the realisability of his approach. These conditions involve different semantics (convex, concave, occluding, external) and labels. The technique assumes that edges and surfaces meeting at a vertex are non-tangential and that the vertices are trihedral. His approach can handle straight as well as curved lines. He has also proved that a 3D wireframe model of a polyhedron with simple trihedral vertices is unambiguous. And he has proposed a novel labeling technique, which involves a number of surfaces in front of and behind each edge.

24.4.2.2 Martin et al. [17]

The technique used by Martin et al. [17] is based upon B-rep, and performs reconstruction from a single view using a labeling scheme. In this research, a new line labeling technique is proposed and a good survey of previous line labeling techniques is supported. The authors claim their approach to be more efficient than other approaches, but lack an analysis of their technique on actual engineering drawings.

24.4.2.3 Feng et al. [18]

This is a top-down approach that deals with a single view, depends on human perception, and requires heavy user interaction. The technique of Feng et al.

[18] does not deal with hidden lines. They have proposed different perceptual constraints, such as axis-alignment, symmetry, parallelism, collinearity, and orthogonal corners.

24.5 Summary

In this paper we have developed a versatile technique to reconstruct solid models of engineering objects from 2D camera perspectives. We have produced output in two formats: DXF and SVG. Different comparison metrics are also discussed, and our approach is compared with existing approaches after a critical review of these approaches. The results of this comparison are depicted in Tables 24.1 and 24.2.

References

1. Mansfield PA, Otkunc CB (2003) Adding another dimension to scalable vector graphics. Proceedings of XML Conference and Exposition
2. Su X, Chu CC, Prabhu BS, Gadh R (2006) Enabling engineering documents in mobile computing environment. In: Proceedings of the 2006 International Symposium on a World of Wireless, Mobile and Multimedia Networks
3. Idesawa M (1973) A system to generate a solid figure from three views. Bull Jpn Soc Mech Eng 16:216–225
4. Aldefeld B (1983) On automatic recognition of 3D structures from 2D representations. Comput-Aided Design 15(2):59–72
5. Wang W, Grinstein GG (1993) A survey of 3D solid reconstruction from 2D projection line drawings. Comput Graph Forum 12(2):137–158
6. Company P, Piquer A, Contero M (2004) On the evolution of geometrical reconstruction as a core technology to sketch-based modeling. EUROGRAPHICS Workshop on Sketch-Based Interfaces and Modeling
7. Fahiem MA, Haq SA, Saleemi F (2007) A review of 3D reconstruction from 2D orthographic line drawings. In: 2nd International Conference on Geometric Modeling and Imaging
8. Liu S, Hu S, Tai C, Sun J (2000) A matrix-based approach to reconstruction of 3D objects from three orthographic views. In: 8th Pacific Conference on Computer Graphics and Applications
9. Cicek A, Gulesin M (2004) Reconstruction of 3D models from 2D orthographic views using solid extrusion and revolution. J Mater Process Technol 152:291–298
10. Dimri J, Gurumoorthy B (2005) Handling sectional views in volume-based approach to automatically construct 3D solid from 2D views. Comput-Aided Design 37:485–495
11. Wesley MA, Markowsky G (1981) Fleshing out projections. IBM J Res Develop 25(6):934–54
12. Aldefeld B, Richter H (1984) Semiautomatic three-dimensional interpretation of line drawings. Comput-Aided Design 8(4):371–80
13. Bin H (1986) Inputting constructive solid geometry representations directly from 2-D orthographic engineering drawings. Comput-Aided Design 18(3):147–55
14. Lee H, Han S (2005) Reconstruction of 3D interacting solids of revolution from 2D orthographic views. Comput-Aided Design 37:1388–1398

15. Gong J, Zhang H, Zhang G, Sun J (2006) Solid reconstruction using recognition of quadric surfaces from orthographic views. Comput-Aided Design 38:821–835
16. Cooper MC (2005) Wireframe projections: Physical realisability of curved objects and unambiguous reconstruction of simple polyhedra. Int J Comput Vision 64(1):69–88
17. Martin RR, Suzuki H, Varley PAC (2005) Labeling engineering line drawings using depth reasoning. J Comput Inf Sci Eng 5:158–167
18. Feng, D, Lee S, Gooch B (2006) Perception-based construction of 3D models from line drawings. In: International Conference on Computer Graphics and Interactive Techniques

Chapter 25
DiProNN: VM-Based Distributed Programmable Network Node Architecture

Tomáš Rebok

Abstract The programmable network approach allows processing of passing user data in a network, which is highly suitable especially for video streams processing. However, the programming of complex stream processing applications for programmable nodes is not effortless, since the applications usually do not provide sufficient flexibility (both programming flexibility and execution environment flexibility). In this paper we present the architecture of our DiProNN node—the VM-based *Distributed Programmable Network Node,* that is able to accept and run user-supplied programs and/or virtual machines and process them (in parallel if requested) over passing user data. The node is primarily meant to perform stream processing; to enhance DiProNN flexibility and make programming of streaming applications for a DiProNN node easier, we also propose a suitable modular programming model which takes advantage of DiProNN's virtualization and makes its programming more comfortable.

25.1 Introduction

Contemporary computer networks behave as a passive transport medium which delivers (or in case of the best-effort service, tries to deliver) data from the sender to the receiver. The whole transmission is done without any modification of the passing user data by the internal network elements.[1] However, especially for small and middle specialized groups (e.g., up to hundreds of people) using computer networks for specific purposes, the ability to perform processing inside a network is sometimes highly desired.

---

[1] Excluding firewalls, proxies, and similar elements, where an intervention is usually limited (they do not process packets' data).

T. Rebok (✉)
Faculty of Informatics, Masaryk University, Botanická 68a, 602 00 Brno,
Czech Republic
e-mail: xrebok@fi.muni.cz

N. Mastorakis et al. (eds.), *Proceedings of the European*
Computing Conference, Lecture Notes in Electrical Engineering 27,
DOI 10.1007/978-0-387-84814-3_25, © Springer Science+Business Media, LLC 2009

The principle called *Active Networks* or *Programmable Networks* is an attempt how to build such intelligent and flexible networks using current "dumb and fast" networks serving as a communication underlay. In such a network, users and applications have the possibility of running their own programs inside the network using inner nodes (called *active nodes/routers,* or *programmable nodes/routers*—all with rather identical meaning) as processing elements.

However, the programming of complex (active) programs used for data processing, that are afterwards uploaded on programmable nodes, may be fairly difficult. And usually, if such programs exist, they are designed for different operating systems and/or use specialized libraries than the programmable node provides as an execution environment. Furthermore, since the speeds of network links still increase and, subsequently, applications' demands for higher network bandwidths increase as well, a single programmable node is infeasible to process passing user data in real-time, since such processing may be fairly complex.

The main goal of our work is to propose a distributed programmable network node architecture with loadable functionality that uses commodity PC clusters interconnected via the low-latency interconnection (so called tightly-coupled clusters), which is also able to perform distributed processing. Since the node is primarily meant to perform stream processing, and to make programming of stream-processing applications for our node easier, we also propose a suitable modular programming model which takes advantages of node virtualization and makes its programming more comfortable.

25.2 DiProNN: Distributed Programmable Network Node

The DiProNN architecture we propose assumes the infrastructure as shown in Fig. 25.1. The computing nodes form a computer cluster interconnected with each node having two connections—one *low-latency control connection* used for internal communication and synchronization inside the DiProNN, and at least one[2] *data connection* used for receiving and sending data.

The low-latency interconnection is necessary because current common network interfaces like Gigabit Ethernet or 10 Gigabit Ethernet provide large bandwidth, but the latency of the transmission is still in the order of hundreds of µs, which is not suitable for fast synchronization of DiProNN units. Thus the use of specialized low-latency interconnects like the Myrinet network, providing latency as low as 10 µs (and even less, if you consider, e.g., InfiniBand with 4 µs), which is close to message passing between threads on a single computer, is very desirable.

---

[2] The ingress data connection could be the same as the egress one.

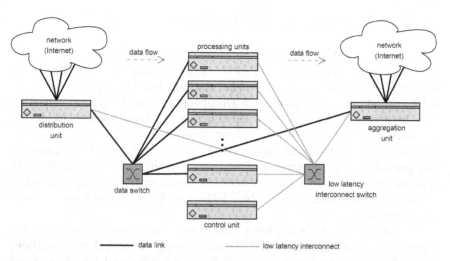

Fig. 25.1 Proposed DiProNN architecture

From the high-level perspective of operation, the incoming data are first received by the distribution unit, where they are forwarded to the appropriate processing unit(s). After processing, they are finally aggregated using the aggregation unit and sent over the network to the next node (or to the receiver). As obvious from Fig. 25.1, the DiProNN architecture comprises four major parts:

- *The distribution unit* takes care of ingress data flow, forwarding it to the appropriate DiProNN processing unit(s), which are determined by the control unit described later.
- *The processing units* receive packets and forward them to the proper active programs for processing. The processed data are then forwarded to the next active programs for further processing or to the aggregation unit to be sent away.
  Each processing unit is also able to communicate with the other ones using the low-latency interconnection. Besides the load balancing and fail over purposes, this interconnection is mainly used for sending control information of DiProNN sessions (e.g., state sharing, synchronization, processing control).
- *The control unit* is responsible for the whole DiProNN management and communication with its neighborhood, including communication with DiProNN users to negotiate the establishment of new DiProNN sessions (details about DiProNN sessions are provided in Section 25.3) and, if requested, providing feedback about their behavior.
- *The aggregation unit* aggregates the resulting traffic to the output network line(s).

25.2.1 DiProNN and Virtual Machines

Virtual machines (VMs) enhance the execution environment flexibility of the DiProNN node—they enable DiProNN users not only to upload the active programs, which run inside some virtual machine, but they are also allowed to upload the whole virtual machine with its operating system and let their passing data be processed by their own set of active programs running inside the uploaded VM(s). Similarly, the DiProNN administrator is able to run his own set of fixed virtual machines, each one with a different operating system, and generally with completely different functionality. Furthermore, the VM approach allows strong isolation between virtual machines, and thus strict scheduling of resources to individual VMs, e.g., CPU, memory, and storage subsystem access.

Nevertheless, the VMs also bring some performance overhead necessary for their management. This overhead is especially visible for I/O virtualization, where the *virtual machine monitor (VMM)* or a privileged host OS has to intervene in every I/O operation. However, for our purposes this overhead is currently acceptable—at this stage we focus primarily on DiProNN programming flexibility.

25.2.2 DiProNN Processing Unit Architecture

The architecture of the DiProNN processing unit is shown in Fig. 25.2. The VM-host management system has to manage the whole processing unit functionality, including uploading, starting, and destroying the virtual machines, communicating with the control unit, and session accounting and management. The virtual machines managed by the session management module could be either fixed, providing functionality given by a system administrator, or user-loadable.

An example of a fixed virtual machine would be a virtual machine providing classical routing as shown in Fig. 25.2. Besides that, the set of other fixed

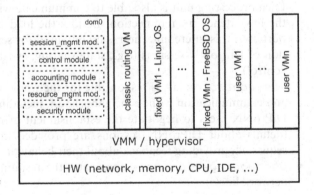

Fig. 25.2 DiProNN
processing unit architecture

virtual machines could be started as an active program execution environment, where the active programs uploaded by users are executed (those not having their own virtual machine defined). This approach does not force users to upload the whole virtual machine, in the case where active program uploading is sufficient.

25.2.3 DiProNN Communication Protocol

For data transmission, the DiProNN users may use both the *user datagram protocol* (UDP) and the transmission protocol called *active router transmission protocol*[3] (ARTP, [1]) which we designed. Depending on the application, the user chooses the transmission protocol he wants to use—whether he wants or needs to use ARTP's extended functionality or not.

25.3 DiProNN Programming Model

In this section we depict a model we use for DiProNN programming.

The model is based on the workflow principles [2] and is similar to the idea of the StreamIt [3], a language and a compiler specifically designed for modern stream programming.

For the DiProNN programming model we adopted the idea of independent simple processing blocks (`filters` in StreamIt), which, composed into a processing graph, carry out the required complex processing. In our case, the processing block is an active program, and communication among such active programs is handled by the virtualization mechanisms provided by a machine hypervisor using common network services. The interconnected active programs then compose the ``DiProNN session'' described by its ``DiProNN session graph,'' which is a graphical representation of a ``DiProNN program'' (an example is given in Fig. 25.3). To achieve the desired level of abstraction, all the active programs as well as the input/output interfaces are referred to by their hierarchical names, as shown in the example in Section 25.3.1.

The DiProNN program defines active programs optionally with the virtual machines they run in, which are neccessary for DiProNN session processing, and defines both data communication and control communication among them. Besides that, the DiProNN program may define other parameters of

---

[3] The ARTP is a connection-oriented transport protocol providing reliable duplex communication channels without ensuring that the data will be received in the same order as they were sent.

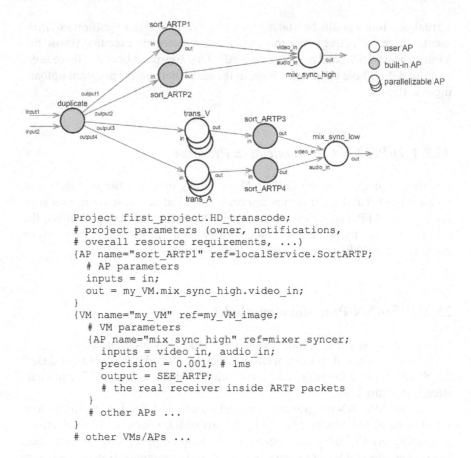

```
Project first_project.HD_transcode;
# project parameters (owner, notifications,
# overall resource requirements, ...)
{AP name="sort_ARTP1" ref=localService.SortARTP;
  # AP parameters
  inputs = in;
  out = my_VM.mix_sync_high.video_in;
}
{VM name="my_VM" ref=my_VM_image;
  # VM parameters
  {AP name="mix_sync_high" ref=mixer_syncer;
    inputs = video_in, audio_in;
    precision = 0.001; # 1ms
    output = SEE_ARTP;
    # the real receiver inside ARTP packets
  }
  # other APs ...
}
# other VMs/APs ...
```

Fig. 25.3 Example of DiProNN session graph together with a fragment of relevant DiProNN program

active programs, or the whole DiProNN session, and/or the resource require-ments they have for proper functionality.

There is one special attribute that can be set for active programs—
parallelizable. If this attribute is set, the DiProNN performs the
active program's processing in parallel.[4] The number of parallel instances
running can be either fixed (set in the DiProNN program and negotiated
during DiProNN session establishment) or variable (controlled by the
control unit, depending on actual DiProNN usage).

---

[4] Note that the DiProNN session must define, on its own, how to distribute data over such parallel instances, or choose such distribution from built-in functions, e.g., round-robin or a simple duplication principle.

25.3.1 DiProNN Program Example

Let's have the following situation: there is one incoming high-bandwidth video stream (e.g., an HD stream at 1.5 Gbps) and one high-quality audio stream, both transferred using the ARTP protocol described before. In the DiProNN, both streams must be transcoded into low quality streams for a specified set of clients behind low-bandwidth lines, and for some clients the streams must remain in the original quality. At the output, there must be both audio and video streams of given quality mixed into just one output stream (thus having two output streams—one in high quality and one in lower quality), and the precise time synchronization between audio and video in both output streams is also required.

The possible DiProNN session graph together with the fragment of relevant DiProNN program are depicted in Fig. 25.3.

25.3.2 DiProNN Data Flow

When a new DiProNN session request arrives at the node, the control unit decides, based on the actual DiProNN usage, whether it can be satisfied or not.

If the request can be satisfied, the session establishment takes place. That means that the control unit decides which processing units each active program will run on, and appropriately sets both their control modules and the distribution unit. Moreover, all the requested resources are reserved, if any.

Since the DiProNN programming model uses symbolic names for communication channels instead of port numbers, the names must be associated with the appropriate port numbers during a DiProNN session startup. This association is done using the control module (a part of each processing unit) where the couple (symbolic name, real port) is using simple text protocol registered. The control module, using this information together with the DiProNN program and, if neccessary, the information set as an option in the ARTP packet coming from an active program, properly sets the IP receiver of passing packets, which are then forwarded to the proper active programs for processing.

However, this approach does not enable active programs to know the real data receiver (each packet is sent by VMM to a given VM address and given the active program's port). Nevertheless, DiProNN users may use the ARTP's extended functionality to make their active programs become aware of the real data receiver (e.g., using the proper option inside each ARTP datagram). In this case, the aggregation unit forwards these packets to the destination given inside the ARTP datagram instead of the one(s) given in the DiProNN program.

The usage of symbolic names in DiProNN doesn't force the active programs to be aware of their neighbourhood,(the active programs processing a given DiProNN session before and after them); they are completely independent of each other so that they just have to know the symbolic names of the ports they want to communicate with and register them at the control module of the processing unit they run in.

25.4 Related work

With the potential for many applications, the active network principles are very popular and investigated by many research teams. Various architectures of active routers/nodes have been proposed; in this section we briefly describe only those most related to our work.

C&C Research Laboratories propose the CLARA (*CLuster-based Active Router Architecture* [4]), providing customizing of media streams to the needs of their clients. The architecture of another programmable network node, LARA (*Lancaster Active Router Architecture* [5]), in comparison with CLARA, encompasses both hardware and software active router designs. The LARA++ (*Lancaster's 2nd-generation Active Router Architecture* [6]), as the name indicates, evolved from the LARA. Compared to the LARA, which provides innovative hardware architecture, the LARA++ lays the main focus on the software design of its architecture.

However, in comparison with the DiProNN, none of these distributed programmable architectures addresses promising virtualization technology and its benefits, or tries to provide enhanced flexibility (both programming flexibility and execution environment flexibility).

25.5 Conclusions and Future work

In this paper, we have proposed a VM-oriented distributed programmable network node architecture named DiProNN. DiProNN users are able to arrange and upload a DiProNN session consisting of a set of their own active programs independent of each other and possibly running in their own virtual machine/operating system, and let their passing data be processed by the DiProNN. Communication among such active programs is provided using standard network services together with a machine hypervisor so that the active programs are not forced to be aware of their neighbourhood. The DiProNN cluster-based design also enables simultaneous parallel processing of active programs that are intended to run in parallel.

Concerning future challenges, the proposed DiProNN architecture is being implemented based on the Xen Virtual Machine Monitor [7]. Further, we want to explore the mechanisms of QoS requirements for assurance and scheduling to be able to utilize DiProNN resources effectively. We want to explore all the three perspectives of DiProNN scheduling: scheduling active programs to VMs (when they do not have their own VM specified), scheduling VMs to appropriate processing units, and scheduling active programs/virtual machines to suitable DiProNN nodes (when there are more DiProNN nodes on the path from a sender to a receiver which are able to process a given DiProNN session).

For efficiency purposes we plan to implement some parts of the DiProNN in hardware, e.g., on FPGA-based programmable hardware cards [8].

Acknowledgments This project has been supported by the research project "Integrated Approach to Education of PhD Students in the Area of Parallel and Distributed Systems" (No. 102/05/H050).

References

1. Rebok T (2004) Active router communication layer. Technical Report 11/2004, CESNET
2. Cichocki A, Rusinkiewicz M, Woelk D (1998) Workflow and process automation: concepts and technology. Kluwer Academic Publishers, Norwell, MA, USA
3. Thies W, Karczmarek M, Amarasinghe S (2002) StreamIt: a language for streaming applications. In: International Conference on Compiler Construction, Grenoble, France
4. Welling G, Ott M, Mathur S (2001) A cluster-based active router architecture. IEEE Micro 21(1):16–25
5. Cardoe R, Finney J, Scott AC, Shepherd D (1999) LARA: a prototype system for supporting high performance active networking. In: IWAN 1999, pp 117–131
6. Schmid S (2000) LARA++ design specification. Lancaster University, DMRG Internal Report, MPG-00–03
7. Dragovic B, Fraser K, Hand S, Harris T, Ho A, Pratt I, Warfield A, m PB, Neugebauer R (2003) Xen and the art of virtualization. In: Proceedings of the ACM Symposium on Operating Systems Principles, Bolton Landing, NY, USA
8. Novotn J, Fučík O, Antoš D (2003) Project of IPv6 router with FPGA hardware accelerator. In: Field-programmable logic and applications. 13th International Conference FPL 2003, Springer Verlag 2778:964–967

Chapter 26
A Semantics of Behavioural Concepts for Open Virtual Enterprises

Mohamed Bouhdadi and Youssef Balouki

Abstract The Reference Model for Open Distributed Processing (RM-ODP) defines a framework for the development of Open Distributed Processing (ODP) systems in terms of five viewpoints. Each viewpoint language defines concepts and rules for specifying ODP systems from the corresponding viewpoint. However, the ODP viewpoint languages are abstract and do not show how these should be represented. We treat in this paper the need for formal notation for behavioural concepts in the enterprise language. Using the Unified Modeling Language and Object Constraints Language UML/OCL, we define a formal semantics for a fragment of ODP behaviour concepts defined in the RM-ODP foundations part and in the enterprise language. We mainly focus on time, action, behaviour constraints, and policies. These concepts are suitable for describing and constraining the behaviour of the ODP enterprise viewpoint specifications.

26.1 Introduction

The Reference Model for Open Distributed Processing (RM-ODP) [1–4] provides a framework within which support of distribution, networking, and portability can be integrated. It consists of four parts. The foundations part [2] contains the definition of the concepts and analytical framework for normalized description of arbitrary distributed processing systems. These concepts are grouped in several categories, which include structural and behavioural concepts. The architecture part [3] contains the specifications of the required characteristics that qualify distributed processing as open. It defines a framework comprising five viewpoints, five viewpoint languages, ODP functions, and ODP transparencies. The five viewpoints are enterprise, information,

M. Bouhdadi (✉)
Department of Mathematics and Computer Science, Mohammed V University,
B.P 1014, Av. Ibn Bettouta, Rabat, Morocco
e-mail: bouhdadi@fsr.ac.ma

N. Mastorakis et al. (eds.), *Proceedings of the European*
Computing Conference, Lecture Notes in Electrical Engineering 27,
DOI 10.1007/978-0-387-84814-3_26, © Springer Science+Business Media, LLC 2009

computational, engineering, and technology. Each viewpoint language defines concepts and rules for specifying ODP systems from the corresponding viewpoint. However, RM-ODP cannot be directly applicable [5]. Indeed, it defines a standard for the definition of other ODP standards. The ODP standards include modeling languages.

In this paper we treat the need for formal notation of ODP viewpoint languages. The languages Z, SDL, LOTOS, and Esterel are used in the RM-ODP architectural semantics part [4] for the specification of ODP concepts. However, no formal method is likely to be suitable for specifying every aspect of an ODP system.

Elsewhere, there has been an amount of research for applying the Unified Modeling Languages UML [6] as a notation for the definition of syntax of UML itself [7–9]. This is defined in terms of three views: the abstract syntax, well-formedness rules, and modeling elements semantics. The abstract syntax is expressed using a subset of UML static modeling notations. The well-formedness rules are expressed in Object Constraint Language OCL [10]. A part of the UML meta-model has a precise semantics [11, 12], defined using a denotational meta-modeling semantics approach. A denotational approach [13] is realized by a definition of the form of an instance of every language element and by a set of rules which determine which instances are and are not denoted by a particular language element.

Furthermore, for testing ODP systems [2–3], the current testing techniques [14] [15] are not widely accepted, especially for the enterprise viewpoint specifications. A new approach for testing, namely agile programming [16, 17] or test first approach [18], is being increasingly adopted. The principle is the integration of the system model and the testing model using the UML meta-modeling approach [19–20]. This approach is based on the executable UML [21]. In this approach, OCL can be used to specify the invariants [12] and the properties to be tested [17].

In this context we used the meta-modeling syntax and semantics approaches in the context of ODP systems. We used the meta-modeling approach to define the syntax of a sub-language for the ODP QoS-aware enterprise viewpoint specifications [22]. We also defined a UML/OCL meta-model semantics for structural concepts in ODP computational language [23]. In this paper we use the same approach for behavioural concepts in the foundations part and in the enterprise language.

The paper is organized as follows. In Section 26.2, we define core behaviour concepts (time, action, behaviour, role, process). Section 26.3 describes behaviour concepts defined in the RM-ODP foundations part, namely, time and behavioural constraints. We focus on sequentiality, nondeterminism, and concurrency constraints. In Section 26.4 we introduce the behaviour concepts defined in the enterprise language and focus on behavioural policies. A conclusion ends the paper.

26.2 Core Behaviour Concepts in the RM-ODP Foundations Part

We consider the minimum set of modeling concepts necessary for behaviour specification. There are a number of approaches for specifying the behaviour of distributed systems, coming from people with different backgrounds and considering different aspects of behaviour. We represent a concurrent system as a triple set, consisting of a set of states, a set of actions, and a set of behaviour. Each behaviour is modeled as a finite or infinite sequence of interchangeable states and actions. To describe this sequence there are mainly two approaches [24]:

1. Modeling systems by describing their set of actions and their behaviours.
2. Modeling systems by describing their state spaces and their possible sequences of state changes.

These views are dual in the sense that an action can be understood to define state changes, and states occurring in state sequences can be understood as abstract representations of actions. We consider both of these approaches as abstractions of the more general approach based on RM-ODP. We provide the formal definition of this approach that expresses the duality of the two mentioned approaches.

We use the formalism of the RM-ODP model, written in UML/OCL. We mainly use concepts taken from the clause 8 "basic modeling concepts" of the RM-ODP part 2. These concepts are: behaviour, action, time, constraints, and state (see Fig. 26.1). The latter are essentially the first-order propositions about model elements. We define concepts (type, instance, pre-condition, post-condition) from the clause 9 "specification concepts." Specification concepts

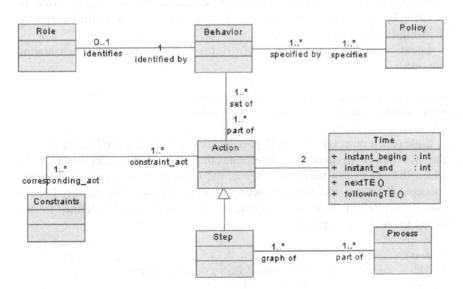

Fig. 26.1 Core behaviour concepts

are the higher-order propositions applied to the first-order propositions about the model elements.

Basic modeling concepts and generic specification concepts are defined by RM-ODP as two independent conceptual categories [25]. The behaviour definition uses two RM-ODP modeling concepts: action and constraints. Behaviour (of an object): "A collection of actions with a set of constraints on when they may occur." That is, a behaviour consists of a set of actions and a set of constraints. An action is something which happens. RM-ODP does not give the precise definition of behavioural constraints. These are part of the system behaviour and are associated with actions. This can be formally defined as follows:

Context c: constraint **inv**:
c.constrained_act - > size > 1
Context m :modelbehaviour **inv**:
m.behaviour- > includesAll(m.Actions- > union(m.constraints))

For any element b from behaviour, b is an action and b has at least one constraint and this constraint is a behaviour element, or b is a constraint and b has at least one action and this action is a behaviour element.

Context b: behaviour **inv**:
m.behaviour- > forall(b |(m.actions- > includes(m.b) and b.constraints- > notempty) or (m.constraints- > includes(m.b) and b.actions- > notempty)

To formalize the definition, we have to consider two other modeling concepts: time and state. We can see how these concepts are related to the concept of action by looking at their definitions. Time is introduced in the following way (RM-ODP, part 2, clause 8.10):

Location in time: An interval of arbitrary size in time at which action can occur.

instant_begin: each action has one time point when it starts
instant_end: each action has one time point when it finishes

State (of an object) (RM-ODP, part 2, clause 8.7): At a given instant in time, the condition of an object that determines the set of all sequences of actions in which the object can take part. Hence, the concept of state is dual with the concept of action, and these modeling concepts cannot be considered separately. This definition shows that state depends on time and is defined for an object for which it is specified.

Context t: time **inv**:
b.actions- > exists (t1,t2| t1 = action.instant_beging - > notempty and
t2 = action.instant_end - > notempty and t1< > t2)

26.3 Meta-Modeling Time and Behavioural Constraints

"Behavioural constraints may include sequentiality, non-determinism, concurrency, real time" (RM-ODP, part 2, clause 8.6). In this work we consider constraints of sequentiality, non-determinism, and concurrency. The concept of constraints of sequentiality is related with the concept of time.

26.3.1 Time

Time has two important roles:

- It serves the purpose of synchronization of actions inside and between processes, the synchronization of a system with system users, and the synchronization of user requirements with the actual performance of a system.
- It defines sequences of events (action sequences).

To fulfil the first goal, we have to be able to measure time intervals. However, a precise clock that can be used for time measurement does not exist in practice but only in theory [26]. So the measurement of time is always approximate. In this case we should not choose the most precise clocks, but the ones that explain the investigated phenomena in the best way. The simultaneity of two events or their sequentiality, and the equality of two durations, should be defined in the way in which the formulation of the physical laws is the easiest" [26]. For example, for the actions' synchronization, internal computer clocks can be used; and for the synchronization of user requirements, common clocks can be used that measure time in seconds, minutes, and hours.

We consider the second role of time. According to [26] we can build some special kind of clock that can be used for specifying sequences of actions. RM-ODP confirms this idea by saying that "a location in space or time is defined relative to some suitable coordinate system" (RM_ODP, part 2, clause 8.10). The time coordinate system defines a clock used for system modeling. We define a time coordinate system as a set of time events. Each event can be used to specify the beginning or end of an action. A time coordinate system must have the following fundamental properties:

- Time is always increasing. This means that time cannot have cycles.
- Time is always relative. Any time moment is defined in relation to other time moments (next, previous, or not related). This corresponds to the partial order defined for the set of time events.

We use the UML (Fig. 26.1) and OCL to define time: Time is defined as a set of time events.

nextTE: defines the closest following time events for any time event. We use the followingTE relation to define the set of the following time events or transitive closure for the time event t over the nextTE relation.

followingTE: defines all possible following time events. Using followingTE we can define the following invariant that defines the transitive closure and guarantees that time event sequences do not have loops:

Context t: time **inv**:
Time- > forAll(t:Time | (t.nextTE- > isempty implies t.follwingTE- > isempty)
and (t.nextTE- > notempty and t.follwingTE- > isempty implies t.follwingTE = t.nextTE) and
(t.nextTE- > notempty and t.follwingTE- > notempty implies t.follwingTE- >
includes(t.nextTE.follwingTE- > union(t.nextTE)) and t. follwingTE- > exludes(t)).

This definition of time is used in the next section to define sequential constraints.

26.3.2 Behavioural Constraints

We define a behaviour as a finite state automaton (FSA). For example, Fig. 26.2 shows a specification that has constraints of sequentiality and nondeterminism.

We can infer that the system is specified using constraints of nondeterminism by looking at state S1 that has a nondeterministic choice between two actions a and b.

Based on RM-ODP, the definition of behaviour must link a set of actions with the corresponding constraints. In the following we give a definition of constraints of sequentiality, of concurrency, and of nondeterminism.

26.3.2.1 Constraints of Sequentiality

Each constraint of sequentiality should have the following properties [28]:

- It is defined between two or more actions.
- Sequentiality has to guarantee that one action is finished before the next one starts. Since RM-ODP uses the notion of time intervals, it means that we have to guarantee that one time interval follows the other one:

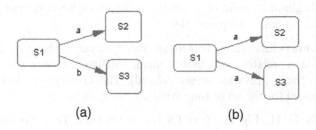

(a)   (b)

Fig. 26.2 (**a**) sequential deterministic constraints; (**b**) sequential nondeterministic constraints

Context sc: constraintseq **inv**:
Behaviour.actions- > forAll(a1,a2 | a1<> a2 and a1.constraints- > includes(sc)
and a2.constraints- > includes(sc) and
((a1.instant_end.followingTE- > includes(a2.instant_begin)   or(a2.instant_end.followingTE- >
includes(a1.instant_begin))

For all SeqConstraints sc, there are two different actions a1, a2; sc is defined
between a1 and a2 and a1 is before a2 or a2 is before a1.

26.3.2.2 Constraints of Concurrency

Figure 26.3 shows a specification that has constraints of concurrency. We can
infer that the system is specified using constraints of concurrency by looking at
state a1 that has a simultaneous choice of two actions a2 and a3.

For all concuConstraints cc there is a action a1, there are two different
internal actions a2, a3, cc is defined between a1 and a2 and a3, a1 is before a2
and a1 is before a3.

Context cc: constraintconc **inv**:
Behaviour.actions-> forAll(a1: Action, a2, a3: internalaction | (a1 <> a2) and
(a2 <> a3) and (a3 <> a1) and a1.constraints-> includes(cc) and
a2.constraints->includes(cc) and a3.constraints-> includes(cc) and
a1.instant_end.followingTE-> includes(a2.instant_begin) and
a1.instant_end.followingTE-> includes(a3.instant_begin))

26.3.2.3 Constraints of Nondeterminism

In order to define constraints of nondeterminism we consider the following
definition given in [24]: "A system is called nondeterministic if it is likely to have
shown a number of different behaviours, where the choice of the behaviour
cannot be influenced by its environment." This means that constraints of
nondeterminism should be defined among a minimum of three actions. The
first action should precede the two following actions, and these actions should
be internal (see Fig. 26.4).

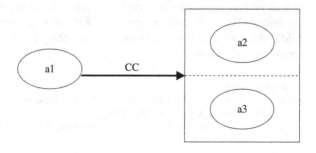

Fig. 26.3 RM-ODP
diagram: Example
constraints of concurrency

Fig. 26.4 RM-ODP
diagram: Example
constraints of
nondeterminism

Context ndc: NonDetermConstraints **inv** :
Behaviour.actions-> forAll(a1 :Action ,a2 ,a3 : internalaction | (a1 <> a2) and
(a2 <> a3) and (a3 <> a1) and a1.constraints->includes(ndc) and
a2.constraints->includes(ndc) and
a3.constraints->includes(ndc) and
a1.instant_end.followingTE-> includes(a2.instant_begin) or
a1.instant_end.followingTE-> includes(a3.instant_begin)) .

We note that, since the choice of the behaviour should not be influenced
by environment, actions a2 and a3 have to be internal actions (not interactions). Otherwise the choice between actions would be the choice of
environment.

26.4 Behavioural Policies in RM-ODP Enterprise Language

The enterprise specification is composed of specifications of the following
elements: the system's communities (sets of enterprise objects), roles (identifiers
of behaviour), processes (sets of actions leading to an objective), policies (rules
that govern the behaviour and membership of communities to achieve an
objective), and their relationships.

The behaviour of an ODP system is determined by the collection of all the
possible actions in which the system (acting as an object), or any of its
constituent objects, might take part, together with a set of constraints on
when these actions can occur. In the enterprise language this can be expressed
in terms of roles or processes or both, policies, and the relationships between
these. That is, the behaviour of an ODP system consists of a set of roles or a set
of processes and a set of their policies. Constraints are defined for actions.
Several authors have proposed different proprietary languages for expressing
ODP policies, usually with formal support (e.g., Object-Z), but with no graphical syntax—hence losing one of the advantages of using UML. We propose
modeling the enterprise viewpoint behavioural concepts using the standard
UML diagrams and mechanisms for modeling behaviour, since policies con-
strain the behaviour of roles.

Context s: System inv:
s.behaviour- > (includesAll(s.Roles) or includesAll(s.Process))- >
union(s.Roles.policy))Context o: object inv:
s.behaviour- > includes(o.behaviour.roles) > -union(o. behaviour.roles.policy)

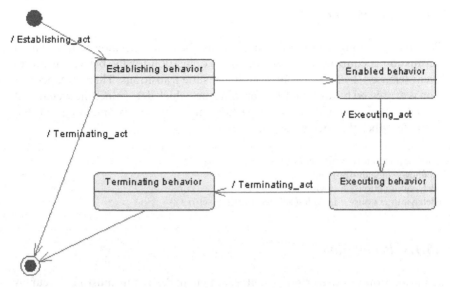

Fig. 26.5 A meta-model for behaviour and policy concepts

We formalize in the following the concepts of policy. Policy is defined as a set of establishing, terminating, and executing actions. Figure 26.5 presents the UML meta-model for behaviour and policy concepts. Establishing actions have to be defined by actions causing communications or processes:

Establishing_act: set of actions which initialize a behaviour.
Terminating_act: set of actions which break some process.
Executing_act: set of actions which execute behaviour.

Context P: Policy **inv**:
P.specified_by - > size > 1

26.4.1 Obligation

To model obligations, we need to specify the actions that the system is forced to undertake as part of its intended behaviour. In fact, an obligation is a prescription that a particular behaviour is required. It is fulfilled by the occurrence of the prescribed behaviour (clause :1 1 . 2 . 4). The actions must be initiated by the **Establishing action**, and be completed by the **Terminating action**.

Context po :policyobligation **inv**:
b.policy- > includes(po) implies
(Behaviour.actions- > (includes(self.Establishing_act) and
(Behaviour.actions- > includes(self.Terminating_act) and
(Behaviour.actions- > includes(self.Executin_act)

26.4.2 Permission

Permission is a prescription that a particular behaviour is allowed to occur. A permission is equivalent to there being no obligation for the behaviour not to occur (clause 1 1 . 2 . 5). Permissions allow state transitions. Therefore, permission is expressed by an action Establishing_act which determines the scenario of the permitted action(s) and their participants, while its Terminating_act diagram describes the effects of such action(s).

Context pp :policypermission inv :
b.policy- > includes(pp) implies
(Behaviour.actions- > (includes(self.Establishing_act) or
(Behaviour.actions- > includes(self.Terminating_act))

26.4.3 Prohibition

A prohibition is a prescription that a particular behaviour must not occur. A prohibition is equivalent to there being an obligation for the behaviour not to occur (clause1 1.2.6.). Prohibitions can be treated in two different ways, depending on their natures. The first way is to express them as conditional statements, explicitly banning an action **Establishing_act**. In this way, the system will automatically prevent the prohibited action from happening. The second way to deal with prohibitions is by using watchdog rules again, which detect the occurrence of the prohibited action and execute the action **Terminating_act**, if possible.

Context ppr: policy Prohibition inv:
b.policy- > includes(ppr) implies
(Behaviour.actions- > (excludes(self.Establishing_act) and
(Behaviour.actions- > excludes(self.Executing_act) and
includes(self.Terminating_act))

26.5 Conclusions

We address in this paper the need of formal ODP viewpoint languages. Using meta-modeling semantics, we define a UML/OCL-based semantics for a fragment of behaviour concepts defined in the Foundations part (time, sequentiality, nondeterminism, and concurrency) and in the enterprise viewpoint language (behavioural policies). These concepts are suitable for describing and constraining the behaviour of open distributed processing enterprise specifications. We are applying the same approach for other ODP enterprise behaviour concepts (real time) and for behaviour concepts in the computational language.

References

1. ISO/IEC (1994) Basic reference model of open distributed processing: Part1: Overview and guide to use. ISO/IEC CD 10746–1
2. ISO/IEC (1994) RM-ODP-Part2: Descriptive model. ISO/IEC DIS 10746–2
3. ISO/IEC (1994) RM-ODP-Part3: Prescriptive model. ISO/IEC DIS 10746–3
4. ISO/IEC (1994) RM-ODP-Part4: Architectural semantics. ISO/IEC DIS 10746–4
5. Bouhdadi M et al (2000) An informational object model for ODP applications. Malaysian J Comput Sci 13(2):21–32
6. Rumbaugh J et al (1999) The unified modeling language. Addison Wesley, New York
7. Rumpe B (1998) A note on semantics with an emphasis on UML. Second ECOOP workshop on precise behavioural semantics. Springer, LNCS 1543:167–188
8. Evans A et al (1998) Making UML precise. Object oriented programming, systems languages and applications (OOPSLA'98). Vancouver, Canada, ACM Press
9. Evans A et al (1999) The UML as a formal modeling notation. UML, Springer, LNCS 1618:349–274
10. Warmer J, Kleppe A (1998) The object constraint language: precise modeling with UML. Addison Wesley, New York
11. Kent S et al (1999) A meta-model semantics for structural constraints in UML. In: Kilov H, Rumpe B, Simmonds I (eds) Behavioural specifications for businesses and systems, chapter 9. Kluwer, Norwell
12. Evans E et al (1999) Meta-modeling semantics of UML. In: Kilov H, Rumpe B, Simmonds I (eds) Behavioural specifications for businesses and systems, chapter 4. Kluwer, Norwell
13. Schmidt DA (1986) Denotational semantics: a methodology for language development. Allyn and Bacon, Massachusetts
14. Myers G (1979) The art of software testing. Wiley, New York
15. Binder R (1999) Testing object oriented systems. Models, patterns, and tools. Addison-Wesley, New York
16. Cockburn A (2002) Agile software development. Addison-Wesley, New York
17. Rumpe B (2004) Agile modeling with UML. Springer, LNCS 2941:297–309
18. Beck K (2001) Column on test-first approach. IEEE Soft 18(5):87–89
19. Briand L (2001) A UML-based approach to system testing. Springer, LNCS 2185:194–208
20. Rumpe B (2003) Model-based testing of object-oriented systems. Springer, LNCS 2852:380–402
21. Rumpe B (2002) Executable modeling UML. A vision or a nightmare? In: Issues and trends of information technology management in contemporary associations. Seattle, Idea Group, London, pp 697–701
22. Bouhdadi M et al (2002) A UML-based meta-language for the QoS-aware enterprise specification of open distributed systems. Collaborative Business EcoSystems and Virtual Enterprises, IFIP Series, Springer, 85:255–264
23. Bouhdadi M, Balouki Y, Chabbar E (2007) Meta-modeling syntax and semantics of structural concepts for open networked enterprises. ICCSA 2007, Kuala Lumpor, LNCS accepted
24. Broy M (1991) Formal treatment of concurrency and time. In: McDermid J (ed) Software engineer's reference book. Butterworth-Heinemann, Oxford
25. Wegmann A et al (2001) Conceptual modeling of complex systems using RMODP based ontology. In: 5th IEEE international enterprise distributed object computing conference—EDOC, USA. IEEE Computer Society, pp 200–211
26. Poincaré H (1983) The value of science. Moscow "Science"

27. Harel D, Gery E (1997) Executable object modeling with statecharts. IEEE Computer 30(7):31–42
28. Balabko P, Wegmann A (2001) From RM-ODP to the formal behaviour representation. In: Proceedings of tenth OOPSLA workshop on behavioural semantics: "Back to Basics." Tampa, Florida, USA, pp 11–23

Chapter 27
A Semantics of Community Related Concepts in ODP Enterprise Language

Mohamed Bouhdadi and Youssef Balouki

Abstract The Reference Model for Open Distributed Processing (RM-ODP) defines a framework within which support of distribution, interoperability, and portability can be integrated. However, other ODP standards have to be defined. We treat in this paper the need for formal notation for community related structural concepts in the enterprise language. Indeed, the ODP viewpoint languages are abstract in the sense that they define what concepts should be supported, not how these concepts should be represented. One approach to define the formal semantics of a language is denotational elaborating of the instance denoted by a sentence of the language in a particular context. Using the denotational semantics in the context of UML/OCL, we define in this paper semantics for the community related concepts defined in the RM-ODP foundations part and in the enterprise language. These specification concepts are suitable for describing and constraining ODP enterprise viewpoint specifications.

27.1 Introduction

The rapid growth of distributed processing has led to a need for a coordinating framework for the standardization of Open Distributed Processing (ODP). The Reference Model for Open Distributed Processing (RM-ODP) [1–4] provides a framework within which support of distribution, networking, and portability can be integrated. It consists of four parts. The foundations part [2] contains the definition of the concepts and the analytical framework for normalized description of (arbitrary) distributed processing systems. These concepts are grouped in several categories. The architecture part [3] contains the specifications of the required characteristics that qualify distributed processing as open. It defines a framework comprising five viewpoints, five viewpoint languages, ODP functions, and ODP transparencies. The five viewpoints, called enterprise, information, computational, engineering, and technology, provide a basis for the

M. Bouhdadi (✉)
Department of Mathematics and Computer Science, Mohammed V University,
B.P 1014, Iv. Ibn Bettouta, Rabat, Morocco

N. Mastorakis et al. (eds.), *Proceedings of the European*
Computing Conference, Lecture Notes in Electrical Engineering 27,
DOI 10.1007/978-0-387-84814-3_27, © Springer Science+Business Media, LLC 2009

specification of ODP systems. Each viewpoint language defines concepts and rules for specifying ODP systems from the corresponding viewpoint. The enterprise viewpoint is the first specification of an open distributed system. It is concerned with the purpose, scope, and policies for the ODP system.

However, RM-ODP can not be directly applicable [3]. In fact, RM-ODP only provides a framework for the definition of new ODP standards. These standards include standards for ODP functions [6–7]; standards for modeling and specifying ODP systems; and standards for programming, implementing, and testing ODP systems. Also, RM-ODP recommends defining the ODP types for ODP systems [8].

In this paper we treat the need for formal notation of ODP viewpoint languages. Indeed, the viewpoint languages are abstract in the sense that they define what concepts should be supported, not how these concepts should be represented. The RM-ODP uses the term "language" in its broadest sense: "a set of terms and rules for the construction of statements from the terms;" it does not propose any notation for supporting the viewpoint languages. A formal definition of the ODP viewpoint languages would permit testing the conformity of different viewpoint specifications and verifing and validating each viewpoint specification. In the current context of software engineering methods and formal methods, we use the UML/OCL denotational meta-modeling semantics to define semantics for structural specification concepts in ODP enterprise language. The part of RM-ODP considered consists of modeling and specifying concepts defined in the RM-ODP foundations part and concepts in the enterprise language. The syntax domain and the semantics domain are defined using UML and OCL. The association between the syntax domain and the semantics domain is defined in the same models. The semantics of the UML model is given by constraining the relationship between expressions of the UML/OCL abstract syntax for models and expressions of the UML/OCL abstract syntax for instances. This is done using OCL.

The paper is organized as follows. Section 27.2 describes related works. In Section 27.3 we introduce the subset of concepts considered in this paper, namely, the object model and main structural concepts in the enterprise language. Section 27.4 defines the meta-model for the language of the model: object template, interface template, action template, type, and role. The meta-model syntax for the considered concepts consists of class diagrams and OCL constraints. Section 27.5 describes the UML/OCL meta-model for instances of models. Section 27.6 makes the connection between models and their instances using OCL. A conclusion and perspectives end the paper.

27.2 Related Works

The languages Z [9], SDL [10], LOTOS [11], and Esterel [12] are used in the RM-ODP architectural semantics part [4] for the specification of ODP concepts. However, no formal method is likely to be suitable for specifying every aspect of an ODP system. In fact, these methods have been developed for hardware design and protocol engineering. The inherent characteristics of ODP systems imply the need to integrate different specification languages and different verification methods.

Elsewhere, this challenge in the formal methods world is the same as in software methods. Indeed, there had been an amount of research for applying the Unified Modeling Languages UML [13] as a notation for the definition of UML itself [14–16]. This is defined in terms of three views: the abstract syntax, well-formedness rules, and modeling elements semantics. The abstract syntax is expressed using a subset of UML static modeling notations, that is, class diagrams. The well-formedness rules are expressed in Object Constraints Language OCL [17]. OCL is used for expressing constraints on object structure which cannot be expressed by class diagrams alone. A part of the UML meta-model itself has a precise semantics [18, 19] defined using the denotational meta-modeling approach. A denotational approach [20] would be realized by a definition of the form of an instance of every language element and a set of rules which determine which instances are and are not denoted by a particular language element. The three main steps of the approach are: (1) define the meta-model for the language of models (syntax domain), (2) define the meta-model for the language of instances (semantics domain), and (3) define the mapping between these two languages.

Furthermore, for testing ODP systems [2–3], the current testing techniques [21, 22] are not widely accepted, especially for the enterprise viewpoint specifications. A new approach for testing, namely, agile programming [23, 24], or the test first approach [25], is being increasingly adopted. The principle is the integration of the system model and the testing model using the UML meta-modeling approach [26, 27]. This approach is based on the executable UML [28]. In this context OCL is used to specify the invariants [19] and the properties to be tested [24]. The OCL invariants are defined using the UML denotational semantics and OCL itself.

In this context we used the UML meta-modeling approach in the context of ODP languages. We defined syntax of a sub-language for the ODP QoS-aware enterprise viewpoint specifications [29]. We also defined a UML/OCL meta-model semantics for structural constraints in the ODP computational language [30].

27.3 The RM-ODP

RM-ODP is a framework for the construction of open distributed systems. It defines a generic object model in the foundations part and an architecture which contains the specifications of the required characteristics that qualify distributed processing as open. We overview in this section the core structural concepts for the ODP enterprise language. These concepts are sufficient to demonstrate the general principle of denotational semantics in the context of the ODP viewpoint languages.

27.3.1 The RM-ODP Foundations Part

The RM-ODP object model [3] corresponds closely to the use of the term data model in the relational data model. To avoid misunderstandings, the RM-ODP defines each of the concepts commonly encountered in object-oriented models. It underlines a basic object model which is unified in the sense that it has to serve each of the five ODP viewpoints successfully. It defines the basic concepts concerned with existence and activity: the expression of what exists, where it is, and what it does. The core concepts defined in the object model are object and action.

An object is the unit of encapsulation: a model of an entity. It is characterized by its behavior and, dually, by its states.

Encapsulation means that changes in an object state can occur only as a result of internal actions or interactions.

An action is a concept for modeling something which happens. ODP actions may have duration and may overlap in time. All actions are associated with at least one object: internal actions are associated with a single object; interactions are actions associated with several objects.

Objects have an identity, which means that each object is distinct from any other object. Object identity implies that there exists a reliable way to refer to objects in a model. When the emphasis is placed on behavior, an object is informally said to perform functions and offer services; these functions are specified in terms of interfaces. It interacts with its environment at its interaction points, which are its interfaces. An interface is a subset of the interactions in which an object can participate. An ODP object can have multiple interfaces. Like objects, interfaces can be instantiated.

The other concepts defined in the object model are derived from the concepts of object and action; those are class, template, type, subtype/supertype, subclass/superclass, composition, and behavioral compatibility.

Composition of objects is a combination of two or more objects yielding a new object.

An object is behaviorally compatible with a second object with respect to a set of criteria if the first object can replace the second object without the environment being able to notice the difference in the object behavior on the basis of the set of criteria.

A type (of an $<x>$) is a predicate characterizing a collection of $<x>$s. Objects and interfaces can be typed with any predicate. The ODP notion of type is much more general than that of most object models. Also, ODP allows ODP to have several types, and to dynamically change types.

A class (of an $<x>$) defines the set of all $<x>$s satisfying a type. An object class, in the ODP meaning, represents the collection of objects that satisfy a given type. ODP makes the distinction between template and class explicit. The class concept corresponds to the OMG extension concept; the extension of a type is the set of values that satisfy the type at a particular time. A subclass is a subset of a class. A subtype is therefore a predicate that defines a subclass. ODP subtype and subclass hierarchies are thus completely isomorphic.

A <x> template is the specification of the common features of a collection x in sufficient detail that an x can be instantiated using it.

Types, classes, and templates are needed for object, interface, and action.

27.3.2 The RM-ODP Enterprise Language

The definition of a language for each viewpoint describes the concepts and rules for specifying ODP systems from the corresponding viewpoint. The object concepts defined in each viewpoint language are specializations of those defined in the foundations part of RM-ODP

An enterprise specification is concerned with the purpose, scope, and policies for the ODP system. Below, we summarize the basic enterprise concepts.

Community is the key enterprise concept. It is defined as a configuration of objects formed to meet an objective. The objective is expressed as a contract that specifies how the objective can be met. A contract specifies an agreement governing part of the collective behavior of a set of objects. A contract specifies obligations, permissions, and prohibitions for the objects involved. A contract specification may also include the specification of different roles engaged in the contract, the interfaces associated with the roles, quality of service attributes, indications of a period of validity, and behavior that invalidates the contract. The community specification also includes the environment contracts that state the policies governing the interactions of this community with its environment.

A role is a specification concept describing behavior. A role may be composed of several roles. A configuration of objects established for achieving some objective is referred to as a community. A role thus identifies behaviors to be fulfilled by the objects comprising the community.

An enterprise object is an object that fills one or more roles in a community.

A policy statement provides an additional behavioral specification. A community contract uses policy statements to separate behavioral specifications about roles.

A policy is a set of rules related to a particular purpose. A rule can be expressed as an obligation, a permission, or a prohibition. An ODP system consists of a set of enterprise objects. An enterprise object may be a role, an activity, or a policy of the system.

27.4 Syntax Domain

We define a model to be the ODP enterprise viewpoint specification: that is, a set of enterprise objects formed to meet an objective. The objective is expressed as a contract which specifies how the objective can be met A system can only be an instance of a single system model. Objects are instances of one or more object templates; they may be of one or more types. The meta-model supports dynamic types.

We define in this section the meta-models for the concepts presented in the previous section. Figure 27.1 defines the context-free syntax for the core object concepts, and Figs. 27.2 and 27.3 define the context-free syntax for concepts in the enterprise language. There are several context constraints between the syntax constructs. In the following, we define some of these context constraints, using OCL for the defined syntax.

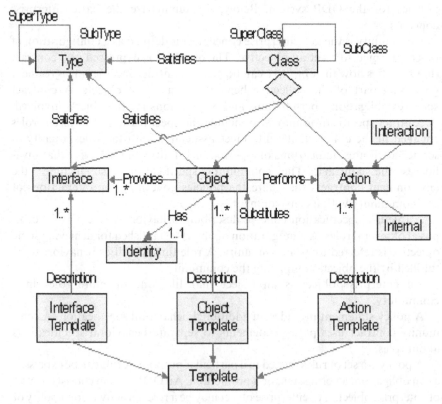

Fig. 27.1 RM-ODP foundation object model

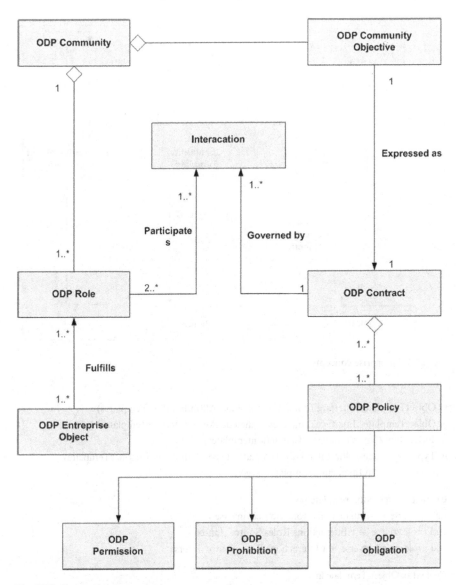

Fig. 27.2 Enterprise concepts

Context m: Model **inv**:
m.Roles- > includesAll(m.Roles.Source - > union(m.Roles.Target)
m.Roles- > includesAll(m.ObjectTemplates.Roles)
m.Roles- > includesAll(m.Interactiontemplate.roles)
m.Roles- > includesAll(m.InterfaceTemplate.roles)

m.InteractionTemplates - > includesAll(m.ObjectTemplates. Interactiontemplates)
m.InteractionTemplates.Types- > includesAll(m.Types)

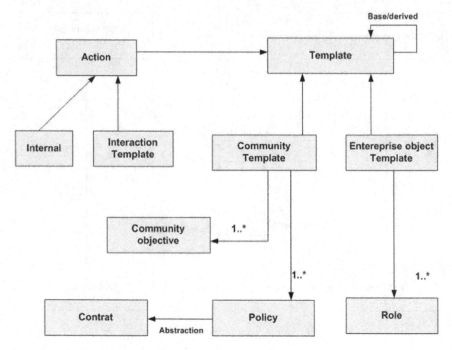

Fig. 27.3 Enterprise concepts

m.ObjectTemplates.InterfaceTemplates- > includesAll(m.InterfaceTemplates)
m.ObjectTemplates.InterfaceTemplates- > includesAll(m.InterfaceTemplates)
includesAll(m.ObjectTemplates.Interactiontemplates)
m.Types- > includesAll(m.InteractionTemplates.Types- > union(m.InterfaceTemplates.
Types)- > union(m.InteractionTemplate.Target)

Context i: Interaction template **inv**:
r.role.inverse = r.Interactions.Roles.Source .inverse
and r.role.source = r.Interactions.Roles.Source .source
and r.role.source.inverse = r.Interactions.Roles.Source .inverse

Context o: Object Template **inv**:
eot (enterprise object template) is not parent of or child of itself
not (eot.parents - > includes(cot) or eot.children- > includes(eot))

27.5 Semantics Domain

The semantics of a UML model is given by constraining the relationship
between a model and possible instances of that model: that is, constraining
the relationship between expressions of the UML abstract syntax for models

and expressions of the UML abstract syntax for instances. The latter constitute the semantics domain, which is integrated with the syntax domain within the same meta-model (Figs. 27.1, 27.2, and 27.3). This defines the UML context-free abstract syntax.

We give in the following a context constraint between the instances of semantics domain.

Context s: system **inv**:
The source and target objects of s'slinks are objects in s
s.objects- > includesAll(s.links.source- > union(s.links.target)
links between two objects are unique per role
s.links- > forAll(l|s.links
- > select(l'|l.'source = l.source&l.'target = l.target&l.'of = l.of) = l)

27.6 Meaning Function

The semantics for the UML-based language defined focuses on the relationship between a system model and its possible instances (systems). Both of the domains are defined using UML/OCL. The association between the instances of the two domains (syntax and semantics) is defined using the same UML meta-models. The OCL constraints complete the meaning function in the context of the UML [20, 31]. We give in the following some constraints which are relatively simple, but which demonstrate the general principle. We use the semantics of invariants as defined in [19].

Firstly, there is a constraint relating to objects. It shows how inheritance relationships can force an object to be of many classes of its parents.

Context o: object **inv**:

The templates of o must be a sign template and all the parents of that template.

o.of- > exists(t | o.of = t- > union(t.parents))

Secondly, there are four constraints which ensure that a model instance is a valid instance of the model it is claimed to be an instance of.

The first and second constraints ensure that objects and interfaces are associated with templates known in the model.

Context s: system **inv**:

The model that s is an instance of includes all object templates that s objects are instances of.

s.of.ObjectTemplates- > includesAll(s.Objects.of)

The model that s is an instance of includes all community templates that s communities are instances of.

s.of.CommunityTemplates- > includesAll(s.Communitys.of)

The third constraint ensures that communities are associated with roles known in the model.

Context s: system **inv**:

The model that s is an instance of includes all the roles that s communities are instances of.

s.of.roles - > includesAll(s.community.of)

The fourth constraint ensures that within the system, cardinality constraints on roles are observed.

Context s: system **inv**:
s.community.of - > forAll(r | let community_in_s be
r.instances - > intersect (s.community) in
(r.upperBound - > notEmpty implies community_in_s - > size < = r.upperBound) and
   coomunity_in_s- > size > = r.upperbound)

27.7 Conclusion

We address in this paper the need of formal ODP viewpoint languages. Using the denotational meta-modeling semantics, we define in this paper the UML/OCL-based syntax and semantics of a language for ODP object concepts defined in the foundations part and in the enterprise viewpoint language. These concepts are suitable for describing and constraining the structure of the ODP enterprise viewpoint specifications. We are applying the same denotational semantics to define semantics for concepts characterizing dynamic behavior in other viewpoint languages.

References

1. ISO/IEC (1994) Basic reference model of open distributed processing-Part1: Overview and guide to use. ISO/IEC CD 10746–1
2. ISO/IEC (1994) RM-ODP-Part2: Descriptive model. ISO/IEC DIS 10746–2
3. ISO/IEC (1994) RM-ODP-Part3: Prescriptive model. ISO/IEC DIS 10746–3
4. ISO/IEC (1994) RM-ODP-Part4: Architectural semantics. ISO/IEC DIS 10746–4
5. OMG (1991) The object management architecture.
6. ISO/IEC (1999) ODP type repository function. ISO/IEC JTC1/SC7 N2057
7. ISO/IEC (1995) The ODP trading function. ISO/IEC JTC1/SC21
8. Bouhdadi M et al (2000) An informational object model for ODP applications. Malaysian J Comput Sci 13(2):21–32
9. Spivey JM (1992) The Z reference manual. Prentice Hall, Englewood Cliffs, NJ
10. IUT (1992) SDL: Specification and description language. IUT-T-Rec. Z.100
11. ISO/IUT (1998) LOTOS: A formal description technique based on the temporal ordering of observational behavior. ISO/IEC 8807
12. Bowman H et al (1995) FDTs for ODP. Comput Stand Inter J 17(5–6):457–479
13. Rumbaugh J et al (1999) The unified modeling language. Addison-Wesley, New York
14. Rumpe B (1998) A note on semantics with an emphasis on UML. Second ECOOP workshop on precise behavioral semantics. Springer, LNCS 1543:167–188

15. Evans A et al (1998) Making UML precise. Object oriented programming, systems languages and applications (OOPSLA'98). ACM Press, Vancouver, Canada
16. Evans A et al (1999) The UML as a formal modeling notation. UML, Springer, LNCS 1618:349–364
17. Warmer J, Kleppe A (1998) The object constraint language: precise modeling with UML. Addison-Wesley, New York
18. Kent S et al (1999) A meta-model semantics for structural constraints in UML. In: Kilov H, Rumpe B, Simmonds I (eds) Behavioral specifications for businesses and systems. Kluwer, New York, chapter 9
19. Evans E et al (1999) Meta-modeling semantics of UML. In: Kilov H, Rumpe B, Simmonds I (eds) Behavioral specifications for businesses and systems. Kluwer, Norwell, chapter 4
20. Schmidt DA (1986) Denotational semantics: a methodology for language development. Allyn and Bacon, Massachusetts
21. Myers G (1979) The art of software testing. Wiley, New York
22. Binder R (1999) Testing object oriented systems. Models, patterns, and tools. Addison-Wesley, New York
23. Cockburn A (2002) Agile software development, Addison-Wesley, New York
24. Rumpe B (2004) Agile modeling with UML. Springer, LNCS 2941:297–309
25. Beck K (2001) Column on test-first approach. IEEE Soft 18(5):87–89
26. Briand L (2001) A UML-based approach to system testing. Springer, LNCS 2185:194–208
27. Rumpe B (2003) Model-based testing of object-oriented systems. Springer, LNCS 2852:380–402
28. Rumpe B (2002) Executable modeling UML. A vision or a nightmare? In: Issues and trends of information technology management in contemporary associations. Seattle, Idea Group, London, pp 697–701
29. Bouhdadi M et al (2002) A UML-based meta-language for the QoS-aware enterprise specification of open distributed systems. IFIP Series, Springer, Boston, 85:255–264
30. Bouhdadi M et al (2007) A UML/OCL denotational semantics for ODP structural computational concepts. First IEEE international conference on research challenges in information science (RCIS'07). Ouarzazate, Morocco, pp 259–264
31. France R, Kent S, Evans A, France R (1999) What does the term semantics mean in the context of UML. In: Moreira AMD and Demeyer S (eds) ECOOP workshops 1999, ECOOP'99 workshop reader. Springer, LNCS 1743:34–36

Chapter 28
Modeling of the Speech Process Including Anatomical Structure of the Vocal Tract

Zygmunt Ciota

Abstract The most important features of voice processing have been presented. The properties of glottal waves have been extracted using recorded microphone signals of the speech. Therefore, it was necessary to solve the inverse problem, of finding the glottis input of the whole vocal tract, having the resulting output waves of the speech process. The frequency parameters of glottal waves have been extracted using a vocal tract model. The autocorrelation and cepstrum methods are also helpful in such extraction. The results are important not only for speaker identification and emotion recognition, but can also be helpful for glottis malfunction diagnosis.

28.1 Introduction

One of the most important aspects of spoken language results from the simplicity of mutual communication which it allows. In the case of computer communication, the speech process becomes an effective tool of computer work control, reducing keyboard and monitor importance. Furthermore, human speech permits exchanging information in a simple and natural way, without special and time-consuming preparation. In other words, the speech signal permits you to express your thoughts by acoustic means using the vocal tract. Articulated sounds transmit some information, therefore it will be possible to find a mutual relationship between the acoustic structure of the signal and the corresponding information. Consequently, natural language can be described as a system of mutual associations between an acoustic structure and transmitted information [2, 3, 6]. The whole speech process is a result of complex transforms present on different levels: the semantic, linguistic, articulator, and acoustic levels.

Z. Ciota (✉)
Department of Microelectronics and Computer Science, Technical University of Lodz, Lodz, Poland

N. Mastorakis et al. (eds.), *Proceedings of the European Computing Conference*, Lecture Notes in Electrical Engineering 27, DOI 10.1007/978-0-387-84814-3_28, © Springer Science+Business Media, LLC 2009

According to the latest researches, the proper behavior of the glottis plays a critical rule in speech phenomena. Even small distortions, caused by diseases or some inborn malformation, have a significant influence on the final voice quality. The anatomical structure of the human vocal tract and the actions of all speech production organs, e.g., the glottis, nostril, tongue, velum, lips, and also the corresponding individual muscle groups, are very complicated [6, 8]. Therefore, the natural vocal tract is a datum point of different mathematical models [1, 6, 7]. The proper model should take into account all elements and phenomena appearing during the speech process, especially including input signal source coming from the larynx, faucal tract, mouth tract, nasal tract, and radiation impedances of both the mouth and nose. An examination of the condition of the larynx and especially the glottis, taking into account only the output speech signals, still remains an important task [6, 7].

28.2 Selection of Feature Vectors

Nowadays, methods of speech processing, based on spectrograms of speech sounds, are not faultless. Therefore, further researches leading to the improvement of such methods, combining spectral parameters with temporal characteristics, are necessary. One of the most important tasks is a proper definition of feature vectors. Each vector can contain several specific features of voice signals, and finally, one must calculate more than a hundred features for each utterance. The following four vectors are especially important and useful [2, 6], covering the most important features of the speech signal: the long-term spectra vector (LTS), the speaking fundamental frequency vector (SFF), the time-energy distribution vector (TED), and the vowel formant tracking vector (VFT).

The LTS vector is one of the most important indicators of voice quality, very helpful for speaker verification. The LTS can be described as a high sensitivity vector for speaker identification based on low quality voice signals. Therefore, proper results can be achieved using signals containing a rather high level of noise and limited frequency passband. Additionally, different emotions of the same speaker, like anger, happiness, and sadness, cannot distort the LTS significantly. The application of speaking fundamental frequency also can be a powerful vector. The efficiency of the SFF depends however, on a processing system which has to track and extract the fundamental frequencies precisely. As a result, statistical behaviors of these frequencies also have to be included. The TED vector is also important, because it characterizes speech prosody, so it is possible to extract an individual speaker's features, like the rate of speaking, pauses, loudness characteristics, etc. Unfortunately, this vector is rather dependent on the speaker's emotion. On the other hand, the VFT vector depends strongly on the individual size and shape of the vocal tract, which means that you can define this vector by the anatomical properties of the tract. The VFT is then stable and independent of almost all types of distortions.

One of the most difficult problems is proper vector construction. In our software environment we applied Matlab and Java platforms to create the above vectors. Particularly, we can create the LTS vector using time domain speech signals up to 20 s in length.

In the proposed method we also applied the features describing the speaking fundamental frequency F_0 and the time-energy distribution of the voice. In the case of the fundamental frequency calculation, two basic methods are available: autocorrelation and the cepstrum method. The first permits obtaining precise results, but we discovered that additional incorrect glottis frequencies have been created. We observed further improper frequencies, especially for ranges lower than 100 Hz and higher than 320 Hz. Additionally, the program indicates some glottis excitations during breaks between phones and in silence regions. There-fore, in this method it is necessary to apply special filters to eliminate all incorrect frequencies.

Another method is based on cepstrum analysis. In this transform the con-volution of glottis excitation and vocal tract is converted, first to the product (after a Fourier transform), separated them finally as the sum. In our method we use cepstrum analysis as less complex, especially when we apply modulo of cepstrum by using modulo of Fourier transform. The following values of glottis frequency F_0 have been taken into account: F_0-minimum, F_0-maximum, and the range and average values.

The parameters describing time-energy distribution have been calculated using fast Fourier transform and dividing speech utterances into 20 ms slots. As a result we obtained seven values of the energy for the following frequency ranges expressed in Hz: 0–400, 400–800, 800–1500, 1500–2000, 2000–3500, 3500–5000, and 5000–8000.

28.3 Model of Vocal Tract

An influence of fundamental frequency changes on the final speech sound can also be verified using a vocal tract model. Having a speech signal recorded using a microphone, and applying the parameters of the vocal tract, it should be possible to solve the inverse problem, obtaining the parameters of the glottal waves. Combining the above methods, it is possible to obtain high efficiency in the characterization of the frequency properties of a vocal excitation signal.

One of the possible modeling methods is replacement of anatomical tracts by coaxial connections of cylindrical tube sections. Each section has to fit as nearly as possible the dimensions of the cross-sectional area and the section length of the natural vocal tract (see Fig. 28.1). Such a vocal tract model should take into account a faucal tract which forks into the nasal tract, and also a mouth tract. It is also important to maintain the dimensions of the natural tract: the length of the faucal tract, the mouth tract, and the nasal tract. Unfortunately, cross-sectional areas cannot be unequivocally calculated, because people have

Fig. 28.1 Signal flow of speech channel and corresponding tube model

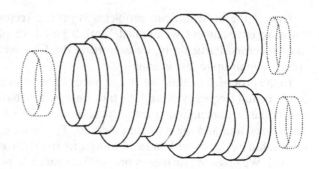

different cross-dimensions of vocal tracts. The complexity of the model depends on the number of tube sections: a bigger number of these sections gives a higher precision to the model, but the difficulty of the parameter calculation can increase significantly.

The behavior of model sections can be analyzed as relationships between the pressures of acoustic wave p_{in}, the volumetric velocity V_{in}, and the corresponding output quantities p_{out} and V_{out} for the current section. Moreover, an acoustic pressure and volumetric velocity correspond to electrical values: voltage and current, respectively.

Afterwards, one can replace each tube by its electrical circuit equivalent. All parameters should be calculated from the geometrical dimensions of the tube:

- inductance L as an equivalent of an air acoustic mass in the tube,
- capacitance C as an equivalent of air acoustic compliance,
- serial resistance, as an equivalent of resistance loss caused by viscotic friction, near the tube walls,
- additional negative capacitance C_N; an equivalent of the inverse acoustic mass of the vibratory tube walls,
- conductance G, an equivalent of acoustic loss conductance of thermal conductivity near the tube walls,
- additional conductance G_S, an equivalent of acoustic conductance of loss conductance of the vibratory tube walls, and
- pulsation ω.

The input signal corresponding to the glottal waves has been simulated using the model presented in Fig. 28.2. According to the above assumptions, we can calculate the equivalent parameters for all sections of the model. However, the calculations are complicated and time-consuming. To avoid these problems, we simplified the model. One can establish that the acoustic wave in the channel is a two-dimensional plane wave. Using such a model, it is possible to obtain the transmittance of the vocal tract. As a source of soundless signals we propose a nozzle model, because a time-domain characteristic of soundless phonemes is random, and the signal can vary for the same phoneme and for the same person. Using this model, we can present different features of noise phonemes. It is possible to calculate the middle frequency of the noise, as well as the total

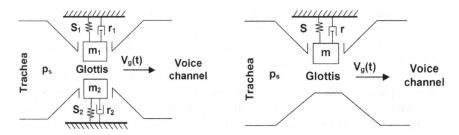

Fig. 28.2 Modeling of glottis behavior

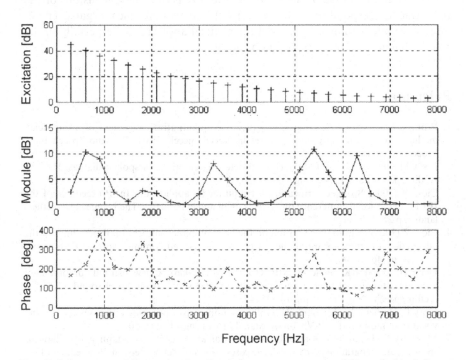

Fig. 28.3 Spectrum of input signal and vocal tract response (module and phase) for a vowel "U"

acoustic power of the channel. It is also possible to obtain the frequency characteristics. More complex analysis, including input excitations and frequency characteristics of the vocal tract, is presented in Fig. 28.3.

28.4 Conclusions

The results of the speech processing system are rewarding, but sometimes we can observe mistakes in the recognition process. However, some of these processes, especially for emotion recognition, make difficulties also for human evaluation. The proposed algorithms not only can be applied for emotion detection, but also can be helpful in the process of people identification. The sensitivity of the program for such emotions as anger or fear is measurable, but the vectors of the properties can be modified in the future.

The frequency parameters of glottal waves have also been extracted using a rather simple vocal tract model. The autocorrelation and cepstrum methods are also helpful in such extraction. The results are important for speaker identification and emotion recognition systems; furthermore, the model can be helpful for a better understanding of speech processes.

The main advantage of the proposed method is that the dimensions of the vocal tract correspond to the dimensions of human anatomical organs. Therefore, the model can also be helpful for medical applications, especially in the case of diagnosis and treatment monitoring.

References

1. Brookes M, Naylor PA, Gudnason J (2006) A quantitative assessment of group delay methods for identifying glottal closures in voiced speech. IEEE T Audio Speech Lang Processing 14:456–466
2. Lee CM, Narayanan SS (2005) Toward detecting emotions in spoken dialogs. IEEE T Audio Speech Lang Processing, 13:293–303
3. Ciota Z (2004) Speaker verification for multimedia application. In: Proceedings of the IEEE international conference on systems, man and cybernetics. The Hague, The Netherlands, pp 2752–2756
4. Ciota Z (2005) Emotion recognition on the basis of human speech. In: Proceedings of the international conference on applied electromagnetics and communications. Dubrovnik, Croatia, pp 467–470
5. Deng H, Ward RK, Beddoes MP, Hodgson M (2006) A new method for obtaining accurate estimates of vocal-tract filters and glottal waves from vowel sounds. IEEE T Audio Speech Lang Processing 14:445–455
6. Gray P, Hollier MP, Massara RE (2000) Non-intrusive speech-quality assessment using vocal-tract models. IEE P-Vis Image Signal Processing 147:493–501
7. Mozzafry B, Tinati MA, Aghagolzadeh A, Erfanian A (2006) An adaptive algorithm for speech source separation in overcomplete cases using wavelet packets. In: Proceedings of the 5th WSEAS international conference on signal processing. Istanbul, Turkey, pp 140–144
8. Santon J (ed) (1996) Progress in speech synthesis. Springer, New York

Chapter 29
Carbon Nanotube FET with Asymmetrical Contacts

Alireza Kargar, Zoheir Kordrostami, and Mohammad Hossein Sheikhi

Abstract We compute the current-voltage transfer characteristics of a new structure of coaxial carbon nanotube field effect transistors and simulate its behavior. In the proposed structure one end of a carbon nanotube is contacted to the metal intrinsically. The other end is doped and then contacted to the metal. Thus, the source nanotube-metal interface forms a Schottky barrier and the drain contact exhibits Ohmic properties. The simulation results of the new device presented show unipolar characteristics but are lower on current than when both contacts are Schottky or Ohmic. The coaxial single Schottky barrier CNTFET (SSB-CNTFET) is proposed for the first time and is modeled based on a semi-classical approach for the calculation of tunneling probability of electrons through the barrier. Both analytical and numerical computations have been used in order to solve Laplace and Schrödinger equations and also to compute tunneling probability for the device current.

29.1 Introduction

Carbon nanotubes (CNTs) are accepted as promising candidates for the channel material of the nanoscale field effect transistors. Such transistors with the nanometer scale channel length have ballistic transport through the channel length. This is the most important advantage of using a single-wall semiconducting carbon nanotube as the channel material in CNT field effect transistors (CNTFETs), which makes them appropriate for high speed applications. The typical diameter of a single-wall carbon nanotube is about 1 nm. Coaxial structures allow for better electrostatics than their planar counterparts [1]. Commonly, two types of the CNTFETs are known [2]. One of them works on the principle of tunneling through the Schottky barriers at the interfaces of the

Z. Kordrostami (✉)
Department of Electrical Engineering, School of Engineering, Shiraz University, Shiraz, Iran; Shiraz Nanotechnology Research Institute, Shiraz, Iran
e-mail: zkrostami@shirazu.ac.it

N. Mastorakis et al. (eds.), *Proceedings of the European*
Computing Conference, Lecture Notes in Electrical Engineering 27,
DOI 10.1007/978-0-387-84814-3_29, © Springer Science+Business Media, LLC 2009

metal-nanotube contacts and exhibits ambipolar characteristics. The other one uses potassium-doped source and drain regions which lead to formation of Ohmic contacts at both ends of the CNT, so that the transistor action is based on the bulk barrier modulation by the gate voltage and has unipolar characteristics and sometimes is called bulk-switched CNTFET. The latter type suffers from a lower on current compared to the Schottky barrier CNFETs. Here we propose a new structure that has unipolar characteristics but a lower on current than bulked-modulated CNTFETS.

Both quantum transport and classical calculations can be used to simulate the device. In this paper we have used a semi-classical approach by which the current of the transistor has been computed after massive computations such as solving Laplace equations analytically, solving Schrödinger equations using the WKB approximation, calculating the transmission probability integral analytically, and computing the CNTFET current by integrating the product of the transmission coefficient and the difference of the Fermi-Dirac distribution functions for each CNT terminal (Launder formula). In the following lines the formulations of the modeling approach are described and the solution results are shown as curves in the figures. We have skipped the explanation of the massive computations, which have not been necessary.

29.2 Single Schottky Barrier CNTFET

In the proposed new CNTFET designation, the metal-intrinsic CNT contact constitutes the source and by selective doping of the CNT in the drain region, the metal-doped CNT contact constitutes the drain. After heavily doping the drain region with potassium in vacuum, a high contact transparency at the interface of the metal (Pd) and the potassium-doped end of the CNT is achieved, suggesting an Ohmic contact structure [2, 3]. This type of CNTFET operates on the principle of Schottky barrier CNTFETs, but have a unipolar current-voltage characteristic like bulk-switched CNTFETs, and we call them single Schottky barrier CNTFETs (SSB-CNTFET). The structure of the device is shown in Fig. 29.1.

Fig. 29.1 Schematic side view physical diagram of the proposed device. The channel region near the drain is heavily potassium (K)- doped

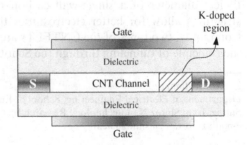

29.3 Simulation Methodology

A semi-classical method is used to calculate the transfer characteristics of the CNTFET. We have calculated the electrostatic potential along the CNT by solving a 2-D Laplace equation analytically in cylindrical coordinates for the device geometry shown in Fig. 29.1. Under appropriate approximations, the band diagram near the source Schottky barrier (solution of the Laplace equation) has an exponential dependence on the gate voltage [4, 5] (Fig. 29.2).

The transmission probability $T(E)$ through the Schottky barrier in the interfaces between the intrinsic CNT and the contact metallic electrodes can be derived by solving the Schrödinger equation using the Wentzel-Kramers-Brillouin (WKB) approximation [6, 7]. The transmission coefficient can be calculated from Eq. (29.1).

$$T(E) = \exp\left(-2\int_{z_1}^{z_2} K(z)dz\right) \qquad (29.1)$$

where $K(z)$ is the parallel momentum (wave number), which is related to the energy through the dispersion relation of the CNT,

$$K(z) = \frac{2}{3a_0 V_{pp\pi}} \sqrt{\left(3a_0 V_{pp\pi} K_n\right)^2 - 4(E - E(z))^2} \qquad (29.2)$$

where $a_0 = 0.142$ nm is the carbon-carbon bond distance, $V_{pp}n$ is the nearest neighbor interaction tight-binding parameter, $E(z)$ is the electrostatic along the CNT, and K_n is the perpendicular momentum.

This integration has been done analytically using exact expanding of the integrand sentences. The result is very long, and we skip the analytical solution

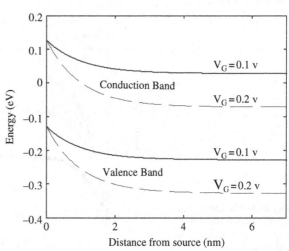

Fig. 29.2 The exponential dependence of the band diagram near the source Schottky barrier on the gate voltage, derived by solving the Laplace equation analytically

here. Under a common assumption of ballistic transport through the channel, the current can be expressed using the Landauer-Buttiker formalism as follows [7]:

$$I = \frac{4q}{h} \int_{-\infty}^{\infty} T(E)(F(E - E_{FS}) - F(E - E_{FD}))dE \qquad (29.3)$$

where $F(E)$ is the Fermi-Dirac distribution function, q is the electronic charge, and h is the Planck constant. The Fermi levels of the drain and source interfaces are represented by E_{FS} and E_{FD}.

The current can not be computed analytically, and a numeric integration is used to calculate the device current. The variation limits of the integrand variable (E) from previous approximations should be considered here, so that the variation range of E may not be infinite.

29.4 Results and Discussion

Using the described semi-classical model, the current of the SSB-CNTFET has been calculated and the dependence of its magnitude on the bias voltages has been investigated. The source contact is assumed to be ground. Tunneling from band to band is not considered in our calculations because there is not a semi-classical methodology for computing the band to band tunneling current similar to the barrier tunneling computation approach which is used here. The transistor current versus the drain voltage for different gate voltages is shown in Fig. 29.3. It is obvious that increasing the gate voltage suppresses the Schottky barrier of the source and thus increases the probability of the tunneling of electrons from the source contact to the conduction band of the intrinsic region. The drain current versus the gate voltage for different drain voltages is shown in Fig. 29.4.

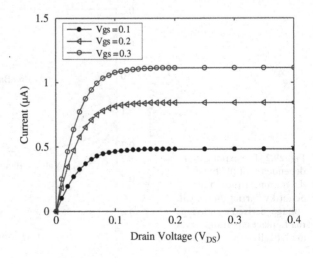

Fig. 29.3 The calculated current versus the drian voltage for different gate voltages

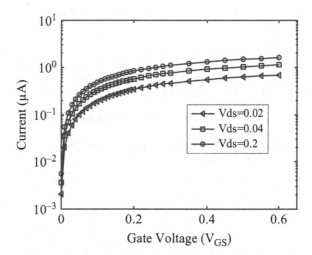

Fig. 29.4 Unipolar current-voltage characteristic of the SSB-CNTFET

The only effect of the negative gate voltage is increasing the barrier at the source and thus decreasing the current. That is why the SSB-CNFET has a near zero current when no gate bias is applied. Low on current is the trade off of the unipolar characteristic. However, in comparison to the conventional SB-CNTFETs, the SSB-CNTFET has a higher on/off ratio, which is another advantage of the proposed structure.

29.5 Conclusion

The new structure of coaxial asymmetrical contacted CNTFET with two different types of contacts (Schottky and Ohmic) at the source and drain was introduced. Because of a lower on current in comparison with conventional CNTFETs, the device has the potential of being used in low bias ultra-high speed integrated circuits. The higher on/off ratio and the unipolar transfer characteristics are the main advantages of the proposed transistor structure. The semi-classical computations have been used to simulate the device behavior. The SSB-CNTFET band diagram and current computation results have been shown as curves under different bias conditions.

References

1. John DL, Castro LC, Clifford J, Pulfrey DL (2003) Electrostatics of coaxial Schottky-barrier nanotube field-effect transistors. IEEE T Nanotechnol 2:175–80
2. Raychowdhury A, Keshavarzi A, Kurtin J, De V, Roy K (2006) Carbon nanotube field-effect transistors for high-performance digital circuits—DC analysis and modeling toward optimum transistor structure. IEEE T Electron Dev 53:2711–7

3. Javey A, Tu R, Farmer DB, Guo J, Gordon RG, Dai H (2005) High performance n-type carbon nanotube field-effect transistors with chemically doped contacts. Nano Lett 5:345–8
4. Jiménez D, Cartoixà X, Miranda E, Suñé J, Chaves AF, Roche S (2006) A drain current model for Schottky-barrier CNT-FETs. Comput Elect 5:361–4
5. Castro LC, John DL, Pulfrey DL (2002) Towards a compact model for Schottky-barrier nanotube FETs. In: IEEE optoelectronic and microelectronic materials and devices. Canada, pp 303–6
6. Mintmire JW, White CT (1998) Universal density of states for carbon nanotubes. Phys Rev Lett 81:2506
7. Datta S (2005) Quantum transport: from atom to transistor. Cambridge University Press, Cambridge, U.K.

Chapter 30
Optimizing Prosthesis Design by Using Virtual Environments

Adrian Zafiu, Monica Dascălu, Eduard Franti, and Annely-Mihaela Zafiu

Abstract This paper contains the presentation of a virtual environment which allows the rapid design and verification of human prostheses. Primarily, the platform has been created for inferior limb prostheses, but it is usable for any kind of prosthesis. The system allows the virtual testing of the prosthesis models, control systems, sensors, and the comparison between a real prosthesis and a virtual one. We can perform, in a virtual environment, tests on larger sets of configurations, in a very short time and with a very low cost.

30.1 Introduction

In this paper we present an interactive application allowing prosthesis development. The application comprises a virtual environment for designing and testing of different kind of prosthesis architectures in order to achieve complex abilities. This simulator allows designs at a low cost. A real prosthesis is expensive in terms of costs, and the existence of a versatile virtual environment replaces the tests with real prostheses. Also, the software environment allows the testing of the interactions between the prostheses, the human body, and the environment. In contrast with the experiments on real prostheses, in a virtual environment we can perform tests on larger sets of configurations in a very short time and with a very low cost.

The virtual environment is designed as a preliminary phase of real experiments. The platform allows the testing of control systems. After the virtual tests and system optimizations, the best logic developed in the virtual environment is implemented on a hardware platform integrated into the real prosthesis.

The results obtained with this platform are certainly different from the results obtained with a real prosthesis and offer new future developments.

A. Zafiu (✉)
Department of Electronics, Communications and Computers, University of Pitesti,
Str. Targul din Vale, nr.1, Pitesti, Arges, Romania
e-mail: adrian_zafiu@yahoo.com

N. Mastorakis et al. (eds.), *Proceedings of the European*   297
Computing Conference, Lecture Notes in Electrical Engineering 27,
DOI 10.1007/978-0-387-84814-3_30, © Springer Science+Business Media, LLC 2009

In the next stages of platform development, the simulation mechanism can be refined, so the results of future experiments are closer to real values. The graphics (including the human body shape [1, 2]) support future improvements. The system allows connections to other additional programs, each of them covering a different aspect of the limb simulation. For example, it is possible to analyze a prosthesis according to the patient's sex, age, height, and weight.

A main goal is designing and simulating in the virtual environment a prosthesis control system and movement system in accordance with all prosthesis facilities and patient assets.

The designed system comprises:

- A virtual environment. It allows generating, configuring, and modifying different work environments for the prosthesis testing.
- A module allowing a virtual prosthesis description. The users can create different prosthesis architectures according to the patient's disability.

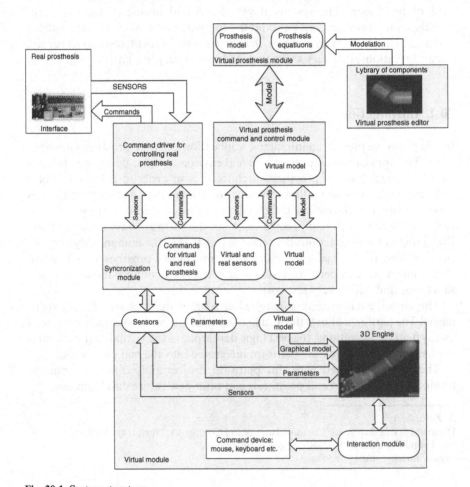

Fig. 30.1 System structure

- A module allowing real prosthesis specification. The module is necessary because the virtual prosthesis must have a behavior identical to the behavior of a real prosthesis, in the same simulated conditions (on different external stimulations). The virtual environment allows the connection to external hardware which reads a set of virtual sensors corresponding to real sensors.
- A command module generating signals to a set of engines used in real prosthesis movements, as a response to different kinds of external stimuli. The module works in real time and performs a continuous monitoring of the parameter states.
- A module implementing the control logic using, usually, a software specification.
- A module for synchronizing and coordinating the environmental components' evolution.

The system allows a comparison between the real and simulated prosthesis behaviors in order to solve the problems of real prostheses. The tests start only in a virtual environment, and, when the virtual prosthesis is free of errors, a real prosthesis is designed and its behavior is compared to the virtual model's behavior (Fig. 30.1).

30.2 The Virtual Module

This software module is also called the virtual environment. This module is used to generate, configure, and actualize the environment.

The virtual environment contains:

- a graphic 3D engine;
- an interactive system between the user and the graphic elements;
- a module for virtual prosthesis parameters administration;
- a module synchronizing graphic scene components according to each new value of the prosthesis parameters; and
- a set of functions for environment settings and design.

The software simulator for human body parts is an interactive open base for prosthesis design.

30.3 The 3D Engine

The 3D engine graphically shows a virtual prosthesis. We adopt two solutions for the engine:

- a powerful engine under DIRECTX [3, 4] to create a high quality scene;
- a less powerful engine using VRML [5] to enable online tests. The engine works under a web browser and it is accessible over the Internet.

30.4 Virtual Sensors

Virtual sensors simulate the interaction between

1. The elements of the prosthesis. This kind of sensor detects collisions between prosthesis elements and blocks illegal movements. The sensors are supposed to be hardware implemented and send values to the control module.
2. The prosthesis and the environment. This type of sensor comprises two categories: real sensor models (like a pressure sensor placed on the comprehension part), and events from the real world (collision sensors that overload the simulated engine effects and don't allow illegal movements).

These sensors capture the new prosthesis parameters as a result of the interaction between the user and the 3D engine, and also manage the collision tests.

The process of movement learning uses these sensors to create databases with movement sequences correlated with a time scale. The sensors are used in the validation process of the movement's equations.

30.5 The Virtual Prosthesis Module

The prosthesis is designed in this module. The design is adapted to each patient. It contains a library of interconnectable subelements and predefined prostheses. The prosthesis contains a set of subelements (link and mobile elements) custom defined or selected from the library, a set of sensors, and a set of action elements (Fig. 30.2).

This module also contains mathematical models of movements for the subelements. The software implemented for this module automatically creates a

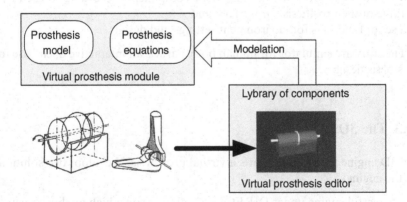

Fig. 30.2 Virtual prosthesis module

cinematic model [6] for the entire prosthesis during the design process. The prosthesis model is used during simulation to create the scene for the 3D engine.

30.6 The Synchronization Module

The middleware of the system is represented by the synchronization module. It interconnects all the other modules.

The entire real prosthesis, or only parts of it, are connected to the system by using a hardware interface. For testing, we use a kit of Velleman K8000 [7] connected to a parallel port. The kit has 16 digital I/Os, 8 analog outputs, and 4 analog inputs. The prosthesis model is implemented by using the virtual prosthesis editor. The system synchronizes the real and the virtual prosthesis. An adjustment of the virtual prosthesis is transmitted to the real prosthesis and vice versa. The movements of the real prosthesis are detected by the system and transmitted automatically to the prosthesis model, which computes the new attributes of the model. The synchronization module sends new attributes as parameters to the graphics engine, which repaints the graphic scene according to the real model.

Just as the interaction between the operator and the graphics engine implies scene modification, the synchronization module sends the new parameters of the virtual prosthesis to the model, the model computes the appropriate commands that are transmitted by using the synchronization module, and the hardware interface modifies the parameters of the real prosthesis.

Thus, all the time, the real prosthesis and the virtual prosthesis are in the same state (Fig. 30.3).

The system allows the storage of a set of movement sequences received from the real or the virtual prosthesis. This set of sequences can be transmitted anytime to the real prosthesis, or to the virtual prosthesis, or to both at the same time.

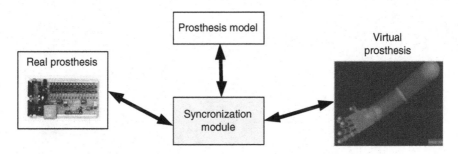

Fig. 30.3 The synchronization module

30.7 Conclusions

The software simulator for human body parts is an interactive open base for prosthesis design. The application allows the design, verification, and validation of different kinds of prosthesis architectures.

The utility of this platform is shown by the following:

- A real prosthesis is more expensive than virtual design and virtual testing.
- A virtual framework allows the testing of the behavior of individual components without integrating them into the system.
- The virtual framework is easy to use for modeling many prostheses.
- The virtual framework allows the testing of prosthesis behavior in different environments.
- It allows the testing of control systems.
- The software environment is used as the first stage of the physical experiments; first the control systems and prosthesis are simulated, then the best model is implemented as a real system.

References

1. Sinelnikov RD (1974) Human anatomic atlas. Russian edition. Medicine Publishing, Moscow
2. Papilian V (1982) The human anatomy. Editura didactica si pedagogica, Bucharest
3. Frank D (2006) Introduction to 3D game programming with DirectX 9.0 Luna. Wordware Publishing, Plano
4. Walsh P (2003) Advanced 3D game programming with DirectX 9.0. ISBN:1556229682 Wordware Publishing, Plano
5. Ames AL, Nadeauand DR, Moreland JL (1997) The VRML 2.0 sourcebook. Wiley Computer Publishing, New York
6. Stănescu A, Curaj A, Catana D (1997) Tehnici de generare automată a mişcării roboţilor. Bucharest
7. Velleman K8000 user manual.

Part IV
Multi-Agent Systems

A multi-agent system is a system composed of multiple interacting intelligent agents. In the most general case, agents will be acting on behalf of users with different goals and motivations. To successfully interact, they will require the ability to cooperate, coordinate, and negotiate with each other, much as people do. Multi-agent systems can be used to solve problems which are difficult or impossible for an individual agent to solve. Multi-agent systems are also referred to as "self-organized systems" because they tend to find the best solution for their problems "without intervention." There is great similarity here to physical phenomena, such as energy minimizing, in which physical objects tend to reach the lowest energy possible within the physically constrained world. The main feature which is achieved when developing multi-agent systems is flexibility, since a multi-agent system can be added to, modified, and reconstructed without the need for detailed rewriting of the application. These systems also tend to be rapidly self-recovering and failure proof, usually because of the heavy redundancy of components and the self-managed features. In this part we present four papers dealing with a Maude-based tool for simulating a DIMA model, a two-level morphology for recognizing derived agentive nouns, an agent framework to support distributed problem solving, and building moderately open multi-agent systems.

Chapter 31
A Maude-Based Tool for Simulating DIMA Model

Noura Boudiaf and Lakhdar Sahbi

Abstract The lack of formal semantics in the existing formalisms describing multi-agents models, combined with the complexity of multi-agents systems, are sources of several problems during their development process. The formal methods are known to bring rigorous and precise descriptions. They offer the possibility of checking the correction of specifications. In this framework and in previous papers (Ferber J, Gutknecht O (2002) MadKit user's guide. Version 3.1) [23], we have proposed a formal and generic framework called DIMA-Maude, allowing formal description and validation of the DIMA model [15] with the Maude language [22]. This language, based on rewriting logic [21], offers a rich notation supporting formal specification, implementation, validation through simulation, and verification through accessibility analysis and model checking of concurrent systems (Koning JL (1999) Algorithms for translating interaction protocols into a formal description. In: IEEE Int Conf Syst Man Cybernet Conf, Tokyo, Japan). In this paper, we propose a rewriting logic-based tool for creating a DIMA model. This tool allows the edition and the simulation of a DIMA system. It allows the user to draw a DIMA system graphically and translates the graphical representation of the drawn DIMA model to a Maude specification. Then the Maude system is used to perform the simulation of the resulting Maude specification. This tool allows preserving the graphic notations offered by the DIMA model for clarity and getting a formal specification in Maude for analysis.

31.1 Introduction

The formalization of multi-agents systems (MAS) is not a very recent idea. Many approaches aiming at formal MAS specification have been proposed in the literature: graphic methods such as Petri nets [2], approaches representing

N. Boudiaf (✉)
Computer Science Department, University of Oum El Bouaghi, 04000 Oum El Bouaghi, Algeria
e-mail: boudiafn@yahoo.com

N. Mastorakis et al. (eds.), *Proceedings of the European* 305
Computing Conference, Lecture Notes in Electrical Engineering 27,
DOI 10.1007/978-0-387-84814-3_31, © Springer Science+Business Media, LLC 2009

an adaptation of object-oriented specification methods like Lotos [8], and more recent approaches based on some kind of logic like temporal logic [13]. In the literature, the proposed approaches to formal MAS specification are often limited to some specific aspects. Several notations are often used to describe the same MAS. Such combinations constitute a serious obstacle to rigorous and founded checking of the properties of the described systems [9]. We showed in previous papers [3, 11] the feasibility and the interest in formalizing the agent's behavior of the DIMA multi-agents model using the Maude language. The constructions offered by this language are rich enough to capture the multiple aspects of the DIMA model. Maude is considered one of the most powerful languages in description, programming, and verification of concurrent systems. Maude is a formal language based on a sound and complete logic called rewriting logic [9]. This logic allows us to reason correctly on non-deterministic concurrent systems in terms of the "true concurrency" semantics. The majority of formal methods used in the framework of formalization of MAS do not bring anything more in terms of expressivity or verification power compared to rewriting logic, because they are integrated into the rewriting logic [9, 10].

The generated DIMA-Maude descriptions have been validated by means of simulation and model checking thanks to the Maude platform. We offer to the user to reuse the obtained model DIMA-Maude core. The user can model his application directly in Maude by importing modules implementing the DIMA-Maude core. However, execution under the Maude system is done by using the command prompt style. In this case, we lose the graphical aspect of DIMA formalism, which is important for the clarity and readability of a MAS description. Moreover, the DIMA model is represented in a graphic way (in a theoretical manner), and for the moment, this model is implemented on a Java platform. There is no tool allowing an automatic translation from the graphic notation to the Java program.

In this paper, we present an interactive tool to create a MAS by using DIMA notations. The tool proposed in this paper allows the user to graphically edit a DIMA system and then converts the graphical representation to its equivalent description in Maude. Thereafter, the tool calls the Maude system for the execution of the obtained code and reconverts the obtained result described in Maude to a graphical representation. With a DIMA system example, we will compare the simulations of the example under the Maude system and our tool.

Let's note that there are many tools for the creation and the manipulation of MAS, most of them implemented in the Java language. The best known ones are AgentTool [5], AgentBuilder [1], and MadKit [6]. These tools are Java-based graphical development environments to help users analyze, design, and implement MAS. With regard to these tools, only the preliminary specification in our application is diagrammatically developed by using the DIMA notations; the rest of the development process phase is made in Maude: formal specification, implementation, and validation by means of simulation. Our application has not yet reached the level of these tools in terms of widely offered services. For the moment, it is a prototype including a minimum of services, to be completed in the future. But, to our knowledge, the tool presented in this paper is the first rewriting logic-based analysis tool for MAS. This tool allows us to benefit from the power of rewriting logic in the specification and analysis of concurrent systems in the context of MAS.

The remainder of this paper is organized as follows: In Section 31.2, we present briefly the DIMA model and an example of a system described by this model. In Section 31.3, we give a short outline on rewriting logic and Maude. Section 31.4 presents in a short way our proposed process for the formalization of the DIMA model by using Maude. We show also in this section how we run the previous example under our application. The most important functionalities of our application are illustrated in Section 31.5. The technical aspect of our simulator is mentioned in Section 31.6. Finally, we discuss our current work and give some conclusions and ideas about future work in Section 31.7.

31.2 The DIMA Multi-Agents Model

The DIMA model [7] agent's architecture proposes to decompose an agent into different components whose goal is to integrate some existing paradigms, notably an artificial intelligence paradigm. An agent can have one or several components that can be reactive or cognitive. An agent is a simple component or a composite component, as illustrated in Fig. 31.1, that manages the agent's interaction with its environment and represents its internal behavior. The environment regroups all other agents as well as entities that are not agents. The basic brick of this architecture is represented by a proactive component on which are founded the different models of agents. A proactive component must describe:

- Goal: implicitly or explicitly described by the method IsAlive ().
- Basic behaviors: a behavior could be implemented as a method.
- Meta-behavior: defines how the behaviors are selected, sequenced, and activated. It relies on the goal (use of the method Step ()).

A communicating agent is a proactive agent that introduces new functionalities of communication. Every communicating agent must have:

- A mailbox to stock its messages.
- A communication component to manage the sending and the reception of messages.
- Possibly, a list of its acquaintances.

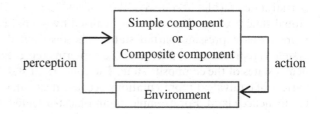

Fig. 31.1 DIMA agent's architecture

The meta-behavior of communicating agents is described by an ATN. This last is defined by a finished set of states and transitions. A state of the ATN describes the agent's state. To define an ATN, we must:

- Define states: There are three types of states, initial (1), final (0 or more), and intermediate (0 or more).
- Define transitions: Every transition has conditions (one or more), actions (one or more), and a target state.
- Attach transitions to states.

The construction of a proactive agent requires the description of its behaviors and its meta-behavior by inheriting the ProactiveComponent class. We must also describe the agent's goal by describing the method IsAlive(). In the case of agents whose meta-behavior would be described by an ATN, such as ATN-based communicating agents, the method IsAlive() tests if the final state is not reached. For the same type of agents, the method Step(), describing a cycle of the meta-behavior of the proactive component, allows us to activate from the current state of an agent a transition whose condition is verified.

31.2.1 Example: Auction Application

This section illustrates a concrete example. It is about a simple example of an auction. We have two kinds of agents: Auctioneer and Bidder.

Each auction involves one Auctioneer and several Bidders. The Auctioneer has a catalog of products. Before starting the auction, the Auctioneer sends the catalog to all the participants. Then he starts the auction of all the products. The products are therefore proposed sequentially to the participants. In this example we use the iterated contract net protocol. Figure 31.2 describes respectively the meta-behavior of the Auctioneer.

31.3 Rewriting Logic and Maude Language

The rewriting logic, having a sound and complete semantics, was introduced by Meseguer [9]. In rewriting logic, the formulas are called rewriting rules. They have the following form: $R: [t] \rightarrow [t']$ if C. The rule R indicates that term t is transformed into t' if a certain condition C is verified. Term t represents a partial state of a global state S of the described system. The modification of the global state S to another state S' is realized by the parallel rewriting of one or more terms expressing partial states. The distributed state of a concurrent system is represented as a term whose subterms represent the different component's states of the composite state. The concurrent state's structure can have a variety of equivalent representations because it satisfies certain structural laws (equivalence class). For example, in an object-oriented system, the concurrent

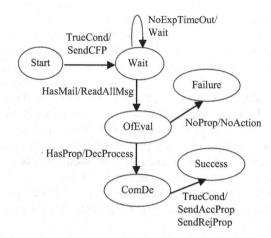

Fig. 31.2 ATNs describing the meta-behavior of the Auctioneer

state that is usually called configuration has the structure of a multiset of objects and messages. Therefore, we can see the constructed configurations as a binary operator applied to binary sets. The portion of the following program gives a definition of three types: *Configuration, Object,* and *Msg. Object* and *Msg* are subtypes of *Configuration*. Objects and messages are configurations containing one element (singleton).

```
mod CONFIGURATION is *** basic object system sorts
sort Configuration . sort Object. sort Msg.
*** Construction of configurations
subsort Object < Configuration. subsort Msg < Configuration.
op null: - > Configuration.
op__: Configuration Configuration - > Configuration [assoc comm id : null].
```

Complex configurations are generated from the application of the union of these multiset singletons. In the case where there are no floating messages or no alive objects, we have an empty configuration (null). The construction of a new configuration in terms of other configurations is done with the operation (__). We can see that this operation has no name and the two under-lines indicate the positions of the two parameters of the configuration type. This operation, which is the multiset union, satisfies the structural laws of associativity and commutativity. It also supports an identity element null.

Maude is a specification and programming language based on rewriting logic [4, 10]. Maude is simple, expressive, and efficient. It is rather simple to program with Maude, considering that it belongs to the declarative programming languages. It is possible to describe different types of applications by using Maude, from proto-typing ones to high concurrent applications. Maude is a competitive language in terms of execution and simulation with imperative programming languages. For a good modular description, three types of modules are defined in Maude.

Functional modules define data types and related operations, which are based on equations theory. System modules define the dynamic behavior of a system. This type of module augments the functional modules by the introduction of rewriting rules, and offers a maximum degree of concurrency. Finally, object-oriented modules can be specified by system modules. The object-oriented module offers a syntax more suitable than the system module to describe the basic entities of the object paradigm. To make the description easier for a user of Maude, the pre-defined object-oriented module *CONFIGURATION* is introduced, encapsulating the basic concepts of object-oriented programming. A part of this module is described in Section 31.4. The remainder of this module will be presented there-after. A typical form of a configuration is: Ob-1 ... Ob-m M-1 ... M-n.

Given that Ob-1 ... Ob-m are objects, and M-1 ... M-n are messages, it does not matter the order (commutativity). In general, a rewriting rule has the following form:

rl Ob-1 ... Ob-k M-1 ... M-n = > Ob-1' ... Ob-j' Ob-K + 1 ... Ob-m M-1' ... M-n'

where Ob-1' ... Ob-j' are updated versions of the objects Ob-1 ... Ob-j if j $\leq$ k, and Ob-k + 1 ... Ob-m are new created objects. If a left part of a rule contains only one object and only one message, then this rule is asynchronous. The rule is synchronous if its left side contains several objects. The remainder of the module *CONFIGURATION* is:

sorts Oid Cid . sorts Attribute AttributeSet . subsort Attribute < AttributeSet . op none : - > AttributeSet. op _,_:AttributeSet AttributeSet - > AttributeSet [ctor assoc comm id: none] . op <_:_ | _> : Oid Cid AttributeSet - > Object [Ctor object] .

In this syntax, the objects have the following general form: < O: C | att-1, ... , att-k > such that O is an identifier of an object, C is an identifier of a class, and att-1, ... , att-k are attributes of objects. Only one rewriting rule makes it possible to express the following at the same time: consumption of certain floating messages, sending of new messages, destruction of the objects, creation of new objects, changes of certain objects states, etc.

31.4 DIMA-Maude Model

For simplicity, we will give only a part of our formalization of the DIMA concepts by using Maude language as proposed in [3, 11]. We explain this formalization through the previous example.

31.4.1 Modeling of the DIMA Model with Maude

We developed a formal framework allowing the construction of communicat-ing agents. Their meta-behavior is described by an ATN. The framework is composed of several modules: the object-oriented module *ATN-BASED-*

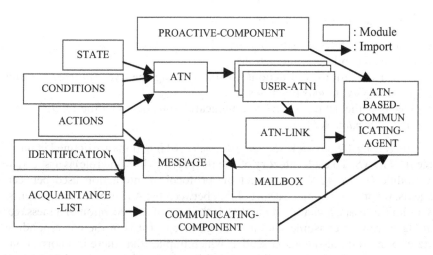

Fig. 31.3 Architecture of the framework DIMA-Maude

COMMUNICATING-AGENT and several functional modules (Fig. 31.3). The functional module *STATE* contains the necessary type declarations for the definition of a state, the definition of operations used for the construction, and the manipulation of a state and equations implementing these operations:

(fmod STATE is sorts State KindState NameState.
ops initial final ordinary : - > KindState.
op AgentState : NameState KindState - > State. op IsInitial : State - > Bool.
op IsOrdinary : State - > Bool . op IsFinal : State - > Bool.
var k : KindState . var ns : NameState .
eq IsInitial(AgentState(ns, k)) = if k = = initial then true else false fi .
eq IsOrdinary(AgentState(ns,k)) = if k = = ordinary then true else false fi .
eq IsFinal(AgentState(ns, k)) = if k = = final then true else false fi . endfm)

To define an ATN, we propose the *ATN* module. This module reuses the *STATE, CONDITIONS,* and *ACTIONS* modules. It includes the definition of two operations: *TargetState* that determines the state destination according to a state source and a condition, and the *AccomplishedAction* operation to determine the action executed according to a state and a condition.

(fmod ATN is protecting STATE . protecting CONDITIONS . protecting ACTIONS . op TargetState : State Condition - > State. op AccomplishedAction: State Condition - > Action . endfm)

For the construction of an ATN for an application, we propose to spread the *ATN* module in another module *USER-ATN:*

(fmod USER-ATN is extending ATN. ***User part*** endfm)

In this module the user must:

- Mention all states constituting the ATN.
- Define all conditions and actions.
- Attach conditions and actions to states, using the two *TargetStates* and *AccomplishedActions* functions.
- Specify if the action is type communication or not, using the *IsInternalActions, IsSendingActions.*

IsReceivingActions and *IsSendingActionToAll* functions are defined in module *ACTIONS*. To each category of agents (playing the same role) is associated a module *USER-ATN*. To respect the protocol of interaction used between agents, we propose to make a sort of link between the ATNs of different agents. This link consists in guaranteeing that at the time of the reception of a message, an agent cannot consume such a message except if it is in the corresponding state to the state of the agent sender, which implies that there is a correspondence between different agent states. As a sending action accomplished by an agent, the sender represents an event for an agent receiver. Therefore, there is a correspondence between sending actions of the sender and events received by the receiver. For that, the user must develop the ATN-LINK module that contains the correspondence on the one hand between different agent states, and on the other hand between actions of senders and conditions of receivers.

(fmod ATN-LINK is protecting USER-ATN.

op CorrspondingAgentState: State - > State.

op CorrspondingCondition: Action - > Condition . ***User part*** endfm)

To describe the identification mechanism of agents, we define the functional module *IDENTIFICATION* :

(fmod IDENTIFICATION is sorts AgentIdentifier . subsort AgentIdentifier < Oid . endfm)

The adopted form in our approach for messages is defined in the functional module *MESSAGE*. This last imports the modules *IDENTIFICATION* and *ACTIONS*

(fmod MESSAGE is protecting IDENTIFICATION . protecting ACTIONS .

sorts Message Content . subsort Action < Content .

op _:_:_ : AgentIdentifier Content AgentIdentifier - > Message . endfm)

The communicating agents are generally endowed with a Mailbox containing the received messages of other agents. For that, we define a functional module *MAILBOX* to manage Mailboxes of agents. This module imports the functional module *MESSAGE*. An agent must be endowed with a list of its acquaintances. In this way, it can exchange messages with other agents. For that, we define a functional module *ACQUAINTANCE-LIST* to manage lists of agent acquaintances. Due to limitation of space and the important size of these last two modules, we don't present them in this paper. The communicating agents are endowed with a module of communication that manages the exchange of messages between agents. In our approach, we define a functional module *COMMUNICATION-*

COMPONENT that imports the two *ACQUAINTANCE-LISTS* and *MESSAGE* modules. In this module, we define three messages: *SendMessages* and *Receive-Messages* that have as parameter a message of the form defined in the module *MESSAGE:*

(fmod COMMUNICATION-COMPONENT is
protecting ACQUAINTANCE-LIST . protecting MESSAGE .
subsort acquaintance < AgentIdentifier . op SendMessage :Message- > Msg .
op ReceiveMessage : Message - > Msg . endfm)

Indeed, a communicating agent is first a proactive agent that has a goal to achieve described by the function *IsAlive*. The *Parameters* and *Void* types are generic. They can be replaced by other types in modules wanting to reuse the *PROACTIVE-COMPONENT* module, provided that these new types are undertypes of *Parameters* and/or *Void*. The basic cycle of the meta-behavior of proactive components described by method *step( )* is modeled by rewriting rules of the module *ATN-BASED-COMMUNICATING-AGENT.*

(fmod PROACTIVE-COMPONENT is sort Parameters Void .
op IsAlive : Parameters - > Bool . op Step : Parameters - > Void . endfm)

The object-oriented module *ATN-BASED-COMMUNICATING-AGENT* is the main module. It imports the *PROACTIVECOMPONENTS, ATN-LINK, COMMUNICATION-COMPONENT* and *MAILBOX* modules. For a formal description of communicating agents, we propose the class Agent (line [1]). The definition of this class has the *CurrentStates, MBox,* and *AccList* attributes, to contain, in this order, the current state of the agent, its Mailbox, and the list of its acquaintances. For ATN-based communicating agents, the description of the goal requires redefining the *IsAlive* function. We use a state as a parameter of this function (line [5]) that returns a *Boolean*. This function returns the *Boolean* value "true" if the current state is final, otherwise it returns "false". The *State* type is an undertype of *Parameters*. The appearance of an event is expressed by the message *Event* (line [2]) having as parameters an agent, a state, and a condition. The execution of the corresponding action for this event is described by the message *Execute* (line [3]) if it is internal, either by one of the two *SendMessages* or *ReceiveMessages* defined in the module *COMMUNICATION-COMPONENT*. The *GetEvent* message (line [4]) is a message used temporarily during the execution of rewriting rules.

(omod ATN-BASED-COMMUNICATING-AGENT is
pr PROACTIVE-COMPONENT. pr ATN-LINK. pr COMMUNICATION-COMPONENT.
pr MAILBOX .subsort State < Paramaters .
class Agent | CurrentState : State, MBox : MailBox, AccList: acquaintanceList ***[1]
Msg Event : AgentIdentifier State Condition - > Msg .   ***[2]
Msg Execute : Action - > Msg .   ***[3]

Msg GetEvent : AgentIdentifier State Condition -> Msg . ***[4]

vars A A1: AgentIdentifier . var NS : NameState . var KS : KindState . vars S S1 : State . var Cond
: Condition . var Act : Action . vars MB MB1 : MailBox . vars ACL ACL1 : acquaintanceList .

eq IsAlive(AgentState(NS,KS)) = IsFinal(AgentState(NS,KS)) ***[5]

***********************Firstcase***********************

crl [StepCaser1] : Event(A, S, Cond) < A : Agent | CurrentState : S, MBox : MB, AccList :
ACL > = > Execute(AccomplishedAction(S, Cond)) GetEvent(A, S, Cond) < A : Agent |
CurrentState : S, MBox : MB, AccList : ACL > if (IsAlive(S) = = false) and
(IsInternalAction(AccomplishedAction(S, Cond)) = = true) .

rl [StepCase11] : GetEvent(S, Cond) Execute(Act) < A : Agent | CurrentState : S, MBox : MB,
AccList : ACL > = > < A : Agent | CurrentState : TargetState(S, Cond), MBox : MB, AccList :
ACL > .

***Other cases . . . endom)

The meta-behavior is expressed in four possible cases, the first, when the
action to execute is a simple internal action. In this case, the meta-behavior is
described by the two rewriting rules of the first case. One of the three
remaining cases is used when it is about an action of communication. If the
action of communication is a sending action, the meta-behavior is described,
either by the two rewriting rules of the second case (if the destination is only
one agent), or by the two rewriting rules of the fourth case (if the message is
sent to all agents).

31.4.2 Simulation of the Example Under Maude

After taking an example of the initial state, we find the final state of the
unlimited rewriting in Maude of this initial configuration in Fig. 31.4.

31.5 Steps of the DIMA Model Simulator

As depicted in Fig. 31.5, the architecture of the simulator is realized in seven
units:

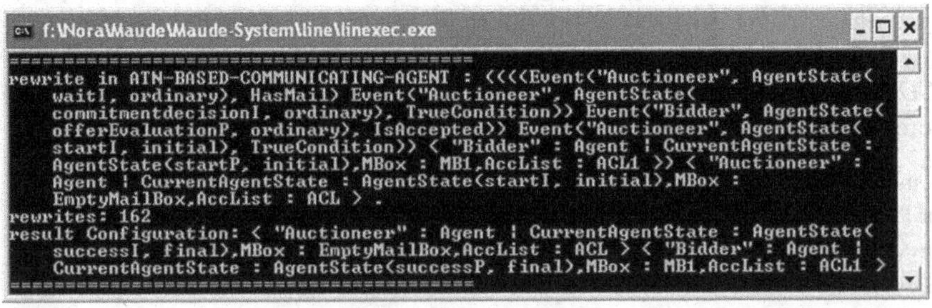

Fig. 31.4 Execution of DIMA-Maude program example under Maude system

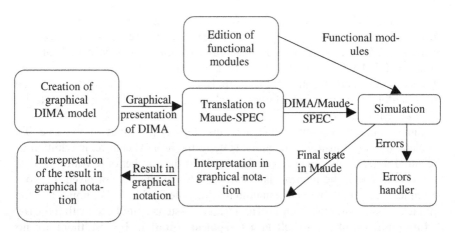

Fig. 31.5 Methodical view of DIMA model simulator

Graphical User Interface. In this step, the user can create a new DIMA system. In Fig. 31.6, we find the principal window, on which the previous example of DIMA has been created. On the right and on the left of the window, we find the two ATNs describing respectively the meta-behavior of the Bidder and the Auctioneer. At the middle of the window, we find a class diagram of the MAS example.

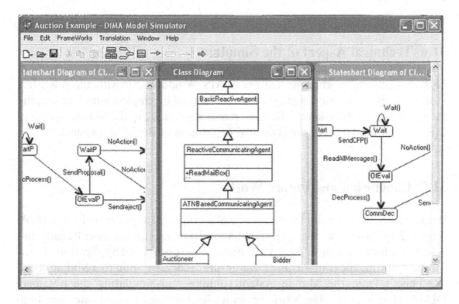

Fig. 31.6 Tool main menu of DIMA simulator with an example

Generation of Code Maude. This step has the graphical representation of a DIMA model as input. It consists of translating this representation into an equivalent Maude code. This representation contains the structure of one model of DIMA (class, collaboration, states-transitions) and the initial state (if necessary). The output of this step is two elements: a code in Maude equivalent to the structure of the translated model, and an initial state by using the Maude syntax.

Simulation. The output of the previous steps is the input of this one. To simulate, the user must give the initial state to the tool. Therefore, in addition to the initial state, the user may give to the simulator the number of rewriting steps, if he wants to check the intermediate states. If this number is not given, the tool continues the simulation operation until reaching the final state. We notice that the infinite case is possible and our tool as a Maude system cannot deal with this case.

Interpretation of the Result in a Graphical Notation. In case there are no errors, the user obtains his result (final configuration) in a dialog box as that of the initial configuration. This result is one returned by Maude after being reinterpreted in our description.

Edition of Functional Modules. Maude provides a library of many data types widely used, like integers, reels, and lists. The user can use these types or define new data types. We propose a simple editor to create new data types. We use the syntax of functional modules as defined. Functional modules and generated system modules (implementing DIMA descriptions) are integrated to form executable Maude code.

Error Handling. If some errors have been detected, we display to the user the errors generated by Maude.

31.6 Technical Aspect of the Simulator

This environment is implemented under MS Windows XP with the following tools: The programming language Delphi 8.0 is used for implementing the graphical editor and the translator. The most recent version 2.2 of the Maude system is used for the simulation of the generated description of the DIMA system.

31.7 Conclusion and Future Work

In this paper, we have shown the outline of a graphical application for a DIMA model. This tool allows using the DIMA model in a simple manner by introducing a graphical interface. This is not new, but our proposed application allows for the first time the translation of this graphic representation to a Maude code. We benefit from the Maude system running and simulating the obtained DIMA-Maude code. The work presented in this paper constitutes the first step in the construction of an environment for edition and analysis of the

DIMA model. We plan integrating into our proposed environment analysis techniques supported by Maude. Such techniques include debugging and troubleshooting, in addition to verification by means of accessibility analysis and model checking. We also plan the integration of a client server execution technique supported by Maude into our environment. Our work concerning the description and verification of MAS is not limited to the DIMA model. We have also created a formal version of Maude for AUML in [12]. We consider this work as an important investigation way, because we can use the same platform (Maude) to allow communication between some MAS applications first developed by using different models.

References

1. AgentBuilder RM (2000). An integrated toolkit for constructing intelligent software agents. Reference Manual
2. Bakam I, Kordon F, Le Page C, Bousquet F (2000) Formalization of a spatialized multiagent model using coloured Petri nets for the study of a hunting management system. In: FAABS2000, Greenbelt
3. Boudiaf N, Mokhati F, Badri M, Badri L (2005) Specifying DIMA multi-agent models using Maude. Springer-Verlag, Berlin, Heideberg LNAI 3371:29–42
4. Clavel M et al. (2005) Maude manual (Version 2.2). Internal report, SRI International
5. DeLoach SA, Wood M (2001) Developing multiagent systems with agent tool. In: ATAL'2000, Berlin
6. Ferber J, Gutknecht O (2002) MadKit user's guide. Version 3.1
7. Guessoum Z (2003) Modèles et architéctures d'agents et de systèmes multi-agents adaptatifs. Dossier d'habilitation à diriger des recherches de l'Université Pierre et Marie Curie
8. Koning JL (1999) Algorithms for translating interaction protocols into a formal description. In: IEEE International conference on systems, man, and cybernetics. Tokyo, Japan
9. Meseguer J (1996) Rewriting logic as a unified model of concurrency: a progress report. Springer-Verlag LNCS 119:331–372
10. Meseguer J (2000) Rewriting logic and Maude: a wide-spectrum semantic framework for object-based distributed systems. In: FMOODS2000
11. Mokhati F, Boudiaf N, Badri M, Badri L (2004) DIMA-Maude: Toward a formal framework for specifying and validating DIMA agents. In: Proceedings of Moca'04, Aarhus, Denmark
12. Mokhati F, Boudiaf N, Badri M, Badri L (2007) Translating AUML diagrams into Maude specifications: a formal verification of agents interaction protocols. J Obj Technol, ISSN 1660–1769, USA
13. Torroni P (2004) Computational logic in multi-agent systems: recent advances and future directions, computational logic in multi-agent systems. Ann Math Artif Intell 42(1–3):293–305

Chapter 32
Using Two-Level Morphology to Recognize Derived Agentive Nouns

Arbana Kadriu

Abstract This paper is about morphophonological modeling of derived agentive nouns. The system is implemented using the two-level morphology model, through the PC-KIMMO environment. The phonological rules are implemented through nine two-level rules, while the morphotactics is implemented using a finite state automaton.

32.1 Introduction

The core task of computational morphology is to take a word as input and produce a morphological analysis for it. Morphotactics is the simple concatenation of the underlying morphemes of an word, and it is usually expressed through finite state automata. But there are situations where the word formation process is not a simple concatenation of morphemes, such as vowel harmony, reduplication, insertion, etc., and it is here that the phonological rules become relevant.

Many linguists have modeled phonological rules, but it is considered that the most successful one is the model called *two-level morphology* (Koskenniemi 1983) [1]. This model includes two components:

- the phonological rules described through finite state transducers,
- the lexicon, which includes the lexical units and the morphotactics.

In this model, the formalism called *two-level phonology* is used. The Koskenniemi model is two-level in the sense that a word is described as a direct, symbol-to-symbol correspondence between its lexical and surface forms. For example, the agentive noun *maker,* is obtained as a concatenation of the verb *make* with

A. Kadriu (✉)
CST Department, SEE-University, b.b. 1200 Tetovo, FYROM
e-mail: a.kadriu@seeu.edu.mk

N. Mastorakis et al. (eds.), *Proceedings of the European Computing Conference*, Lecture Notes in Electrical Engineering 27, DOI 10.1007/978-0-387-84814-3_32, © Springer Science+Business Media, LLC 2009

the suffix *er*. But when adding the suffix, the last character of *make* (*e*) is deleted. The two-level representation for this case will be:

Lexical Form: m a k e + e r
Surface Form: m a k 0 0 e r

This approach is used in developing a system for general morphological analyses of the English, Hebrew, Japanese, Turkish, Finnish, Tagalog, and a few other languages. As far as we know it is not done for a particular grammar category.

The subject of the research presented in this paper is to build a system that will recognize derived agentive nouns. A noun that marks an active participant in a situation is called an agentive noun. Usually these nouns are created as a concatenation of a verb and/or noun (object or prepositional object) with an element from a predefined set of suffixes [2]. For example: administrator, activist, adapter, airman, etc.

32.1.1 Two-Level Rules

The two-level rules consist of three components: the correspondence, the operator, and the environment [3].

Every pair lexical symbol–surface symbol is called a correspondence pair. The notation for this correspondence is:

lexical symbol : surface symbol

For the first character of the previous example, we would write *m:m,* while for morpheme boundary we have +:0.

The operator defines the relation between the correspondence and the environment. There are four operators, illustrated as follows:

- =>—the correspondence appears only in this environment, but not necessarily always;
- <=—the correspondence appears always in this environment, but not necessarily only in this one;
- <=>—the correspondence appears always and only in this environment;
- /<=—the correspondence never appears in this environment.

The third component relates to the environment and specifies the phonological context where a specific phenomenon happens. The notation for this component is realized through the underline sign "_," called the environment line, and its general form is LC_RC, where LC denotes the left context, while RC denotes the right context.

32.2 Two-Level Rules for Morphological Recognition of Derived Agentive Nouns

In total, nine two-level rules are implemented to recognize the derived agentive nouns. The rules are implemented for those cases that come into view more than twice, that is, there are no rules implemented for situations that appear only once or twice. All implementations are done using state transition tables.

Because of the fact that every valid correspondence used in the system must be declared explicitly, rules have to be implemented for the standard correspondence. So, the first and the second rules implemented in this system relate to the standard correspondence of vowels and consonants. These transducers have only one state, that state is finite and every transition ends in it.

32.2.1 Remaining Rules

Before going further with the description of the other rules, it should be noted that there are situations when two or more rules can conflict with each other. For example, let us consider the following rules:

1. x:y => S __ S
2. x:y => t __

It is obvious that these two rules are in conflict with each other, because both say that "the correspondence can appear only in this environment." In these cases, the rules are combined in a single one, so that the left side of the new rules shows the correspondence, while the right side shows the new environment that is the disjunction of the two previous contexts:

3. x:y => [S __ S | t __]

Second, a rule can apply to more than one symbol. In these conditions, a whole *subset* of symbols that relate to the same particular rule is defined. This subset is declared prior to declaring the transducer, with a name that does not contain white spaces, and then, in the transducer, we refer to that name, instead of referring to every symbol individually.

Third, the phonological processes which delete or insert symbols (or strings), in the two-level model are expressed through the correspondence with the NULL symbol, here written as 0.

RULE 3 "[[e:0<=>_+ :0 S]| [a:0<=_+ :0]| [a:0 t:0 e:0<=>_+ :0 a n t]]"

This rule is in fact an intersection between rules "e:0 <=>_+ :0 S," "a:0 t:0 e:0 _+ :0 a n t" and "a:0 <=_ + :0," but they are combined in one since there is a conflict.

(a) RULE "e:0 <=>_+ :0 S"

The rule says that "if *e* is the last character of the word to which a suffix that starts in *a, o, i* or *e*, is added, then the character *e* will be removed." [1].

Example:

Lexical Form: s o l v e + e r
Intermediate Form: s o l v 0 0 e r
Surface Form: s o l v e r

(b) RULE "a:0 < = __ + :0"

The rule says: if a word ends in *a*, then it will be deleted in every suffixation.

Example:

Lexical Form: p r o p a g a n d a + i s t
Intermediate Form: p r o p a g a n d 0 0 i s t
Surface Form: p r o p a g a n d i s t

(c) RULE "a:0 t:0 e:0 < = > __ + :0 a n t"

For this rule, the following principle applies: if a word ends in *ate*, and the suffix *ant* it is added to it, the characters *a, t and e* will be removed.

Example:

Lexical Form: p a r t i c i p a t e + a n t
Intermediate Form: p a r t i c i p 0 0 0 0 a n t
Surface Form: p a r t i c i p a n t

There is a conflict between rules (a) and (c) in relation to the fact that the last character of *ate is e*, while the first character of the right context of the second rule is an element of the set of symbols related to the right context of the first rule.

The second conflict relates to the rules (b) and (c). Namely, rule (c) says that it stands only for this environment, and that is true (*ate* is deleted only when it precedes *ant*). But, the deletion of *ate* includes also the correspondence *a:0*, and this leads to a conflict situation with rule (b), which also contains this correspondence, but in another context.

RULE 4 "y:i < = > S4 __ + :0 e r"

The rule says: in the words that end in *y*, that follows symbols *f, l, n, p, r* or *v* (left context), it changes into *i*.

Example:

Lexical Form: n o t i f y + e r
Intermediate Form: n o t i f i 0 e r
Surface Form: n o t i f i e r

---

[1] The sign + represents the morpheme boundary, and 0 the NULL character.

For this rule the operator is $<=>$, meaning that this situation is valid only, and always, in this con-text. For example, for the surface form of the agentive noun *employer* (lexical form *employ + er*), there is no reflection of *y* to *i*, because of the fact that it does not fulfill the left context condition.

RULE 5 "0:S1 = > S1__ + :0 e r"

While the previous rules were about deletion and change of symbols, this rule is used to insert characters using the NULL character. In fact, when a character from the subset *{b, d, g, k, l, m, n, p, t}*, is at the end of a word to whom the suffix *er* is added, it's doubled.

Example:

Lexical Form: g r a b 0 + e r
Intermediate Form: g r a b b 0 e r
Surface Form: g r a b b e r

This doubling appears only in this context, but not always: the agentive noun *climber* is gained as a result of concatenation of the verb *climb* and suffix *er*, and it ends in *b*, but there is no doubling of this symbol. That's why the operator $=>$ is put here.

RULE 6 "[[s:0 | ism:0 | ize:0] <=> __ + :0 i s t] | [s:0 <=> __ + :0 ian]]"

All previous rules were about suffixation of verbs,[2] except subrule 2 of the third rule. This rule and the subsequent ones are about suffixation of nouns.

This is the most complex rule, and it is a combination of two rules:

- if the morpheme boundary and the suffix *ist* are preceded by the character *s*, or the strings *ism* or *ize*.[3] these are removed.
- if the morpheme boundary and the suffix *ian* are preceded by the character *s*, that character is removed.

Example for the first rule:

Lexical Form: p h y s i c s + i s t
Intermediate Form: p h y s i c 0 0 i s t
Surface Form: p h y s i c i s t

Example for the second rule:

Lexical Form: c o s m e t i c s + i a n
Intermediate Form: c o s m e t i c 0 0 i a n
Surface Form: c o s m e t i c i a n

---

[2] The rule for deleting the last e states also for the nouns, for example, from prose + er, it is derived proser. The doubling rule relates also to the nouns, for example, the surface form gunner, it is derived from the lexical form gun + er.

[3] This rule is valid for the verbs, not for the nouns.

The string *ism* is always deleted in this environment and only in this environment. It is the same for the string *ize*. The character *s* is always deleted when it appears before the suffix *ist*, but also when it comes before the suffix *ian*. So, the rule for this character is the intersection of two disjunct rules.

RULE 7 "y:0 = > S2__+ :0 [er | ian | ist]"

This rule is in fact a combination of three rules: the character y is deleted, if it supercedes one of the characters *d, g, h, r, l, m, n, p,* and it precedes the suffixes *er, ian* or *ist.*
Example:

Lexical Form: p h i l o s o p h y + e r
Intermediate Form: p h i l o s o p h 0 0 e r
Surface Form: p h i l o s o p h e r

This rule is valid only for this environment, but not always. For example, essayist is gained from the suffixation process of the noun *essay* with *ist,* so there is no deletion. That's why the operator = > stays here.

RULE 8 "[[0:s = > __ + :0 man] | [0:i = > __ + :0 cide]]"

This rule represents in fact a composition of two seemingly disjunctive rules, but with two common elements: it represents an insertion (in the first case it is the character *s,* in the second it is the character *i*); and it has only a right context (for the first case it is the suffix *man,* for the second one it is the suffix *cide*).
Example:

Lexical Form: c r a f t 0 + m a n
Intermediate Form: c r a f t s 0 m a n
Surface Form: c r a f t s m a n

This insertion is valid only under this environment, but not always. The agentive noun *birdman* is created by adding the suffix *man* to the noun *bird,* but there is no insertion of *s.*

RULE 9 "o:0 <= > __ + :0 ist"

According to this rule, if an agentive noun is obtained by adding the suffix *ist* to a noun that ends in *o,* then this character is removed.
Example:

Lexical Form: p i a n o + i s t
Intermediate Form: p i a n 0 0 i s t
Surface Form: p i a n i s t

This happens always and only in this environment.

32.2.2 The Lexicon

The lexicon consists of one main file (.LEX) together with additional files of lexical units. In the main lexicon, the description of the morphotactics of the implemented system is given. It also contains the information about the included supplemental files. These files hold the lists of lexical units, such as nouns, verbs, and adjectives.

The morphotactics implemented for this system is simple. First, the set of used suffixes is defined:

- suffixes that are added to the nouns: er, ian, ist, or, ee, ier, arian, man, cide, monger, smith, ster, boy, 0;
- suffixes that are added to the verbs: ant, ent, ar, er, ee, ist, or, man, 0.
- After this set is outlined, it is declared that an agentive noun can be:
- verb + verbal suffix,
- noun + noun suffix,
- noun + verb + verbal suffix,
- adjective + verb + verbal suffix.

32.2.3 Results and Examples

After all the phonological and morphotactical rules are implemented, together with the noun, verb, and adjective files, the system for morphological recognition of the agentive nouns is tested. Figure 32.1 gives a

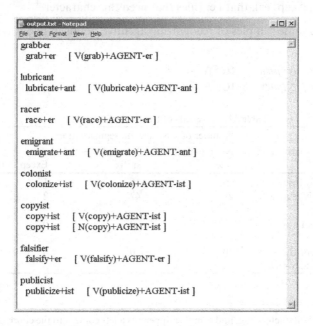

Fig. 32.1 Examples of morphological segmentation of agentive nouns

picture of some examples of morphological segmentation using the PC-KIMMO environment [4].

The system is tested on: 639 agentive nouns that are derived only from a verb, 504 agentive nouns that are derived only from a noun, 124 agentive nouns that are derived from a verb and a noun, and 5 agentive nouns that are derived from an adjective and a verb.

Table 32.1 shows the statistics of the achieved results.

No implemented rule for the agentive nouns that are derived only from a verb or only from a noun is a consequence of the fact that phonological rules are not implemented. While, for those that are derived from a noun and from a verb, that is a consequence of the fact that morphotactics rules are not implemented. Namely, the morphotactics states that the verb always comes after the noun. But there are two exceptions: *cutpurse (cut + purse)* and *makepeace (make + - peace)*, where the verb precedes the noun.

There are only three cases as exceptions:

- The agentive noun that is derived from a verb: *flyer (fly + er)*, as an exception to rules 4 and 7, i.e., the only case when *y* is preceded by the character *l*, and precedes *+ er*, but does not change to *i* and is not deleted.
- The agentive nouns that are derived from a noun: *sorcerer (sorcery + er)*, *adulterer (adultery + er)*, that are not recognized because a conflict forms between rules 4 and 7. According to rule 4, when *y* follows *r* and precedes *+ er*, it changes to *i;* while according to rule 7, it is deleted. The character *r* must be an element of the subset used by rule 7, because it contains other suffixes (except *er*), that i.e. rules that need this character.[4]

Here are some samples of recognizing agentive nouns in two ways:

jurist
jura + ist [N(jura) + AGENT-ist]
jury + ist [N(jury) + AGENT-ist]

Table 32. 1 Results of the implemented system

Type of the agentive noun that:	Number of unsuccessful segmentations			Success rate
	No implemented rule	No implemented suffix	Exceptions	
Is derived only from a verb	6	6	1	98%
Is derived only from a noun	29	42	2	86%
Is derived from a verb and a noun	2	0	0	98%
Is derived from an adjective and a verb	0	0	0	100%

---

[4] Namely, the characters h, g, and double r appear as a left context to the combination y + er, the other characters in the subset are part of the rule because of the other suffixes.

lyricist
lyric + ist [N(lyric) + AGENT-ist]
lyrics + ist [N(lyrics) + AGENT-ist]
organist
organize + ist [V(organize) + AGENT-ist]
organ + ist [N(organ) + AGENT-ist]
impressionist
impression + ist
[N(impression) + AGENT-ist]
impressionism + ist
[N(impressionism) + AGENT-ist]
systematist
systematize + ist
[V(systematize) + AGENT-ist]
systematics + ist
[N(systematics) + AGENT-ist]

The output also contains two morphologically, but not semantically, correct segmentations:

warrantor
warrant + or [N(warrant) + AGENT-or]
warrant + or [N(war)V(rant) + AGENT-or]
prosecutor
prosecute + or [V(prosecute) + AGENT-or]
prosecut + or [N(prose)V(cut) + AGENT-or]

These two cases are, in fact, the only cases of semantically non-correct segmentations that are produced in the process.

32.3 Conclusions

This paper was about morphological segmentation of derived agentive nouns. These can also be of the form that has white space in it, like *candy maker,* or hyphenation, like *script-writer.* In the future the system can be extended to handle these cases as well.

The second route for a possible expansion of this system is to use it for machine translation of the agentive nouns.

References

1. Mitkov R (2003) The Oxford handbook of computational linguistics. Oxford University Press, Oxford

2. Kadriu A (2006) Analysing agentive nouns. In: Proceedings of the 28rd international conference on information technology interfaces. Cavtat, Croatia
3. Antworth EL (1995) Developing the rules component. North Texas natural language processing workshop. University of Texas, Arlington, USA
4. Antworth EL (1995) Introduction to PC-KIMMO. North Texas natural language processing workshop, University of Texas, Arlington, USA

Chapter 33
Towards an Agent Framework to Support Distributed Problem Solving

G.A. Mallah, Z.A. Shaikh, and N.A. Shaikh

Abstract With the development of multi-agent systems, agent-oriented programming is getting more and more attention. An abstract level view is being provided to programmers to develop agent-based applications to solve distributed problems. Many multi-agent systems have been developed, but the least importance has been given to .NET. Because of the unavailability of the agent platform based on .NET, programmers are unable to exploit .NET services to build agent applications for the solution of the grid-like applications that are distributed in nature.

An agent collaborative environment based on .NET (ACENET) has been developed that provides the programming platform to develop application agents at the application layer. At present, agents have been developed that solve distributed problems such as matrix multiplication.

33.1 Introduction

Software agents are computational systems that inhabit some complex dynamic environment and that sense and act autonomously in that environment to achieve the tasks for which they are designed. Agent-oriented software is the appropriate paradigm with which to model, design, and implement complex and distributed software systems [2]. It is viewed as an abstraction with intelligent behavior to cope with the complexity of system design. Agent communication languages, interaction protocols, and agent markup languages are designed to provide high-level abstraction. In distributed systems agents are viewed as autonomously migrating entities that act on behalf of network nodes [3]. Agent platforms are the development environments that provide agent builders with a sufficient level of abstraction to allow them to implement intelligent agents with the desired attributes, features, and rules. When agents socially interact with

G.A. Mallah (✉)
National University of Computer and Emerging Sciences, ST-4, Sector-17/D,
Shah Latif Town, Karachi, Pakistan

N. Mastorakis et al. (eds.), *Proceedings of the European*
Computing Conference, Lecture Notes in Electrical Engineering 27,
DOI 10.1007/978-0-387-84814-3_33, © Springer Science+Business Media, LLC 2009

each other, they need some environment. Such an environment is called a multi-agent system (MAS), and involves coordination among agents to solve complex problem. A grid can be viewed as an integrated computational and collaborative system where users interact with the grid application (at a high level) to solve problems through the underlying distributed grid resources that handle resource discovery, scheduling, and processing tasks [4].

The .NET platform has been exploited for multi-agent systems since 2004. CAPNET [5] and MAPNET [6] were the initial multi-agent systems developed in .NET.

We propose a new agent platform, ACENET (Agent Collaborative Environment based on .NET), that supports distributed problem solving. As .NET is the emerging technology, therefore the proposed ACENET platform exploits the advanced features of .NET.

The platform is completely distributed, because there is no server or controlling machine, as normally happens in other platforms [7]. The disadvantage of client-server architecture is that if the server goes down, the complete system and all agents will be shut down. In our system, all platforms have the ability to work as the server. In the same way, when the server needs some services from the other machines (the agents), it will work as a client and can be connected to any machine just by sending a request. All machines can independently connect to or disconnect from the framework. If one node goes down, all the other nodes remain intact and function without disturbance. The entire system will not shut down; only the agents of that machine will not be available to provide services.

We have created agents that solve the application of matrix multiplication, which is a good example of parallel computing.

33.2 Agent Standardization

Agent technology has gained a huge amount of attention in the research community during the last few years. This new area of computing, which has its roots in software engineering, distributed computing, and artificial intelligence, has been used in a large number of commercial applications, ranging from e-commerce to a large number of defense applications. In order to facilitate the agent-based approach to software development, a large number of agent frameworks like JATLite, JADE, Aglets, etc., have been developed. With the increasing number of these frameworks, two parallel agent development standards have evolved: FIPA and MASIF.

The Foundation for Intelligent Physical Agents (FIPA) is a standard to promote interoperable agent applications and agent systems. FIPA specifications only talk about the interfaces through which agents may communicate with each other. They do not describe the implementation details. FIPA specifications are divided into five categories: applications, abstract architecture, agent communication, agent management, and agent message transport [8, 9].

The FIPA reference model considers an agent platform as a set of four components: the agents, the directory facilitator (DF), the agent management system (AMS), and the message transport system (MTS). The DF and the AMS support the management of the agents, while the MTS provides a message delivery service.

The FIPA standard doesn't talk about mobility as a mandatory requirement for a FIPA-compliant system. FIPA provides some guidelines for how implementation providers can support mobility on their own. In addition, the specification that talks about this is in its preliminary stages [10].

Parallel to FIPA, another agent standard, MASIF, has also been developed by the Object Management Group (OMG), whose main intention was to integrate all mobile technology. This gave rise to a standard model for distributed computing. MASIF standardizes several terminologies related to the agent execution environment [11]. MASIF does not define any communication language. It does address the mobility issue, but it does not address advanced level of mobility features like continuations, etc.

33.3 Implementation Details

33.3.1 ACENET

The ACENET architecture (illustrated in Fig.33.1) has been designed so that developers may create and integrate complete distributed agent applications. Application agents are the main component of the framework. They are autonomous and have their own thread of execution. They can utilize the services provided by the platform, such as the directory service, the message transport service, the agent transport service, etc. ACENET agents have the capability to interact with each other using FIPA interaction protocols. Messages are encoded in the FIPA agent communication language.

Agents can register, de-register, and modify their services to the directory facilitator. An agent can find other agents, within the same ACENET or a remote ACENET, by querying the directory facilitator with the AgentID, AgentName, or ServiceType. That makes agents able to create an emerging virtual society, where they are able to find and interact with each other for cooperation and collaboration to solve problems.

An abstract agent class is provided with all the necessary functionality. The developer should derive its agent's class from the abstract agent class. Abstract functions are provided in order to notify the derived agent classes of essential events such as message notification and agent notification before, during, and after migration notifications, etc.

The AMS takes agent assemblies and creates instances of the agents by using the services of the reflection API provided by the .NET platform. Each agent is loaded in a separate application domain in order to isolate the agent code and

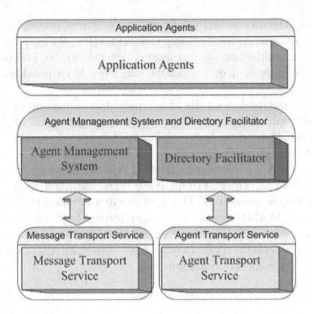

Fig. 33.1 ACENET architecture

data from other agents. The great advantage associated with this approach is that once an agent assembly is loaded in an application domain it can be unloaded. The agents are accessed via a proxy that is created by the .NET framework, thereby providing a balance between isolation and security of code and data.

The AMS handles all the communication, interaction, and conversation mechanisms at the platform and agent level. The AMS is able to handle tasks like creating agent, suspending agent, resuming agent, and killing agent. When an agent is started, the AMS assigns a globally unique identifier, called the AgentID, to the agent and stores its instance in the local agent container.

The DF provides a "yellow pages" service to the agents. Any agent can search any other agent by AgentID, Agent Name, or Agent Service type. Only those agents can be searched that are registered to the DF. An agent can register/de-register itself to the DF and can also modify its registration details. The DF can also be used to query for current and remote platform addresses respectively.

A socket-based implementation has been done to control the transportation of messages and agents. Both client and server sockets have been implemented, enabling the platform to send and receive messages and agents to and from the remote ACENET. RemoteReceiver and RemoteSender are the classes made for this purpose. RemoteReceiver and RemoteSender follow the publish-subscribe mechanism. If an agent needs to be notified of a message destined for it, it should subscribe to the RemoteReceiver service of the platform. This service asynchronously calls the subscribers' MsgProc() method with the message in the argument list of that function. Only those subscribers' MsgProc() will be called for which the message is destined. If an agent arrives into the platform, the service asynchronously sends that agent to the AMS.

The RemoteSender service, when it receives an agent or a message to send to a remote ACENET, first checks the availability of the remote ACENET. If the ACENET is available, the message or the agent will be sent to that ACENET, otherwise it waits for a certain period of time in order to reconnect to that ACENET. This repeats a predefined number of times, and if that remote ACENET is still out of reach, the RemoteSender sends that agent or the message back to the sender, specifying the error of "unreachable host".

33.3.2 Grid-Like Distributed Architecture

We have recommended the usage of the master-slave design pattern for the development of the agent-based grid computing environment. The server actually manages the grid, whereas the nodes are dynamic. The master agent has the logic for the distribution of processes into tasks and the recompilation of the results. The tasks are then sent to nodes with idle CPU cycles, in order to utilize the free CPU cycles.

We propose a programming model that is a derivation of the master-slave pattern. The master class has all the controlling logic. It divides the problem into small computable tasks which are called slaves. Only one master class, but one or more slave files, can exist. During the execution, the master thread will create slave threads, which will be executing the slave code. These threads will be mapped to the agents. When a call to create a slave thread is made, a new task will be created and assigned to any available agent, which will execute it. The slave code will report back to the master after finishing the task assigned to it. The messaging is supported through the agent framework.

We propose a system where agents will explore in a distributed environment that will provide autonomy, dynamic behavior, and a robust infrastructure.

33.4 Matrix Multiplication over ACENET

We deployed ACENET (Agent Collaborative Environment based on .NET) and created two agents: a manager agent and a task agent (shown in Fig. 33.2).

We chose matrix multiplication as the trial application. The reason for selecting this application is that it is the best understood application of parallelism; it truly needs parallel algorithms to solve the problem. After the successful execution of this application, we will also address some other distributed applications.

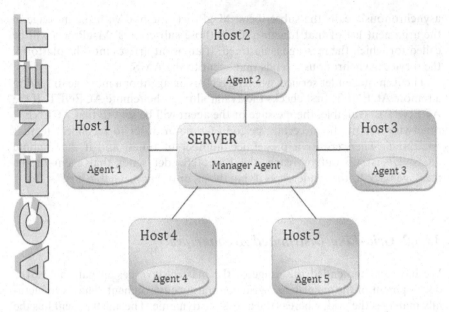

Fig. 33.2 ACENET and application agents for distributed problem solving

We followed the following steps to achieve the task:

1. Manager agent (MA) analyses the matrices and divides the first matrix into a block of rows.
2. MA then creates task agents (TAs) equal to the number of rows of the matrix.
3. MA broadcasts in full the second matrix and the row specific to the TAs.
4. TAs perform computations on the assigned matrix at the idle nodes.
5. Each TA uses its block of the first matrix to multiply by the whole of the second matrix.
6. TAs return results to the MA, which places the value-specific location in the resultant matrix.

33.5 Conclusion and Discussion

A multi-agent system execution environment, the ACENET agent platform, has been developed that supports distributed problem solving. The platform provides a foundation to work on agent-oriented programming for .NET developers. The ACENET provides a framework in which to develop the next generation of distributed applications enriched with the robust features provided by the .NET platform. The framework provides a test bed for future ideas in distributed and parallel computing. The main focus during research was on agents. The research can be tested and even extended to more complex computations, where large grid systems may be deployed over which agents may run.

The ACENET will be used as a foundation for evaluating different coordination algorithms. Additional services will be added to the ACENET if required for the implementation of the distributed problem-solving framework.

In future, the performance of the ACENET will be compared with other agent platforms in which some distributed and parallel computations will be performed over a few selected agent platforms.

References

1. Franklin S, Graesser A (1996) Is it an agent or just a program? A taxonomy for autonomous agents. In: Proceedings of the third international workshop on agent theories, architectures, and languages. Springer-Verlag, New York
2. Jennings NR (2001) An agent-based approach for building complex software systems. Commun ACM 44(4):35–41
3. Wooldridge M (1997) Agent-based software engineering. IEE Proc Software Eng 144:26–37
4. Foster I, Kesselman C (1999) The grid: blueprint for a future computing infrastructure. Morgan Kaufmann, San Fransisco, CA
5. Miguel C, Ernesto G, et al (2004) Design and implementation of a FIPA-compliant agent platform in .NET. In: .NET Technologies'2004 workshop proceedings
6. Staneva D, Dobrev D (2004) MAPNET: A .NET-based mobile-agent platform. International conference on computer systems and technologies—CompSysTech'2004
7. Farooq Ahmad H, Arshad A, et al (2004) Decentralized multi-agent system, basic thoughts. In: 11th assurance system symposium. Miyagi University, Sendai, Japan, pp 9–14
8. FIPA specifications grouped by year of publication, last accessed on 6th Jan, 2007
9. FIPA interaction protocol specifications, last accessed on 17th June, 2007
10. Georgousopoulos C, Rana Omer F (2002) An approach to conforming MAS into a FIPA-compliant system. In: AAMAS'02, Bologna, Italy
11. Milojicic D, Campbell J, et al (1998) The OMG mobile agent system interoperability facility. Lecture notes in computer science. Stuttgart, Germany, September 1998, Springer-Verlag, Berlin, pp 50–67

Chapter 34
Building Moderately Open Multi-Agent Systems: The HABA Process

A. Garcés, R. Quirós, M. Chover, J. Huerta, and E. Camahort

Abstract In this work we propose a development process for Moderately Open Multi-Agent Systems. Our process has the advantage that it allows both high- and low-level behavior specifications. The HABA.DM methodology, the HABA.PL programming language, and the HABA.PM project manager are the structural elements of the HABA development process we propose. As an example, we use a real-world application. It implements a prototype for the collection and analysis of seismic data.

34.1 Introduction

Recently a lot of progress has been made in agent-oriented technology. GAIA [7] was the first complete methodology for the analysis and design of multi-agent systems (MAS). GAIA offers a conceptual framework to model agent-based systems using an organizational metaphor.

From a practical point of view GAIA has certain disadvantages. Its high level of abstraction is the most important of them. Its design methodology is so generic that it does not provide a clear relationship between the design specification and the final system implementation [3]. The reason is that GAIA avoids tying its methodology to any particular programming language, technique, or architecture [3].

In this work, we propose a GAIA-based framework for the development of MAS. It is called Homogeneous Agents-Based Architecture (HABA). This framework is suitable for Moderately Open Multi-Agent Systems (MOMASs) [2]. There is a large class of systems that can benefit from our model; for example, electronic commerce systems [3]. We describe the use of our development process with a real-world application example. It implements a prototype for the collection and analysis of seismic data.

A. Garcés (✉)
Department of Computer Systems, Jaume I University, Castellón, Spain

N. Mastorakis et al. (eds.), *Proceedings of the European*
Computing Conference, Lecture Notes in Electrical Engineering 27,
DOI 10.1007/978-0-387-84814-3_34, © Springer Science+Business Media, LLC 2009

Our paper is organized as follows. In the background section we present related work. In Section 34.3 we introduce the architecture of our system and the HABA process. In Sections 34.4 and 34.5 we present the development stages in HABA. We conclude our paper with a discussion.

34.2 Background

34.2.1 Related Work

In the literature one may find many different methodologies for the development of agent-oriented systems like BDI [5] or MAS-CommonKADS [4]. GAIA [2] was the first complete methodology for the analysis and design of MAS. It is based on an organizational metaphor. The model is based on agents; each has certain roles in the organization. The agents cooperate to achieve the common objectives of the system.

There have been theoretical studies of GAIA [1, 8]. They extend the development of MAS to complex open environments. Still, none of them proposes how to implement MAS on real platforms.

The Moderately Open Multi-Agent Systems (MOMASs) model constitutes an alternative to model MASs. This model was presented for the first time in [2]. In [3], its application to e-commerce transaction modeling using moderately open multi-agent systems was described. These works constitute the basis of the complete process of HABA.

34.2.2 An Application to Seismic Monitoring

In 1998 the Cuban Seismic Service replaced its old system by a network of digital sensor stations. The system evolved from a centralized system to a corporate distributed information system. It features a hierarchical structure, with a coordinating agency that handles data collection and analysis and report generation and delivery. We use it as our example application. The reader can find a complete description in [2].

34.3 Moderately Open Multi-Agent Systems

We describe how to organize agents in communities and how to architect a MOMAS. The HABA process is suitable for building this kind of MAS.

34.3.1 Multi-Agent System Architecture

The architecture of MOMAS comprises four fundamental components: *agents, packages,* a *management module,* and a *social state* (Fig. 34.1). Agents are clustered into communities called packages, each managed by an agent community coordinator. A *management module* handles the life cycle of packages and agents, their communication, and the social state's resources. The system includes the social state, the management module, and all the packages handled by the management module.

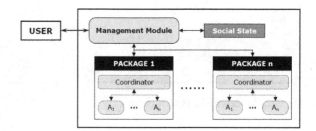

Fig. 34.1 Architecture of a moderately open multi-agent system

34.3.2 The HABA Development Framework for Moderately Open Multi-Agent Systems

To overcome the problems of GAIA and its extensions we propose a new MAS development framework called HABA. This process can be used for the analysis, design, and implementation of moderately open multi-agent systems. The main differences of HABA with respect to GAIA are:

- HABA is used to model MOMAS.
- HABA incorporates models related to the environment of the system in each stage of development.
- The whole development process, from the analysis to the implementation, is performed by refining the roles step by step.
- For the implementation of the system, the HABA framework provides the HABA.PL programming language and the HABA.PM tool.

HABA models are shown in Fig. 34.2. The HABA.DM methodology, the HABA.PL programming language, and the HABA.PM project manager are the structural elements of the HABA development process we propose.

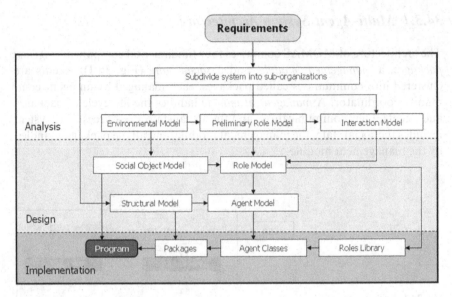

Fig. 34.2 Models in the HABA process

34.4 The HABA.DM Methodology for Agent-Oriented Analysis and Design

The whole development process, from the analysis to the implementation, is performed by refining the roles step by step. Modeling the environment, architecture, and communications is another important issue in HABA.

34.4.1 The Analysis Stage

In the analysis stage we obtain a set of models that represent the organizational structure, the environment, a preliminary description of the roles, and the interaction protocols of the MOMAS.

The environmental model contains the computational resources of the MAS. Figure 34.3(a) shows how the environment of the seismic service has been described. The environment contains two indexed object sets: *Seisms* and *Trace*. A constraint has also been defined. The constraint is expressed in Z notation [6]. It restricts *seisms* to having one or more traces.

Roles define the generic behavior of entities within an organization. Figure 34.3(b) shows the *monitor* role. This role monitors a seismometer that takes local seismic measurements in real time. The role has three basic tasks: ignore, register, and evaluate. Using the *AwaitCallReport* protocol, the role communicates with other roles for measurement reports. The permissions

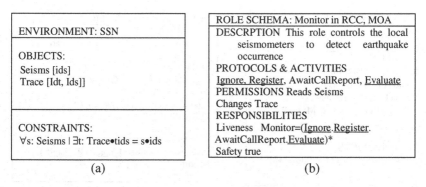

Fig. 34.3 (a) Environmental model (b) Preliminary role model

Fig. 34.4 Interaction model:
AwaitCallReport protocol

AwaitCallReport		
Analysis	Monitor	
The goal is to send the seismic measurements		t : Trace

authorize the monitor role to read the values of the *Seisms*. It also produces changes in the *Trace* records.

We limit the interaction between roles to asynchronous and controlled ask-reply communications. Figure 34.4 shows the *AwaitCallReport* protocol, part of the *Monitor* role. In this protocol, the *Analysis* role is the client and the *Monitor* role is the server.

34.4.2 The Design Stage

In HABA.DM, the design stage produces the social objects model, the role model, the agent model, and the structural model. A social object model describes the information resources that belong to the social state of the MOMAS. Figure 34.5(a) shows an instance of such a model applied to our seismic monitoring system.

The role model describes each role of the MOMAS using four basic components: the activities, the services, the properties, and the permissions. This role model derives directly from the preliminary role model and the interaction model. Figure 34.5(b) shows the specification of the *Report* service, obtained from the *AwaitCallReport* protocol and the interactions model.

The agent model specifies the agent classes used in the MOMAS. Figure 34.6 shows two agent classes: *LocalStationAgent* and *CentralObservatoryAgent,* defined with the roles *Monitor* and *Analysis,* respectively. Figure 34.7 shows how the MOMAS system is built using packages.

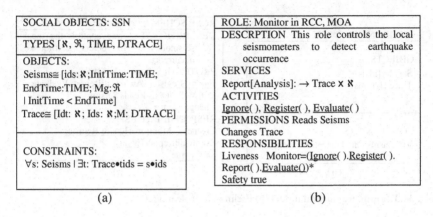

SOCIAL OBJECTS: SSN
TYPES [ℵ, ℜ, TIME, DTRACE]
OBJECTS: Seisms≅ [ids: ℵ;InitTime:TIME; EndTime:TIME; Mg:ℜ I InitTime < EndTime] Trace≅ [Idt: ℵ; Ids: ℵ;M: DTRACE]
CONSTRAINTS: ∀s: Seisms I ∃t: Trace•tids = s•ids

ROLE: Monitor in RCC, MOA

DESCRPTION This role controls the local
  seismometers to detect earthquake
  occurrence

SERVICES Report[Analysis]: → Trace x ℵ
ACTIVITIES Ignore( ), Register( ), Evaluate( )
PERMISSIONS Reads Seisms Changes Trace
RESPONSIBILITIES Liveness Monitor=(Ignore( ).Register( ). Report( ).Evaluate())* Safety true

(a)   (b)

Fig. 34.5 (a) Social objects model (b) Role model

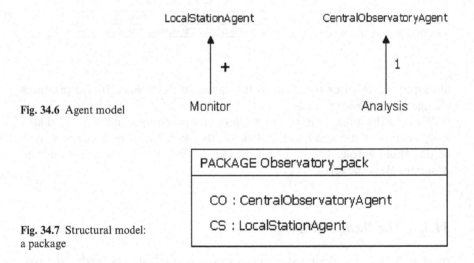

LocalStationAgent   CentralObservatoryAgent

+   1

Fig. 34.6 Agent model   Monitor   Analysis

PACKAGE Observatory_pack
CO : CentralObservatoryAgent CS : LocalStationAgent

Fig. 34.7 Structural model:
a package

34.5 Programming Environment

To illustrate the implementation of a MOMAS, we describe how to program the
monitor role using the HABA.PL language of our development process. We focus
on the general structure of the system, omitting unnecessary low-level details.

34.5.1 Programming in HABA.PL

The monitor role has the following attributes: an array x with the seismic
measurements, a variable *xactual* with the current measurement value, and
two logical variables *event* and *saturated*. *Event* signals the triggering of a
seismic event and saturated signals when array x has been filled.

The experience of the *Monitor* role is declared in the knowledge section. Procedures *ignore* and *register* are role activities. *Report* is a public service that can be accessed by any agent with the *Analysis* role. Its goal is to send the seismic measurements.

```
Role Monitor
{ real x[200], actualX = 0; int n = -1; boolean event =
   false, saturated = false;
/* Other attributes come here . ... */
knowledge
{ trigger(X) :- X > = 10.
stop_register(X) :- X < 5.
/* Other knowledge come here . ... */}
ignore() : not event
{ while(not ?(trigger(actualX))) {get(com1,actualX)};
   event = true;}
register() : event and not saturated
{ while (not ?(stop_register(actualX)) AND n<199)
{ n = n + 1 ; x[n] = actualX ; get(com1,actualX); }
saturated = true; }
/* Other tasks or activities come here . ... */
service report(real &xtrace[200] ; int &nmax): [Analysis]
{ xtrace = x ; nmax = n ; n = -1; event = false ; saturated
   = false;}
/* Other services come here . ... */ }
```

Having the monitor role, we can build an agent class Local_Station with the monitor role and other functions.

```
Agent Local_Station
{ extend Monitor;
string Station_name ; code = "RCS";
knowledge { ... }
Local_Station(string name) { Station_name = name; }
init() { print("Station ," Station_name); }
start() { init(); ... } }
```

34.5.2 Creating and Running a MAS

Now we show how to create and run an MAS executing three types of console commands: instance control commands, execution control commands, and communication commands. Imagine a simple seismic service with a central observatory and a single local station.

```
# Observ_pack = NEW PACKAGE (observ_host)
# RCS_pack = NEW PACKAGE (rcs_host)
# Central_Observ CO = NEW Central_Observ ("Senén") IN
   Observ_pack
# Local_Station RCS = NEW Local_Station("Río Carpintero")
   IN RCS_pack
# CO.start()
# RCS.start()
```

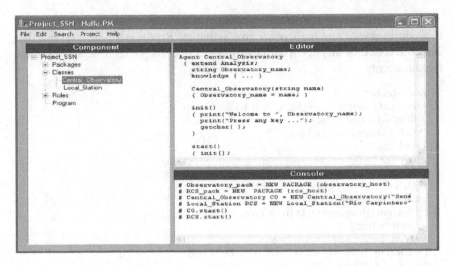

Fig. 34.8 MOMAS development tool

These commands create two packages: *Observ_pack* and *RCS_pack*. They are located in the servers of the *Central_Observ* and the *Local_Station,* respectively. We instantiate two agents: CO and RCS. Finally, commands CO.start() and RCS.start() run agents CO and RCS by executing their respective activation routines.

34.5.3 The HABA.PM Project Manager

Figure 34.8 shows a screenshot of our development tool. The left-hand side of the window contains the system's main components organized in a tree. They include packages, agent classes, roles, and a console program that supports batch execution. The console is used to prototype and debug the MOMAS, and launch its execution. The application can be compiled to a program written in JAVA or C++.

34.6 Conclusions

In this work we present the HABA process for the development of MAS. Our methodology is a variant of the GAIA methodology. The goal of HABA is to reduce the gap between the abstract modeling of MAS and its practical implementation. To achieve this goal, we reduce the scope of our process to a specific class of MAS that we have called Moderately Open Multi-Agent Systems (MOMAS). The HABA process supports not only the analysis and design stages of GAIA, but also the implementation stage.

HThe ABA process allows fast prototyping of an agent-based model in an incremental process based on roles. The HABA.DM methodology provides models to capture the complexity of the information of the environment and the system architecture. For the implementation of the system, we use the HABA.PL programming language and the HABA.PM project manager.

The HABA.PM project manager has a visual user interface and can generate JAVA and C++ code. We have used it to build a system for seismic monitoring. Our immediate goal is to apply our framework to other problems suited for moderately open agent-based modeling.

References

1. Cernuzzi L, Juan T, Sterling L, Zambonelli F (2004) The Gaia methodology: basic concepts and extensions. In: Methodologies and software engineering for agent systems. Kluwer, Dordrecht
2. Garcés A, Quirós R, Chover M, Camahort E (2006) Implementing moderately open agent-based systems. In: Proceedings of the IADIS international conference WWW/Internet. Murcia, Spain
3. Garcés A, Quirós R, Chover M, Huerta J, Camahort E (2007) E-commerce transaction modeling using moderately open multi-agent systems. In: Proceedings of the 9th international conference on enterprise information systems. Funchal, Madeira, Portugal, pp 12–16
4. Iglesias C, Garito M, González J, Velaso J (1998) Analysis and design of multiagent systems using MAS-CommonKADS. In: Singh MP, Rao A, Wooldridge M (eds) Intelligent agents IV, LNAI 1365:313–326
5. Kinny D, Georgeff A, Rao A (1996) A methodology and modelling technique for systems of BDI agents. In: Van de Velde W, Perram JW (eds) Agents breaking away: proceedings of the seventh European workshop on modelling autonomous agents in a multi-agent world, LNAI 1038:56–71.
6. Spivey JM (1992) The Z notation: a reference manual, 2nd ed. Prentice Hall, Upper Saddle River, NJ
7. Wooldridge M, Jennings N, Kinny D (2000) The Gaia methodology for agent-oriented analysis and design. Auton Agent Multi Agent Syst 3(3):285–312
8. Zambonelli F, Jennings NR, Wooldridge M (2003) Developing multi-agent systems: the Gaia methodology. ACM Trans Soft Eng Methodol 2(3):317–370

Part V
Educational Software and E-Learning Systems

Recently e-learning has become one of the most important forms of education. E-learning systems include learning content, but also the infrastructure that allows content to be created, stored, accessed, and delivered, and the learning process to be managed. The architecture of these e-learning systems is a crucial aspect. Architectures define structures that connect an e-learning software and information system with its instructional and educational context. In this part we present some papers that deal with the principles and components of the architectures of e-learning systems and how they operate in terms of the instructional and content aspects involved. The papers include using mathematics for data traffic modeling within an e-learning platform, developing statistics learning at a distance using formal discussions, distance teaching in the technology course in senior high school, electronic exams for the 21st century, effects of the Orff music teaching method on creative thinking abilities, knowledge application for preparing engineering high school teachers, a collaborative online network and culture exchange project, a study on mobile devices in education, the solutions of mobile technology for primary school, a study of verifying the knowledge reuse path using structural equation modeling, a web-based system for distance learning of programming, new software for the study of the classic surfaces from differential geometry, data requirements for detecting student learning styles, and an ontology-based feedback e-learning system for mobile computing.

Chapter 35
Using Mathematics for Data Traffic Modeling Within an E-Learning Platform

Marian Cristian Mihăescu

Abstract E-Learning data traffic characterization and modeling may bring important knowledge about the characteristics of that traffic. Without measurement, it is considered impossible to build realistic traffic models. We propose an analysis architecture employed for characterization and modeling using data mining techniques and mathematical models. The main problem is that real data traffic usually has to be measured in real time, saved, and later analyzed. The proposed architecture uses data from the application level. In this way the data logging process becomes a much easier task, with practically the same outcomes.

35.1 Introduction

The Tesys e-Learning platform was developed and deployed. It has a built-in mechanism for monitoring actions performed by users and data traffic transferred during usage. In this paper we study the possibility of modeling data traffic using data mining techniques and mathematics based on performed actions. This would have great benefits regarding the overhead within the platform. The Introduction presents the Tesys e-Learning platform in brief. In the second section are presented the methods employed: actions and data monitoring, the process of clustering users. The clustering process groups similar users based on a specific similarity function. At the cluster level, the self-similarity of data traffic is then examined. This is accomplished by estimating the Hurst parameter.

In the third section of the paper is presented the proposed architecture and the analysis process. In brief, the architecture and employed process will try to estimate the data traffic self-similarity within a cluster of users. In the fourth section are presented the results obtained. Finally, the conclusions and future works are presented.

M.C. Mihăescu (✉)
Software Engineering Department, University of Craiova, Craiova, Romania

N. Mastorakis et al. (eds.), *Proceedings of the European*
Computing Conference, Lecture Notes in Electrical Engineering 27,
DOI 10.1007/978-0-387-84814-3_35, © Springer Science+Business Media, LLC 2009

The platform has a built-in capability of monitoring and recording a user's activity at the application level. The activity represents valuable data, since it is the raw data for our machine learning and modeling process. The user's sequence of sessions makes up his activity. A session starts when the student logs in, and finishes when the student logs out. Under these circumstances, a sequence of actions makes up a session.

35.2 Methods and Materials

There are many different ways for representing patterns that can be discovered by machine learning. Out of all of them we choose clustering, which is the process of grouping a set of physical or abstract objects into classes of similar objects [1]. For our platform we create clusters of users based on their activity.

As a product of the clustering process, associations between different actions on the platform can easily be inferred from the logged data. In general, the activities that are present in the same profile tend to be found together in the same session. The actions making up a profile tend to co-occur to form a large item set [2].

There are many clustering methods in the literature: partitioning methods such as [4], hierarchical methods, density-based methods such as [7], grid-based methods, or model-based methods. Hierarchical clustering algorithms like the single-link method [8] or OPTICS [9] compute a representation of the possible hierarchical clustering structure of the database in the form of a dendrogram or a reachability plot from which clusters at various resolutions can be extracted, as has been shown in [10]. From all of these we chose to have a closer look at partitioning methods.

The EM algorithm [11] takes into consideration that we know none of these things: not the distribution that each training instance came from, nor the parameters μ, σ, or the probability. So we adopt the procedure used for the k-means clustering algorithm and iterate. Start with an initial guess for the five parameters, use them to calculate the cluster probabilities for each instance, use these probabilities to estimate the parameters, and repeat. This is called the EM algorithm, for "expectation-maximization." The first step, the calculation of cluster probabilities (which are the "expected" class values), is "expectation;" the second step, calculation of the distribution parameters, is "maximization" of the likelihood of the distributions, given the data [2].

The EM algorithm is implemented in the Weka package [12] and needs the input data to be in a custom format called *arff*.

Self-similarity and long-range dependence of data traffic are discussed in detail in [13–15]. A process is considered to be self-similar if the Hurst parameter satisfies the condition:

$$Y(t) = a^{-H} Y(at) \quad t>0, \ a>0, \ 0<H<1 \tag{35.1}$$

where the equality is in the sense of finite-dimensional distributions.

Parameter H can take any value between 1/2 and 1, and the higher the value the higher the degree of self-similarity. For smooth Poisson traffic, the value is H = 0.5. There are four methods used to test for self-similarity. These four methods are all heuristic graphical methods, they provide no confidence intervals, and they may be biased for some values of H. The rescaled adjusted range plot (R/S plot), the Variance-Time plot, and the Periodogram plot, as well as the theory behind these methods, are described in detail by Beran [13] and Taqqu et al. [16]. Molnar et al. [17] describe the index of dispersion for the counts method and also discuss how the estimation of the Hurst parameter can depend on estimation technique, sample size, time scale, and other factors.

35.3 Proposed Analysis Process

The analysis process starts from logged data about actions and data traffic and comes up with an estimation of the Hurst parameter. This estimation of self-similarity represents important knowledge in characterizing and modeling data traffic. In Fig. 35.1 the employed analysis process is presented.

The analysis process starts by creating clusters of users based on their activity. For this, only the data regarding performed actions is used. Once the clusters of users are obtained, the data traffic transferred within each cluster is taken into consideration by the H Parameter Estimator module. This module will produce three plots: the R/S plot, the Variance-Time plot, and the Periodogram plot.

35.4 Results

The EM algorithm is implemented in the Weka package [19] and needs the input data to be in a custom format called *arff*. Under these circumstances we have developed an offline Java application that queries the platform's database and creates the input data file called *activity.arff*. This process is automated and is

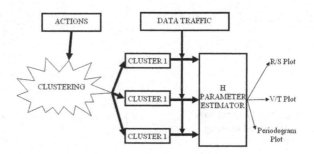

Fig. 35.1 Analysis process

driven by a properties file in which there is specified what data will lie in the activity.arff file.

Running the EM algorithm created three clusters. The procedure clustered 91 instances (34%) in cluster 0, 42 instances (16%) in cluster 1, and 135 instances (50%) in cluster 3. The final step is to check how well the model fits the data by computing the likelihood of a set of test data given the model. Weka measures goodness-of-fit by the logarithm of the likelihood, or log-likelihood; and the larger this quantity, the better the model fits the data. Instead of using a single test set, it is also possible to compute a cross validation estimate of the log-likelihood. For our instances the value of the log-likelihood is -2.61092, which represents a promising result in the sense that instances (in our case, students) may be classified into three disjoint clusters based on their activity.

The clustering process produced the following results (Table 35.1):

For the obtained clusters a study of self-similarity of traffic was performed. More precisely, self-similarity was studied for cluster 0, formed of 91 students (34%).

In six months of functioning on the platform, there were executed over 10,000 actions of different types: course downloads, messaging, self tests, exams. For computations a packet was considered to have 1,000 bytes.

For estimation of the Hurst parameter a 3 h interval, between 18:00 and 21:00, was chosen, which is considered to be a heavy traffic period. This may be observed from the general traffic statistics presented in Fig. 35.2.

Table 35.1 Distribution of users in clusters

Cluster	No. of users
0	91 (34%)
1	42 (16%)
2	135 (50%)

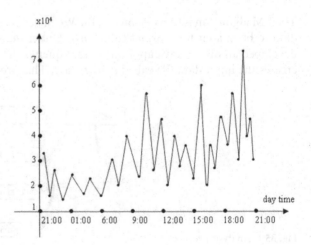

Fig. 35.2 General data traffic on e-Learning platform

The interval from 18:00 to 21:00 was chosen for close analysis. The R/S plot estimated the H parameter to a value of 0.89. The Time-Variance plot showed a slope of -0.320, which means a value of H of $1 + \text{slope}/2 = 0.84$. The IDC (Index of Dispersion for Counts) shows an H parameter of 0.88. In the Periodogram plot there may be observed a value of $H = 0.85$. These methods do not obtain exactly the same values, but the values are over 0.5, which is a good indication of the traffic's self-similarity Figs. 35.3, 35.4, 35.5, and 35.6.

The self-similarity of byte traffic presents similar values for the H parameter. The number of bytes transferred in each bin was computed and the results presented in Table 35.2. In this table there are presented estimations of the H parameter for different dimensions of the time bin.

Having in mind that non-stationary traffic may be easily taken as self-similar stationary traffic, smaller intervals of time bins were also examined. The H parameter was estimated for each of the 6 intervals of 30 min between 18:00 and 21:00. The results are presented in Fig. 35.7.

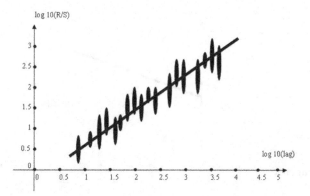

Fig. 35.3 H parameter—R/S plot

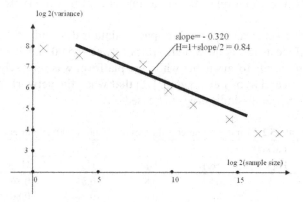

Fig. 35.4 H parameter—V-T plot

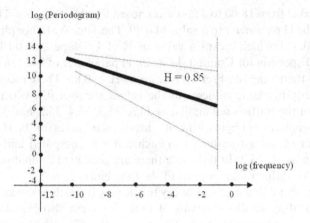

Fig. 35.5 H parameter–P plot

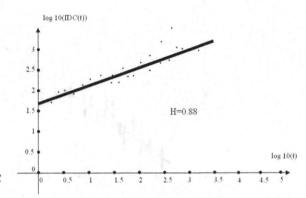

Fig. 35.6 H parameter–IDC
plot

In this way, there was estimated an H parameter for 3 h from a complete interval of 24 h. Estimation of the H parameter for other intervals is presented in Fig. 35.8.

Estimations were accomplished for packet data traffic and a time interval of 15 min. All three methods show high values between 15:00 and 20:00. This time interval corresponds to moments when the platform was intensely used, confirming the researches of Leland et al. [18] that when the network load is high, then the degree of self-similarity is increased.

Table 35.2 Hurst parameter estimates for 18:00–21:00 time interval

Bin size	Packets			Bytes		
	$H_{R/S}$	H_{VT}	H_P	$H_{R/S}$	H_{VT}	H_P
2 h	0.85	0.84	0.87	0.83	0.86	0.88
4 h	0.82	0.78	0.85	0.81	0.86	0.88
6 h	0.84	0.83	0.89	0.77	0.79	0.85

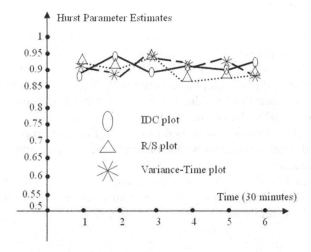

Fig. 35.7 Hurst parameter from 18:00 to 21:00

The fact that traffic is found to be self-similar does not change its behavior, but it changes the knowledge about real traffic and also the way in which traffic is modeled. It has led many [19] to abandon the Poisson-based modeling of network traffic for all but user session arrivals. Real traffic, well described as self-similar, has a "burst within a burst" structure that cannot be described with the traditional Poisson-based traffic modeling.

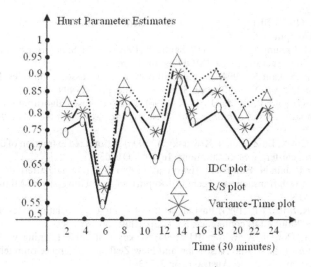

Fig. 35.8 Hurst parameter for each hour of traffic monitoring

35.5 Conclusions

Data analysis is done using an EM clustering algorithm implemented by the Weka system and mathematical traffic modeling.

Mathematical modeling estimates the self-similarity of data traffic. This is accomplished by heuristic graphical methods: R/S plot, Variance-Time plot, IDC plot, Periodogram plot. The analysis is performed rigorously for a 3 h interval, from 18:00 to 21:00, but also for the whole day.

All the analysis follows a proposed process that has data regarding actions executed and bytes transferred within the platform as input, and estimates of the Hurst parameter as output.

The values found for the Hurst parameter are very promising. All the calculations showed values above 0.7, and many above 0.8, which indicate a good level of self-similarity.

References

1. Han J, Kamber M (2001) Data mining—concepts and techniques. Morgan Kaufmann, San Fransisco, CA
2. Witten IH, Eibe F (2000) Data mining—practical machine learning tools and techniques with Java implementations. Morgan Kaufmann, San Fransisco, CA
3. MacQueen J (1967) Some methods for classification and analysis of multivariate observations. In: 5th Berkeley symposium on mathematical statistics and probability, pp 281–297
4. Nasraoui O, Joshi A, Krishnapuram R (1999) Relational clustering based on a new robust estimator with application to web mining. In: Proceedings of the international conference of the North American fuzzy information processing society (NAFIPS 99), New York
5. Sibson R (1973) SLINK: an optimally efficient algorithm for the single-link cluster method. Comput J 16(1):30–34
6. Ankerst M, Breuing M, Kriegel HP, Sander J (1999) OPTICS: ordering points to identify the clustering structure. In: SIGMOD'99, pp 49–60
7. Agrawal R, Srikant R (1994) Fast algorithms for mining association rules. In: Proceedings of the 20th VLDB conference, Santiago, Chile, pp 487–499
8. Ester M, Kriegel HP, Sander J, Xu X (1996) A density-based algorithm for discovering clusters in large spatial databases with noise. In: Proceedings of the KDD'96, Portland, OR, pp 226–231
9. Sander J, Qin X, Lu Z, Niu N, Kovarsky A (2003) Automated extraction of clusters from hierarchical clustering representations. In: PAKDD'03
10. Mobasher B, Jain N, Han E-H, Srivastava J (1996) Web mining: pattern discovery from world wide web transactions. Technical Report 96–050, University of Minnesota, Minneapolis, MN
11. Agrawal R, Srikant R (1994) Fast algorithms for mining association rules. In: Proceedings of the 20th VLDB conference, Santiago, Chile, pp 487–499
12. Holmes G, Donkin A, Witten IH (1994) Weka: a machine learning workbench. In: Proceedings of the second Australian and New Zealand conference on intelligent information systems, Brisbane, Australia, pp 357–361
13. Beran J (1994) Statistics for long-memory processes. Chapman & Hall, New York

14. Willinger W, Taqqu MS, Sherman R, Wilson D (1995) Self-similarity through high-variability: statistical analysis of Ethernet LAN traffic at the source level. In: Proceedings of SIGCOMM '95, pp 100–113

15. Willinger W, Paxson V, Taqqu MS (1998) Self-similarity and heavy tails: structural modeling of network traffic. In: Adler R, Feldman R, Taqqu MS (eds) A practical guide to heavy tails: statistical techniques and applications, Birkhauser, Boston

16. Taqqu MS, Teverovsky V (1998) On estimating the intensity of long-range dependence in finite and infinite variance time series. In: Adler R, Feldman R, Taqqu MS (eds) A practical guide to heavy tails: statistical techniques and applications, Birkhauser, Boston

17. Molnar S, Vidacs A, Nilsson A (1997) Bottlenecks on the way towards fractal characterization of network traffic: estimation and interpretation of the Hurst parameter. In: International conference on the performance and management of communication networks, Tsukuba, Japan

18. Leland WE, Taqqu MS, Willinger W, Wilson DV (1994) On the self-similar nature of Ethernet traffic (extended version). IEEE-ACM T Network 2(1):1–15

20. Paxson V, Floyd S (1995) Wide-area traffic: the failure of Poisson modeling. IEEE-ACM T Network 3(3):226–244

Chapter 36
Developing Statistics Learning at a Distance Using Formal Discussions

Jamie D. Mills

Abstract The purpose of this paper is to report some preliminary empirical results of students enrolled in two graduate-level hybrid courses. A new design feature, the discussion board, was recently implemented in one section in order to increase overall interaction as well as to better monitor and assess student learning. The preliminary results of this study indicate that students who were more actively involved with the course materials, discussions, and others in the class performed better academically than students who were less involved. The results support formal asynchronous discussions as one teaching strategy that might facilitate the learning of statistics concepts online. There have been no empirical studies that focus on how effective asynchronous discussions might be utilized in an online/hybrid statistics course.

36.1 Introduction

Today, teaching and learning using technology is utilized more than ever before in higher education. According to the Council for Higher Education Accreditation, many traditional colleges and universities are now offering courses and complete degree programs in a wide variety of disciplines at a distance [1]. This alternative form of course delivery is a fast-growing trend and has the potential to change and revolutionize teaching and learning at every level of education, perhaps forever.

Teaching statistics at a distance is also becoming a popular course offering [2, 3]; however, there is a lack of empirical results and discussions about teaching and learning in this environment. In particular, many questions might be of interest: How is instruction "best" delivered? What specific technologies seem to be helpful for learning specific statistics concepts online? How does student-to-student

J.D. Mills (✉)
Department of Educational Studies, University of Alabama, 316-A Carmichael Hall, Box 870231, Tuscaloosa, AL 35487–0231, USA
e-mail: jmills@bamaed.ua.edu

N. Mastorakis et al. (eds.), *Proceedings of the European Computing Conference*, Lecture Notes in Electrical Engineering 27, DOI 10.1007/978-0-387-84814-3_36, © Springer Science+Business Media, LLC 2009

interaction and student-to-teacher interaction take place? Is there an optimal course design? Which design features (i.e., whiteboard, chat feature, discussion board) in the course development system appear to be most effective for students learning statistics? Although there are researchers who are studying how to deliver statistics courses in this new technological environment, there is still much to learn about how to effectively implement these courses and what practices are best.

Another common problem in many online courses is a lack of teacher-to-student interaction, as well as student-to-student interaction [2]. Students may feel not only isolated from the teacher, but also isolated and deprived of the "normal" social interaction and cognitive learning processes that take place in a face-to-face class [4, 5]. Particularly in a statistics course, where students often feel anxious and insecure about learning with an instructor face-to-face each week [6], additional online support may be necessary and critical for student academic success.

In an effort to improve interaction in an online course, the discussion board might provide one avenue in which to accomplish this task. There is substantial evidence to indicate that students learn more when they are actively engaged with the course materials, their classmates, and the instructor [7, 8]. Therefore, how can students in statistics courses utilize the discussion board? In other disciplines, asynchronous discussion groups have been very valuable. The following authors [9] found that asynchronous discussions in their psychology and educational sciences courses reflected high phases in knowledge construction; while others have reported that discussion boards can provide an interactive venue where students can reflect, evaluate, solve problems, and exchange ideas [10]. Is it possible to effectively discuss statistics concepts online through the use of asynchronous discussions and assist students through different levels of learning?

36.2 The Hybrid Course

36.2.1 Course Modules

The WebCT course management system was used to deliver the hybrid courses. Although many students were familiar with WebCT or have used it in other courses, there was an animated "talking head" on the front page of the site to welcome students to the course, provide brief announcements and course logistics, and encourage students to begin the course with the training video module, which illustrates how to use relevant modules in the course (i.e., how to submit assignments, how to read the calendar, how to use the discussion board). The training videos also explained all of the links in the course (i.e., where assignments are posted, where the lecture and SPSS movies are located, examples of how files should "look" before being submitted). The streaming videos on the site require a high-speed Internet connection, which is available in the labs on campus, if working off-campus is not feasible for students. Technical difficulties were handled through the university help desk.

An interactive SMART Board, SMART pen, projector, and a computer were used for the recording of all course materials. The SMART pen was used to write text and formulas as well as drawings that were all captured as streaming audio and video clips. The clips were segmented into smaller chunks of related concepts/topics to allow for greater flexibility in viewing the videos and to encourage step-by-step learning. Fig. 36.1 provides an example of a typical video the students might see in each chapter.

36.2.2 Collaborative Learning

For both the spring and fall 2005 courses, discussion board problems (without solutions) and practice problems (with solutions) were posted for specific topics in their respective discussion areas. The discussion board problems, which consisted of 5–6 questions to answer for each problem, were context-related research problems which required students to apply the concepts they were learning to a specific research scenario. The practice problems had similar objectives but were not as comprehensive (i.e., one question). The discussion board problems were reserved for specific students involved in group work, while the practice problems could be discussed by all students. During the spring semester, the discussion board and practice problems were available to all students for discussions, but postings were not required. The students enrolled in the fall semester, however, were required to make contributions and participate in group work. Although students in both courses had access to the same discussion board and practice problems, the major difference was the

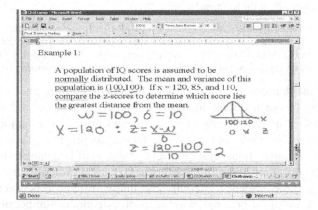

Fig. 36.1 Students watch streaming videos about the normal distribution

requirement to participate in posting solutions to the discussion board and practice problems.

The students in the fall course received a post grade for their contributions, which was weighted at 35% of their final course grade. A contribution could be (1) a question about a practice or discussion board problem, (2) a discussion of a solution to a practice problem, or (3) some other content-related contribution on the specific topic of interest. Students were generally given full credit if they posted by the deadline. Therefore, accuracy was generally not considered. As long as the student demonstrated a concerted effort to make a contribution, full credit was granted. Students in the spring course posted their questions to the discussion board and practice problems, which were answered primarily by the instructor.

There were 5 topics considered for the discussion board problems: normal distribution, hypothesis testing using z and t, independent samples t-test, dependent samples t-test, and correlation/regression. An example of a discussion board assignment, the independent samples t-test, required a group to analyze the data using SPSS, address assumptions, set up the null and alternative hypotheses, interpret confidence intervals within the context of the example, make a decision about the null hypothesis, and make a final conclusion within the context of the example. For this topic of interest for this study (independent and dependent samples), the group in the required-post course were responsible for posting the solutions and SPSS files for the rest of the class; while other students (not in the group) and the instructor could ask questions either related to the discussion board assignment or participate in another eligible manner (i.e., discuss practice problems, ask questions, or make some other content-related contribution).

Approximately 52 optional postings were observed for students enrolled in the spring course, compared to 268 required postings for the fall students. A typical posting from each course that discussed differences in one- and two-tail tests are presented for each course below:

> I know that we talked about 2-tailed tests in our notes but how will the question be worded to let us know that we need to do a 2-tail test? Are there any key words or phrases to look for in the problem? (Spring 2005)

> I came up with the hypothesis as a two tail test because there was no difference between the groups. When there is no difference between the groups, then you use the two tail test which is Ho: Mu1 = Mu2 and Hi: Mu1 is not equal to Mu2. I got this from your lecture notes on Hypotheses of interests, so I hope I interpreted it correctly (Fall 2005)

Other contributions for independent/dependent samples and other topics were similar for each course in terms of the type, quality, and content of questions asked as well as any other comments posted. The majority of the postings were comments and questions related to practice problems, SPSS, verification of assumptions, and questions about what language to use regarding the decision about the null hypothesis, statistical evidence, and the interpretation of the confidence interval. The major difference between the two courses, in terms of the postings, was the disproportionate number of contributions made (i.e., required vs. optional postings).

36.3 Method

36.3.1 Sample

The study was conducted during the spring (n = 22) and fall (n = 14) of 2005 at a large research university in the South. The same instructor, course design, content, textbooks/software, discussion board assignments and practice problems, computer assignments, optional lectures, and tests were used for both sections. One difference was the way in which the discussion boards were utilized. Students in the fall course were required to participate and received a post grade for their contribution, while the spring students were allowed to work on the discussion board assignments, practice problems, and post questions/comments as an option.

All graduate students in the College of Education were required to take this introductory statistics course, and all students enrolled in both sections elected to take this course online (there was an optional face-to-face course also available). The course is normally offered in the evenings (i.e., 6:00–9:00 pm) to primarily full-time working graduate students progressing toward the Doctor of Philosophy degree. Approximately 82% (n = 18) of the students in the spring course indicated that their GPA was between 3.5 and 4.0, while 85% for students enrolled in the fall course indicated the same average range. As indicated previously, there was no random assignment of students to either section; the students elected to enroll in the online section of this course during the spring and fall semesters of 2005.

36.3.2 Student Performance and Evaluation

In both courses, there were 3 computer assignments, a midterm examination (in-class), and a final examination (in-class). Assignment 1 was designed to introduce students to SPSS; therefore, this assignment only required that students input data into SPSS and generate the output. Assignment 2 required students to have an understanding of basic descriptive statistics, through computing hand calculations and using SPSS to interpret graphs and measures of central tendency and variability. Assignment 3 presented two research scenarios: one for independent samples and one for dependent samples. This assignment required students to input data and generate the output, answer questions related to assumptions, write null and alternative hypotheses, and read, report, and understand related statistics from the SPSS output. Both Assignments 2 and 3 followed the postings, comments, and questions of the related discussion board assignments and practice problems for both courses. Therefore, these two assignments were two of the dependent measures for this study. Other variables of interest for this study included a midterm, which covered topics from the introductory material up through the normal distribution.

This test included objective (i.e., true/false, multiple choice) and short answer (i.e., hand calculations) problems, as well as excerpts from the SPSS output, in which the students were required to demonstrate their ability to report and interpret the statistics. The final examination, directly related to the overall objectives for the course, was a series of research scenarios related to the following tests: one-sample z-test, one-sample t-test, independent samples t-test, dependent samples t-test, correlation, regression, and chi-square. For each scenario, the students were required to address the assumptions, set up the null and alternative hypotheses, find the calculated statistic by hand, make a decision about the null hypothesis, and make a final conclusion within the context of the research scenario. This was also an in-class test (done by hand—no SPSS) and followed a comprehensive review of each test in the discussion area as well.

Student performance was measured by the following variables: Assignments 2, 3, and the final examination grade, where scores for all measures were assigned from 0 to 100. The midterm grade was used as a covariate for both groups. The end-of-semester student evaluations, 1-to-5 Likert-scaled items (1 = strongly disagree and 5 = strongly agree) that measured student attitudes toward the course and the instructor, were also considered.

ANCOVA was used to determine if differences existed between the two classes for all variables of interest. The assumptions were investigated for all measures. Although the normality assumption was violated for Assignments 2, 3, and the final examination, according to the Wilk's statistic ($p < 0.001$), the F-statistic is generally robust with regard to this violation [11]. When examining the distributions, all three were negatively skewed. In addition, the homogeneity of variance tests, examined at the 0.10 level of significance, was tenable for all three measures.

Using the student's midterm score as the covariate, the results revealed a statistically significant difference between average adjusted scores for Assignments 2 [$\bar{x}_s = 85.2$, $s_s = 2.5$; $\bar{x}_f = 94.1$, $s_f = 2.0$; $F(1, 33) = 7.3$, $p<0.011$; $\eta^2 = 0.182$] and 3 [$\bar{x}_s = 66.1$, $s_s = 6.4$; $\bar{x}_f = 94.2$, $s_f = 8.1$; $F(1, 33) = 7.1$, $p<0.012$; $\eta^2 = 0.177$], but not for the final examination [$\bar{x}_s = 85.1$, $s_s = 3.3$; $\bar{x}_f = 88.1$, $s_f = 4.2$; $F(1, 33) = 0.307$, $p>0.05$; $\eta^2 = 0.009$]. Both statistically significant results represent large effects, according to [12]. It appears that the students in the fall course performed better statistically on Assignments 2 and 3 than the students enrolled in the spring course. In addition, although student attitudes in the fall course were generally higher than those in the spring course, student attitudes were positive in both courses and revealed no statistically significant differences between sections (attitude results are presented in the full paper).

36.4 Summary and Concluding Remarks

The preliminary results indicate that the students enrolled in the required asynchronous discussion section performed better than students not involved in formalized discussions. Specifically, students who were required to participate in required discussion board assignments and discussions performed better statistically than students who were not required to participate. In addition, these students performed better on the final examination, but not better statistically. Because there have been no empirical studies that focus on how effective asynchronous discussions might be utilized in an online statistics course, this study presents some evidence that indicates that discussions might offer one way to help students learning statistics concepts online.

Although students in both courses were exposed to the same content and course materials, the students in the fall course contributed more to the discussion area, because their postings were required. This increased involvement alone is potentially why the results may have differed. The required contributions were implemented as a way to ensure that everyone stayed involved with the course materials and to determine whether these kinds of contributions might make a difference in student performance. In terms of the content of the contributions, there appeared to be no differences in the type and quality of contributions made between the two sections, despite the fact that the students in the fall course also received a grade for their posts. This could be because the fall students were not evaluated based on a "right" or "wrong" solution, since all of the solutions were provided for both courses. Thus, requiring participation appeared to influence attitudes and academic performance in this study. The lack of randomization and the fact that the sections were not taught simultaneously in the same semester are also limitations of this study.

The results of this study not only lend support to previous work regarding the notion that students benefit academically when they are actively involved with others and engaged with the course materials [7, 8]; but they also extend this notion to the online course; they further corroborate the previous research that claims asynchronous discussions can assist students in progressing through different levels of learning online [9, 10]; and the results might also be extended to learning in an online statistics course. However, additional descriptive and empirical studies and results will be needed in order to further advance our knowledge and understanding of these online teaching and learning practices. Although there are many more questions than answers at this point about teaching statistics online, it is hoped that our results and experiences might contribute to the scarce research literature in this area, as well as encourage further pedagogical dialogue and empirical results about how to effectively and successfully deliver these kinds of courses online. We hope our study begins this much-needed exchange.

References

1. Accreditation and assuring quality in distance learning (2002) Council for Higher Education Accreditation, CHEA Monograph Series, 1, 5
2. Utts J, Sommer B, Acredolo C, Maher M, Matthews H (2003) A study comparing traditional and hybrid internet-based instruction in introductory statistics classes. J Stat Educ 11(3), http://www.amstat.org/publications/jse/v11n3/utts.html

3. Ward B (2004) The best of both worlds: a hybrid statistics course. J Stat Educ 12(3), http://www.amstat.org/publications/jse/v12n3/ ward.html

4. Arnold N, Ducate L (2006) Future foreign language teachers' social and cognitive collaboration in an online environment. Lang Learn Technol 10(1), http://llt.msu.edu/vol10num1/arnoldducate/default.html

5. Pawan F, Paulus T, Yalcin S, Chang C (2003) Online learning: patterns of engagement and interaction among in-service teachers. Lang Learn Technol 7(3), http://llt.msu.edu/vol7num3/pawan/

6. Gal I, Ginsburg L, Schau C (1997) In: Gal I, Garfield J (eds) The assessment challenge in statistics education. IOS Press, Amsterdam, pp 37–51

7. Mills J, Johnson E (2004) An evaluation of ActivStats® For SPSS® for teaching and learning. Am Stat 58(3):254–258

8. Moore D (1997) New pedagogy and new content: the case of statistics. Int Stat Rev 65(2):123–137

9. Schellens T, Valcke M (2005) Collaborative learning in asynchronous discussion groups: what about the impact on cognitive processing? Comput Human Behav 21:957–975

10. DeWert M, Babinski L, Jones B (2003) Safe passages: providing online support for beginning teachers. J Teacher Educ 54(4):311–320

11. Keppel G, Wickens T (2004) Design and analysis: a researcher's handbook. Pearson/Prentice Hall, Upper Saddle River, NJ

12. Cohen J (1977) Statistical power analysis for the behavioral sciences (rev ed). Academic Press, Orlando, FL

Chapter 37
Distance Teaching in the Technology Course in Senior High School

Shi-Jer Lou, Tzai-Hung Huang, Chia-Hung Yen, Mei-Huang Huang, and Yi-Hui Lui

Abstract In the general outline of the technology course in senior high school in Taiwan, several topics are included. These curricula emphasize inspiring the students' capacity for creative design, and developing their ability at the same time. However, faced with teaching so many different categories, the science and technology teacher may lay particular stress on or attach undue importance to one thing, but neglect the other categories to some extent, because of being limited by his own specialty. In order to improve these problems, this research adopts the distance-education method of coordination, combining teachers of different specialties and using teaching equipment which the ordinary high school has at present. After teaching this way in practice, researchers interview the teachers in coordination and conduct a questionnaire investigation of students. The results of this research are as follows: (1) Students give positive opinions about distance team teaching of the class in science and technology. (2) Teachers approve students' performance in distance team teaching of the class in science and technology. (3) The present science and technology classrooms' lack of equipment, students, and teachers indicates that sufficient equipment and space should be made available. (4) Teachers and students hold positive opinions about the practice of distance team teaching.

37.1 Introduction

In the 2006 general outline of the living technology senior high school course in Taiwan, the following topics are included: technology of communication, construction, production, transportation, power and energy, and biology. These curricula emphasize inspiring the students' capacity for creative design, and developing their ability at the same time. However, faced with teaching about so

S.-J. Lou (✉)
Institute of Technological and Vocational Education, National Pingtung University
of Science and Technology, Taiwan, ROC
e-mail: lou@mail.npust.edu.tw

N. Mastorakis et al. (eds.), *Proceedings of the European*
Computing Conference, Lecture Notes in Electrical Engineering 27,
DOI 10.1007/978-0-387-84814-3_37, © Springer Science+Business Media, LLC 2009

many different categories, the science and technology teacher may lay particular stress on or attach undue importance to one thing, but neglect the other categories to some extent, because of being limited by his own specialty.

Researchers also face those problems in the real teaching field. In order to resolve the problems mentioned above, researchers adopt the distance-education method of coordination, combine teachers of different specialties, and make use of teaching equipment which the ordinary high schools already own to develop a concrete plan as the basis of transnational team teaching.

Based on the reasons stated above, the goals of this study include:

1. To provide an example of distance team teaching.
2. To encourage teachers to use basic computer equipment at school for conducting distance education.
3. To understand students' learning effectiveness via distance team teaching.
4. To provide suggestions for conducting different team teaching.

37.2 Literature Review

37.2.1 Team Teaching

In Taiwan, the general outline of the first- through ninth-grade curriculum alignments, the content of that indicates the conducting of learning domains should contain the spirit of integration; teachers and governors should refer the character of the class to conduct team teaching (Department for Education, 2006). Therefore, team teaching has become a cynosure in education during recent years.

Team teaching means teachers working in coordination. In order to understand the meaning of team teaching objectively, a detailed description follows.

Shaplin defines team teaching as a kind of teaching style, two or more teachers working together to teach all or parts of courses [1].

Li indicates that team teaching means two or more teachers should integrate their own specialties to teach students in one or several domains [2].

Buckley identifies team teaching as a group of teachers who aim at designing a curriculum, schedules, and lesson plans in coordination. Then these teachers have to teach students, to evaluate the results, to share their views, and to discuss things together [3].

Jeng says that team teaching means two or more teachers forming a teaching team, and being responsible for teaching a group of students. Teachers have to plan, to teach, to assess students, and to evaluate their instruction in one or several domains [4].

Chen indicates that team teaching means two or more teachers, or teachers of two or more subjects, forming a teaching team. These teachers have to make use of their specialties to design lesson plans, to help students via different teaching

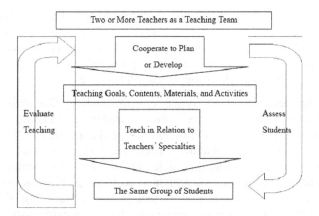

Fig. 37.1 The model of team teaching

methods, and to assess their students' learning and activities in one or several domains [5].

Because these definitions of team teaching are very similar, researchers have collected them into a model, as shown in Fig. 37.1.

Team teaching is most suitable for teachers who have just a single specialty to apply in their teaching, therefore it needs many specialized teachers working together. Because science and technology includes many learning domains, team teaching is especially appropriate for teachers who cannot be proficient in each specialty.

According to the developmental history of team teaching, this is not a new teaching theory. Many schoolteachers have known this before, but few can conduct this kind of teaching in the class. The reasons for this situation are imputed to the subjective and objective viewpoints in the school environment [6]. On the other hand, the difficulties of team teaching are shown in the following comments [7]: "It's not easy to arrange co-teaching time." "Media teaching equipment, individual learning facilities, classrooms, and learning spaces are very insufficient." "There is deficient motivation for and experience of attending related conferences and workshops." "There is a need to spend more time conferring with co-teachers." "There is a lack of teaching assistants." "It is difficult to look for commune help with team teaching." "The teaching plan is always changeable because the commune and assistants' schedules are not easy to match up." Therefore, this study tries to coordinate team teaching with distance education, then to find the best approach to solving the problems.

37.2.2 Distance Education

The concept of distance education means teachers and students teach and learn in a separated space, so they make use of man-made broadcast media to send information. In a word, distance education provides the interaction platform for teachers and students [8]. Regarding the history of distance education, Moore indicates that correspondence education is the foundation of distance education, and this foundation has evolved into several styles, such as broadcast, Internet, and videotex, with the development of media technology [9].The current types of distance education are as follows [10]:

1. The instantaneous broadcast system of team teaching: One main broadcast room and several distance classrooms are used in this teaching system. The teacher conducts instruction in the broadcast classroom, and students have to attend the class in the distance classrooms. A teaching assistant captures the picture and sound to let the teacher and the students communicate with each other instantaneously via a high-speed Internet transmission.
2. The virtual classroom teaching system (network teaching): This system applies software to design and make use of a manageable system to create a real classroom context. Network teaching is not only an assisted teaching, but also a substitute for a teacher's real lecture in a classroom.
3. The teaching system for instantaneous course taking: Students use computers and television set-top boxes to get the teaching materials they need; then they can follow their own learning speed by controlling the transmission process of distance learning.

The instantaneous broadcast system of team teaching provides vivid and lively class quality; but its shortcoming is that it requires a high level of material resources, so some ordinary organizations cannot afford it. In addition, it needs extra professionals to manipulate the teaching equipment.

Comparing an ordinary learner with a distance education learner, the latter is more active and intentional than the former.

To conclude, most junior and senior high school students fail at independent leaning. But the instantaneous broadcast system of team teaching is a synchronous distance education. In this system, a teacher can guide students to learn instantaneously, and the effectiveness of this should be superior to other systems. For that reason, this kind of distance education is adopted for this study.

Nevertheless, some junior and senior high schools cannot afford the pay for professional equipment and specialists. Therefore, this study also investigates whether or not the advantages of the instantaneous broadcast system of team teaching can be achieved with the available teaching staff and equipment.

37.3 Research Method

37.3.1 Research Subjects

The subjects of this study were 43 students of one class at National Pingtung Girl's Senior High School, and 40 students of one class at Taipei Municipal Jianguo High School. All together, there were 83 students in the study. The schedules of the science and technology classes were synchronous.

37.3.2 Method of the Study

The study was based on distance team teaching, and adopted the synchronous method to conduct the class. For that reason, the researchers had to look for two teachers who lived in different counties but who could conduct the class at the same time. The designated teachers had to conduct the two classes in 3 weeks (6 h). The researchers had to interview the teachers and to carry out a questionnaire survey to understand the effectiveness of synchronous distance education.

37.3.3 Research Tools

The research tools of this study were questionnaires and structural questions for interviews; the following are descriptions of the content and the amendment of them.

(1) Questionnaire and the content of the interview question: According to the literature review and researchers' discussions,,the factors of synchronous distance team teaching are as follows:

1. The degree of learning effectiveness due to the teacher's team teaching.
2. The degree of learning effectiveness due to the synchronous distance video-tex class.
3. The overall opinion of the class regarding distance team teaching.

Next, the researchers draw up the items for each question, and set the question totals to 21. Because the students' backgrounds are very clear and definite, the researchers list only each student's school name and gender to indicate the student's file. Finally, the researchers use the factors as the unit and apply the items of questions as the descriptions to be the structural questions for the teachers to conduct interviews with their students.

After finishing the first draft of the questionnaire, the researcher requests the adviser to revise any useless and unclear questions. Then those revised questions are arranged into "The Learning Effectiveness Questionnaire about Distance Team Teaching in the Area of Science and Technology in Senior High School," and "The Interview Questions for Teachers about Distance Team Teaching in the Area of Science and Technology in Senior High School."

(2) The method for editing questionnaires:

The questions for the interviews with students about learning effectiveness should match the factors of the questions and the students' understanding with regard to vocabulary (subjects) as the main principle. Eighteen questions make use of the Likert five-point scale, and the responses to the questions are "strongly agree, agree, neutral, disagree, strongly disagree." Another three questions are open-ended, and their content is based on the overall process.

37.3.4 Data Collection, Processing and Analysis

The subjects of this study come from northern and southern Taiwan. In order to conduct the questionnaire easily, the researchers adopt online questionnaires, transforming the rough data into SPSS for setting up the file, and use SPSS to carry out statistical analysis via the method of descriptive statistics. The researchers would also interview teachers and record their thoughts and feelings about this teaching to compare with the students' responses to the questions.

37.4 Curriculum Design

37.4.1 The Title of the Class

Information Communication Technology (ICT)—A Story Fan-Tan.

37.4.2 Purpose of the Curriculum

1. The available teaching equipment at the senior high school is the foundation on which to conduct synchronous across-school teaching so that specialized teachers at different schools can teach in coordination.
2. Students who are located in different areas can study and work with each other using the Internet and videotex to shorten the distance between urban and rural, and achieve the goal of multiple learning.

37.4.3 Goal of Teaching

1. Students can be able to shoot a video and use image editing software to revise or to capture the frames from the video.
2. Students can carry out the system model of ICT (the input-process-output model).
3. Students can apply ICT to the teaching environment in order to achieve the purpose of assimilating information into their instruction.
4. Students can make use of distance team teaching to promote their learning.

37.4.4 Location and Place

The distance team teaching was held in two schools. One school is located in southern Taiwan: the scientific and technological classroom at National Pingtung Girl's Senior High School. The other school is located in northern

Taiwan: the scientific and technological classroom at Taipei Municipal Jianguo High School. It is about 400 km from Pingtung to Taipei.

37.4.5 The Plan of Curriculum

37.4.5.1 Teaching Equipment

Hardware

The researchers list the equipment at one school in Table 37.1, but they also set up the same equipment at the other school while conducting the class. Figure 37.2 illustrates the structure model for distance team teaching.

Software

The following software was used in this system; some can be downloaded online. The teachers may need to adjust the settings (Table 37.2).

37.4.5.2 Contracted Lesson Plan

The lesson plan is developed for a three-week period. Input (week one)→ process (week two)→ output (week three) (Table 37.3).

Table 37.1 The equipment at the researched school

Title/name	Function and use	Quantity
FTP server	Teachers can use the original FTP sever at the school or use the available computer to set up an FTP server with a Linux/Windows system. Teachers can open the space for students to upload videos or to save related files before lectures begin.	1.
Media server	Teachers can use the available computer or the professional server computer with a Linux/ Windows system to conduct the class. If the expense is too great, the researchers can use the FTP sever. This machine is used for broadcasting the video.	1
Equipment for Internet	The main factor in videotex quality is the frequency channel, so an optical fiber line is the best tool for that. If the expense is too great, ADSL, cable, and the school network can also be utilized for this teaching.	1
Computer with videotex	The functions of computer and notebook are to project the image onto the screen and to connect the webcam.	1
Internet video camera	To video the teachers and students, and to send the video to the other school.	At least 2

Table 37.1 (continued)

Title/name	Function and use	Quantity
Microphone	To record the voices of teachers and students, and to send them to the other school.	1
Projector and screen	To show the video from the other school.	1
Speaker	To broadcast the voices from the other school.	1
Students' computer	For students to make the video.	Several or provided by students
Digital video (DV)	For students to produce their own work. Suggestions: The DV with hard discs and memory stick is how the students capture the files after they are connected via a computer. If the users have a Mini-DV, they have to use a IEEE1394 device to capture the files. Avoid using DV recorded on DVD, because the process of formatting is very complicated.	Several or provided by students
Digital camera	For students to produce their own work. Suggestions: The auto-camera is how the students shoot still pictures and moving pictures. A single-lens reflex camera is better, but is just for shooting still pictures.	Several or provided by students
IEEE1394 and USB cards	To capture the video. A recent PC and NB are regarded as basic equipment; new computers should be equipped with both 1394 and USB.	Several

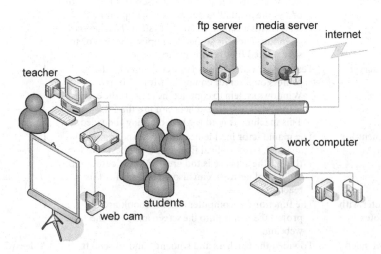

Fig. 37.2 The structure model for distance team teaching

Table 37. 2 The software used in this study

Title / Name	Function and Description	Memo
Yahoo Messenger	Software for videotex that can be downloaded on the Internet.	
MSN	Software for videotex that can be downloaded on the Internet.	Choose one
Skype	Software for videotex that can be downloaded on the Internet.	
Microsoft Movie Maker	Software for making videos or images; the controlling interface is easy to utilize. Windows XP has this software in it, but its capabilities are too simple.	
Gold Wave	The program for recording and editing voice messages. It can be downloaded from http://www.goldwave.com	Shareware
WinAvi Converter	The program for transforming media files, such as avi, mpg, wmv, mov so they can be used interchangeably. It can support the formats of DC and DV. This program can be downloaded from http://www.winavi.com.	Shareware
Filezilla	FTP software for both server and client, for transferring files. This program can be downloaded from http://filezilla.sourceforge.com.	Shareware

Table 37. 3 The lesson plan for this study

Week	Content	Equipment	Memo
One 20 Min	1. Description of activity: (1) Students at National Pingtung Girl's Senior High School and Taipei Municipal Jianguo High School to choose the location and the materials for the shoot.	DC or DV FTP Server	The specification for pictures is 1024×768, and the video file should be wmv format.
20 Min	(2) The subject matter for the shoot will be uploaded to the FTP server, and the works of students at the two schools will be saved in document style.	PC/ Notebook The video editing software.	Students can freely add asides, captions, and background music for their video.
80 Min	(3) Students at the two schools have to use the present materials to make a video via the processing of images. The length of the video is 2 min. The subject matter and the script of the video should be produced by the students. 2. Start to collect data.		
Two 100 Min	Video editing. Upload the works to the Internet.		The format of the video should be wmv.

Table 37. 3 (continued)

Week	Content	Equipment	Memo
Three 100 Min	Sharing moment. Students at the two schools have to present their works on the stage.	PC. webcam, communications software, microphone	After broadcasting the video, students have to present their thoughts or discuss online.

37.5 Findings and Discussion

37.5.1 The Questionnaire on Learning Effectiveness

After processing the data, the results of the closed questions are as follows:

(1) The impact on learning effectiveness of teachers' team teaching

According to the average (M>4, SD<1) and mode values, most students gave positive opinions on the atmosphere in the class, the content of the class, and different teachers' lecturing. Not a few students chose "agree" (4) for the content of the class and the support for producing works. Most students chose "strongly agree" (5) for the atmosphere in the class, and interesting feelings in the class. Just one or two students gave negative opinions, the percentage being 1–2%. The subject of this class is more interesting than in an ordinary class; none of the students gave negative opinions.

(2) The impact on learning effectiveness of distance videotex synchronous classes

In this part, students agreed with the method of distance videotex synchronous classes. Because students could encourage and interact with the other group of students via the Internet, their opinions on learning motivation, works, and the atmosphere in the classroom presented a positive result (M>4,SD<1). Hence, most students chose "agree" (4) and "strongly agree" (5) to this section. Only 1–3 students chose "disagree," the percentage being 1–3%. On the item "The method of distance education makes me understand other students' thinking in the same class," nobody expressed a negative opinion. This clearly shows that this distance videotex synchronous class inspired students to improve their abilities.

(3) The entire experience of the distance team teaching program

The results of the students' questionnaires indicate that students do learn a lot from this class and like this class (M>4, SD<1), with most students expressing "agree" (4) and "strongly agree" (5). The negative opinion for this section was only 1–4 students choosing "disagree," the percentage being around 1–4%. Nobody gave negative opinions on the question of "Overall, I like this class." This indicates that all students liked the method of this class.

(4) The open-ended questions on the whole perception of the class of distance team teaching

In the process of this learning program, what do you think that you learned best? The students' answers in this section fell into the following categories:

a. To interact with students at the other school (68.67%).
b. To learn the technique for video editing (45.78%).
c. To like and understand science and technology (8.43%).

What do you think was the biggest difficulty you faced in the process of this learning program? The students' answers in this section fell into the following categories:

a. The ability to shoot and produce video was insufficient (43.37%).
b. The equipment was insufficient and of low quality (31.33%).
c. The class time was insufficient (30.12%).
d. Conflict within the group was a difficulty (4.8%).

In this learning process, what do you think could be improved further (facility, program, etc.)? The students' answers in this section fell into the following categories:

a. The equipment (74.7%).
b. The technique and ability to shoot and edit (9.6%).
c. The class time—too insufficient. (14.56%)
d. The content of the class (7.2%).
e. Everything was pretty good (6.02%).

37.5.2 Interview with Participating Teachers

The interviewees are three teachers in this distance education, and the researchers used A, B, and C to represent them.

(1) The impact on learning effectiveness of teachers' team teaching

Teacher A: In creating the learning atmosphere, lesson plans and preparation are very important for the class, except for the class management. However, the lesson plan should be developed by the team of teachers, then the process of the class would be smoother and more successful.

In the process of this class, the teachers at National Pingtung Girl's Senior High School and Taipei Municipal Jianguo High School had to draw up the lesson plan and correct its shortcomings by long-distance discussions three times. After that, the teachers could reach a consensus and then conduct the actual teaching process accordingly. Consequently, the students would have an authority to follow, and disarticulation between teachers and students would not occur.

Finally, I know students like this class from their attitude and responses in the class; they do get involved in this class. Meanwhile, they can also finish the task on time. Thus I think this distance team teaching raises the students' learning effectiveness.

Teacher B: Students are curious about new things, and also interested in the appearance of teachers or female classmates. Therefore, is this teaching method able to help students understand the content or not? I hold a neutral opinion on this. Besides, owing to the limitation in bandwidth and equipment, both sides cannot have an international atmosphere in the class. Thus if we had more high-quality DV and bandwidth, that would improve the students' ability in producing their works.

(2) The impact on learning effectiveness of distance videotex synchronous classes

Teacher A: I never taught this kind of class, but it is very good for students. It is very good for promoting students' learning motivation. Students like to interact with students at the other school, and they have confidence in their works. Nevertheless, when students can understand the different thoughts and viewpoints of others, these thoughts and viewpoints inspire their competitiveness and actualization. Beside, after brainstorming, the students' achievement is really good.

Students attend the class with students at the other school, and this kind of environment makes a good atmosphere for the class. Maybe this kind of novel class affects the students' learning motivation and atmosphere; therefore, if the teachers can make their instruction innovative, the students' motivation and positive atmosphere can be continued.

Finally, is the distance videotex synchronous class better than an ordinary class? I hold a qualified attitude, because different teaching methods make for different leaning effectiveness. This teaching method is not superior to other teaching methods, but it is still a good teaching method for students.

Teacher B: This is cooperation between a school for boys and a school for girls, and this method drives boys to present a good learning attitude toward the class and makes a good atmosphere in the class. This kind of class is more interesting than the atmosphere in an ordinary class, and the students' learning effectiveness is better than the learning effectiveness in an ordinary class.

However, the class time is too short; there is no time for students to share their feelings and thoughts in the class. Otherwise, students could understand more and different thoughts from others.

About the cooperation of a school for boys and a school for girls, boys are eager to present good learning motivation in the class, the class atmosphere is very good, and this class is more interesting than others.

(3) The section on the complementariness to teachers' specialties

Teacher A: In this class, I think different teachers' specialties are complementary to each other, because a teacher cannot be an expert on everything. Science and technology includes many professional domains, but a teacher can only be conversant with certain domains. By enabling students to study this field comprehensively, team teaching corrects this deficiency.

Besides, team teaching promotes my professional ability. Teaching benefits teachers as well as students; the key point is the interaction between teachers and students. Team teaching can inspire new thinking and creative teaching methods for teachers, and it is really good for the teachers' profession. Thus, I think I really do learn a lot from this class.

Teacher B: In the process of team teaching, participating teachers have a chance to learn from each other's work, which can also achieve the function of complementariness. Advance communication and preparation are very important for team teaching; teachers can dispose of problems by creative brainstorming and problem solving, and thus achieve the function of complementariness.

I am sure this kind of teaching can promote my profession, and the thoughts and attitudes of teacher A are much appreciated by me. Teacher A has knowledge about network management; he helps us directly and opens a new teaching field for us.

(4) The entire experience of the distance team teaching program

Teacher A: In this teaching process, because the bandwidth is not enough, the effect of different versions and voices is terrible. This problem obstructs the process of interaction. Next time, the transmission equipment (webcam, computer, and transmission speed), the completed lesson plans, the students' video editing (students use Windows Movie Maker with only basic functions), the governmental support, and the costs should be improved and reformed; the results will be better and better.

In general, the results of this teaching process are satisfactory. Students are very happy with the learning process; the outcome of the class is really good. If future teachers and researchers want to promote this kind of teaching, the equipment must first be improved.

Teacher B: In this teaching process, I think the most difficult part is the lack of manpower and equipment; this kind of teaching needs better equipment for videotex and bandwidth, and requires sufficient manpower to manipulate the machines as director. Therefore, improved bandwidth and capture of DV are imminently required in the future.

As a man sows, so shall he reap. Without diligence, there is no opportunity for gain. Teachers think that the identify and change of this class are more important than the degree of difficulty and practicability. In other words, if a teacher disapproves of this kind of class, the effectiveness of teaching will diminish.

37.6 Conclusion and Suggestions

37.6.1 Conclusion

In line with the findings of this study, our conclusions are as follows:

1. Students have positive overall feelings about distance team teaching for the science and technology class.

2. Teachers approve the students' performance in the distance team teaching of the science and technology class.
3. Both students and teachers indicated that insufficient equipment and space should be improved.
4. Both teachers and students hold positive and affirmative views about the practice of distance team teaching.
5. Teachers agree with the method of distance education, which provides an opportunity for different specialty teachers in different locations to conduct the class. The outcome of this method is very affirmative and successful, and it improves the teachers' professionalism.

37.6.2 Suggestions

1. Suggestions for teachers of distance team teaching:

   a. With regard to creating a learning atmosphere, lesson plans and preparation are very important for the class, except for the class management. However, the lesson plan should be arranged by teachers in the team, in order to make the process of the class smoother and more successful.
   b. A good class plan helps the teaching team achieve a fine effectiveness, despite the lack of adequate equipment.

2. Suggestions for future researchers:

   a. This time, the researchers used basic and simple equipment to conduct synchronous distance team teaching. Future researchers should use superior equipment to investigate whether or not teaching effectiveness will be improved.
   b. In this study, coeducation (boys and girls) was adopted in the process of distance team teaching. Future researchers could arrange for students of the same gender to study together, and to investigate whether or not the degree of achievement will be the same.
   c. This study was based on a unit of science and technology for senior high school students. Future researchers could conduct other units or investigate the learning effectiveness of team teaching in the first- through ninth-grade curricula. Transnational team teaching could also be investigated and examined by future researchers.

References

1. Shaplin JT (1964) Team teaching. Harper & Row, New York
2. Li KC (2001) The teaching materials and methods of integrated learning. Psychological Publishing, Taipei
3. Buckley FJ (2000) Team teaching: what, why, and how? Sage, Thousand Oaks, CA
4. Jeng BJ (2002) The study of team teaching implementing in elementary schools. Dissertation, National Kaohsiung Normal University, Kaohsiung, Taiwan

5. Chen FM (2003) A study of teachers' problems with team teaching in elementary school. Thesis, National Taitung University, Taitung, Taiwan
6. Kao PC (2003) The conception and practice of team teaching in middle school education. Ministry of Education, Taipei, pp 124–139
7. Lin YP (2004) A study of the types of team teaching in language arts in Kaohsiung County. Thesis, National Pingtung University of Education, Pingtung, Taiwan
8. Moore MG, Kearsley G (1996) Distance education: a system view. Wadsworth, Belmont, CA
9. Moore MG (1996) Theory of transactional distance. In: Keegan ED (ed) Theoretical principles of distance education. Rutledge, New York, pp 22–38
10. Chang YF (2000) A study of distance learning on policy and execution in junior high school. Thesis, National Chengchi University, Taipei, Taiwan

Chapter 38
Electronic Exams for the 21st Century

Antonios S. Andreatos

Abstract Information and communication technology (ICT) has deeply affected the way people in modern societies become educated and learn new things. Several "digital skills" are needed for young people in order to live and work in the 21st century. During the years to come, a tremendous development and spread of e-learning, and consequently e-assessment, is expected. Common question types used today will soon prove inadequate to assess the required "digital skills." In this paper we deal with this problem and we propose some solutions.

38.1 Introduction

Information and communication technology (ICT) has deeply affected the way people in modern societies communicate, work, transact, get informed, entertain, and learn. Teachers and students use the Internet in many different ways: as a huge online encyclopaedia and dictionary, as a medium for communication, as a teleworking or collaborating medium, as a distance learning medium, etc. Learning and education have been deeply affected and enriched by ICT. The introduction of computers and the Internet in education has radically changed the way people become educated and learn new things [1].

We are currently in a transitional stage as far as ICT skills are concerned. Many of us were born before the introduction of PCs in the 1980s or the invention of the World Wide Web (WWW) in the 1990s. Many older people attend lessons in order to be able to use ICT for their jobs or everyday life, but not all of them succeed. Adults have difficulty in learning new things not related to their previous knowledge or skills [2]. However, our children grow up with computers and ICT and they are familiar with their use.

A.S. Andreatos (✉)
Division of Computer Engineering & Information Science, Hellenic Air Force Academy, Dekeleia, Attica, TGA-1010, Greece
e-mail: aandreatos@hafa.gr; aandreatos@gmail.com

N. Mastorakis et al. (eds.), *Proceedings of the European Computing Conference*, Lecture Notes in Electrical Engineering 27, DOI 10.1007/978-0-387-84814-3_38, © Springer Science+Business Media, LLC 2009

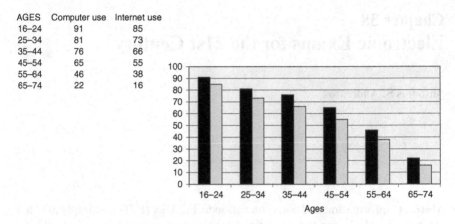

AGES	Computer use	Internet use
16–24	91	85
25–34	81	73
35–44	76	66
45–54	65	55
55–64	46	38
65–74	22	16

Fig. 38.1 Use of PCs and the Internet by individuals in the European Union (25 members). Legend: *Black* = computer use, *Gray* = Internet use

Figure 38.1 shows age-related statistics for individuals' use of PCs and the Internet in the European Union [Source: Eurostat, 2004].

The new generation has incorporated technology in its everyday life, where many activities are supported by computers and presuppose a technophile attitude. People nowadays prefer to photocopy, scan, or photograph a document instead of writing down information by hand on a piece of paper. Digital photography is not only superseding analog; it is also replacing other paper works such as photocopies and paper notes. Students use their mobile phone cameras to copy each other's notes or even to cheat! Delegates in conferences use their zoom-capable digital cameras to copy interesting slides or whole presentations. Digital technology is everywhere nowadays, even in our pockets; and it is handy and easy to use.

Our children spend a lot of time in front of several kinds of screens, such as monitors, handheld devices, TV screens, etc. However, they spend less time reading books; they are anxious, and they find it difficult to sit in a chair for a long period of time and concentrate on books or notebooks.

Web usability studies have shown that mature people who were educated reading paper books have difficulty in reading from computer screens. It has been measured [1, 3–7] that:

- 79% of users scan web pages for keywords because they find it painful to read too much text online.
- Only 16% of the newcomers to a web page read it word-by-word.
- Reading from a screen is 25–30% slower and much more boring than reading from paper with comparable size and fonts [1].
- Web content should have half the word count of the corresponding paper version.

– Web text should be written following journalists' style, i.e., using the model of the pyramid: since the reader is exposed to the title and abstract at first sight, the most important information should be placed on the home page and the details (or the balance of the article) elsewhere, connected to the abstract by a hyperlink.

Should the same research be repeated some years from now, different statistics may come out; but for now we have to take these figures into account when using educational material in electronic format.

From the aforementioned we can easily draw the following conclusions:

1. Our children grow up with computers and read information from screens.
2. Our children use ICT as a tool for many activities, such as games, education, information, etc.
3. Young people form the "critical mass" of Internet and ICT users, and therefore will drive the evolution of those technologies.

This paper is structured as follows: in Section 38.2 we point out the advantages of electronic documents; in Section 38.3 we stress the need for new types of e-tests; in Section 38.4 we show some sample problems testing the students' "digital skills;" in Section 38.5 we present types of problems especially for computer science students. Finally, in Section 38.6, we conclude.

38.2 Electronic Documents

Today a lot of information is produced in electronic format from the beginning; entire encyclopaedias, conference proceedings, telephone directories, online journals, and books are published on CD-ROMs or as PDF files. At the same time, traditional paper documents are converted into electronic format. Clearly, the electronic format has some distinct advantages, such as:

– It may easily be distributed, stored, reproduced, processed, encrypted, digitally signed, etc.
– It may easily be searched for specific keywords.
– It is digital, thus it may easily fit in a handy digital storage medium such as a CD, DVD, flash memory stick, PDA, mobile phone, etc., and be transported anywhere, even uploaded to the Internet, making it available anytime.
– We can easily incorporate multimedia in it as well as hyperlinks, in order to connect it to related documents.
– It is economical and ecological.

Let us examine a case study. The Technical Chamber of Greece numbers about 102,000 members, increasing by about 3,000–4,000 new members yearly. For many years, they have issued a weekly News Bulletin (about 100 pages), which is sent by mail to all members. Beause of the increase in the price of paper, and especially that of postage costs, the circulation of the News Bulletin is facing economic problems. The annual costs for paper, layout, printing, and posting the News Bulletin ran to 3.9 million euros for 2006, and it is estimated that the respective entry for the year 2007 will exceed 4.26 million euros. Therefore, they have recently decided (June 2007) to circulate the News Bulletin electronically [8].

We claim that today, digital is the format of choice. We frequently give our students additional educational material in electronic format (e-books, links, etc.). Hence, we would like to educate our students to use electronic documents fluently. The ability to use electronic documents is a prerequisite skill needed in various activities as well as in many disciplines and, as such, it must be assessed as well.

38.3 Need for New Types of E-Tests

E-assessment has been widely used since the development of e-learning. However, what most e-tests do is to transform paper exams into electronic format. This is clear from the types of questions used by most e-learning and e-assessment platforms. Common question types include: multiple choice, multiple correct, true/false, correspondence, fill in the blanks, simple arithmetic or numeric results, etc.

Transforming paper journals into electronic ones is good but not enough, if the journals do not take full advantage of the new format; we may easily use online journals in PDF, but the use of hyperlinks is limited (except emails and references from the web). However, there are other documents in electronic format which take full advantage of the medium by using hyperlinks throughout the text. Wikipedia is a good example: all underlined keywords link to related lemmas, while there is usually an extensive bibliography at the end of each lemma, where the reader can find more information.

Similarly, transferring the common question types from paper to computer is only half the job. The other half is to assess the students' "digital skills." This is important not only for computer science students, but for all candidate professionals in the 21st century, because computers are tools for all disciplines, just as mobile phones are for practically everyone. Thus, instructors also have to verify that their students can answer the questions using the right tools for the 21st century.

Instructors creating test questions should take the aforementioned screen readers' behaviour, as well as web usability rules [9], into account.

38.4 Towards a New Generation of E-Tests

Most modern e-assessment tools, either part of e-learning platforms or autonomous [10], support a general set of rather naïve types of questions. The problem with multiple choice and similar questions is that they transfer the tedious work from grading to properly formulating the questions [11]. The problem with naïve and generic (hence limited-capability) types of questions is that they are not a panacea; more specifically: first, they are not adequate for all majors, and second, they are less applicable to higher education [10].

In order to overcome such limitations and gain more control over types of questions, we have developed an experimental e-assessment software, called "e-Xaminer" [10, 12]; this has been used in our institution, the Hellenic Air Force Academy (HAFA), during the past and current academic years. The main reason for developing "e-Xaminer" was to experiment with new question types and, of course, to take advantage of all the merits of e-assessment [12]. e-Xaminer incorporates some new, advanced types of questions [10]. However, we realize that it is not enough to transfer questions from paper to electronic exams; a new philosophy is required, not only because the skills of the 21st century are different [13, 14], but also because ICT has significantly changed the way contemporary people learn new things and assimilate new skills [1].

38.4.1 New Possibilities

In order to assess the aforementioned "digital education" skills, we have employed questions like the following:

"Given the specification of HTTP 1.1 [RFC 2616] in electronic format, answer the following questions:

1. What signalling mechanism is used to terminate a persistent connection?
2. Who may terminate such a connection? (a) the client, (b) the server, (c) both of them.
3. What does error code 409 signify?
4. Which are the idempotent methods of HTTP?"[12].

The HTTP standard is described in RFC 2616, a 176-page document. Students are obviously not asked to memorize this text; besides, it is impractical to copy the entire document for each student just for the purpose of the exam. They are however expected to be able to extract information from electronic documents using various text/word processing tools (such as "Search" or "Find"). Testing this skill is only possible via electronic exams. During a recent exam (January 2007), it was proven that students with adequate "digital skills" had an advantage and finished earlier.

Another example activity is: "Use the service provided by www.traceroute.org, to find all the intermediate servers between your server and a given server" (somewhere abroad).

As already mentioned, it is wrong to try to transfer paper exams to the computer; the correct approach is to try to take advantage of the new possibilities offered by ICT and include them in our tests. Some examples are:

1. Allow students to use additional programs or utilities (such as the calculator, print screen/screen capture tools, multimedia processing/editing utilities, dictionaries, etc. [15].

2. Allow students to use the web during the exams; the web may be used in many different ways, even as an alternative source of information, like an encyclopaedia, etc. [1].

38.4.2 Facilitating e-Assessment

Because of the high demand for distance learning courses and the large number of students [15], it is imperative to devise methods for automatic assessment. The employment of auxiliary programs is of great use. Let us give some examples:

1. In a signal processing course using MATLAB or other special tools, the instructor may easily check whether the student has constructed a correct filter for clearing the noise, just by hearing the output waveform.
2. In an image signal processing course using special tools such as MATLAB, the instructor may easily verify whether the student has cleared the noise, just by looking at the resulting image.
3. In an electronic design the instructor may automatically verify the correctness of a schematic using the Design Rules Check (DRC) tool of the software (e.g., OrCAD).
4. In a programming language course the correctness of a piece of code may easily be verified by the compiler/assembler [12].

Telecommunication engineering students in HAFA are required to be able to write, assemble, and run machine language programs. In another type of question, examinees were required to assemble their program and submit the assembler listing file. This is another example showing how e-assessment can test skills that would have been impossible to measure with paper-and-pencil exams.

In another course, students were asked to construct their own blog and upload some specific information. It is obvious that blogs today are for everyone, not just for computer scientists or students. Blogs may be considered to be the digital equivalent of paper diaries, or personal announcements, or news bulletins, or even photo albums [17]. By asking students to create their own blogs we try to encourage them to develop a "digital culture" and at the same time, acquire "digital skills."

Another important "digital skill" required of our students is the ability to find correct and precise information on the Internet. The ability to use search engines is not adequate, because often the results number in the thousands or millions, and it is impossible for the user to check them all.

Instructors writing e-test questions should ask themselves the following things:

1. "Does my e-assessment software support this type of question?"
2. "How can I formulate the question/problem so that it will be easily graded by the software without sacrificing test quality?"
3. "Does the resulting test assess the required knowledge, skills, or attitudes?"

38.5 Teaching Telematics via Telematics

Some years ago we introduced the "Teaching Telematics via Telematics" approach in our classes [18]. The concept is that instead of trying to teach telematics theoretically, we do it practically on the computer, or on the network, or on the Internet, or in the lab. To give an example: instead of trying to explain in the classroom what an email message looks like, we may ask them to send email messages (a general skill), but also ask them to identify the address of a specific email server by searching the raw email text message (an exercise for computer scientists).

Other such tests require students to perform tasks that are automatically done through the HTTP protocol and the browser (e.g., communicate with an HTTP server and ask for an HTML document) manually, via "telnet."

Other examples concern the use of software such as MATLAB/Simulink and various toolboxes, related to courses such as telecoms, signal processing, etc.

Obviously such skills are electronically tested in the lab and not with paper-and-pencil exams.

Therefore, a good procedure to use for electronically testing computer science courses is to combine available e-assessment tools with assignments to be carried out in the lab during exam time, following the "TeachingTelematics via Telematics" approach.

38.6 Conclusion

In this paper we have argued that modern e-assessment should be based on ICT and not just a transfer of questions from paper to computer. We have demonstrated some applicable questions; we have also demonstrated a new series of questions for CS students, based on the "TeachingTelematics via Telematics" approach.

Electronic tests should be written keeping in mind user behaviour when reading from screens, as well as web usability rules [9]. They should also take into account the possibilities of new technologies, as well as their use by modern people.

References

1. Avouris N (2004) Quality issues on the structure and content of the Internet, used as a teaching tool. In: Vlachavas I, Dagdilelis V et al (eds) Information and communication technologies in Greek education. University of Macedonia Press, Thessalonica, Chapter 1, pp 16–33 (in Greek).
2. Rogers A (1996) Teaching adults. Open University Press, Buckingham
3. Bernard M (2003) Criteria for optimal web design. psychology.wichita.edu/optimalweb.
4. Nielsen J (1996) Top ten mistakes in web design. www.useit.com/alertbox/ 9605.html

5. Nielsen J, Writing for the web. www.useit.com/papers/webwriting
6. Nielsen J, Schemenaur PJ, Fox J, Writing for the web. http://www.sun.com/980713/webwriting
7. Nielsen J, Schemenaur PJ, Fox J Difference between paper and online presentation. www.sun.com/980713/webwriting/wftw1.html
8. Technical Chamber of Greece (2007) News Bulletin 4 June 2007, No. 2442, p 7
9. Nielsen J (2000) Designing web usability. New Riders, Indianapolis
10. Andreatos A, Doukas N (2006) The 'e-Xaminer' approach: a proposed electronic examination system. WSEAS Trans Adv Eng Educ 3(5):431–438
11. Bush M (1999) Alternative marking schemes for online multiple choice tests. In: 7th annual conference on the teaching of computing. CTI Computing, Belfast
12. Doukas N, Andreatos A (2007) Advancing electronic assessment. IJCCC 2(1):56–65
13. Salpeter J (2003) 21st century skills: will our students be prepared? Available online at: www.techlearning.com/story/showArticle.jhtml?articleID = 152020 90 (retrieved on March 22, 2006)
14. The New London Group (1996) A pedagogy of multiliteracies: designing social futures. Harv Educ Rev 66(1):60–92
15. Jesshope CR (2003) Towards the dynamic publication of multimedia presentations—a strategy for development. In: Proceedings of ALT-C
16. Burns E (2006) Continuing education drives distance learning enrollment. Available online at: www.clickz.com/stats/sectors/education/article.php/3605 321 (retrieved on May 25, 2006)
17. Angelopoulos C (2006) Blogs change the landscape of communication. Special edition of Kathimerini (newspaper): New media: the alternative choice. 28:78–79 (in Greek)
18. Andreatos A, Stefaneas P (1997) Teaching telematics via telematics. In: Proceedings of Neties '97, Ancona, Italy

Chapter 39
Effects of the Orff Music Teaching Method on Creative Thinking Abilities

Rong-Jyue Fang, Hung-Jen Yang, C. Ray Diez, Hua-Lin Tsai, Chi-Jen Lee, and Tien-Sheng Tsai

Abstract This study starts with the area of arts and humanities, integrates sentiment and rationality, and emphasizes fostering the abilities of creative thinking. Quasi-experimental research was adopted to examine the effects of the Orff music teaching method on creative thinking abilities, and the method was implemented on the experimental group. The Torrance Test of Creative Thinking was used for the pre-test and post-test, and it was found after analyzing the covariance data that the Orff music teaching method has positive influences on children's creative thinking abilities. The fluency, flexibility, and originality of their creative thinking increased significantly. There are suggestions for future research.

39.1 Introduction

The grade 1–9 curriculum emphasizes subject integration, team teaching, and cooperative learning to develop students' self-learning abilities. In this technological society, people need not only to adapt to impacts from society but also to possess the ability of creative thinking. Therefore, much attention has been paid on how to encourage students' learning motivation, their self-learning ability, and their creative thinking ability [1, 5, 7]. This study starts with the area of arts and humanities, integrates sentiment and rationality, and emphasizes fostering the abilities of creative thinking. The Orff music teaching method guides children's intelligence with their feelings [4]. Children have their sense of sight, hearing, touch, and physical acts like crawling, walking, etc. The Orff music teaching method inspires children's basic understandings of music using these natural experiences.

R.-J. Fang (✉)
Department of Information Management, Southern Taiwan University
of Technology, Taiwan, ROC
e-mail: rxf26@mail.stut.edu.tw

N. Mastorakis et al. (eds.), *Proceedings of the European
Computing Conference*, Lecture Notes in Electrical Engineering 27,
DOI 10.1007/978-0-387-84814-3_39, © Springer Science+Business Media, LLC 2009

39.2 Purpose of the Study

The purpose of this study is to explore whether the creative thinking abilities of children who have been trained with the Orff music teaching method are superior to the abilities of those who have not been trained with the method. The content, method, process, and characteristics of the Orff music teaching method are as follows.

39.2.1 Content

The curriculum (Orff Schulwerk) is designed starting from childhood and uses children's musical experience as the teaching material. It is aimed at impromptu expression and emphasizes rhythm [4, 6].

39.2.2 Characteristics

A pentatonic scale is included in the first volume of Orff Schulwerk. Ostinato and bass are repeated in the teaching materials [3, 2]. Folk songs and music related to students' experiences are adopted in the Orff music teaching method and the Kodaly teaching method. Special Orff musical instruments are used. It is taught starting from sol-mi and the concept of elemental music is presented in Orff Schulwerk [8]. The language of rhythm starts from words and develops as sentences and canon [1, 3].

39.3 Activities of the Orff Music Teaching Method

39.3.1 Language of Rhythm

Language is a part of musical experience, which is also a special part of Orff Schulwerk, from clapping to playing instruments [1]. Orff transfers the students' experiences of talking and chanting into musical experiences [5, 7]. Some music concepts, such as accent and meter signature, can be introduced in languages, enhanced in other activities, and learned in music.

39.3.2 Singing

The Orff music teaching method makes students listen to many sounds and the five-note scale mode is considered to be suitable for children.

39.3.3 Performing

In Orff Schulwerk, performing includes both solo and ensemble. Children use different skills to perform and learn the composition when they perform together.

39.4 Research Method

The pre-test and post-test design for two groups:

The two groups are given the Torrance Test of Creative Thinking to determine the standard of the subjects' creative thinking abilities.

After one term of music training, the two groups are tested again using the Torrance Test of Creative Thinking to examine the differences.

39.4.1 Sampling

Students are classified according to their language ability and their parents' professions and 30 subjects are taken from each class for testing.

The parents' professions are classified into six categories: business, military service, civil service, industry, freelance, and other. Six students are chosen from each category in each group. After excluding the highest and the lowest scores in Chinese in the class, 30 students are chosen as subjects according to their language ability. In the sampling process, 36 students are chosen according to their parents' professions, and 30 are chosen from them according to their language ability.

39.4.2 Research Tools

This study adopts the Torrance Test of Creative Thinking, published by the Chinese Behavioral Science Corporation and revised by Ingmao Liu.

The test examines the subjects' creative thinking ability for solving problems and discovering new relationships. It is an open-ended questionnaire with a standard answer for each item to test creative thinking ability.

The test is revised from the Torrance Test of Creative Thinking, edited by E.P. Torrance in the U.S.A.

Students are scored according to their fluency, flexibility, and originality on the Torrance Test. The mean and standard deviation are calculated. A one-way analysis of covariance is used to exclude the influence of the pre-test. Scores from both groups are analyzed.

39.5 Results Analysis

39.5.1 Influence of the Orff Music Teaching Method on Children's Creative Thinking

The study adopts a one-year teaching experiment and uses the teaching method as the independent variable with the experimental group and the control group. The students' pre-test scores on the Torrance Test of Creative Thinking constitute the covariance and their post-test scores constitute the dependent variable. The influence of the teaching method (the independent variable) on the students' creative thinking ability is examined with a one-way analysis of covariance.

To learn the differences between the two subjects groups' entry behaviors, the homogeneity of within-class regression coefficient test was used to examine the differences of creative thinking ability between the two groups before the teaching experiment. The results are shown in Table 39.1. The mean in the post-test is shown in Table 39.2.

In Table 39.1, $F = 2.941$ and $P > 0.05$ for the homogeneity of regression slope test in the total scale; this does not reach the standard of significance. The statistical testing accepts the null hypothesis which the homogeneity of within-class regression coefficient hypothesis establishes in the total scale, and the one-way analysis of covariance is then processed. Tables 39.3 and 39.4 indicate that

Table 39.1 Summary of homogeneity of within-class regression coefficient test in Torrance Test of Creative Thinking

Item	F	p
Total scale	2.941	0.920
Fluency	0.076	0.784
Flexibility	0.343	0.560
Originality	2.680	0.107

$p < 0.5$

Table 39.2 Mean in the post-test of both groups in the Torrance test of Creative Thinking

	Experimental group	Control group
Total scale	82.778	71.889
Fluency	23.408	14.392
Flexibility	30.342	26.258
Originality	39.872	20.395

Table 39.3 Mean and standard deviation in the pre- and post-test of the Torrance Test of Creative Thinking

	Experimental group (n = 30)		Control group (n = 30)	
	Pre-test	Post-test	Pre-test	Post-test
M	72.93	99.67	44.27	55.00
SD	17.338	22.068	6.731	7.254

Table 39.4 Covariance analysis of the pre- and post-test in the Torrance Test of Creative Thinking

Variation source	SS	df	MS	F	Sig.
Between groups (teaching method)	797.985	1	797.985	26.432	0.000*
Within groups (error)	1720.836	57	30.190		

*$p < 0.001$

there are differences between the two groups. After excluding the influence of covariance in the pre-test, the differences are significant for the scores on the Torrance Test of Creative Thinking between the two groups ($F = 26.432$, $P < 0.001$). There are significant differences between the experimental group and the control group in creative thinking ability.

From Table 39.2, we can see that the mean of the experimental group is higher that that of the control group. This shows that the experimental group made better progress after the experiment.

39.5.2 Influence of the Orff Music Teaching Method on Children's Fluency in Their Creative Thinking

In Table 39.1, $F = 0.076$ and $P > 0.05$ for the homogeneity of regression slope test in the "fluency" subscale; this does not reach the standard of significance. The statistical testing accepts the null hypothesis which the homogeneity of the within-class regression coefficient hypothesis establishes in the subscale of both groups' "fluency," and a one-way analysis of covariance is then processed. Tables 39.5 and 39.6 indicate that there are differences between the two groups. After excluding the influence of covariance in the pre-test, the differences are

Table 39.5 The mean and standard deviation for both groups in the "fluency" subscale in the Torrance Test of Creative Thinking

	Experimental group (n = 30)		Control group (n = 30)	
	Pre-test	Post-test	Pre-test	Post-test
M	16.17	26.23	9.80	11.57
SD	5.790	6.816	2.905	3.002

Table 39.6 Covariance analysis of the "fluency" subscale for the pre- and post-test in the Torrance Test of Creative Thinking

Variation Source	SS	df	MS	F	Sig.
Between Groups (Teaching Method)	912.970	1	912.970	71.305	0.000*
Within Groups (Error)	649.871	57	11.401		

*$p < 0.001$

significant between the two groups on the scores for "fluency" in the Torrance Test of Creative Thinking (F = 26.432, P < 0.001).

From Table 39.2, we can see that the mean of the experimental group is higher that that of the control group. This shows that the experimental group made better progress after the experiment.

39.5.3 Influence of the Orff Music Teaching Method on Children's Flexibility in Their Creative Thinking

In Table 39.1, F = 0.343 and P > 0.05 for the homogeneity of regression slope test in the "flexibility" subscale; this does not reach the standard of significance. The statistical testing accepts the null hypothesis which the homogeneity of the within-class regression coefficient hypothesis establishes in the subscale of both groups' "flexibility," and a one-way analysis of covariance is then processed. Tables 39.7 and 39.8 indicate that there are differences between the two groups. After excluding the influence of covariance in the pre-test, the differences are significant between the two groups on the scores for "flexibility" in the Torrance Test of Creative Thinking (F = 20.259, P < 0.001).

From Table 39.2, we can see that the mean of the experimental group is higher that that of the control group. This shows that the experimental group made better progress after the experiment.

39.5.4 Influence of the Orff Music Teaching Method on Children's Originality in Their Creative Thinking

In Table 39.1, F = 2.941 and P > 0.05 for the homogeneity of regression slope test in the "originality" subscale; this does not reach the standard of significance. The

Table 39.7 The mean and standard deviation for both groups on the "flexibility" subscale in the Torrance Test of Creative Thinking

	Experimental group (n = 30)		Control group (n = 30)	
	Pre-test	Post-test	Pre-test	Post-test
M	25.57	33.30	19.10	23.30
SD	7.537	7.675	4.780	5.181

Table 39.8 Covariance analysis of the "flexibility" subscale for the pre- and post-test in the Torrance Test of Creative Thinking

Variation source	SS	df	MS	F	Sig.
Between groups (teaching method)	196.854	1	196.854	20.259	0.000*
Within groups (error)	553.855	57	9.717		

*p < 0.001

Table 39.9 The mean and standard deviation for both groups on the "originality" subscale in the Torrance Test of Creative Thinking

	Experimental group (n = 30)		Control group (n = 30)	
	Pre-test	Post-test	Pre-test	Post-test
M	16.37	40.13	15.37	20.13
SD	3.624	10.129	4.222	4.524

Table 39.10 Covariance analysis of the "originality" subscale for the pre- and post-test in the Torrance Test of Creative Thinking

Variation source	SS	df	MS	F	Sig.
Between groups (teaching method)	5597.162	1	5597.162	95.983	0.000*
Within groups (error)	3323.900	57	58.314		

*p < 0.001

statistical testing accepts the null hypothesis which the homogeneity of the within-class regression coefficient hypothesis establishes in the subscale of both groups' "originality," and a one-way analysis of covariance is then processed. Tables 39.9 and 39.10 indicate that there are differences between the two groups. After excluding the influence of covariance in the pre-test, the differences are significant between the two groups on the scores for "originality" in the Torrance Test of Creative Thinking ($F = 95.983$, $P < 0.001$).

From Table 39.2, we can see that the mean of the experimental group is higher that that of the control group. This shows that the experimental group made better progress after the experiment.

39.6 Conclusion and Suggestions

The analysis and discussion in this study shows, first, that the Orff music teaching method is suitable for teaching music; second, that the method promotes students' interests in music and helps cultivate talented children; and third, that the method benefits the development of students' creative thinking ability.

It is thus concluded that teaching music with this method has positive influences on students' creative thinking ability. Such abilities can be taught through music teaching. From the students' feedback, it appears that they like to be taught using the Orff method, and that it promotes their learning motivation and thus enhances their creative thinking ability.

It is hoped that this study can provide a reference for music teachers. It is noted also that innovative teaching should be encouraged, in order to promote efficient learning with diverse teaching methods. Students' individual differences should be valued for positive learning. And differences due to subjects, courses, and time periods in teaching "creative thinking" can be considered for future research. Diverse teaching methods need to be developed for innovative teaching.

References

1. Li D (1983) Research on teaching music. Compulsory Educ Couns 23(8):9–12
2. Lin R (1983) Teaching with language rhythm. Children and Music Magazine 2(1):49
3. Liu Y (1979) Instruction for Torrance Test of Creative Thinking. Chinese Behavioral Science
   Corporation, Taipei
4. Lu S, Yang M (1982) Teaching viewpoints of
5. Orff and Kodaly, Children and Music Magazine 1(6):17
6. Orff C (1982) Elemental music (trans A-lang). Children and Music Magazine, 1(4):1
7. Su E (1972) Children's music, 1. Huaming Publisher, Taipei
8. Su E (1979) Children's music, 2. Huaming Publisher, Taipei
9. Wu D (1982) About Music Magazine 1:21

Chapter 40
Knowledge Application for Preparing Engineering High School Teachers

Kuo-Hung Tseng, Ming-Chang Wu, and See-Chien Hou

Abstract Educators view knowledge as their most valuable resource. For many years, researchers have been suggesting that preparing engineering high school teachers should use information and communication technologies to facilitate learning, critical thinking, and discussions. Therefore, the main subject of this research is to develop new applications and capabilities of knowledge management (KM) technologies to support the digital capture, storage, retrieval, and distribution of explicitly teacher-documented knowledge. The research data were gathered by using literature reviews and self-evaluated questionnaires filled out by teachers. Participants who join the study should consider the knowledge application (KA) platform as a useful tool, and consequently, the knowledge platform revealed that its potential development to the teacher when it was implemented appropriately. In addition, after the questionnaire survey, the participants demonstrated positive and affirmative attitudes towards the KA platform.

40.1 Introduction

In today's world, the capabilities for knowledge creation and knowledge application (KA) are considered the most important resources of an organization if it wants to maintain a competitive advantage [1–3]. Due to environmental factors such as competitive conditions, rapid changes in technology, etc., knowledge nowadays is considered to be a fundamental asset of an organization. In fact, there are many tools available to support knowledge management (KM) and there are many tools being developed currently by researchers [4–6]. The most valuable aims are the expansion of knowledge perimeters, the increase of knowledge transfer speed, and the support of knowledge codification [7].

It takes time for teachers to let students spend time to be familiar with the teaching tools and sometimes it is too complex to utilize. Researchers [6, 8] have

K.-H. Tseng (✉)
Meiho Institute of Technology, Taiwan, ROC
e-mail: ken@meiho.edu.tw

N. Mastorakis et al. (eds.), *Proceedings of the European Computing Conference*, Lecture Notes in Electrical Engineering 27, DOI 10.1007/978-0-387-84814-3_40, © Springer Science+Business Media, LLC 2009

observed that it is difficult to create a design for developing the skills to effectively utilize these technologies in a meaningful manner. The challenges faced by those preparing to be engineering high school teachers are summarized in the following statement:

> When teachers attempt to implement a technology innovation in the classroom, they naturally face the complex challenge of fitting it together with deep-rooted pedagogical beliefs and practices [9, p 39].

The reasons people still doubt that technology can be an effective tool are given above. Those reasons affect teachers' teaching practices and students' learning processes. The motivation for the establishment and development of a platform was derived from the aspiration to create a more effective teaching tool for learners. The platform design was based on strong theoretical grounds and on previous experience and knowledge. The development of a platform can not only foster teaching skills but also enable teachers to apply and share their knowledge. This paper attempts to investigate a KA process, particularly focusing on the process by which knowledge is transmitted to, and applied or absorbed by, a user in a platform. As a result, the establishment of a platform in this study also supports development skills such as information searching, problem solving, critical thinking, communication, and collaborative team-work [10]. The study intends to establish a useful and powerful platform for preparing engineering high school teachers to enhance KA and better improve teaching materials.

40.2 Literature Reviews

40.2.1 The Applications of Knowledge

In fact, the literature of knowledge is very rich. There are many fields such as psychology, pedagogy, philosophy, etc., that have made knowledge itself a common research topic. Knowledge is complicated, situational, and multidimensional [11]. The complex nature of knowledge is considered through an examination of its dynamic qualities, and its inner meaning is related to its specific social and cultural contexts. Therefore, before discussing the inner meaning of knowledge, the differences among data, information, and knowledge must be distinguished. Data inputs are defined as series of observations, measurements, or facts in the form of numbers, words, sounds, and images. Data have no meaning, but provide the raw material to become information, which is classified according to different patterns. Data may result from the conduct of a survey, and information results from the analysis of the meaningful data in the form of reports or graphs [12]. In this research, knowledge is here defined as applicable and productive information.

40.2.2 Knowledge Technologies

KM attempts to develop a strategy for the capture, use, and transfer of knowledge in order to improve efficiency and increase a competitive edge [13]. The scope of KM includes interaction between people and people, or between people and organizations. Knowledge technologies not only facilitate the rapid collection, collation, storage, and dissemination of data, but also assist the process. Furthermore, recent rapid changes in the macro-environment have made a significant and positive impact on organizations' abilities and intentions. A lot of web-based educational material developers have a tendency to use the powerful and new knowledge technologies in traditional ways [1, 14].

As technological developments become more and more advanced in applications and utilizations, it is imperative for educators to think more broadly to utilize the materials from knowledge technologies. Therefore, it is necessary to provide interesting settings for learners, such as experiential and problem-solving tasks, which allow learners to learn more actively. Knowledge technologies themselves will not lead educators to apply knowledge effectively, but when KM technologies and the teacher are aligned and integrated, they can provide a structure as a scaffold to enhance the abilities of educators to create and exploit knowledge. Besides, with well-designed, standardized, technical support and good management of knowledge, information-capturing abilities, knowledge discoveries, and knowledge creation can also be enhanced.

The KA platform in this study can impact knowledge in many ways. First, the platform allows for collection, collation, storage, and dissemination of data. Second, the platform facilitates knowledge transfer and application through data and information. The researchers actually don't want to emphasize knowledge technologies too much, because the technological approach toward KA is often unsatisfactory. The researchers wish to explore the process of KA during usage of the platform by teachers as a tool to apply the knowledge most needed in the class and transfer it to students.

40.3 Research Methods

The method used in the present study is different from those used in previous studies. Most of the past studies utilized only paper and pencil questionnaires, scales, or surveys to collect data on participants' perceptions or attitudes. Instead, in this study, the researchers tended to use the modified Delphi technique to explore the richer and deeper views of the participants. In addition, after using the KA platform and questionnaires to explore the participants' perceptions, surveys of feelings and intentions were utilized in this study. The descriptions of the methods are as follows.

40.3.1 Modified Delphi Technique

In this study, the modified Delphi technique was designed to elicit information and estimates from participants to facilitate decision-making. After reviewing the literature and the discussions of the research team members, the first round of the modified Delphi questionnaire was developed. It consisted of several structured questions, including multiple-choice questions, instead of unstructured questions. All the questionnaire items in this study were evaluated using the Likert 5 scale, with measurements ranging from very important to not very important, with average scores above 3 meaning the experts' opinions could be regarded as needed. The initial topics are shown as follows:

1. Integrating knowledge into regular teaching curricula
2. Using knowledge to manage the classroom
3. Integrating knowledge into the curriculum and content of courses
4. Using knowledge to design multimedia teaching materials
5. Integrating knowledge into general programs preparing engineering educational teaching
6. Using knowledge to develop teachers' research activities
7. Using knowledge to cooperate with enterprise
8. Using knowledge to guide students during the contexts

40.3.2 The Development of the Questionnaire

To access the views of a teacher, a questionnaire was used to extract relevant information on the teacher's perceptions, feelings, and actions toward using the KA platform to apply the information he or she needed. The questionnaire for the teachers consisted of 11 items on perceptions, 17 items on feelings, and 19 items on intensity of action statements related to KA; these indicated levels of agreement or disagreement on a 5-point Likert-type scale with 5 standing for "strongly agree" and 1 standing for "strongly disagree." The detailed descriptions of the three scales are presented below:

1. *Teacher's perception scale:* To measure perceptions of the extent to which the teachers can explain the usage and implementation of the KA platform.
2. *Teacher's feeling scale:* To measure feelings of the extent to which the teachers see the advantages of applying knowledge from the KA platform.
3. *Teacher's action scale:* To measure the intensity of action of the extent to which teachers can actively utilize knowledge from the KA platform.

40.3.3 Knowledge Application Platform

The KA platform was established by using SharePoint Team Services software. The purpose of setting up the platform for teachers was to cultivate the mode and integrated functions of content knowledge (CK), pedagogical knowledge

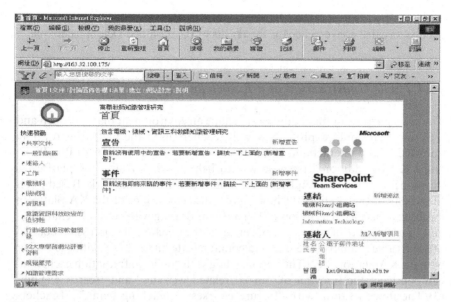

Fig. 40.1 Knowledge application (KA) platform website

(PK), and pedagogical content knowledge (PCK). Researchers also enriched the website by including plentiful and various content, including PK (teaching materials, teaching methods, administrative knowledge, counseling knowledge, and classroom management), PCK, and CK.

40.4 Results

40.4.1 Investigation Results Analysis for the 1st Round and the 2nd Round Using the Modified Delphi Technique

According to the statistical results, the identification of KA perspective from the professional domain experts, eight items had average scores above 4. The following lists proceed from high to low as follows: "teaching activities," "problem solving," "teaching media," "management and maintenance of experimental lab or intern factory," "guidance of student's project," "students joining the technical competition," "teaching research development activities," and "science fair." Two items had average scores of between 3 and 4: "media of teaching" and "management of classroom." The mean score above 3 were evaluated by the experts and their opinions, which were regarded as needed to construct the KA Platform for this study. After doing chi-square examination, the results show that the χ^2 values of the items are between 0 and 1.05, with a critical value ($\chi^2 = 3.841$, df $= 1$, p $= 0.05$), which means these items that were evaluated by the experts had homogeneity. In other words, the opinion of the experts revealed convergence for the results of the modified Delphi technique.

40.4.2 The Construction of a Knowledge Application Platform

The main purpose of this study was to shape an environment that was open and trusting, that could explore how well teachers used the platform to apply knowledge, and interact and communicate within the constructs of PK, PCK, and CK. An additional purpose was to help students realize the platform's purpose and use it to complete, solve, and share their homework. Based upon the results of the modified Delphi survey, the significance of the KA platform has been constructed as follows: (1) The items dealing with "Q&A," "chatting," and "qommunities of practice" can construct the function of "discussion board." Students able to upload the files and communicate interactively with each other with the KA platform. (2) The items dealing with "online information exchange" can construct the function of "linked related websites" and "search engine." (3) The items dealing with "lecture courses," "teaching team," "Teachers Research Association," "Courses Development Committee," "meeting between departments," and "teaching achievement exhibition," construct the function of "teaching database," the contents of which include PK, PCK, and CK. (4) The items dealing with "study club" and "diary" can construct the function of "electronic newspaper."

40.4.3 The Results of the Questionnaire

The Cronbach alpha coefficient of the survey was quite acceptable for an overall score ($\alpha = 0.94$). There were six expert reviewers, including three industry professionals and three KM experts. This indicated that there were acceptable internal consistency and validity for each subscale. From the results of the descriptive analysis of the teachers' perceptions, the participants (teachers) revealed that the information system could reinforce the KM activities. The information system was designed for this platform, which easily allowed teachers to acquire, apply, share, and create the knowledge. The teachers' feelings scale indicated that the platform could help them to acquire a method of professional teaching capability. The KA platform not only can form a wonderful learning atmosphere and an interactive learning environment between teachers and students, but also can help teachers and students to file and manage their teaching and learning processes to share with other teachers. Finally, the participants intend to learn how to actively utilize the platform. Teachers were able to explore new information through the internal hyperlinks in order to share their capability using the platform with students from the

Table 40.1 The usage frequency of KA platform

	Week	PK	CK	PCK	Total
Knowledge application	One	12	0	15	27
	Two	15	0	13	28
	Three	13	0	14	27
	Four	18	2	10	30
	Five	11	1	12	24
	Six	29	0	20	49
SUM		98	3	84	185

discussion board. The researchers used self-evaluated scales to examine the attitudes of two teachers toward the KA platform and found out that they had shown positive attitudes. The usage frequencies of the KA platform of PK, PCK, and CK are shown in Table 40.1. This shows that during the experiments, which lasted for six weeks, the teachers were more likely to use the platform for PK and PCK, but they seldom used CK to utilize as teaching materials.

40.5 Discussion

In today's complicated and rapidly changing educational environment, the need for applying technological capabilities and creating knowledge have increased. As [1, 15] have mentioned, preparing engineering high school teachers have evolved from "sage on the stage" to "guide on the side;" the students have switched from a dependent, passive role to a self-directed, discovery-oriented role. It is suggested that teachers need to prepare themselves for these changes. In this study, teachers expressed positive attitudes toward the utility of a platform for information resources and self instruction, which is capable of supporting teachers' instructions. The KA platform is based on the strong theoretical grounds of KM the and friendly interface of the SharePoint Team Services package. This KA platform was designed with both human and technology components, and teachers can easily use it without long practice. One example of this is providing downloaded documents from the discussion board. Teachers can download the teaching materials from the knowledge platform. Generally speaking, the platform is being used primarily as an information resource for teachers, and for the storage of instructional material, including class notes and linked related websites. The most critical purpose of KA is to encourage the teachers to internalize other teachers' knowledge into their individual instruction, to acquire information resources to improve the learning environment. It is clear that KA related to the platform is necessary for preparing engineering high school teachers.

40.6 Conclusion

From this study's results it can be seen that KA is highly dependent on the capabilities of knowledge technologies. In fact, there are many tools that have been developed, but the impacts of these technologies on the applications are always unevenly distributed because of their inappropriate forms of design or human interaction. Finally, after six weeks of experimentation with integrating KA into the curriculum, the conclusions of this research were as follows:

1. We suggest using the platform (SharePoint) to achieve a process of KA. After the participants of this study experienced the process of information searching, using the application, and sharing their knowledge, the PK, PCK, and CK were enhanced and preparing engineering high school teachers were more likely to apply their knowledge through this platform.
2. The KA platform enhances the practical experience of the preparing engineering high school teachers.
3. Because of the unique characteristics of knowledge, it is vital to merge knowledge platforms into teaching tools in order to expand and spread KA.
4. Establish teaching material database and electrical designated topic learning database to enhance the teaching systematic strategy. The role of the platform is to apply knowledge. We should focus on the educational goals and learning environment, not just pay attention to the functions or human-like responses of the platform. Furthermore, after designing the platform or so-called knowledge technologies, future research should emphasize evaluations of the platform from students.

References

1. Kearsley G (1998) Educational technology: a critique. Educ Technol 38:47–51
2. Nelson RR (1991) Why do firms differ, and how does it matter? Strateg Manage J (Winter Special Issues) 12:61–74
3. Nonaka Chishiki-SouZou no keiei (1990) A theory of organizational knowledge creation. Nihon Keizai Shimbun-Sha, Tokyo
4. Davenport TH, De Long DW, Beers MC (1998) Successful knowledge management projects. Sloan Manage Rev 39:43–57
5. Davenport T, Prusak L (1998) Working knowledge—how organizations manage what they know. Harvard Business School Press, Boston, MA
6. Rickards T (2003) Technology-rich learning environments and the role of effective teaching. In Khine, SM, Fisher, D (eds) Technology-rich learning environments. World Scientific, Hackensack, NJ, pp 115–132
7. Hansen MT, Nohira N, Tierney T (1999) What's your strategy for knowledge management? Harv Bus Rev 77:106–116
8. Ruggles RL (1997) Knowledge management tools. Butterworth-Heinemann, Boston, MA
9. Russell LD, Schneiderheinze A (2005) Understanding innovation in education using activity theory. Educ Technol Soc 8:38–53
10. Ashcroft K, Foreman-Peck L (1994) Managing teaching and learning in further and higher education. The Falmer Press, London

11. Margerum-Leys J, Marx R (2002) Teacher knowledge of educational technology: a case study of student teacher/mentor teacher pairs. J Educ Comput Res 26:427–462
12. Roberts J (2000) From know-how to show-how? Questioning the role of information and communication technologies in knowledge transfer. Technol Anal Strateg Manage
13. Demerest M (1997) Understanding knowledge management. J Long Range Plann 30:374–84
14. Dehoney J, Reeves T (1999) Instructional and social dimensions of class web pages. J Comput Higher Educ 10:19–41
15. Fischer G (1999) Lifelong learning: changing mindsets. In: Cumming G, Okamoto T, Gomez, L (eds) Advanced research in computers and communications in education, ICCE99. IOS Press, Amsterdam, pp 21–30

Chapter 41
Collaborative Online Network and Cultural Exchange Project

Rong-Jyue Fang, Hung-Jen Yang, C. Ray Diez, Hua-Lin Tsai, Chi-Jen Lee, and Tien-Sheng Tsai

Abstract This study starts with the idea of dealing with digital learning and the information explosion. English plays a major role in world communication. All trade areas as well as other diversified fields emphasize English proficiency. Students enhance their global vision via English comprehension. Therefore, students' interactions with their peers in other nations through digital multimedia on the web is a very important skill in this digital era. This project attempts to understand course material while interacting through online web learning. It stimulates students' cultural experiences and expands their learning fields. The project provides a problem solving model and encourages students to try to solve problems by themselves. It recruits high school and primary school students to make international contacts. Four hundred teams of parents, teachers, and students are involved in a web platform to discuss international issues. The topics can be related to natural resources, people and society, and human and self-renunciation, and there are many more topics for multicultural learning. The project also intends to do cross-cultural learning exchanges and experience sharing, and will submit findings to related educational administrators and web learners for further reference.

41.1 Introduction

While much attention has been paid to delivering learning objects through existing systems, and similarly to the use of mobile devices in learning, connecting the two has taken a back seat.

Due to the liberalization of the global economy, all trades and professions emphasize the promotion of English comprehension in order to expand their international markets, and to keep themselves at the same pace with global

R.-J. Fang (✉)
Department of Information Management, Southern Taiwan University of
Technology, Taiwan, ROC
e-mail: rxf26@mail.stut.edu.tw

N. Mastorakis et al. (eds.), *Proceedings of the European*
Computing Conference, Lecture Notes in Electrical Engineering 27,
DOI 10.1007/978-0-387-84814-3_41, © Springer Science+Business Media, LLC 2009

economic development. To improve the competitiveness of the country, the government has proposed enhancing students' English comprehension as an important policy [1]. The Ministry of Education suggests the perspective of "Creative Taiwan, Connecting Globally," in which "fostering talents with foreign language abilities" is one of the action plans [2]. Due to the universal availability of the Internet and the rapid development of multimedia techniques, e-learning has become an important learning tool. Digital teaching materials produced with multimedia films and pictures provide a varied and more active course content. Meanwhile, there is less restriction on space and time using digital teaching materials, and that provides a more interactive and convenient learning environment. Therefore, using multimedia digital teaching materials in international interactive teaching is a necessary way of learning in this digital era [3,4].

The Ministry of Education has provided many resources for learning, such as the teaching resources center, the Six Major Learning Systems, and seeded schools to encourage teachers applying computers in their teaching, and to encourage students using the Internet. The digital plan for augmenting manpower will bring wireless Internet to each school, city, and county. Thus, instructing teachers and students in using online information to connect with the world has currently become an important issue.

41.2 Background of the Project

On behalf of Taiwan K-12 teachers and students, we would like to invite our international partners to join our CONCEPT Project in 2006—Collaborative Online Network and Culture Exchange Project in Taiwan.

In the global village connected by the Internet, industries from different walks of life utilize information technology to reform their enterprises, to promote their work efficiency, and to improve the quality of their service [5]. The application of information technology facilitates the prosperous development of modern society. There is no exception for education. Both teachers' teaching and students' learning should incorporate the advantages of modern technology. Consequently, continuous changes should be made in accordance with the facilitation modern technology can provide.

Recognizing the importance of information technology to students' learning, the government in Taiwan is establishing an online environment to prepare students for the information society in the 21st century and to promote their problem-solving abilities [6]. Meanwhile, it is crucial that students need to understand the world more, and their international awareness and perspective should be fostered to cope with globalization. With strong support from MOE [7], we have begun to design and create an online learning environment for students in Taiwan. The core spirit lies in our students' learning collaboratively with international partners [8, 9] synchronously through the Internet. Based on the WebQuest theory, one team from Taiwan will match with another team from abroad as follows:

1. They plan their own projects and formulate the topics.
2. They discuss the problems online and finally come to the solutions.

Such a process-oriented project includes learning:

1. to formulate topics,
2. to decide learning tasks,
3. to establish the exploring processes and procedures,
4. to seek the resources,
5. to conclude the project.

To know more about this activity, please e-mail to jerome@mail.stu.edu.tw.

Echoing this mission, we plan to conduct a project-based learning (PBL) project—the Collaborative Online Network and Culture Exchange Project in Taiwan. In this plan, 400 teams from Taiwan and 400 teams from other countries will be invited to the project on the basis of a one-to-one team model. These 800 teams together will learn asynchronously.

41.2.1 Content

The curriculum (Orff Schulwerk) is designed starting from childhood and uses children's musical experience as the teaching material.

41.2.2 Characteristics

A pentatonic scale is included in the first volume.

This project will recruit 400 teams with students from the elementary schools and high schools in Taiwan. Each group consists of 2–5 people coached by 1–3 teachers. The details will be published on the following website: http://nt.stu.edu.tw/PBL

For more information about PBL, please refer to the following websites:

1. http://course.fed.cuhk.edu.hk/s031705/EDD5169F/
2. www.berksiu.k12.pa.us/webquest/bowman/index.htm
3. http://et.sdsu.edu/APaxton-smith/eslwebquest/WebQuest.htm
4. www.geocities.com/SiliconValley/Mouse/8059/CurriculumQuest.html
5. http://projects.edtech.sandi.net/grant/oceanpollution/introduction.html
6. http://inshow.idv.tw/webs/oldstreet/default.htm

We look forward to seeing you on line soon!
Truly yours, Committee of CONCEPT, Taiwan

41.3 Discussions of Web-Enhanced Project-Based Learning

41.3.1 PBL

Recently there have been teachers trying project-based learning (PBL), which is different from traditional teaching methods. PBL emphasizes helping students apply what they have learned in their daily lives innovatively. PBL is a constructivist approach which provides learners a complex and authentic project in order to have students find subject matter, design questions, draw up action plans, collect information, solve problems, set up policies, complete the research process, and present a learning mode for the project. This learning mode integrates learning from life experiences and from interdisciplinary courses.

41.3.2 Discussing Project-Based Learning with Constructivism

Constructivists consider that knowledge is acquired by the learner's construction. Students should express the learning results by applying surface features of diversified knowledge. The knowledge and techniques already possessed by a learner may influence their learning of something new. Learning cooperatively and learning to interact with communities will help in depth learning [2]. PBL also emphasizes a learner's construction of knowledge, and thus adopts a learner-centered model. Instructors will not tell the answer but instruct students in exploring questions and trying to solve problems. Learners construct their knowledge by themselves in the process of exploring. Products of PBL can be presented as oral presentations, websites, briefings, etc.

41.3.3 Internet Courses and Problem-Based Learning

First, interdisciplinary courses are based on important issues in real life, and thus they fit the nature of project-based learning. Second, interdisciplinary courses apply knowledge of context rather than being limited to knowledge of subject matter. This conforms to the interdisciplinary feature of project-based learning, and that it is designed to explore a question which combines contexts in the real world. Third, interdisciplinary courses do research on current issues, not on a subject. Project-based learning also does research.

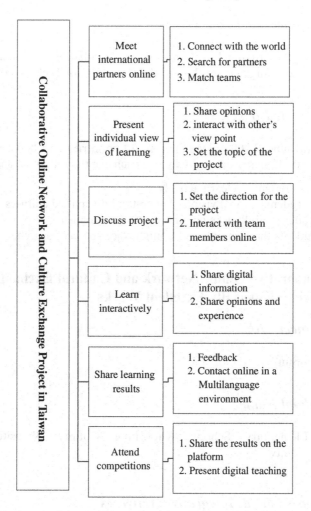

Fig. 41.1 The structure of the project

41.4 Contents

This project is aimed at building a high-quality online learning environment, integrating learning resources, improving the online learning environment, and stimulating the teachers' motivation for learning. It is designed to promote e-learning at domestic schools and to achieve the perspective of "Creative Taiwan, Connecting Globally."

41.4.1 Members of This Project

This project will be attended by the teachers and students in the Department of Applied Foreign Languages at the Shu-Te University of Technology.

41.4.2 Expected Benefits

400 teams (around 2,000 students) from 40 countries and 800 teachers are expected to attend this project. We hope to achieve the following goals:

1. Encourage students to learn actively.
2. Broaden students' global vision. Understand diversified cultures.
3. Foster students' abilities in communication in English.
4. Teach students the techniques of doing project-based research.

41.5 Collaborative Online Network and Cultural Exchange Project—Especially Excellent Works

41.5.1 Project Title

"Me-Hu-Hu-Sun!"

41.5.2 School Name

Taoyuan Elementary School, Kaohsiung County. Website: http://www.typ.ks.edu.tw

41.5.3 Course Field: Integrative Activities

41.5.3.1 Attendants

All students of our school and the kindergarten.

41.5.3.2 Project Website

http://khc.k12.edu.tw. Account: typbear. Password: 123123.

41.5.4 Summer

This proposal is to be worked out and applied to the teaching characteristics of our school—international interaction on education. Hopefully, our students can contact students from urban areas and foreign countries with regard to sharing their learning through the network, and to extending their vision. We have cooperated with Yanagi Elementary School of Japan on the online seminar and learning sharing since September 2005. Our school lies on a remote mountain with only six classes, and insufficient manpower and budget. The only way to bring our students off the mountain and into interactions with cities and other countries is by means of the network, easy webcams, and microphones. Students are guided to the palace of cultural sharing by virtual characters named "Taoyuan bear, Taoyuan monkey, Taoyuan pig." The activity is just an access to cities and other countries. We expect the students can discover the beauty of the world by expanding their learning.

41.5.5 Curriculum and Innovative Teaching

The curriculum is designed around theme activities for students by project learning. The virtual characters for cultural sharing are toys named "Taoyuan animals" for kindergarten, "Taoyuan bear" for the first and second grades, "Taoyuan pig" for grade 5, and "Taoyuan monkey" for grade 6. All the characters join the students' lives when school begins. Students learn to keep their friendships with the "Taoyuan animals" by bringing them home and recording their daily lives. A month later, the activity of exchanging the students' "Taoyuan animals" with those from cities and other countries began. We sent the virtual characters to Jhih Kai Elementary School by ourselves, and mailed to Fuyoh Elementary and Kanazawa Municipal Ougidai Elementary Schools in Japan. Now, the "students" are proceeding with their learning at the cooperating schools. Meanwhile, "students" from Chi Kai Elementary and Japanese schools are learning at Taoyuan Elementary.

41.6 Conclusion and Analysis

The virtual students are looked after by assigned students every day; their learning situation is also recorded in diaries in various ways (each animal has a diary and a notebook prepared by the schools at both ends). We expect all "students" to bring their learning records back to their school after the project, and share their learning experiences (Taoyuan animals) with their classmates or schoolmates.

We transmit information simultaneously via a web platform to promote the students' self-understanding, build their confidence, and cultivate their active learning attitude; to cultivate their ability for teamwork and their trust in and assistance for each other; to promote their understanding of the differences among various peoples and their interaction with others; to inspire their

Members of our school in Disaster briefing by Yanagi
2006NDYS   Elementary, Nagoya

Fig. 41.2 Face to face sharing: Taoyuan Elementary and Yanagi Elementary (1)

Fig. 41.3 Report by Liberty Times on our students' attending 2006NDYS

Fig. 41.4 Student's learning sharing after international interaction on education (1)

Fig. 41.5 We exchange the students with Japan using their virtual characters named "Taoyuan bear," "Taoyuan monkey," and "Taoyuan pig." On the lower right are Doraamo and baby elephant from Japan

imaginations to express their ideas accurately; and to improve their writing ability and cultivate their ability for independent thinking and problem solving.

41.6.1 WHY (Why We Insist on Proceeding with Educational Communication by ICT)

The weak people are usually the information poor in digital society, too. Most of them are from families of low income, residents of remote area, women, the old, children, aborigines, and foreign spouses. Because of the unfriendly cultural and economic conditions, there are difficulties and discrimination for them to purchase computers and related accessories. Though the difficulty and discrimination have become less, the socioeconomic status of the family still controls students' informational learning (Wei-Chu Chen 2003). Digital discrepancies between cities and countries keep growing, but are not getting better, as shown by Hung-Chih Chan on the investigation of Taiwanese consumers in 2004. The number of "country-folk" remind us of the obvious insufficiency of information equipment and applications (May-Yi Chiang 2004).

All citizens in an information society are authorized to use computers and networks equally. The U.S., France, Japan, Finland, and Germany not only focus on the basic building of information, but also on the development and application afterwards. This shows that an individual's information equipment in an information society is related to his/her employment and living quality, and even to the competence of the whole nation.

Owing to our efforts on educational interaction with other countries and the cultivation of aboriginal students' worldview, wireless Internet equipment in this project is the key expected by all parents, teachers, and students of our school. Students can be sharing their learning with foreigners like the urban students by video conferencing simultaneously or asynchronously any time and anywhere, and by applying the skills of online interaction to their daily lives. If online working is necessary in the future, the student's right to learn can't be sacrificed. Thus, promoting learning visions and effects for aboriginals is a target for the future.

41.6.2 What, When, Who, Where, How (What are We Doing? When We Do It? Whom Do We Do It With? Where We Do it? How We Do It?)

ICT instruction is carried out gradually over three years. In the first year, students share their learning about culture with urban and foreign students through web video conferencing; in the second year, they share their learning about ecology through discussion boards on platforms and web video conferences by teamwork automatically; in the third year, students expand their learning individually.

Regarding the curriculum development, the learning direction of our students is based on the globalization of local school-based curricula in the first year.

References

1. Bean JA (1996) On the shoulders of giants! The case for curriculum integration. Middle Sch J 28:6–11
2. Krajcik JS, Czerniak CM, Berger C (1999) Teaching children science: a project-based approach. McGraw-Hill, New York
3. Bereiter C, Scardamalia M (1993) Surpassing ourselves: an inquiry into the nature and implications of experts. Open Court, Chicago
4. Brown JS, Collins A, Dugid P (1989) Situated cognition and the culture of learning. Educ Res 18(1):32–42
5. Ertmer PA (1999) Addressing first- and second-order barriers to change: strategies for technology integration. ETR&D 47(4):47–61
6. Moursund D (1999) Project-based learning using information technology. International Society for Technology in Education Books, Eugene, OR
7. Lave J, Wenger E (1991) Situated learning: legitimate peripheral participation. Cambridge University Press, Cambridge
8. Brown AL, Ash D, Rutherford M, Nakagawa K, Gordon A, Campione JC (1993) Distributed expertise in the classroom.
9. Linn MC (1995) The effects of self-monitoring on students course performance, use of learning strategies, attitude, self-judgment ability, and knowledge representation. J Exp Educ 64(2):101–115

Chapter 42
A Study of the Project on Mobile Devices in Education

Rong-Jyue Fang, Hung-Jen Yang, Hua-Lin Tsai, Chi-Jen Lee, Chung-Ping Lee, and Pofen Wang

Abstract This study intends to investigate the Personal Handyphone System (PHS) phone and its application in an elementary school. The interview process focused on the director who managed the planning for the system from an experimental elementary school located in Kaohsiung County, southern Taiwan. She joined of the whole digital experimental project and observed interactions between the parents and their own children in that specific elementary school. In the experimental period, the director observed each parent using a PHS phone for message transmission and mobile net communication, and the parents discussed and shared feelings with the teacher and other parents by using a PHS phone. The research summarizes the director's ideas and provides references, opinions, and information to instruction designers and mobile learning directors for developing further innovative instruction.

42.1 Introduction

People increasingly access audio, video, and animation data, or search for information via the World Wide Web, while the out-of-date way of connecting to the Net by cable is inconvenient. As a result, "information appliances" discard the PC's complicated architecture, presenting a simple, low-price, and consumer-oriented substitute [4]. At the same time, we also need a convenient environment for e-learning.

In 1996, the Ministry of Communication in Taiwan opened the telecommunications business up to the local people, caused the liberalization and internationalization of telecommunications in Taiwan, and promoted market competition. Therefore, mobile phones became convenient and necessary in our daily life, and made the wireless communication market flourish. Currently

R.-J. Fang (✉)
Department of Information Management, Southern Taiwan University
of Technology, Taiwan, ROC
e-mail: rxf26@mail.stut.edu.tw

N. Mastorakis et al. (eds.), *Proceedings of the European*
Computing Conference, Lecture Notes in Electrical Engineering 27,
DOI 10.1007/978-0-387-84814-3_42, © Springer Science+Business Media, LLC 2009

the Personal Handyphone System's (PHS) smartphone comes with the follow-ing standard features [6]: large-volume phone number and email address book; "emoji" (i-mode picture symbol) support; multiple typefaces/fonts; fast digitalized telecommunication key, and diverse recording functions. Its main feature consists of a powerful standard transmission and communication pro-tocol in its core technology and Java support that brings excellent performance with wireless networking.

With the advantages of mobility, mobile wireless technologies help improve efficiency and effectiveness in teaching and learning [5]. We will search for solutions that this equipment and its relevant infrastructure are going to bring about. This research intends to demonstrate the director's ideas for PHS phone applications in education, in order to provide references, opinions, and infor-mation to instruction designers and mobile learning directors for developing further innovative instruction. In the future, educational authorities will work with PHS mobile telecommunication suppliers to carry out an experiment that will help determine the feasibility and effect of PHS mobile phone utilization in education.

42.2 Definitions and Conditions

42.2.1 Content

Generally speaking, m-learning is defined as e-learning through mobile devices. Mobile wireless technology is defined as any wireless technology that uses the radio frequency spectrum in any band to facilitate transmission of text data, voice, video, or multimedia services to mobile devices with freedom from time and location limitations [3]. In Taiwan, the production of WLAN exceeded 90 percent of the world total and our rate of mobile phone usage also surpassed 100 percent. We are number one in the world. The plan for "the demonstration and application of wireless broadband networks" of the Ministry of Economic Affairs has helped 25 cities, nearly 400 schools, 16 wireless networks industries, and the 1 telecommunications industry connect with the international roaming organizer iPass [7]. This will be helpful in implementing the public wireless regional network service in Taiwan schools.

42.2.2 Conditions

Mobile phones are the most popular and common mobile wireless technology among personal communication tools. They use the wireless application pro-tocol (WAP) to enable access to the Internet from mobile phones. Although it is possible to deliver content to WAP phones, reading it is rarely easy enough. The mobility of the devices used in m-learning scenarios involves a new context data to be considered—location. The service providers had extremely limited opportunities to offer interactive data services. Interactive data applications are required to support now commonplace activities such as e-mail by mobile phone, news headlines, activity messages, and music downloads.

42.3 Types and Characteristics

PHS is basically a wireless telephone, with the capability to handover from one cell to another. The PHS smartphone's standard is not unified now, but their characteristics are as follows [5, 9].

Many value-added services: Modern PHS phones can also support high-speed wireless data transfers, Internet connections, WWW access, e-mailing, text messaging, and even color image transfers.

Polyphonic sound support plus stereo widening and adjustable brightness: Polyphony chord ringtones are the basic component of the handset. There are stereo widening sounds, incoming call ringtones to select; and tunes can be downloaded, composed, and even recorded. Additionally, you can adjust the built-in brightness mode, make other levels yourself, or even choose an alternately flashing mode.

Large volume phone number and e-mail address book: There are entries for contact phone books and group management, each entry with up to three names and two email addresses, an email inbox, and a sent messages/drafts outbox. It is easy to manage e-mails and thumb information, coupled with an offline reading function.

MiMi (Mobile Information Mobile Internet) Thumb service: PHS MiMi thumb information is just like an Internet home page. When your PHS mobile phone is connected to the home page, you can freely browse any value-added services on the MiMi menu, such as news flashes, monthly billing services, categorized menus, and search engines.

WiWi Netspeed service: You can avoid troublesome fixed-line connections, and get online anywhere with your mobile phone at anytime. WiWi connection equipment has the widest choices, so you can get access via a notebook PC or any kind of PDA. The simple operating method and connection settings allow you to easily expand your wireless vision.

Fast digitalized telecommunication key and diverse record functions: there are many services, such as call blocking, automatic call return, calendar and scheduler, alarm clock, passthrough hole for straps, interchangeable voice-control, as radio telephone within short distance.

42.4 PHS Applications in Education

According to the American web site K12 Handhelds [2], there are different ways to use handheld computers in education. We can find great educational uses for handheld information devices; some new ideas include:

42.4.1 Administrative Applications

Keep your schedule. Track student progress on specific skills. Instantly access student information. Take notes at a meeting or in a class. Record and tabulate grades. Store and access lesson plans. Use a rubric to assess and score student work. Evaluate teacher performance and record observation notes. Let students have constant access to their current grades (very motivating!). Keep emergency procedures and checklists readily accessible.

42.4.2 Communication and Collaboration Applications

Send an email. Group schedule school meetings. Distribute school activity information to students and parents. Get parents' sign-offs. Send and receive instant messages. Access online educational events and news.

42.4.3 Teaching and Learning Applications

Keep track of your class schedules, assignments, and grades. Record observations on a field trip. Take notes in class. Practice handwriting. Take part in a collaborative simulation. Give students step-by-step instructions or visual plans for projects. Manage a collaborative project. Have classes create their own mobile information channels to share information with other classes or the community.

42.5 Study Method

This research went through the relevant literature and interviewed the director who managed the PHS phone plan from an experimental elementary school located in Kaohsiung County which was participating in this digital learning special case. The director's feelings and opinions were discussed, and conclusions were reached after the experiment. Later those reference materials will provide opinions and information to instruction designers and mobile learning directors to help them develop further innovative instruction.

42.5.1 Procedure

In the experiment, the research team composed a semi-structured interviewing outline, revised by professionals and experts. The interview was carried out to let the director with experience in utilizing PHS express the teachers' and the parents' afterthoughts. We would understand what background the director had.

42.5.2 Subject Background

This study involved a class in the director's school participating in the digital learning special case. Every pair of parents was assigned a PHS handset. There were six parents and teachers who stayed for the whole process. However, the director is an educational expert with experience in using personal digital assistants. She is the director of the educational affairs division in Cheng-Jeng Primary School in Kaohsiung County.

42.5.3 Interviewer's Background

The interviewer is an educational expert with experience in using personal digital assistants. He was a professional consulted in the interview outline.

42.5.4 Interview Data

The keynotes of the director's comments and encodings were as follows and as shown in Table 42.1:

(1) As you work as a project initiator in this PHS mobile learning case, what is your motivation?

Many kinds of artificial intelligence or technological products are designed for people to process data more efficiently and to make our work more effective. Therefore, it is very important to teach our students to learn how to adapt to the artificial environment early via technological implements (Director IDI01082007_1–1).

(2) What were users' difficulties and opinions in the beginning? How did you overcome the hurdles to continue promoting their learning?

It is very important to learn technological knowledge and scientific thought while a child, but this fails to give our children new messages about science and technology to study today. When users face new technology in the beginning, they feel embarrassed about operating it. In addition, it is not easy to change their learning habits. In the face of this predicament, we do our best to communicate with them and offer them assistance with the software and hardware, in order for them to get familiar with the various operational functions of the PHS (Director IDI01082007_1–2).

Table 42.1 The illustration of encoding

Source	Code	Example
In-depth Interview	IDI	Director IDI01082007 represents what the director said on January 8, 2007.

(3) As time goes by, how many concrete changes have taken place among the users?

We see some changes in the users, including improving their technological knowledge about mobile devices, accepting PHS hand-helds in education, sharing their opinions with others, and applying cooperative learning in this course (Director IDI01082007_1–3).

(4) What did they find were concrete educational applications after using the PHS mobile learning device?

The PHS basically serves as an individual helper. It is not necessary to use mobile devices in class, so it takes a long time to teach parents and teachers to know how important it is. Nevertheless, we have learned much about education in science and technology from the "PHS handset case," and we need to teach what our students want to learn in the future (Director IDI01082007_1–4).

(5) What did you accomplish in this PHS mobile learning case? What would you suggest that other directors do for further research in the future?

In this course, we have already seen that the users have changed their attitudes about behavior in science and technology, including their attitudes toward using the PHS and their behavioral intention to use mobile devices. In the future, we may design questionnaires to collect data from experimental schools which carry on this special project in order to design mobile learning targets. In addition, we should encourage new scientific and technological products, to give our students a chance to accept education in science and technology in their basic early education (Director IDI01082007_1–5).

42.5.5 Analysis of Data

The author encoded categories from the literature and developed concepts from content analysis [8]. The results are shown in Table 42.2.

42.5.6 Formulas

1. For computing the coefficient of reliability, the mutual agreements formula (Pi) is as follows:
   To compute the coefficient of reliability, we use this formula:

$$\text{Intercoder agreement formula (Pi)} = \frac{2M}{N_1 - N_2}$$

where M = number of times the two coders agree and N_1 and N_2 = number of coding decisions each coder made. (Pi is 0.894.)

Table 42.2 The register of the interview's opinions

Category	Concept register	Text	Code
Efficiently and effectively	Explicit register: **efficiently and effectively**	• to process data more **efficiently** and to make our work more **effective**.	Director IDI01082007_1–1
Artificial environment	Explicit register: **artificial environment**	• to learn how to adapt to the artificial environment early via technological implements.	Director IDI01082007_1–1
	Implicit register: **new messages, a chance**	• to give our children new messages about science and technology to study today.	Director IDI01082007_1–2
		• to give our students a chance to be able to accept the education of science and technology in basic education early!	Director IDI01082007_1–5
Mobile devices	Explicit register: **mobile devices**	• how to adapt to the artificial environment early via technological **implements**.	Director IDI01082007_1–1
	Implicit register: **PHS' technological implements**	• to offer them assistance with the software and hardware to get familiar with various operational functions in **PHS**.	Director IDI01082007_1–2
		• improving their technological knowledge about **mobile devices**, accepting **PHS hand-helds** in education.	Director IDI01082007_1–3
		• **PHS** basically serves as an individual helper.	Director IDI01082007_1–4
		• including attitude toward using **PHS** and behavioral intention to use **mobile devices**.	Director IDI01082007_1–5
Education	Explicit register: **education**	• to **teach** our students to learn how to adapt to the artificial environment early.	Director IDI01082007_1–1
	Implicit register: **teach**	• It is very important to **learn** technological knowledge and scientific thought while a child accepting PHS hand-helds in **education**.	Director IDI01082007_1–2
			Director IDI01082007_1–3
		• we need to **teach** what our students want to learn in the future.	Director IDI01082007_1–4
		• to be able to accept the **education** of science and technology in basic education early!	Director IDI01082007_1–5

Table 42.2 (continued)

Category	Concept register	Text	Code
Change	Explicit register: **change** Implicit register: **feel embarrassed, improving, accepting, sharing, applying**	• When the users face new technology in the beginning, they feel embarrassed to operate it. In addition, it is not easy to change their learning habits.	Director ID101082007_1–2
		• We see some changes in the users, including improving their technological knowledge about mobile devices, accepting PHS hand-helds in education, sharing their opinions with others, and applying cooperative learning in this course.	Director ID101082007_1–3
		• we have already seen that the users change attitude about their behavior in science and technology.	Director ID101082007_1–5
		• attitude toward using PHS and behavioral intention to apply mobile devices.	Director ID101082007_1–5
		• to give our students a chance to be able to accept the education of science and technology in basic education early!	Director ID101082007_1–5

2. The complete agreements formula:

(P) is as follows : Complete agreement formula $= \dfrac{\sum_{i=1}^{n} Pi}{N}$

where N = number of coding decisions each coder made.
(Here N is equal to 1 and P is 0.894).

3. The coding reliability formula (P) is as follows:

$$\text{Coding reliability (P)} = \frac{nP}{1 + [(n - 1)P]}$$

where N is the number of coders. (Here N is equal to 2).

$$\text{Composite reliability} = \frac{N \text{ (average intercoder agreement)}}{1 + [(N - 1)(\text{average intercoder agreement})]}$$

where N is the number of coders. (P is 0.8) [1].

Note that the composite reliability coefficient is larger than the average intercoder agreement (0.894 compared to 0.65). The composite reliability coefficient (0.944) assumes that the more coders there are, the more valid their combined observations will be.

42.6 Conclusion and Suggestions

Many educational opportunities will be made possible in the future because of mobile technologies' unique characteristics and positive impacts identified progressively in education. Using technology products brings additional value for teachers, parents, and their children. In the beginning, lack of proficiency may cause inconvenience for users, but making up their minds to try new things is really rewarding. We were happy to hear that PHS has been gradually accepted in education, and was used to encourage good interactions between parents and teachers. We would also offer the m-learning environment of choice in elementary schools.

In this article, we examined many resources and cited studies to answer the practicability of mobile wireless technologies in basic education. This research intends to demonstrate the director's ideas and the application of PHS phones in education, in order to provide references, opinions, and information to instruction designers and mobile learning directors to develop further innovative instruction.

We summarize from documents of the interviews we conducted as follows:

1. It is very important to teach our children to learn how to adapt to the artificial environment early via PHS hand-helds.

2. Only when we do our best to communicate with parents and teachers, will they become familiar with various operational functions in PHS, by our offering assistance with the software and hardware.
3. PHS has been gradually accepted in education. At the same time, it is used to encourage interactions and communication between parents and teachers.
4. We learn many concepts of mobile learning from the "PHS handset case" and we will know what our students want to learn in technological education.
5. We will strive for the new mobile products to give our students many chances to learn science and technology in elementary school.

We hope that our contribution to the use of mobile devices will encourage other researchers to look at the big picture of how we presently plan to interact and communicate between parents and teachers using small devices. Through this experiment, the director and the research team received good responses and effects that were proven to be useful in education. These enhancements will be crucial for supporting the growth of mobile devices in education.

References

1. Berelson B (1952) Content analysis in communication research. The Free Press, Glencoe, IL
2. K12 Handhelds (2005) 101 great educational uses for your handheld computer. Retrieved January 3, 2007 from: http://www.k12handhelds.com/101list.php
3. Kim SH, Mims C, Holmes KP (2006) An introduction to current trends and benefits of mobile wireless technology use in higher education. AACE J 14(1):77–100
4. Lynch P, Horton S (2005) Web style guide, 2nd ed. Retrieved December 15, 2006, from: http://www.webstyleguide.com/index.html?/index.html
5. Maginnis F, White R, Mckenna C (2000) Customers on the move: m-commerce demands a business object broker approach to EAI. EAI J, pp 58–62
6. McCarthy J (2005) The PalmGuru report on palm and pocket PC. PC 2000 Magazine
7. Ministry of Economic Affairs (2005) The new era of wireless M in Taiwan. Retrieved December 30, 2006, from: http://www.gov.tw/PUBLIC/view.php3id=131407&sub=60&main=GOVNEWS
8. Newman WL (1997) Social research methods—qualitative and quantitative approaches, 3rd ed. Allyn and Bacon, Boston, MA
9. PHS.com (2005) The function of PHSR. Retrieved January 04, 2007, from: http://www.phs.com.tw/en/index.asp

Chapter 43
The Solutions of Mobile Technology
for Primary School

Rong-Jyue Fang, Hung-Jen Yang, C. Ray Diez, Hua-Lin Tsai, Chi-Jen Lee, and Tien-Sheng Tsai

Abstract It is very important to understand the mobile technology content in order to bring up the new century students' technology equipment, as it is the headstone of technology development. The mobile technology includes handle mobile devices, digital transmission modes, mobile information software and so on. The mobile technology realized the perpetual contact, the network tools of teenagers' social intercourse. In the science and technology curriculum, the students had to answer questions such as: What is mobile technology? How do these things execute? What kinds of value and belief effect influenced the mobile technology? What kind of ideas fostered the mobile technology? How did we put these ideas into practice? What is the interaction between mobile technology and sociality? What is very important, is to promote the students' adaptation to the new environment via technology instruction. The mobile technology content integrated into instruction could be applied to instructional strategies of dissemination, facilitation, inside and outside collaboration, apprenticeship, generative development and so on. Therefore, perplexities and countermeasures that have risen from theory exploration and experience of teaching can be shown in this chapter. The result will provide a resource for administrations and teaching.

43.1 Introduction

Since 1993, a new economical scenery has been shaped, with the United States continuously affecting the global economy and the competitive power development. In 2000, the OECD noted that the new economy is attributed to the rapid advances in information and communication science and technology (information and communication technology, ICT), and eight country leaders summit in

R.-J. Fang (✉)
Department of Information Management, Southern Taiwan University of Technology, Taiwan, ROC
e-mail: rxf26@mail.stut.edu.tw

N. Mastorakis et al. (eds.), *Proceedings of the European Computing Conference*, Lecture Notes in Electrical Engineering 27, DOI 10.1007/978-0-387-84814-3_43, © Springer Science+Business Media, LLC 2009

Ryukyu Islands and pass a resolution for information science and technology (information and technology, IT). The charter recognized IT as creating one of the 21st Century's formidable forces [1]: the education and the technical union were important subjects acknowledged by the audiences.

Encouraging the use of mobile technology may result in encouraging continuous studying, in multiple places, and adopting the lifelong study custom [2]. When the learning environment is able to receive the information from science and technology at any given time and place, we achieve a greater impact on the learner. As science information and technology assistance are adopted by students, the majority has also adopted the motive for intense studying.

At the present time, information system applications are growing within the education systems. Mobile technology is about to change the present situation; no longer must the teaching process be limited to the classroom. The teacher may soon become obsolete; therefore, in order to realize the immediate information transport, it is necessary for the academic community to develop a motion research system. Mobile technology is an indispensable, important item for technical education and its applications.

Therefore, knowledge of mobile technology is also very important for the contemporary educational system. Mobile technology is planted deep in our life; we already were using it.

In order to upgrade the new generation of student science and technology accomplishments we have to be emphatic on the essence of the mobile technology, which, in any case, will be the future science and technology development cornerstone.

There are two reasons for studying the mobile technology:

1. The connotations of mobile technology integrate into the science and technology curriculum for primary school.
2. The solutions of mobile technology integrate into the science and technology curriculum for primary school.

43.2 The Meaning of Mobile Technology

The science and technology community must now act upon the following aspects of the mobile technology study.

43.2.1 The Connotations of Mobile Technology

Teachers must understand that their role now has changed. Nowadays, teachers are evaluated on how well they manage technology and science educational contents. The teacher wants to attract the student's interest in learning about technological advantages [3]. Along with the mobile technology (e.g., cell phone, PDA), business studies can be a helpful educational direction [4].

In the third generation of mobile technology, everyone personalizes his/her communication (the Mobile Portal). Communication abilities of each user in its own personalized form depend on the attributes of the terminal equipment. Every one of the users can create his/her own content regarding image, animation, photograph, speech and writing. In a word, people would be willing to pay to use these functions and contents which, in fact, are digital messages.

Taking part in the information society may not only be affective on one, but also provide a way to make a profit. Here are some of the most vivid aspects of the information and mobile technology life: content of digital (the digitalization of all content), the Internet as a universal message media carrier, flexibility in wireless equipment and cell phones (mobility).

Larger in the life of freedom at present, in spite of the body can enjoy personalization what place of service. Such service further turns a passive service content into the dynamic state function, and the service promoter also, therefore, makes a profit plenteous.

The digital content has to be adequately applicable on other devices, such as multimedia telephones (the media phones), notebooks (laptops), personal digital assistants (PDA) and other equipment any time, anywhere. The cell phone has become everyone's personal equipment of choice, satisfying everyone's special needs.

Any user that has handheld mobile equipment (such as cellular phone, PDA, the Smart phone, etc.) can get information, arrange business or shopping using third generation wireless communication (3G), send and receive messages, e-mails, etc.

One should not forget the amusement applications that are available to mobile equipment users. 3G wireless communication possesses many functions such as walkman, camera, personal certification, remote control, E-Wallet, long distance operation and surveillance, etc. They can also connect to the Internet anywhere, anytime.

43.2.2 Characteristics of Mobile Technology

Mobile technology made "perpetual contact" possible, thereby giving rise to the sociologic phenomenon, the everlasting contact. New social behaviors are structured on these applications; however, the first goal of the contact is to interact and exchange ideas (the pure communication), hoping to share intelligence (mind) [5]. Katz thinks the image of pure communication is rooted in the logic of everlasting contact supported by how people judge, invent and use communication science and technology; personal development with technological communication are the basis for the lasting contact.

Oksman and Rautiainen [6] studied Finnish children, and, in particular, the relationship between the teenager and the cell phone, and found that until the person is 20-years-old the cell phone has already been a very important everyday medium, and truly a part of their lives. To a Finnish teenager, the function of the cell phone is, among others, to create their social set, and help in the personal definition of the individual.

Rautiainen [7] points out that the teenager uses a vernacular speech within his/her written messages very broadly. Grinter and Eldridge [8] detected that the teenager's message is mainly used among peers. The length restriction of the message makes the user change conversation quickly. Using the message can be a discreet, quiet form of communication. The message has changed the teenager's communication motive, but not the content. The messages used are shorter than other means of communication, but more abundant.

Communication by messaging is instant, putting the teenager in the position to monitor every change in his/her plans, when someone notifies him/her, instantly. Their research went on to show that, even when a land line phone is at the teenager's disposal, free of charge, the teenager still prefers to contact by text messages. Slang and abbreviations are the rule in the teenager's cell phone. A certain kind of language is continuously evolving through the use of messages forming a certain messaging culture.

Holmes and Russell [9] think that the teenager adopts the communication and information science and technology (the communication and information technology, CIT) quickly due to the fact that this gives them a medium to communicate that is not controlled by the family or the school. They think this has already produced a new kid type. The teenager perceives the world as an extension of his/her cell phone antenna, and is able to keep contact no matter where he/she is.

The fast adoption of the cell phone promoted an alternative society shaping, and had an impact on the school learner's properties. They point out that the telephone may be one media that is being used in order to construct ego. The importance of these media is that they can combine science and technology to postpone the exhibition to the educational environment.

A characteristic example is the global antenna that will increasingly break the geography Jiang boundary, weakening the value of what is taught by conventional teaching. As far as culture is concerned, this trend is subjected to the proliferation that the digital world rubs.

The research of Ling and Yttri [10] among teenagers showed that the cell phone, apart from its use as a communication tool, is also used to declare the social group that one belongs to: "super moderate" (hypercoordination).

The so-called "super moderate" constitutes the fact that mobile communications surmounted the mere business contact, expanded in the fields of social intercourse, interactive coordination, and emotion. As far as the teenage user of the mobile technology is concerned, he/she tries to maintain a contact status with his/her peers while keeping distance from his/her parents.

Considering what has just been presented above under terms of science and technology illustrates an ideal realization of everlasting contact. The teenager's tool of social intercourse interacts with others; that is, to some extent, what Yun teaches about the teenager's language: it is the limitless boundary extension trend, the super coordination of the peers by contacting each other with the provided science and technology applications and tools.

43.3 The Connotations of Mobile Technology Integrate into the Science and Technology Curriculum for Primary School

43.3.1 Our Country's Future National Elementary School Nature and Life Science and Technology Curriculum Mobile Technology Connotation

The traditional course content has been unable to deal with the fast vicissitude of the social environment. The teaching plan and form must be greatly revised if it is to keep pace with a largely changing environment. In order to achieve "effective teaching," apart from the teacher's idiosyncrasy and behavior, mobile technology must be introduced into the learning process. In what way, and how effectively, is it going to be introduced in the learning process, is a matter of great importance.

The Ministry of Education on 30 September 1987 declared "the compulsory nine year education curriculum general outlines" and, the ten items about "the compulsory education's purpose to raise the basic capabilities." The eighth item is "the utilization science and information technology: Correct, safe, and effective use of the science and technology. Collection, analysis, evaluation, conformity and utilization of the information; the promotion of studying efficiency and the life quality."

The general outline summary divides the compulsory educational curriculum (language, health and sports, society, art and humanism, mathematics, nature, science and technology, synthesis and so on) to seven big study domains. In the "nature, science and technology" domain, material about the terrestrial environment, ecology care, information science and technology, and so on, is included.

The emphasis on scientific research raises the grasp of energy issues, respect for life and the natural environment, which is a good way for science and technology to be used in everyday life.

The "nature, science and technology curriculum" should cover the following essential topics.

43.3.1.1 Content

Systems and procedures concerning technical issues are always topics to be dealt with (e.g., material or energy processing). This implies the following: Material processing constitutes the change of the material with regard to its shape or nature; material processing may include the coding, storing, transmitting, receiving and decoding, whereas transformation procedures are involved in the energy elaborations (e.g., mechanical energy – electrical energy).

The study content must be such, and have the kind of pattern required that will promote the student's exploration and perception of such issues.

43.3.1.2 Procedure

When considering a technical question many pieces of information can be examined, including the development solution, the manufacture, the test, the appraisal and the improvement solution procedure, to propose solutions. The student selects one kind of solution and implementation.

43.3.1.3 System Vein

The learning activity must deal broadly with the real world so it doesn't "see only the tree, and not the forest."

Therefore, the technical question needs to be confirmed in terms of family, school, community, leisure, business circle data, etc., in the system vein.

43.3.1.4 Progress

When the technical knowledge is built gradually from the simple to the more sophisticated and complex topics, the student develops an overall perception about local and global technical science.

43.3.1.5 Making Uniform

In the lower grades, the nature, science and technology curriculum emphasizes subject (thematic) teaching; thus, makes different domains uniform and enriches each child's learning experience.

In higher grades, the nature, science and technology curriculum may continue with more complex science and technology functions, giving the student the opportunity to utilize his/her acquired information on these topics.

43.3.1.6 Document

The mobile technology tutorial content and procedures cannot just determine the student's memorization abilities. In order to assure a greater depth in the student's understanding of the content we should encourage the student to elaborate and analyze certain processes.

In other words, the student's main content of study within the nature, science and technology curriculum must be:

What is mobile technology? How does it progress? Does mobile technology receive values and belief influence? Do these concepts encourage mobile technology? How is a concept caused to move? Does mobile technology and society share mutual influences?

It is very important to promote the student's opportunities in the science and technology world through science and technology teaching.

43.3.2   The Teaching Strategies for Mobile Technology Integrate into Science and Technology Curriculum

Integrated types of teaching involving mobile technology connotations are encouraged. They may involve dissemination, facilitation, inside and outside collaboration, apprenticeship, generative development, etc.

These teaching strategies are:

43.3.2.1   Teaching of Strategy Dissemination

The strategy of the teaching is dissemination. Its goal is to use the ultra chain to tie with the homepage. The primary teaching activities include issuing the curriculum news, organizing/reorganizing the network resources and the chain tie, and providing a larger amount of material than the traditional classroom teaching.

43.3.2.2   Teaching of Strategy Facilitation

The teaching strategy to help the student understand mobile technology connotations will provide the directions, the guidance discussion and suggest related resources.

43.3.2.3   Teaching of Strategy Inside Collaboration

The inside collaboration strategy focuses on the students' interaction and communication, and uses inside collaboration in order to understand the mobile technology connotations. Its main teaching activity provides support to the student to answer the question; directions and answering plans are facilitated by an inside corporation.

43.3.2.4   Teaching of Strategy Outside Collaboration

The outside collaboration strategy offers outside assistance to students in order to understand the mobile technology connotations. Its main teaching activities are:

- to invite exterior personnel to participate in the classroom instruction,
- to tie the homepage chain to exterior resources, and
- to participate in other website social groups.

43.3.2.5 Teaching of Strategy Apprenticeship

The apprenticeship strategy of teaching includes the scholar expert who will provide instructions on mobile technology connotation. Its main teaching activity is:

- The scholar expert instructs by the school district nearby in view of the mobile technology connotation learning activity.

43.3.2.6 The Teaching Strategy Generative Development

The generative development teaching strategy is aimed at publishing mobile technology connotations by the student himself/herself. Its main teaching activity is: As the mobile technology connotations remain the subject, the penetration creation, the organization, the reorganization and assimilation achievement are topics for the article to be composed.

The above six kinds of teaching strategies may be applied to the mobile technology connotations and can be included in the teaching process to maximize effectiveness.

Each kind of teaching strategy may be utilized alternately and supplementary, keeping the student study as the main axle, and further facilitate the teaching process.

43.4 The Solutions of Mobile Technology Integrate into Science and Technology Curriculum for Primary School

The student, teacher and administration stratification plane can be a bitter and difficult experience.

43.4.1 Student Stratification Plane

The emerging science and technology level in the student community may not yet be high enough because the national elementary school students are still under-age, and do not practice independent use with their belongings. The high tech knowledge level is low as far as the mobile technology is considered, as well.

The counterplan to respond to such a problem is applied science and technology that accepts the mode theories (Davis, Bagozzi, & Warshaw, 1989); this mode puts forward consciousness to the Perceived Ease of Use (EOU) and the Perceived Usefulness (U), two special convictions with consciousness.

The consciousness uses (EOU) to easily guide an individual toward a particular information system that he/she perceives to be easy to use; the consciousness is useful (U) so that an individual can promote using a certain information system and results of cognition degree.

As far as EOU is concerned, when an individual finds it easy to use a particular information system, he/she is more likely to present a positive attitude toward that system. Similarly, when the individual has a positive evaluation to the particular information system, he/she is more likely to adopt a positive attitude towards that kind of system. When science and technology accept the mode to be inquiring for a particular information system and speech, the individual consciousness is useful.

If a student is unfamiliar with the technology, it may present difficulty in inducting the mobile technology connotations. In this instance, the counterplan is to teach related knowledge derived from ordinary items, and give the student the opportunity to personally experience on his/her own, cognize the aiming at action science and technology content after measuring before carrying out teaching.

Another consideration is whether all students possess personal characteristics suitable to receive education.

Everyone is put into ordinary classes when they are enrolled in elementary school. Everyone attends normal classes, although teaching sophisticated action science and technology isn't suitable for all students. The specific counterplan in response to this need is experimental teaching of progressive contents to more advanced students, that can show the possibility of teaching such content to the entire class.

The different resources available at the local and national level are a widespread problem in the educational circles; the government has been working intensively to reduce this margin.

As far as mobile technology content teaching is concerned, in addition to the government's efforts, the following counterplan can be applied: The education administration conducts on-the-job training of the specialty teachers and the subsidy of the budget at ordinary items, even the national region, can fight for project subsidies from the budget; Private cars can be granted to specialized teachers who can tour the country giving lectures to encourage and strengthen the teaching power of the country's regional schools.

As mobile technology equipment is a product of the emerging technology, cost is an issue. Therefore, acquiring auxiliary equipment and applications is an issue to be dealt with.

If students are given the possibility to physically act and experiment, the learning results would ascend to much better levels. So, the correspondent counterplan would be: Fight for an upper grade organization subsidy and obtain a sponsor manufacturer from which one can buy –arranging a discount – equipment; this should increase the teaching effectiveness.

43.4.2 Teacher Stratification Plane

The teacher psychology accepts the possibility of emerging science and technology; therefore, integrating instruction for mobile technology, is easy to create.

The counterplan: Applied science and technology acceptance of mode theories by the school administration, through education or guide.

Encourage a teaching professional to permit and invite action science and technology views in education; a teacher who is not afraid to step aside and let technological novelties be introduced.

Among the teacher's official duties should be studying the introduction of a new culture in the curriculum and their personal self-study which will involve the connotations of mobile technology.

The current position of the teacher is in the office, ought to be such to promote the correlation of his job description with expanding the technical knowledge that is obtained at this time.

Currently, the curriculum and teaching material,do not yet include mobile technology curriculum or related teaching material.

The correspondent counterplan would be: The curriculum programming of the system organization should favor developing the teaching material or the teaching activity so that science and technology is integrated with teaching in the future.

The present state in which a teacher uses auxiliary equipment, which is expensive and vulnerable to technological advances, is not easy to overcome.

The corresponding counterplan is: Combined efforts to obtain a subsidy from community resources, research a sponsor – manufacturer – or use second-hand equipment should all be taken into consideration.

43.4.3 Administrative Stratification Plane

Administrative personnel have a limited acceptance of emerging science and technology, due to restricted application in every day life; thus it is not easy for them to accept it.

The correspondent counterplan is: Administrative personnel should undergo related seminars, in order to obtain a homogenous grasp on the topic and promote the practical technological applications on an everyday basis. This will help raise the administrative staff's level on science and technology.

Science and technology information changes daily; even if something was once considered to be the best solution, it must constantly evolve in order to keep up with the continuous increasing demands. But, for actual technological cognition, which nowadays is in relatively short supply, particularly when applied to education, sufficient support must be given to the teaching system.

Firstly, the school tutoring system has not degraded the teacher to the level of carrying the mere teaching burden, so the teacher's utilization also waits for its enhancement.

Secondly, the school information system using many teaching procedures is unable to complete them as a whole procedure which should be done in units.

Thirdly, the teachers are unable to reach the teaching target and the necessary resources (e.g., individual student progress, average and attendance rate, or perhaps curriculum progress and program).

Thus, the administrative support of the emerging science and technology depends on the hardware level.

Here is the correspondent counterplan:

Because of the need to purchase hardware equipment, a reward system will be established by the administration for the teachers; research and educational motivation are encouraged in this way.

On the other hand, the administrative personnel should invest in new ways of science and technology teaching. Not doing so will yield high costs and poor results.

The correspondent counterplan is:

Invest in the newly arisen science and technology products, with regard to administrative personnel training, which will result in promoting national competitive ability.

43.5 Conclusion

Much technical progress and development has taken place in recent years. The personalization of applications is gradually reaching more and more people. The Internet is already widespread and constitutes an indispensable element of the consumers' practices.

The use of applied science and technology special features have been introduced in communications. Along with the occurrence and development of 3G technology, not only was progress made in regard to independent products of science and technology, but the potential for a vast information industry has been expanded.

At present, the science and technology will go on to form a large market in the business community.

Large international information corporations, one after another, seek a way to break into this market and invest under various types of strategy, in order to control a piece it.

It is believed that in the very near future we will see applied science and technology being used in our everyday life. If education is to meet the challenges of today, school teaching should include new science and technology information, and consider behaviors produced by these new social activities.

The country's education should evaluate the new science and technology content, including the universal motion ability of the cellular phone and the stability of the Internet.

It is, therefore, very important to be educated about the mobile communication science and technology; this can actually contribute to a better quality of life.

Mobile technology is the key for a competitive national presence. It can help people deal with communication inconvenience, save time and money and result in a better quality of life.

Various methods of teaching the applied science and technology strategies are presented here, along with analyses of the difficulties observed in various educational applications and proposed counterplans for each one of them. Data is collected from study theory and actual teaching experience, and combines them to propose a new subject curriculum and teacher profile.

References

1. eCollege.com (2002) Retrieved form World Wide Web: http://eCollege.com.
2. Sharples M (2000) The design of personal mobile technologies for lifelong learning. Comput Educ 34:177–193.
3. Mitchell DL, Hunt D (1997) Multimedia lesson plans-help for preserve teachers. J Phys Educ Recreation Dance, 17–20.
4. Barnes SJ (2000) The mobile commerce value chain analysis and future developments. Inf Manag, 91–108.
5. Katz JE, Aakhus M (2002) Perpetual contact: Mobile communication, private talk, public performance. Cambridge University Press, Cambridge.
6. Oksman V, Rautiainen P (2002) 'Perhaps it is a body part' How the mobile phone became an organic part of the everyday lives of children and adolescents: A case study of Finland. In Katz JE (ed) Machines that become us. Transaction Publishers, New Brunswick, NJ.
7. Rautiainen P (2001) The role of mobile communication in the social networks of Finnish teenagers. In: 'Machines that become us' conference, Rutgers University, New Jersey.
8. Grinter RE, Eldridge MA (2001) y do tngrs luv 2 txt msg? In: Proceedings of the European conference on computer-supported cooperative work. Bonn, Germany.
9. Holmes D, Russell G (1999) Adolescent CIT use: Paradigm shifts for educational and cultural practices? Br J Soc Educ 20(1):69–78.
10. Ling R, Yttri B (2000) 'Nobody sits at home and waits for the telephone to ring:' Micro and hyper-coordination through the use of the mobile telephone. In Katz JE and Aakhus M (eds) Perpetual contact: Mobile communication, private talk, and public performance. Cambridge University Press, Cambridge.

Chapter 44
A Study of Verifying Knowledge Reuse Path Using SEM

Hung-Jen Yang, Hsieh-Hua Yang, and Jui-Chen Yu

Abstract The purpose of this study is to apply SEM to identify the path structure of knowledge reuse of senior vocational high IT major students. An investigation research method was applied to collect data for creating a theoretical model and then was tested with LISREL 8.52. The test result of the SEM fit well. Based on the SEM, a knowledge reuse model was revealed. Both the theory model and the verified model are discussed. These results suggest to us that the knowledge reuse path of senior vocational high students is a two-path model, from capture to both packaging and distributing. In the second level, packaging would affect distributing. The packaging knowledge would act as an intermediate variable to be a transferring path from collecting knowledge to distributing knowledge.

44.1 Purpose

The purpose of this study was applying SEM to identify the path structure of knowledge reuse of senior vocational high information technology (IT) major students. To fulfill that purpose, we set out to (1) examine possible knowledge reuse paths of senior vocational students; (2) create a theory model of knowledge reuse; (3) investigate how IT major students conducted knowledge reuse; (4) verify our theory model.

The following discussions cover relevant theories and empirical evidence for this research. The major topics are learning and transfer, knowledge reuse, and SEM.

H.-J. Yang (✉)
Department of Industrial Technology Education, National Kaohsiung Normal University, Kaohsiung City, Taiwan, ROC
e-mail: hjyang@nknucc.nknu.edu.tw

N. Mastorakis et al. (eds.), *Proceedings of the European Computing Conference*, Lecture Notes in Electrical Engineering 27, DOI 10.1007/978-0-387-84814-3_44, © Springer Science+Business Media, LLC 2009

44.2 Literature Review

44.2.1 Process and Transfer

We hope that students will transfer their learning experience from one problem to another within a course, from one domain to another, from the school learning environment to the real world, and from the information technology industry to professional development. There are three dimensions of the technology learning model; those are knowledge, process, and context. Processes of learning and the transfer of learning are central to understanding how people develop important competencies. Learning is important, because no one is born with the ability to function competently as an adult in a certain community. It is also important to understand the kinds of learning experiences that lead to transfer. Transfer is defined as the ability to extend what has been learned in one context to a new context [1]. Assumptions about transfer accompany the belief that it is better to broadly "educate" people than to simply "train" them to perform particular tasks [2].

Paths of knowledge reuse play an important role in preparing the learning environment for students. Different paths of reusing knowledge provide students different opportunities. Learners value paths of transfer differently, according to their experiences.

Thorndike and his colleagues were among the first to use transfer tests to examine assumptions about learning [3]. One of their goals was to test the doctrine of "formal discipline" that was prevalent at the turn of the century. According to this doctrine, practice by learning Latin and other difficult subjects had broad-based effects, such as developing general skills of learning and attention. But these studies raised serious questions about the fruitfulness of designing educational experiences based on the assumptions of a formal discipline. Rather than developing some kind of "general skill" or "mental muscle" that affected a wide range of performances, people seemed to learn things that were more specific. Early research on the transfer of learning was guided by theories that emphasized the similarity between conditions of learning and conditions of transfer. Thorndike, for example, hypothesized that the degree of transfer between initial and later learning depended upon the match between elements across the two events [4].

The essential elements were presumed to be specific facts and skills. By such an account, the skill of writing letters of the alphabet is useful for writing words (vertical transfer). The theory posited that transfer from one school task to a highly similar task (near transfer), and from school subjects to non-school settings (far transfer), could be facilitated by teaching knowledge and skills in school subjects that have elements identical to activities encountered in the transfer context [5]. Reusing knowledge could also be negative, in the sense that experience with one set of events could hurt performance in related tasks [6]; A major goal of schooling is to prepare students for flexible adaptation to new problems and settings. The ability of students to reuse knowledge provides an important index of learning that can help teachers evaluate and improve their instruction.

Finally, a meta-cognitive approach to teaching can increase transfer by helping students learn about themselves as learners in the context of acquiring content knowledge. One characteristic of experts is an ability to monitor and regulate their own understanding in ways that allow them to keep learning adaptive expertise: this is an important model for students to emulate.

44.2.2 *Structural Equation Modeling*

Structural equation modeling, or SEM, is a very general, chiefly linear, chiefly cross-sectional statistical modeling technique. Factor analysis, path analysis, and regression all represent special cases of SEM. A researcher is more likely to use SEM to determine whether a certain model is valid, rather than to "find" a suitable model—although SEM analyses often involve a certain exploratory element [7]. As Garson suggested, SEM is usually viewed as a confirmatory rather than exploratory procedure, using one of three approaches [8]:

1. Strictly confirmatory approach: A model is tested using SEM goodness-of-fit tests to determine if the pattern of variance and covariance in the data is consistent with a structural (path) model specified by the researcher. However, as other unexamined models may fit the data as well or better, an accepted model is only a not-disconfirmed model.
2. Alternative models approach: One may test two or more causal models to determine which has the best fit. There are many goodness-of-fit measures, reflecting different considerations, and usually three or four are reported by the researcher. Although desirable in principle, this AM approach runs into the real-world problem that in most specific research topic areas, the researcher does not find in the literature two well-developed alternative models to test.
3. Model development approach: In practice, much SEM research combines confirmatory and exploratory purposes: a model is tested using SEM procedures, found to be deficient, and an alternative model is then tested based on changes suggested by SEM modification indexes. This is the most common approach found in the literature. The problem with the model development approach is that models confirmed in this manner are post hoc ones which may not be stable (may not fit new data, having been created based on the uniqueness of an initial data set). Researchers may attempt to overcome this problem by using a cross-validation strategy under which the model is developed using a calibration data sample and then confirmed using an independent validation sample.

SEM grows out of and serves purposes similar to multiple regression, but in a more powerful way which takes into account the modeling of interactions, nonlinearities, correlated independents, measurement error, correlated error terms, multiple latent independents each measured by multiple indicators, and one or more latent dependents also each with multiple indicators. SEM may be used as a more powerful alternative to multiple regression, path analysis, factor analysis, time series analysis, and analysis of covariance. That is, these procedures may be seen as special cases of SEM, or, to put it another way, SEM is an extension of the general linear model (GLM) of which multiple regression is a part [8].

44.3 Methodology

The following sections mainly focus on how the subjects were chosen for this study, how the study was designed and carried out, and what coding schemas were developed.

44.3.1 Theory Model

A nominal group technique, NGT, was used to create a theory model of a knowledge reuse path. Nine experts were invited as NGT members. Through NGT procedures, the components and paths of the theory model were concluded with high consensus.

In the theory model, there are three latent variables: capturing knowledge as Capturing; packaging knowledge as Packaging, and distributing knowledge as Distributing. For the latent variable industry, there are three manifest variables. There are two manifest variables for the latent variable capturing and three manifest variables for representing the latent variable packaging.

44.3.2 Subjects

There are around 20,000 IT major senior vocational students in Taiwan. A random sample technique was applied for selecting 1,500 students.

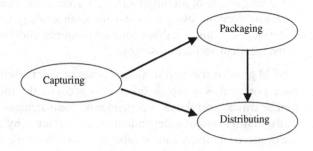

Fig. 44.1 Theory model of this study

Table 44.1 Reliability coefficients alpha value of questionnaire

Knowledge reuse paths	Reliability coefficients alpha
Capturing	0.92
Packaging	0.97
Distributing	0.94

44.3.3 Instrument

After designing the investigating questionnaire based on the NGT conclusion, a pilot test was conducted. A total of 56 subjects were invited to answer the questionnaire. The reliability was calculated and is listed in Table 44.1. This instrument has professional structure validity and high reliability.

After the pilot test, this questionnaire was put online and invitations were sent to the randomly selected subjects with their personal codes, so they could access the questionnaire anytime within a two-week time period.

44.4 Findings

The findings of this study are mainly from the statistical test results of the data collected by our investigation.

In Fig. 44.2, the t-tests of all 8 manifest variables are significant. It is concluded that variables ind1, ind2, and ind3 are significant contributors to the latent variable of industry knowledge. The manifest variables of both sch1 and sch2 are significant contributors to the school knowledge. The per1, per2,

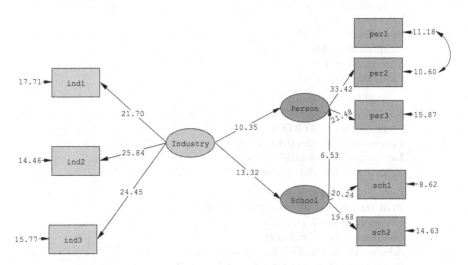

Fig. 44.2 T test of knowledge reuse structure equation model

and per3 are significant contributors to personal knowledge. Industry knowledge contributes to both school knowledge and personal knowledge. School knowledge also contributes to personal knowledge.

44.4.1 The Model-Fitting Process

In Table 44.2, the goodness-of-fit statistics are listed and the chi-square value is 24.23 (p>0.05), thereby suggesting that the hypothesized model is adequate.

Table 44.2 Goodness-of-fit statistics

Degrees of Freedom = 15
Minimum Fit Function Chi-Square = 24.65 (P = 0.055)
Normal Theory Weighted Least Squares Chi-Square = 24.23 (P = 0.061)
Estimated Non-Centrality Parameter (NCP) = 9.23
90 Percent Confidence Interval for NCP = (0.0 ; 26.79)

Minimum Fit Function Value = 0.026
Population Discrepancy Function Value (F0) = 0.0099
90 Percent Confidence Interval for F0 = (0.0 ; 0.029)
Root Mean Square Error of Approximation (RMSEA) = 0.026
90 Percent Confidence Interval for RMSEA = (0.0 ; 0.044)
P-Value for Test of Close Fit (RMSEA < 0.05) = 0.99

Expected Cross-Validation Index (ECVI) = 0.071
90 Percent Confidence Interval for ECVI = (0.061 ; 0.090)
ECVI for Saturated Model = 0.077
ECVI for Independence Model = 7.24

Chi-Square for Independence Model with 28 Degrees
    of Freedom = 6723.46
Independence AIC = 6739.46
Model AIC = 66.23
Saturated AIC = 72.00
Independence CAIC = 6786.16
Model CAIC = 188.82
Saturated CAIC = 282.14

Normed Fit Index (NFI) = 1.00
Non-Normed Fit Index (NNFI) = 1.00
Parsimony Normed Fit Index (PNFI) = 0.53
Comparative Fit Index (CFI) = 1.00
Incremental Fit Index (IFI) = 1.00
Relative Fit Index (RFI) = 0.99

Critical N (CN) = 1155.80
Root Mean Square Residual (RMR) = 0.014
Standardized RMR = 0.014
Goodness-of-Fit Index (GFI) = 0.99
Adjusted Goodness-of-Fit Index (AGFI) = 0.98
Parsimony Goodness-of-Fit Index (PGFI) = 0.41

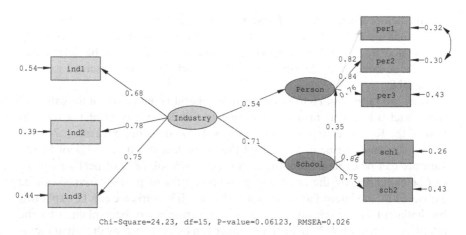

Chi-Square=24.23, df=15, P-value=0.06123, RMSEA=0.026

Fig. 44.3 LISREL parameter estimates of knowledge reuse model

The first goodness-of-fit statistic to be reported is the root mean square error of approximation (RMSEA).

The RMSEA takes into account the error of approximation in the population and asks the question, "How well would the model, with unknown but optimally chosen parameter values, fit the population covariance matrix if it were available?" This discrepancy, as measured by the RMSEA, is expressed per degree of freedom, thus making it sensitive to the number of estimated parameters in the model; values less than 0.05 indicate good fit, and values as high as 0.08 represent reasonable errors of approximation in the population. The RMSEA value of the model is 0.026, indicating good fit. Other indexes were listed for reference purposes.

The estimation process yields parameter values such that the discrepancy between the sample covariance matrix S and the population covariance matrix implied by the model is minimal. In Fig. 44.3, the standardized covariance values are listed.

Model assessment is listed in Table 44.3. Of primary interest in structural equation modeling is the extent to which a hypothesized model "fits" or, in other words, adequately describes the sample data.

Table 44.3 Structural equations model of knowledge reuse

Person $= 0.25$*School $+ 0.54$*Industry, Errorvar. $= 0.32$, $R^2 = 0.68$
(0.038)   (0.052)   (0.038)
6.53   10.35   8.19
School $= 0.98$*Industry, Errorvar. $= 1.00$, $R^2 = 0.51$
(0.076)
13.32

The measurement model is listed in Table 44.4. The second step in assessing model fit is to examine the extent to which the measurement model is adequately represented by the observed measures. This information can be determined from the squared multiple correlation (R2) reported for each observed variable in the table.

They can range from 0 to 1, and serve as reliability indicators of the extent to which each adequately measures its respective underlying construct. Examination of the R2 values reported in Table 44.4 reveals strong measures. 74% of Sch1's variance can be explained by the latent factor school. 57% of Sch2's variance can be explained by the latent factor school. 68% of per1's variance can be explained by the latent factor school. 70% of per2's variance can be explained by the latent factor school. 57% of Sch3's variance can be explained by the latent factor school. 57% of Sch2's variance can be explained by the latent factor school. 46% of ind1's variance can be explained by the latent factor school. 61% of ind2's variance can be explained by the latent factor school. 56% of Sch3's variance can be explained by the latent factor school.

Table 44.4 LISREL estimates (maximum likelihood) measurement equations of knowledge reuse

$$Sch1 = 0.60*School, \ Errorvar. = 0.26 \ , \ R^2 = 0.74$$
$$(0.030) \qquad\qquad\quad (0.030)$$
$$20.24 \qquad\qquad\qquad 8.62$$

$$Sch2 = 0.53*School, \ Errorvar. = 0.43 \ , \ R^2 = 0.57$$
$$(0.027) \qquad\qquad\quad (0.029)$$
$$19.68 \qquad\qquad\qquad 14.63$$

$$Per1 = 0.82*Person, \ Errorvar. = 0.32 \ , \ R^2 = 0.68$$
$$(0.029)$$
$$11.18$$

$$Per2 = 0.84*Person, \ Errorvar. = 0.30 \ , \ R^2 = 0.70$$
$$(0.025) \qquad\qquad\quad (0.028)$$
$$33.42 \qquad\qquad\qquad 10.60$$

$$Per3 = 0.76*Person, \ Errorvar. = 0.43 \ , \ R^2 = 0.57$$
$$(0.035) \qquad\qquad\quad (0.027)$$
$$21.48 \qquad\qquad\qquad 15.87$$

$$Ind1 = 0.68*Industry, \ Errorvar. = 0.54 \ , \ R^2 = 0.46$$
$$(0.031) \qquad\qquad\quad (0.030)$$
$$21.70 \qquad\qquad\qquad 17.71$$

$$Ind2 = 0.78*Industry, \ Errorvar. = 0.39 \ , \ R^2 = 0.61$$
$$(0.030) \qquad\qquad\quad (0.027)$$
$$25.84 \qquad\qquad\qquad 14.46$$

$$Ind3 = 0.75*Industry, \ Errorvar. = 0.44 \ , \ R^2 = 0.56$$
$$(0.031) \qquad\qquad\quad (0.028)$$
$$24.45 \qquad\qquad\qquad 15.77$$

44.5 Conclusions

Initial learning is necessary for transfer, and a considerable amount is known about the kinds of learning experiences that support transfer. The path of transfer should be noticed by the learner. It will provide the necessary connection between the original and the target domains.

Knowledge that is excessively contextualized can reduce transfer; abstract representations of knowledge can help promote transfer. Transfer is best viewed as an active, dynamic process rather than a passive end-product of a particular set of learning experiences. It is meaningful to find the path of knowledge reuse. All new learning involves transfer based on previous learning, and this fact has important implications for the design of instruction that helps students learn.

The major components of industry knowledge could be divided into publisher, software development, and 3C services. The major components of school knowledge could be categorized into Internet and communications media. The major components of personal knowledge are code and protocols, system developing techniques, and professional ethics. According to the estimation process, a measurement model was also reported. It provides a way to understand the relation among industry knowledge, school knowledge, and personal knowledge.

Based on the LISREL output, we examined the model of knowledge reuse as a whole. Our theory model was tested through the model-fitting process. The process yielded a χ^2 value of 24.23 (p = 0.06123). It was concluded that the model is adequate. In the well fitting model, industry knowledge contributes to both school knowledge and personal knowledge. Industry knowledge also contributes to personal knowledge via school knowledge.

These results suggested to us that the knowledge reuse path of senior vocational high students is a two-path model, one from school and one from industry. The school knowledge acts as an intermediate variable to be a transferring path from industry knowledge to personal knowledge.

References

1. Byrnes JP (1996) Cognitive development and learning in instructional contexts. Allyn and Bacon, Boston, MA
2. Broudy HS (1977) Types of knowledge and purposes in education. In: Anderson RC, Spiro RJ, Montague WE (eds) Schooling and the acquisition of knowledge. Erlbaum, Hillsdale, NJ, pp 1–17
3. Thorndike EL, Woodworth RS (1901) The influence of improvement in one mental function upon the efficiency of other functions. Psychol Rev, 8:247–261
4. Thorndike EL (1913) Educational psychology. vol. 1, 2
5. Klausmeier HJ (1985) Educational psychology, 5th ed. Harper and Row, New York
6. Luchins AS, Luchins EH (1970) Wertheimer's seminar revisited: problem solving and thinking, vol. 1. State University of New York, Albany, NY
7. Rigdon EE (2005) What is structural equation modeling? Retrieved July 12, 2007, from http://www2.gsu.edu/~mkteer/sem.html
8. Garson GD (2005) Structural equation modeling. Retrieved July 12, 2007, from http://www2.chass.ncsu.edu/garson/pa765/structur.htm

Chapter 45
A Web-Based System for Distance
Learning of Programming

V.N. Kasyanov and E.V. Kasyanova

Abstract Web-based education is currently an important research and development area, and it has opened new ways of learning for many people. In this
paper, we present the web-based system WAPE that is under development at
the A.P. Ershov Institute of Informatics Systems as a virtual environment for
the distance learning of programming.

45.1 Introduction

Web-based education is currently a hot research and development area. The
benefits of web-based education are clear: classroom independence and platform independence. Web courseware installed and supported in one place can
be used by thousands of learners all over the world that are equipped with any
kind of Internet-connected computer. Thousands of web-based courses and
other educational applications have been made available on the web within the
last ten years. The problem is that most of them are nothing more than a
network of static hypertext pages.

A challenging research goal is the development of advanced web-based
educational applications that can offer some amount of adaptivity and intelligence [1–9]. These features are important for web-based education applications
since distance students usually work on their own (often from home). The
intelligent and personalized assistance that a teacher or a peer student can
provide in a normal classroom situation is not easy to get. In addition, being
adaptive is important for web-based courseware, because it has to be used by a
much wider variety of students than any "standalone" educational application.
Web courseware that is designed with a particular class of users in mind may not
suit other users.

V.N. Kasyanov (✉)
A.P. Ershov Institute of Informatics Systems / Novosibirsk State University,
Novosibirsk, 630090, Russia

N. Mastorakis et al. (eds.), *Proceedings of the European*
Computing Conference, Lecture Notes in Electrical Engineering 27,
DOI 10.1007/978-0-387-84814-3_45, © Springer Science+Business Media, LLC 2009

Two kinds of systems have been developed to support the user in his/her tasks. In *adaptable* hypermedia the user can provide some profile (through a dialog or questionnaire). The adaptable hypermedia system provides a version of the hypermedia application that corresponds to the selected profile. Settings may include certain presentation preferences (colors, media type, learning style, etc.) and user background (qualifications, knowledge about concepts, etc.) On the web there are several such sites that use a questionnaire to tailor some part of the presentation to the user (usually the advertisement part).

In *adaptive* hypermedia the system monitors the user's behavior and adapts the presentation accordingly. Adaptive hypermedia systems build a model of the goals, preferences, and knowledge of the individual user and employ this throughout the interaction for adaptation of the hypertext to the needs of that user: for example, to adapt the content of the hypermedia page to the user's knowledge and goals, or to suggest the most relevant links to follow. The evolution of the user's preferences and knowledge can be (partly) deduced from page accesses. Sometimes the system may need questionnaires or tests to get a more accurate impression of the user's state of mind. Most of the adaptation, however, is based on the user's browsing actions, and possibly also on the behavior of other users. A comprehensive review of adaptive hypermedia techniques and systems can be found in [2, 3, 8].

In this paper the WAPE project [10, 11] under development at the Institute of Informatics Systems is presented. The web-based WAPE is intended to be an intelligent and adaptive virtual environment for supporting the distance learning of programming.

The rest of the paper is structured as follows. Section 45.2 gives a general description of the WAPE system. Sections 45.3 and 45.4 describe two main subsystems, CLASS and PRACTICE, of the WAPE system. Section 45.5 presents a knowledge model of any course supported by the WAPE system. Section 45.6 describes a knowledge model of a student and shows how the student model is used and updated. Using Bayesian networks in the system is considered in Section 45.7. How the system tests student's conceptual knowledge concerning theoretical material learned is considered in Section 45.8. Section 45.9 is our conclusion.

45.2 The WAPE System

The WAPE system supports users of four types: students, instructors, lecturers, and administrators. Users access WAPE through a standard web browser, which presents HTML documents provided by an HTTP server.

After authorization of the user as a student, the appropriate menu shell is opened. The WAPE system supports the following three levels of the learning process.

1. When a student learns theoretical material in a specific domain with the help of hypertext textbook.

2. When the system tests student's conceptual knowledge concerning the theoretical material learned.
3. When a student under the control of the system solves the practical educational problems: tasks and exercises.

The third level is assumed to be the main one in using the WAPE system; in order to learn a course supported by the WAPE system, a student has to perform several individual tasks and exercises.

Goal orientation is an important aspect of our WAPE environment. Since we do not want to determine the learning path of a student or a student group from beginning to end, the students are free to determine their own learning goals and their own learning sequence. At each step they can ask the system for relevant material, teaching sequences, and hints for practice examples and projects. If they need advice to find their own learning path, they can ask the system for the next suitable learning goal.

The WAPE system is intended to serve many students with different goals, knowledge, and experience. In our system the focus is on the knowledge of the students, which may greatly vary. Moreover, the knowledge state of a student changes during the work with the system. So we are using adaptivity concepts in our project.

The WAPE system provides monitoring facilities to follow the students' interaction with the system. It is possible to define the student actions that will be selected for tracking. These actions will be reflected in the student's working history. It is also possible to define some actions that will be needed in monitoring. Whenever a student performs a task (or an exercise), messages are sent to the instructor responsible for monitoring the student's defined actions.

Open discussions supported by the WAPE system provide a full virtual teleclassroom atmosphere, including cooperative learning with other students and tutoring facilities for instructors and lecturers.

The CLASS and PRACTICE systems are the main subsystems of the WAPE system.

45.3 The CLASS Subsystem

The CLASS subsystem is a virtual classroom that allows students to gain expertise in high-level programming. It is a problem-based learning environment in which students learn to write correct and effective programs for solutions of relatively simple problems.

Any course supported by the CLASS system includes problems of two kinds: *exercises* and *tasks*.

An exercise is a question (like a text test); so the expected solution for an exercise is not a program. A task is a problem of programming; so to solve a task means to write a program. Tasks are problems drawn from real applications, which students can readily understand and model.

Each project consists of 30 similar problems (individual variants of the project for all the students of a group) and includes an example (description and analysis of different solutions of one of these problems), which is used for example-based problem solving support.

There are projects of two types. The first type of project consists of problems that test thestudent's understanding of theoretical material; as a rule they relate directly to examples in the textbook. In particular, some problems are chosen in order to teach how to use new algorithmic operations in conjunction with elements already known. Another type of project includes the problems that add new and thought-provoking information to the material. Such tasks encourage a student to think about an important concept that is related to the theoretical material in the textbook, or to answer a question that may have occurred to the student when he/she read the textbook.

Students write programs using a standard general-purpose programming environment such as Turbo Pascal. The CLASS subsystem provides students with help at the stage of understanding a given problem statement, supports the example-based problem solving technology, and makes an intelligent analysis of student solutions by comparing actual program results with the ones expected.

The CLASS system accepts a program as a correct solution if the program works correctly on all tests that are stored by the system, but the final decision about the program is made by the instructor. In particular, to make a decision, the instructor should evaluate the properties of the program's solution that are hard to quantify, such as reliability, and so on.

45.4 The PRACTICE Subsystem

The PRACTICE subsystem is a virtual laboratory that provides students with a collection of 500 projects composed of tasks on six themes: graphs, grammars, languages and automata, formulas and programs, geometry, and games.

In these tasks students have to integrate all that they have learned before and develop the skills which are fundamental to software development at all levels and to programming as a discipline. These include the use of more or less formal methods of problem analysis and program design, with an emphasis on creating efficient and reliable programs which meet given specifications and provide friendly user interfaces.

For convenience, all tasks are divided into three groups based on their complexity: average (or normal), higher, and reduced. The tasks were rated using the following three metrics: complexity of model, complexity of algorithm, and "ingenuity." Measured "ingenuity" is based on such properties of the task as the complexity of understanding the application from which the task is drawn, the complexity of the design model and algorithm for the task, and the complexity of the interconnections between the control and data structures which should be represented by the program.

An example of a task with average complexity is the following: "Write a program which for a given graph G with n nodes, where $n<11$, finds a maximum number m such that the graph G has a clique consisting of m nodes."

An example of a task with reduced complexity is: "Write a program that for a given context-free grammar G determines the emptiness of the language $L(G)$."

An example of a task with higher complexity is: "Write a program that for any two given state transition diagrams DB_1 and DB_2 with nonempty intersection L' of their languages $L(DB_1)$ and $L(DB_2)$, find a string x from L' such that x has the shortest length in L'."

The PRACTICE subsystem can also be used as a hypertext manual on algorithm design, providing both a catalog of basic algorithmic problems (with the algorithms known for them) and several fundamental algorithmic techniques, including data structures, dynamic programming, depth search, backtracking, and so on.

45.5 Knowledge Model

Any course supported by the system is based on a *knowledge model* which is a triple (S, U, W), where S is a finite set of *concepts* (or *knowledge units*), and U and W are two binary relations on S such that the following properties hold for any $p, q \in S$:

- $(p, q) \in U$ denotes the fact that the concept q is a component of the composite concept p;
- $(p, q) \in W$ denotes the fact that p has to be learned before q, because understanding p is a prerequisite for understanding q;
- (S, U) is a forest;
- (S, W) is a directed acyclic graph (or DAG).

On the basis of the knowledge model a *course glossary* is constructed. In the glossary each knowledge unit is provided with two sets: a set of keywords (or phrases) that are used for textual representation of this knowledge unit, and a set of references to the elementary information resources of the course whose contents are connected with this knowledge unit.

Each information resource of a course is indexed by some set of knowledge items describing the content of the resource. These resources can be general HTML pages, examples, projects, etc. The origin of an information resource is not relevant for indexing; only the content defines the index.

Let $<$ denote a partial order on the set S where $p < q$ if and only if either $(p, q) \in W$ or $(q, p) \in U$. The relation $<$ represents all learning dependencies defined by the knowledge models. $p < q$ denotes the fact that p has to be learned before q, because understanding p is a prerequisite for understanding q or because p describes an aspect of the composite concept q.

45.6 Student Model

The knowledge of a student is modeled as a *knowledge vector*

$$K(x) = (p_1, \ldots, p_n),$$

where n is the number of knowledge units in the knowledge model and each p_i is a conditional probability which describes the system's estimation that a student x has knowledge about a topic s_i, on the basis of all observations the system has about the student x.

Each p_i expresses the grade of knowledge the student has on a topic s_i. We use four grades which divide all students into *experts, advanced students, beginners,* and *novices* with respect to s_i. Thus, the elements of the knowledge vector $K(x)$ are, on the one hand, concepts describing the domain model of a course;, on the other hand, they are random variables with the four discrete values $E, A, B,$ and N, coding corresponding knowledge grades.

The evidence we obtain about the student's work with the system changes with time. Normally, the student's knowledge increases while working with the system, although lack of knowledge is equally taken as evidence. Since every kind of observation about a student is collected as evidence, the knowledge vector gives—at each time—a snapshot of the student's current knowledge.

Many adaptive systems detect the fact that the student reads some information to update the estimate of his knowledge. Some of them also include reading time or the sequence of read pages to enhance this estimation. While this is a viable approach, it has the disadvantage that it is difficult to measure the knowledge a student gains by "reading" an HTML page. In the current state of our development, we decided to take into account neither the information about visited pages nor the student's path through the hypertext books. Instead, we use only the tests and the projects for updating the student knowledge model. This is motivated by the problem-based approach for courses supported by our system.

The student model is used to provide a student with the most suitable individually planned sequence of knowledge units to learn and projects to solve. For example, every time a student is going to solve a project, the system checks whether all prerequisite knowledge units are sufficiently known by the student. If not, the student cannot begin to solve the project.

The WAPE system uses three problem-solving support technologies: intelligent analysis of student solutions, interactive problem-solving support, and example-based problem-solving support. All these technologies can help a student in the process of solving a project, but they do it in different ways.

For intelligent analysis of student solutions, a set of program tests with inference rules can be assigned to every task. The program tests are used to decide whether or not the solution of the task is correct, and to find out what exactly is wrong or incomplete. The inference rules are used to identify which missing or incorrect knowledge may be responsible for the error and to update the student model when the student's solution of the project is incorrect.

Instead of waiting for the final solution of a task, the WAPE system can provide a student with intelligent help for each step of problem solving. The level of help can vary: from signaling about a wrong understanding of the statement of the task, to giving a hint, to executing the next step for the student. The system can watch some actions of the student, understand them, and use this understanding to provide help and to update the student model.

45.7 Bayesian Network

Bayesian networks are useful tools for inferring in graphs with dependent vertices. A Bayesian network is a directed acyclic graph, such that the following properties hold:

- each node in the graph represents a random variable;
- there is an edge from a node p to another node q, whenever q is dependent on p;
- each node is labeled with a conditional probability table that quantifies the effect of its predecessors.

We use such a Bayesian network to calculate a probability distribution for each concept, and thus for calculating a knowledge vector of a student.

Bayesian networks are very useful in user modeling because they enable us to manage uncertainty in our observations and our conclusions.

To construct a Bayesian network which calculates the probability distribution for each concept of a course for a particular student, there are two main steps to take:

The first step is generating some acyclic graph which contains the knowledge items as nodes and the immediate learning dependencies between them as edges. There is an edge from p to q in the graph if $p < q$ and there exists no w with $p < w < q$.

The second step is defining probability tables for all the nodes. After generating the directed acyclic graph from the dependency graph, we have to add for each node of the graph a probability table containing the conditional probabilities that a successor node has with respect to its predecessor node.

Because we have already used conditional probabilities for describing the student's knowledge, it is obvious to look for inferring mechanisms which allow us to handle networks with dependent random variables.

It is known that exact inference in Bayesian networks is NP-hard [12]. Indeed, a general Bayesian network can represent any propositional logic problem (if all probabilities are 1 or 0), and propositional logic problems are known to be NP-hard [13].

Linear time algorithms exist for Bayesian networks which have no cycles in the underlying undirected graph. There are several methods to deal with such not continuously directed cycles: clustering, conditioning, and stochastic simulation algorithms.

Clustering algorithms glue two or more nodes together to avoid not continuously directed cycles (see for example [5, 14]).

Conditioning methods transform the network into several simpler networks (see for example [15, 6]). Each of these networks contains one or more of the random variables instantiated by one of their values.

Stochastic simulation methods repeatedly run simulations of the network for calculating approximations of the exact evaluation (see for example [16, 17]).

45.8 Testing

Tests are problems that the system uses for testing a student's conceptual knowledge concerning theoretical material learned. There are three types of tests: *single-choice* tests, *multiple-choice* tests and *textual* tests. In contrast with choice tests that present sets of alternative answers, every textual test assumes some text as a correct answer.

We use computer-based testing with random positioning of answers for single-choice and multiple-choice tests, so the position number can mislead students if they learn the position numbers by heart. Thus, instead of memorizing the questions and the line numbers of the answer options, students are forced to learn the relations between questions and answers, i.e., concepts and other skills. Therefore the approach used in our system includes the random generation of tests in a given area (see test space, below) and random positions of the answer options. Each test measures the verbal, quantitative, and analytical skills related to a specific field of the course studied. A different time constraint is associated with each question. We distinguish three classes of tests: *verbal, quantitative,* and *analytical* questions.

A *verbal* question defines a specific concept and has a time constraint of 60 s. The verbal exercise tests the ability to analyse and evaluate written material and to synthesize information obtained from it, to analyse relationships among component parts of sentences, and to recognize relationships between words and concepts.

In the case of *quantitative* questions, where a more sophisticated concept needs to be explained, an answer is usually expected in 120–240 s. The quantitative exercise tests basic skills and understanding of elementary concepts, as well as the ability to reason quantitatively and to solve problems in a quantitative setting, or to explain more sophisticated concepts.

For an *analytical* question, in which a rather difficult concept has to be explained, the answer should be given within a time constraint of 360–480 s. The analytical exercise tests the ability of the student to understand structured sets of relationships, deduce new information from sets of relationships, analyse and evaluate arguments, identify central issues and hypotheses, draw sound inferences, and identify plausible causal explanations. Questions in the analytical section usually measure the reasoning skills developed in almost all fields of the course studied.

All tests which aim to check the student's knowledge relative to the same concept are grouped into a so-called *test space*. The test space for a concept *s* is a directed acyclic graph (DAG) whose nodes are tests related to *s*, and any edge between nodes represents the possibility of generating one test just after another. Nodes without predecessors are called *input* tests of the test space, and nodes without successors are called its *output* tests. The test space is used for random generation of a sequence of tests for concept *s* as a path from an input test to an output test.

45.9 Conclusion

In this paper the project WAPE, under development at the Institute of Informatics Systems, has been presented. The WAPE system is intended to be an intelligent and adaptive virtual environment for the distance learning of programming. Adaptive presentation can improve the usability of the course material. Adaptive navigation support and adaptive sequencing can be used for overall course control and for helping the student in selecting the most relevant tests and assignments. Problem-solving support and intelligent solution analysis can significantly improve the work by means of tasks providing both interactivity and intelligent feedback, while taking a serious grading load from the teachers' shoulders.

One of the main issues in the development of an advanced technology learning environment is a gap between pedagogues and technicians. The WAPE project is aimed to overcome the gap. Lecturers without programming skills will be able to create adaptive educational hypermedia courses supported by the WAPE system.

At present the WAPE system includes an introductory course on programming based on the Pascal language [18], and an introductory course on programming for the Zonnon language is under development. Zonnon [19] is a new programming language in the Pascal, Modula-2, and Oberon family. It retains an emphasis on simplicity, clear syntax and separation of concerns. Although rather smaller than languages such as C#, Java, and Ada, it is a general-purpose language suited to a wide range of applications. Typically this includes component-oriented composition, concurrent systems, algorithms and data structures, object-oriented and structured programming, graphics, mathematical programming, and low-level systems programming. Zonnon provides a rich object model and may be used to write programs in traditional or object-oriented styles. Zonnon is also well suited for teaching purposes, from basic principles right through to advanced concepts.

Acknowledgments This work was partially supported by grants from the Ministry of Science and Education of Russia and from Microsoft Research Ltd.

References

1. Brusilovsky P (1999) Adaptive and intelligent technologies for web-based education. Kunstliche Intelligenz, Special Issue on Intelligent Systems and Teleteaching 4:19–25
2. Brusilovsky P (2001) Adaptive hypermedia. User Model User-Adapt Interact 11:87–110
3. De Bra P (2002) Adaptive educational hypermedia on the web. Commun ACM 45:60–61
4. De Bra P, Stash N (2002) AHA! A general-purpose tool for adaptive websites. Lect Notes Comput Sci 2347:381–384
5. Dechter R, Pearl J (1989) Tree clustering for constraint networks. Artif Intell 38:353–366
6. Horvitz E, Suermondt H, Cooper G (1989) Bounded conditioning: flexible inference for decisions under scarce resources. In: Fifth conference on uncertainty in artificial intelligence, Windsor
7. Kinshuk, Han B, Hong H, Ratel A (2001) Student adaptivity in TILE. In: IEEE international conference on advanced learning technologies. IEEE Computer Society, Washington, DC, pp 297–300
8. Kasyanov VN, Kasyanova EV (2004) Distance education: methods and tools of adaptive hypermedia. In: Program tools and mathematics foundations of informatics (In Russian). Novosibirsk, pp 80–141
9. Wu H, De Kort E, De Bra P (2001) Design issues for general-purpose adaptive hypermedia systems. In: Proceedings of the ACM conference on hypertext and hypermedia. Aarhus, pp 141–150
10. Kasyanov VN, Kasyanova EV (2003) An environment for web-based education of programming. In: HCI International 2003. Proceedings of the 10th international conference on human-computer interaction. Crete University Press, Heraklion, pp 179–180
11. Kasyanova EV (2007) Adaptive methods and tools for support of distance education in programming (In Russian). IIS Press, Novosibirsk
12. Cooper G (1990) The computational complexity of probabilistic inference using Bayesian belief networks. Artif Intell 42:393–405
13. Russell S, Norvig P (1955) Artificial intelligence: A modern approach. Prentice-Hall, Upper Saddle River, NJ
14. Lauritzen S, Spielgelhalter D (1988) Local computations with probabilities on graphical structures and their application to expert systems. J R Stat Soc B 50:157–224
15. Dechter R (1989) Enhancement schemes for constraint processing: backjumping, learning and cutset decomposition. Artif Intell 41:273–312
16. Henrion M (1988) Propagation of uncertainty in Bayesian networks by probabilistic logic sampling. Uncertain Artif Intell 2:149–163
17. Shachter R, Peot M (1989) Simulation approaches to general probabilistic inference on belief networks. In: Fifth conference on uncertainty in artificial intelligence, Windsor
18. Kasyanov VN (2001) An introductory course of programming in Pascal in tasks and exercises (In Russian). NSU Press, Novosibirsk
19. Gutknecht J, Zueff E (2003) Zonnon for .NET—a language and compiler experiment. Lect Notes Comput Sci 2789:132–143

Chapter 46
New Software for the Study of the Classic Surfaces from Differential Geometry

Anca Iordan, George Savii, Manuela Pănoiu, and Caius Pănoiu

Abstract Important changes are demanded by the informatics society as far as educational programs are concerned. Learning processes, programs, manuals, methods and organization forms of the didactic activities must be reconsidered taking into account novelties such as computer-assisted instruction and self instruction. This chapter presents a software package which can be used for educational purposes. A graphical user interface, implemented in Java, and useful for computer-based learning is presented in this chapter. It is a study about helical and revolution surfaces.

46.1 Introduction

Considering the current condition of informatics society, whose principal goal in socioeconomic development is to produce and consume information, the complex and fast awareness of the information in order to make rational, opportune and effective decisions is the desideratum. This generates the necessity to form a superior level of habituation in information management for the entire population. Computers and their programs have many powerful applications to offer to their users, as far as information manipulation is concerned: image and text visualization on the screen; memory storage of an important quantity of information, selectively accessing and processing data; conduction of a great volume of computation; equipment control and fast decisions; computer-based training [3].

A. Iordan (✉)
Engineering Faculty of Hunedoara, Technical University of Timişoara, Revolutiei 5, 331128 Hunedoara, Transylvania, Romania
e-mail: anca.iordan@fih.upt.ro

N. Mastorakis et al. (eds.), *Proceedings of the European*
Computing Conference, Lecture Notes in Electrical Engineering 27,
DOI 10.1007/978-0-387-84814-3_46, © Springer Science+Business Media, LLC 2009

46.2 Computer-Based Training as a Didactic Method

Many modifications in education programs are proposed by the informatics society. In this scope, the school should be able to provide the preparative knowledge for programmers, maintenance technicians, etc.

At the same time, it is necessary for the teacher to introduce the use of the computer in the education process. These informational techniques require the reorganization of the contents, programs, course books and manuals of the education process, the reconsideration of the methods and organization forms of the didactic activities, which are central to individualizing the teaching process [2].

As a method of informational didactic, computer-based training is based on the programmed teaching. N. Crowder has been working on a new programming type – the branch programming – which is characterized by the content's division into smaller units. His successive presentation is tailored according to the student's needs and corrective feedback, and use of author language.

The programmed teaching consists of a presentation of the information in small units, logically structured, units that compose a program, the teaching program [3]. After each section the user can review the information and challenge his/her comprehension of the material. The programmed teaching method organizes the didactic action applying the cybernetic principles to the teaching-learning-evaluating activities level, considering them to be a complex and dynamic system, composed as an elements ensemble. The programmed teaching method is developing its personal principles valid on the strategic level in any cybernetic organization form of teaching.

On the other hand, programmed teaching assumes some principles which the teaching program must respect [4]:

- The small steps principle consists of progressive penetration, from simple to complex, in a subject content which is logically divided in simple units. This unit series leads to minimal knowledge, which will later form an ensemble. This principle of subject division in contents/ information units gives the user the chance to succeed in his/her teaching activities;
- The principle of personal rhythm of study regarding mannerism is observance and capitalization of each user of the program. The user will be able to manage and control the sequences of knowledge, in a personal rhythm appropriate to his/her psycho-intellectual development, without time limits. As the user accomplishes the respective sequence requirement, he/she can progress further in the program;
- The active participation principle, or active behavior, involves user effort, understanding and applying the necessary information to determine the correct answer. At each step the user must actively participate to resolve each step;

- The principle of inverse connection, with regard to positive or negative inputs of user competence, refers to the success or breakdown in the task performed;
- The immediate and direct control of the task constitutes precisely in the possibility to proceed to the next sequence, in case of success;
- The repetition principle, which is based on the fact that the programs are based on returning to the users' initial knowledge.

The combined programming interposes the linear and branch sequence according to the teaching necessities.

After linear and branch programming, the computer-aided generative teaching has appeared, where the exercises are gradually presented with different difficulty steps and answers on the student's questions.

Computer-based, programmed teaching realizes the learning process with an input flow – the command, an executive controlled system, an output flux – control and control system functions.

In such a system, three stages are present for a teacher to perceive: teaching, evaluating and the feedback loop closing; the computer is present in all three of the stages.

46.3 Application Present

This chapter presents a software package, which can be used as educational software to present helical and revolution surfaces in the course of differential geometry. The application is implemented in Java and Mathematica, under Microsoft Windows operating system. The graphical user interface consists of four parts:

- the theoretical presentation;
- the presentation of solved problems;
- the solution of representative types of problems;
- the evaluation of knowledge.

In the third part of the software entering the parametric equations of helical or revolution surface is enabled. One of the following options can be selected from the main menu of the application window:

- the determination of the first quadratic fundamental form of helical or revolution surface;
- the determination of the second quadratic fundamental form of helical or revolution surface;
- the determination of the asymptotic lines of helical or revolution surface;
- the determination of the lines of curvature of helical or revolution surface;
- the determination of the geodesic lines of helical or revolution surface;
- the graphic representation of helical or revolution surface.

46.3.1 Theoretical Presentation of Classic Surfaces

A surface of revolution is a surface generated by rotating a two-dimensional curve about an axis. The resulting surface, therefore, always has azimuthal symmetry [7]. The standard parameterization of a surface of revolution is given by:

$$\begin{cases} x(u, v) = \varphi(u) \cdot \cos\, v \\ y(u, v) = \varphi(u) \cdot \sin\, v \\ z(u, v) = \psi(u) . \end{cases} \qquad (46.1)$$

In order to generate a helical surface let us consider a curve having a kinematics composed by a rotational movement around a fixed axis Oz, characterized by angular velocity, and a simultaneous, translation movement parallel to Oz axis [5].

We can write the surface parametric equations as follows:

$$\begin{cases} x(u, v) = \varphi(u) \cdot \cos\, v \\ y(u, v) = \varphi(u) \cdot \sin\, v \\ z(u, v) = \psi(u) + k \cdot v. \end{cases} \qquad (46.2)$$

The general name for the quadratic differential forms of the surface is given in coordinates on the surface and satisfying the usual transformation laws under transformations of these coordinates. The fundamental forms of a surface characterize the basic intrinsic properties of the surface and the way it is located in space in a neighbourhood of a given point; one usually singles out the so-called first, second and third fundamental forms.

The first fundamental form characterizes the interior geometry of the surface in a neighbourhood of a given point [6]. This means that measurements on the surface can be carried out by its means.

The first quadratic fundamental form of helical surface is:

$$ds^2 = E(u, v) \cdot du^2 + 2 \cdot F(u, v) \cdot du \cdot dv + G(u, v) \cdot dv^2 \qquad (46.3)$$

where:

$$\begin{aligned} E(u, v) &= x_u'^2 + y_u'^2 + z_u'^2 \\ F(u, v) &= x_u' \cdot x_v' + y_u' \cdot y_v' + z_u' \cdot z_v' \\ G(u, v) &= x_v'^2 + y_v'^2 + z_v'^2 \end{aligned} \qquad (46.4)$$

The second fundamental form characterizes the local structure of the surface in a neighbourhood of a regular point. The second quadratic fundamental form of the helical surface is:

$$\frac{ds^2}{R_n} = D(u, v) \cdot du^2 + 2 \cdot D'(u, v) \cdot du \cdot dv + D''(u, v) \cdot dv^2 \qquad (46.5)$$

where:

$$D(u, v) = \frac{1}{H(u, v)} \left(\vec{r}_u', \vec{r}_v', \vec{r}_{u^2}'' \right)$$

$$D'(u, v) = \frac{1}{H(u, v)} \left(\vec{r}_u', \vec{r}_v', \vec{r}_{uv}'' \right)$$

$$(46.6)$$

$$D''(u, v) = \frac{1}{H(u, v)} \left(\vec{r}_u', \vec{r}_v', \vec{r}_{u^2}'' \right)$$

$$H(u, v) = \sqrt{E(u, v) \cdot G(u, v) - F^2(u, v)}.$$

46.3.2 Examples

Ellipsoid. The general ellipsoid is a quadratic surface which is given in cartesian coordinates by [6]:

$$\frac{x^2}{a^2} + \frac{y^2}{a^2} + \frac{z^2}{c^2} = 1. \tag{46.7}$$

The parametric equations of an ellipsoid can be written as:

$$x(u, v) = a \cdot \sin u \cdot \cos v$$
$$y(u, v) = a \cdot \sin u \cdot \sin v \tag{46.8}$$
$$z(u, v) = c \cdot \cos u$$

The first quadratic fundamental form of ellipsoid, as well as its graphic representation, is presented in Fig. 46.1.

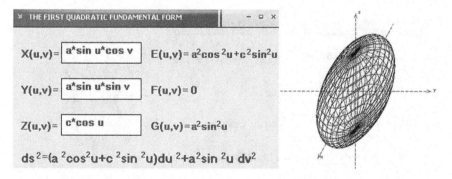

Fig. 46.1 The first fundamental form and the graphic representation of an ellipsoid

Hyperboloid. A hyperboloid is a quadratic surface which may be one-sheeted or two-sheeted [5]. The one-sheeted hyperboloid is a surface of revolution obtained by rotating a hyperbola about the perpendicular bisector to the line between the foci, while the two-sheeted hyperboloid is a surface of revolution obtained by rotating a hyperbola about the line joining the foci. The one-sheeted hyperboloid is given in cartesian coordinates by:

$$\frac{x^2}{a^2} + \frac{y^2}{a^2} - \frac{z^2}{c^2} = 1. \tag{46.9}$$

The parametric equations of a one-sheeted hyperboloid can be written as:

$$x(u, v) = a \cdot \sqrt{1 + u^2} \cdot \cos v$$
$$y(u, v) = a \cdot \sqrt{1 + u^2} \cdot \sin v \cdot \tag{46.10}$$
$$z(u, v) = c \cdot u$$

The first quadratic fundamental form of a one-sheeted hyperboloid, as well as its graphic representation, is presented in Fig. 46.2.

Paraboloid. A paraboloid is a quadratic surface which can be specified by the cartesian equation [6]:

$$z = a \cdot (x^2 + y^2). \tag{46.11}$$

The parametric equations of a paraboloid can be written as:

$$x(u, v) = a \cdot \sqrt{u} \cdot \cos v$$
$$y(u, v) = a \cdot \sqrt{u} \cdot \sin v \tag{46.12}$$
$$z(u, v) = a^3 \cdot u$$

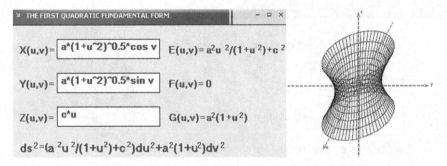

Fig. 46.2 The first fundamental form and the graphic representation of a hyperboloid

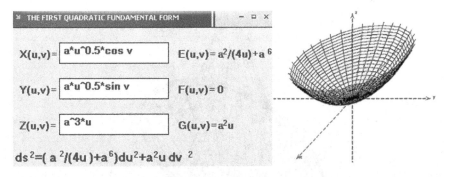

Fig. 46.3 The first fundamental form and the graphic representation of a paraboloid

The first quadratic fundamental form of a paraboloid, as well as its graphic representation, is presented in Fig. 46.3.

Pseudosphere. Half the surface of revolution generated by a tractrix about its asymptote forms a tractroid. The surfaces are sometimes called the antisphere or tractrisoid [5]. The cartesian parametric equations are:

$$x(u, v) = a \cdot \sin u \cdot \cos v$$
$$y(u, v) = a \cdot \sin u \cdot \sin v$$
$$z(u, v) = a\left(\ln tg\frac{u}{2} + \cos u\right)$$

(46.13)

The first quadratic fundamental form of a pseudosphere is presented in Fig. 46.4, and its graphic representation is given in Fig. 46.5.

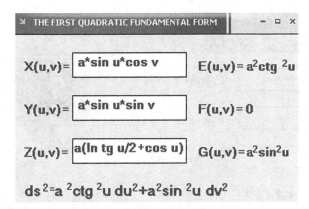

Fig. 46.4 The first fundamental form of a pseudosphere

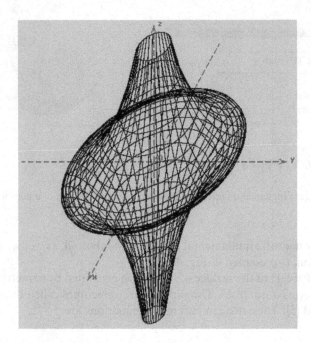

Fig. 46.5 The graphic representation of a pseudosphere

Helicoid. Let us consider the following equations of a surface which represent the helicoid:

$$x(u, v) = u \cdot \cos v; y(u, v) = u \cdot \sin v; z(u, v) = k \cdot v. \qquad (46.14)$$

The first quadratic fundamental form of a helicoid, as well as its graphic representation, is presented in Fig. 46.6.

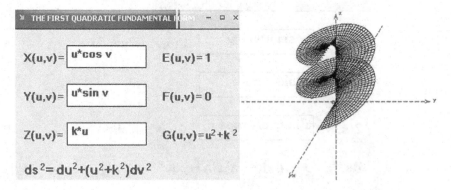

Fig. 46.6 The first fundamental form and the graphic representation of a helicoid

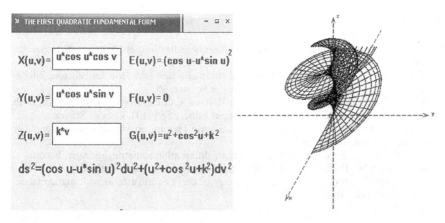

Fig. 46.7 The first fundamental form and the graphic representation of the previous surface

Let us consider the following equations of a helical surface:

$$x(u, v) = a \cdot \sin u \cdot \cos v$$
$$y(u, v) = a \cdot \cos u \cdot \sin v \qquad (46.15)$$
$$z(u, v) = k \cdot v.$$

The first quadratic fundamental form of the previous surface, as well as its graphic representation, is presented in Fig. 46.7.

46.4 Conclusion

In this application, the authors took into consideration the conditions, which a courseware must accomplish in order to make the necessary steps. So, for the elaboration and utilization of this application one must consider the following criteria:

- Follow up the curriculum for a specific domain;
- Plan some teaching and learning strategy. In this kind of self-instruction and evaluation program, basic representation and scanning notions must be found. Animation and graphic modeling must represent the graphical construction way and scanning of them;
- Give the opportunity to use parameterized variables (User-defined variables condition);
- Present a method informing the user how he/she can use the graphical module, i.e. a user-computer interaction.

The presented application accomplishes these criteria. Thus, we consider it to be a good example of how educational software must be realized.

References

1. Nicaud J, Vivet M (1998) Les Tuteurs Intelligents réalizations et tendances de recherché. TSI
2. McDougall A, Squires A (1995) Empirical study of a new paradigm for choosing educational software. Computer Education, Elsevier Science, 25.
3. Scalon E, Tosunoglu C, Jones A, Butcher P, Ross S, Greenberg J (1998) Learning with computers: experiences of evaluation. Comput Educ, 25:93–103, Elsevier Science.
4. Tridade J, Fiolhais C, Almeida L (2002) Science learning in virtual environments. Br J Educ Technol, 33:477–488.
5. Blaschke W, Leichtweiss K (1975) Elementare differentialgeometrie. Springer, Berlin.
6. Klingenberg W (1978) A course in differential geometry. Springer, Berlin.
7. Gray A (1997) Modern differential geometry of curves and surfaces with mathematica. CRC Press, Boca Rotan, FL.

Chapter 47
Data Requirements for Detecting Student Learning Style in an AEHS

Elvira Popescu, Philippe Trigano, and Mircea Preda

Abstract According to educational psychologists, the efficiency and effectiveness of the instruction process are influenced by the learning style of the student. This paper introduces an approach which integrates the most relevant characteristics from several learning style models proposed in the literature. The paper's focus is on the data requirements for the dynamic learner modeling method, i.e., the learner's observable behavior in an educational hypermedia system; based on the interpretation of these observables, the system can identify various learning preferences.

47.1 Introduction

The ultimate goal of adaptive educational hypermedia systems (AEHS) is to provide a learning experience that is individualized to the particular needs of the learner from the point of view of knowledge level, goals, or motivation. Lately, learning styles have also started to be taken into consideration, given their importance in educational psychology [9, 10].

According to [6], learning style designates a combination of cognitive, affective, and other psychological characteristics that serve as relatively stable indicators of the way a learner perceives, interacts with, and responds to the learning environment. During the past two decades, educational psychologists have proposed numerous learning style models, which differ in the learning theories they are based on, and the number and description of the dimensions they include.

Most of today's AEHS that deal with learning styles are based on a single learning style model, such as Felder-Silverman [4] (used in [1, 2, 3]), Honey and Mumford [5] (used in [7]), and Witkin [12] (used in [11]). We take a different approach by characterizing the student by a set of learning preferences, which we included in a unified learning style model (ULSM) [8], rather than directly by a particular learning style. An overview of this ULSM and its advantages are

E. Popescu (✉)
Software Engineering Department, University of Craiova, Romania; Heudiasyc UMR CNRS 6599, Université de Technologie de Compiègne, France

N. Mastorakis et al. (eds.), *Proceedings of the European Computing Conference*, Lecture Notes in Electrical Engineering 27, DOI 10.1007/978-0-387-84814-3_47, © Springer Science+Business Media, LLC 2009

presented in the next section. Subsequently, Section 47.3 details the data requirements for the detection of student learning preferences in an AEHS. Finally, in
Section 47.4 we draw some conclusions, pointing towards future research directions.

47.2 The Unified Learning Style Model Approach

In [8, 10] we introduced an implicit, dynamic learner modeling method, based on
monitoring the students' interactions with the system and analyzing their behavior.
The novelty of our approach lies in the use of a unified learning style model, which
integrates characteristics from several models proposed in the literature. Moreover,
it includes e-learning specific aspects (technology-related preferences), and it is
stored as a set of learning characteristics, not as a stereotyping model. More
specifically, ULSM integrates learning preferences related to: perception modality
(visual vs. verbal), field dependence/field independence, processing information
(abstract concepts and generalizations vs. concrete, practical examples; serial vs.
holistic; active experimentation vs. reflective observation, careful vs. not careful
with details), reasoning (deductive vs. inductive), organizing information (synthesis
vs. analysis), motivation (intrinsic vs. extrinsic; deep vs. surface vs. strategic vs.
resistant approach), persistence (high vs. low), pacing (concentrate on one task at a
time vs. alternate tasks and subjects), social aspects (individual work vs. team
work; introversion vs. extraversion; competitive vs. collaborative), study organization (formal vs. informal), and coordinating instance (affectivity vs. thinking).

The advantages of this approach include: (i) it solves the problems related to
the multitude of learning style models, the concept overlapping, and the correlations between learning style dimensions; (ii) it removes the limitation imposed
by traditional learning in the number of learning style dimensions that can be
taken into consideration in face-to-face instruction; and (iii) it provides a

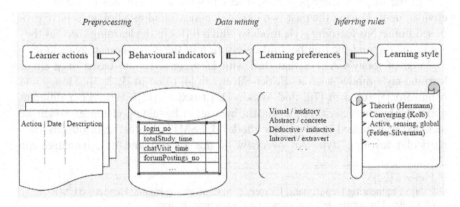

Fig. 47.1 Schematic representation of the proposed approach

simplified and more accurate student categorization (*characteristic-level modeling*) which in turn allows for a finer granularity of adaptation actions.

Obviously, this characteristic-level modeling does not exclude the use of traditional learning style models. Indeed, starting from the identified learning preferences on the one hand and the description of the desired learning style model on the other, the system can easily infer the specific categorization of the student. Our approach thus provides the additional advantage of not being tied to a particular learning style model. A schematic representation of the mechanism is depicted in Fig. 47.1.

47.3 Data Requirements for Detecting Learning Preferences

The first step towards dynamic learner modeling is to track and monitor student interactions with the system. Learner observable behavior in an educational hypermedia system includes: (i) navigational indicators (number of hits on educational resources, navigation pattern); (ii) temporal indicators (time spent on different types of educational resources proposed); and (iii) performance indicators (total learner attempts on exercises, assessment tests). Based on the interpretation of these observables, the system can infer different learning preferences. Table 47.1 summarizes some student actions that can be used to

Table 47.1 Examples of student actions that could be used as indicators of a particular learning preference

Learning preference	Behavioral indicators
Perception modality	
Visual preference (see)	High amount of time spent on content with graphics, images, video
	High performance in questions related to graphics
Verbal preference (read/ write)	High amount of time spent reading text
	High performance on questions related to written text
	High number of visits/postings in forum/chat
Verbal preference (hear)	High amount of time spent on text and audio content
	High participation in audio conferences
Processing information	
Abstract concepts and generalizations	Access of abstract content first (concepts, definitions)
	High amount of time spent on abstract content
	High performance on questions regarding theories
Concrete, practical examples	Access of concrete content first (examples)
	High amount of time spent on concrete content
	High performance on questions regarding facts
Serial	Linear navigation (intensive use of Next/Previous buttons)
	Infrequent access of additional explanations (related concepts)
Holistic	Nonlinear navigation pattern (frequent page jumps)
	High amount of time spent on outlines, summaries, table of contents
	Frequent access of additional explanations (related concepts)

Table 47.1 (continued)

Learning preference	Behavioral indicators
	High performance on questions related to overview of concepts and connections between concepts
Active Experimentation	Access of practical content (simulations, exercises, problems) before theory
	High number of accesses to exercises
	High amount of time spent on simulations and exercises
Reflective Observation	Access of theoretical content before practical content
	Higher time spent on reading the material than on solving exercises or trying out simulations
Reasoning	
Deductive	Access of abstract content first (concepts, definitions)
	High performance on exercises requiring direct application of theory
Inductive	Access of concrete content first (examples)
	High performance on exercises requiring generalizations
Organizing information	
Synthetic	High performance on exercises requiring synthesis competency
	Breadth first navigation pattern
Analytic	High performance on exercises requiring analysis competency
	Depth first navigation pattern
Persistence	
High persistence	High amount of time spent on studying
	High number of test retakes
	High number of returns to educational material
Low persistence	Low number of test retakes correlated with low number of returns to educational material
	Frequent use of hints and answer keys
Pacing	
Concentrate on one task at a time	Low number of web browsers opened at a time
	Linear navigation path (few jumps and returns)
Alternate tasks and subjects	Frequent passages from one section of the course to another (educational material, communication tools, tests) and from one course to another
	High number of web browsers opened at a time and frequent passages between them
	High nonlinearity degree of the navigation path
Social aspects	
Introversion	Passive participation in communication channels
	Higher number of visits/postings in forum versus chat
Extraversion	Active participation in synchronous communication channels (chat, audio conference, etc.)
Individual work	Choice of individual assignments
	Infrequent use of ask/offer peer help facility
Team work	Choice of group assignments
	Frequent use of ask/offer peer help facility
	High number of visits/postings in forum/chat

identify the characteristics of the ULSM discussed in the previous section, as a result of various researches reported in the literature. As can be seen, the main behavioral indicators refer to the relative frequency of learner actions, the amount of time spent on a specific action type, and the order of navigation, all of which can be obtained from the system log, either directly or after some preprocessing (see also Fig. 47.1).

47.4 Conclusions

Accommodating learning styles in AEHS represents an important step towards providing individualized instruction. In this note we presented an approach which is not tied to a particular learning style model, but instead integrates the most relevant characteristics from several models. We showed how these characteristics can be identified from monitoring and analyzing learner behavior in the system, by providing examples of student actions that can be interpreted as indicators for a particular learning preference.

Obviously, the validity of this interpretation can only be assessed through experimental research, which is the subject of our future work. Currently, an AEHS based on the proposed approach is under development and a course module is being implemented in order to conduct experiments.

Acknowledgments This research has been funded partly by the Romanian Ministry of Education and Research, the National University Research Council, under grants CNCSIS TD 169/2007 and CNCSIS AT 102/2007.

References

1. Bajraktarevic N, Hall W, Fullick P (2003) Incorporating learning styles in hypermedia environment: empirical evaluation. In: Proceedings of the workshop on adaptive hypermedia and adaptive web-based systems, pp 41–52
2. Carver CA, Howard RA, Lane WD (1999) Enhancing student learning through hypermedia courseware and incorporation of student learning styles. IEEE T Educ 42(1):33–38
3. Cha HJ, Kim YS, Park SH et al (2006) Learning styles diagnosis based on user interface behaviors for the customization of learning interfaces in an intelligent tutoring system. In: Proceedings of the ITS 2006, LNCS, vol 4053. Springer, New York
4. Felder RM, Silverman LK (1988) Learning and teaching styles in engineering education, Eng Educ 78(7)
5. Honey P, Mumford A (2000) The learning styles helper's guide. Peter Honey Publications Ltd, Maidenhead, UK
6. Keefe JW (1979) Learning style: an overview. NASSP's student learning styles: diagnosing and prescribing programs, pp 1–17
7. Papanikolaou KA, Grigoriadou M, Kornilakis H, Magoulas GD (2003) Personalizing the interaction in a web-based educational hypermedia system: the case of INSPIRE. User Model User-Adap 13:213–267

8. Popescu E, Trigano P, Badica C (2007) Towards a unified learning style model in adaptive educational systems. In: Proceedings of the ICALT 2007. IEEE Computer Society Press, Washington, DC, pp 804–808

9. Popescu E, Trigano P, Badica C (2007) Evaluation of a learning management system for adaptivity purposes. In: Proceedings of the ICCGI'2007. IEEE Computer Society Press, Washington, DC, pp 9.1–9.6

10. Popescu E, Trigano P, Badica C (2007) Adaptive educational hypermedia systems: a focus on learning styles. In: Proceedings of the 4th IEEE international conference on the computer as a tool, EUROCON 2007. IEEE Press, Piscataway, NJ, pp 2473–2478

11. Triantafillou E, Pomportsis A, Demetriadis S (2003) The design and the formative evaluation of an adaptive educational system based on cognitive styles. Comput Educ 41:87–103

12. Witkin HA (1962) Psychological differentiation: studies of development. Wiley, New York

Chapter 48
Ontology-Based Feedback E-Learning System for Mobile Computing

Ahmed Sameh

Abstract An E-Learning system that provides vast quantities of annotated resources (fragments or learning objects) and produces semantically rich feedback is very desirable. It is an accepted psychological principle that some of the essential elements needed for effective learning are custom learning and semantic feedback. In this paper we are making use of a collection (ontology) of metadata for the design of a custom E-Learning system that also provides learners with effective semantic educational feedback support. The learning domain is "mobile computing." We define various concepts in the domain and the relationships among them as the ontology, and build a system to utilize them in customizing the E-Learning process. The ontology is also used to provide informative feedback from the system during learning and/or during assessment. The focus in this research is on the representation of ontology using languages/grammars, grammar analysis techniques, algorithms, and AI mechanisms to customize the learning and create effective feedbacks. The proposed mechanisms, based on ontology, are used to assemble virtual courses and create a rich supply of feedbacks, not only in assessment situations but also in the context of design-oriented education. We are targeting feedbacks similar to ones in programming environments and design editors.

48.1 Virtual Courses

The proposed system aims to expose vast quantities of annotated resources (fragments or learning objects) that have been distributed over time and space by educators and instructional designers to data-mining end users in order for the latter to assemble sequences of "learning objects" (virtual classes). Towards this goal we are proposing an ontology-based feedback model to achieve a number of

A. Sameh (✉)
Department of Computer Science, The American University in Cairo, 113 Kaser Al Aini Street, P.O. Box 2511, Cairo, Egypt
e-mail: sameh@aucegypt.edu

N. Mastorakis et al. (eds.), *Proceedings of the European Computing Conference*, Lecture Notes in Electrical Engineering 27, DOI 10.1007/978-0-387-84814-3_48, © Springer Science+Business Media, LLC 2009

high-level objectives: dynamically generating on-demand virtual courses/services, providing component-based fragments, and facilitating rich semantic feedbacks.

An organization repository is a repository of course components (fragments) at various levels of detail. As such, these fragments can be reused for several courses and contexts and can be distributed over a number of sites. A virtual course authoring process points to various learning components (fragments). A learning fragment is a self-contained, modular piece of course material. It can be either passive or active (e.g., a live or recorded lecture). These fragments are annotated (for example, by the author) according to an ontology metadata schema that provides efficient mechanisms to retrieve fragments with respect to the specific needs of a virtual course. With similar details, administrators cooperate in building various service repositories. This follows the new direction in what is called the "semantic web" [3].

An organization's local architecture consists of repositories, a mediator, concept storage fragments, storage systems, and clients. The mediator provides transparent access for all client requests to all distributed repositories fragments. The huge amount of metadata we have to deal with indicates the need for efficient storage and query mechanisms. The repositories can be accessed either synchronously or asynchronously. Customers can access their composed virtual courses/services either online or offline, in distance and/or conventional education.

For example, specific course instructors in an organization like AUC can join their efforts to build a course repository. In the computer science department, a CSCI106 repository would contain all the CSCI106 fragments. Actually, CSCI106 (Introduction to Computer Science) is offered in at least 6–8 sections every semester and is taught by at least that many full- and part-time instructors. The CSCI106 repository fragment material would come from these instructors. A CSCI106 repository moderator would be one of the instructors, who takes care of fragmenting the deposited material, annotating it, and maintaining the currency of the repository by updating it with new fragments all the time. The repository would also contain all material similar to that which can be found in professional accreditation reports. This would include, for example, three samples of the best, average, and worst student answer sheets for each midterm and final exam for the last x semesters; also samples of projects, reports, assignments, etc. Other accreditation material such as course objectives, and the course's role in achieving the total degree objectives, etc., would also be included as separate fragments within every course repository. In fact, we see a great gain for students in having access to such accreditation material. They will have a better understanding of their course objectives, and where they fit in the global scheme of their target degree. In fact, repositories are living things; they are updated and incremented. For example, embedded student comprehension and track performance fragments can be included; also content development and management tools, as well as TA online sessions.

The development of electronic course material suitable for different learners takes much effort and incurs high costs. Furthermore, professional trainers have huge expenses in order to keep course content up to date. This problem occurs especially in areas where knowledge and skills change rapidly, as in the computer science domain. Thus, we need new approaches to support *(semi-)automatic virtual* course generation in order to keep up with current knowledge and perhaps even to adapt materials to individual user needs. In this proposal *repositories* are used as infrastructure to support the automatic generation of virtual courses from similar course repositories, while also aiming to achieve a high degree of reusability for course content. A popular, promising approach is to dynamically compose virtual courses "on-demand" within the course repositories. The idea is to segment existing course material (e.g., slides, text books, animations, videos) into so-called learning fragments. Such learning fragments represent typically self-contained units that are appropriately annotated with metadata.

48.2 Semantic Feedbacks

In a classroom, learners and teachers can easily interact, i.e., students can freely ask questions and teachers usually know whether their students understand (basic) concepts or problem-solving techniques. Feedback is an important component of this interaction. Furthermore, educational material can be continually improved using information from the interaction between the lecturer and the learners, which results in a more efficient and effective method of course development.

Feedback can be given to authors during virtual course development and to learners during learning. In the current generation of E-Learning systems, automatically produced feedback is sparse, mostly hard-coded, not very valuable, and used almost only in question-answer situations. In this paper we are introducing mechanisms—based on ontologisms—to create a rich supply of feedback, not only in question-answer situations but also in the context of virtual course composition. Ontologisms are formal descriptions of shared knowledge in a domain. With ontologisms we are able to specify (1) the knowledge to be learned (domain fragments and task knowledge) and (2) how the knowledge should be learned (education). In combining instances of these two types of ontologisms, we hope that we (1) are able to create (semi-) automatically valuable generic feedback to learners during learning and to authors during virtual course development, and (2) are able to provide the authors with mechanisms to (easily) define domain- and task-specific feedback to learners.

Feedback describes any communication or procedure given to inform a learner of the accuracy of a response, usually to an instructional question. More generally, feedback allows the comparison of actual performance with some standard level of performance. In technology-assisted instruction, it is information presented to the learner after any input, with the purpose of shaping the perceptions of the learner. Information presented via feedback in instruction might include not only answer correctness, but also other information such as precision, timeliness, learning guidance, motivational messages, background material, sequence advisement, critical comparisons, and learning focus. Feedback is given in the form of hints and recommendations. Both a domain conceptual/structural ontology as well as a task/design ontology is used. The ontologisms are enriched with axioms, and on the basis of the axioms, messages of various kinds can be generated when authors violate certain specified constraints.

In our research we are generating generic, domain, and task feedback mechanisms that produce semantically rich feedback to learners and authors. We distinguish two types of feedback: (1) feedback given to a student during learning, which we call *student feedback*, and (2) feedback given to an author during course authoring, which we call *author feedback*. The generic feedback mechanisms use ontologisms as arguments of the feedback engine. This is important, because the development of feedback mechanisms is time-consuming and specialized work, and can be reused for different ontologisms. Besides generic feedback mechanisms, we will provide mechanisms by means of which authors can add more domain and/or task specific feedback. In this research, we focus on the "mobile computing" domain.

We designed an E-Learning environment for mobile computing courses, in which: (1) learners are able to design artifacts of certain domains using different types of languages, and (2) authors are able to develop virtual courses. Learners as well as authors receive semantically rich feedback during learning, designing artifacts, and developing virtual courses. For example, a student first has to learn the concept (communication) network. Assume that a network consists of links, nodes, a protocol, and a protocol driver. Each of these concepts consists of sub-concepts. The domain ontology "communication technology" represents these in terms of a vocabulary of concepts and a description of the relations between the concepts (see Figs. 48.1, 48.2, and 48.3). On the basis of an

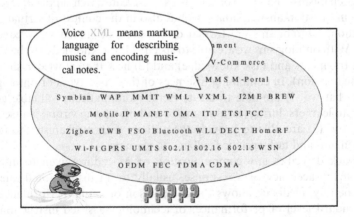

Fig. 48.1 Part of the domain concept ontology of "mobile computing" in acronym terminology with semantic feedback (by clicking on an acronym you get the balloon feedback with colored links to further explanation)

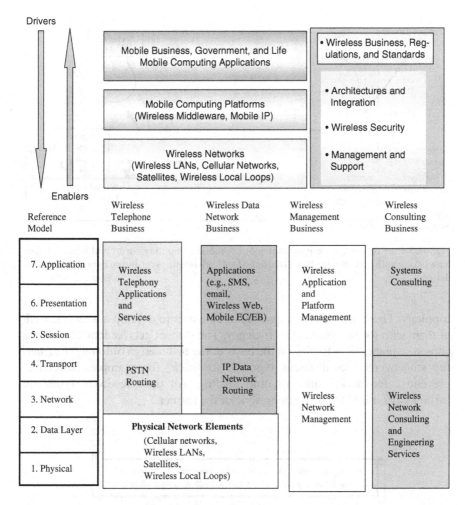

Fig. 48.2 Part of the domain structural ontology of "mobile computing" in framework building blocks with semantic feedback (by clicking a framework block you get the star feedback)

education ontology, which describes the learning tasks, the student is asked to list the concepts and relate the concepts to each other (see Fig. 48.1). Feedback is given about the completeness and correctness of the list of concepts and relations using different balloon dialog patterns.

In a second step the learner is asked to design a part of a local area network (LAN) using the network model developed during the first step (see Figs. 48.4 and 48.5). Instead of concepts, concrete instantiations must be chosen and related to each other. The learner gets feedback about the correctness of the instantiations and the relations between the concepts using different star/lamb/ scroll dialog patterns. Some protocols for example need a specific network

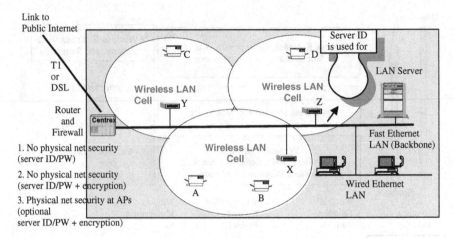

Fig. 48.3 Part of domain design ontology of "mobile computing" in hybrid wired/wireless networking with semantic feedback (the lamb feedback is always on during hybrid networking design)

topology. There are various sequences of activities to develop a network, each of them with its own particular efficiency. The student gets feedback about the chosen sequence of activities on the basis of the task/design ontology. Further, the student receives different types of feedback, for example, corrective/preventive feedback, criticisms, and guiding. All these feedback types are further customized to the learning style of the learner.

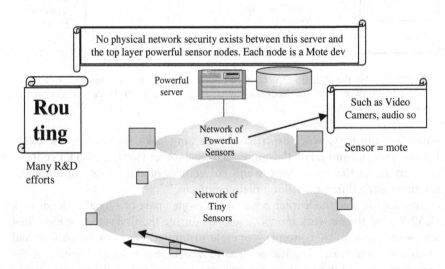

Fig. 48.4 Part of domain task ontology of "mobile computing" in sensor networking with semantic feedback (by clicking on any element you will get the scroll feedback)

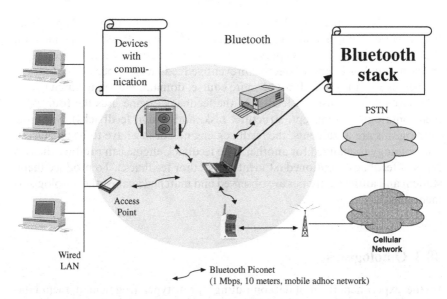

Fig. 48.5 Part of the domain design ontology of the "mobile computing" in Bluetooth networking with semantic feedback (the scroll feedback is always on during a Bluetooth network design)

An author develops and optimizes a virtual course from learning fragments. He/she has to choose, develop, and/or adapt particular ontologisms and develop related fragmented material like examples, definitions, etc. (see Fig. 48.1). Based on analyses of the domain, education, and feedback ontologisms, the author gets feedback, for example, about: (1) Completeness: A concept can be used but not defined. Ideally, every concept is introduced somewhere in the course, unless stated otherwise already at the start of the course. This error can also occur in the ontology for the course. (2) Timeliness: A concept can be used before its definition. This might not be an error if the author uses a top-down approach rather than a bottom-up approach to teaching, but issuing a warning is probably helpful. Furthermore, if there is a large distance (measured, for example, in number of pages, characters, or concepts) between the use of a concept and its definition in the top-down approach, this is probably an error. (3) Synonyms: Concepts with different names may have exactly the same definition. (4) Homonyms: A concept may have multiple, different definitions, sometimes valid, depending on the context.

The E-Learning environment consists of four main components: a player for the student, an authoring tool, a feedback engine, and a set of ontologisms as pluggable components (see Fig. 48.1). The player consists of a design and learning environment in which a student can learn concepts, construct artifacts, and solve problems. The authoring tool consists of an authoring environment where the author develops and maintains courses and course-related materials like ontologisms, examples, and feedback patterns. The feedback engine automatically produces feedback to students as well as to authors. It produces both generic feedback and specific feedback. Generic feedback is dependent on the ontologisms used and is applicable to all design activities and artifacts (e.g., criticism, guidance, and corrective/preventive feedbacks). Specific feedback is defined by the author and can be more course, domain, modeling language, or task specific. To construct feedback, the feedback engine uses the four argument ontologisms (concept, structure, task, and design feedbacks). Since the ontologisms are arguments, the feedback engine doesn't have to be changed if one ontology is changed for another. The feedback engine can produce the two types of feedback mentioned (student and author feedback). To produce them, student and author activities are observed and matched against the ontologisms mentioned.

48.3 Ontologisms

In the experimental "mobile computing" prototype, fragmented metadata-based repositories make use of four standards—resource description framework (RDF), IEEE LOM metadata, learning material markup language (LMML), and XML-based metadata—to build the prototype. The proposed prototype provides gateways among these standards. RDF/RDFS [4] is deployed as one of the underlying modeling languages to express information about the learning objects (fragments or components) contained in the repository, as well as information about the relationships between these learning objects (the ontologisms).

The mobile computing open learning repositories provide metadata-based course portals (also called virtual courses), which structure and connect modularized course materials over the Web. The modular content is distributed anywhere on the Internet, and is integrated by explicit metadata information in order to build virtual courses and connected sets of learning fragment materials. Modules can be reused for other courses and in other contexts, leading to a course portal that integrates modules from different sources, authors, and standards. Semantic annotation is necessary for authors to help them choose modules and to connect them into course structures. Each repository can, for example, store RDF (resource description framework) metadata from arbitrary RDF schemas. Initial loading for a specific virtual course is done by importing an RDF metadata file (using XML syntax, for example) based on this course's RDFS schema. A simple cataloguing (annotation) of its fragments can be deployed using the Dublin Core metadata set [4]. We can also port these metadata to the LOM standard, using the recently developed LOM-RDF binding [4]. With RDF, we can describe for our purposes how modules, course units, and courselets are related to each other, or which examples or exercises belong to a course unit. RDF metadata used in this way are called structural or relational metadata.

The IEEE LOM metadata standard specifies the syntax and semantics of learning object metadata, defined as the attributes required to fully/adequately describe a learning object. A learning object is defined here as any entity, digital or nondigital, which can be used, reused or referenced during technology-supported learning. Examples of learning objects include multimedia content, instructional content, learning objectives, instructional software and software tools, and persons, organizations, or events referenced during technology-supported learning. The learning object metadata standard focuses on the minimal set of attributes needed to allow these learning objects to be managed, located, and evaluated. The standard accommodates the ability to locally extend the basic fields and entity types, and the fields can have a status of obligatory (must be present) or optional (may be absent). Relevant attributes of learning objects to be described include type of object, author, owner, terms of distribution, and format. Where applicable, learning object metadata may also include pedagogical attributes such as teaching or interaction style, grade level, mastery level, and prerequisites. It is possible for any given learning object to have more than one set of learning object metadata.

LMML [4] proved itself as a pioneer in this field providing component-based development, cooperative creation and re-utilization, as well as personalization and adaptation of E-Learning contents. Considering both economic aspects and the aim to maximize the success of learning, LMML meets the new requirements of supporting the process of creation and maintenance for E-Learning contents, as well as supporting the learning process itself. Each E-Learning application has its own specific requirements structuring its contents.

The ARIADNE [4] knowledge pool standard is a distributed repository for learning objects. It encourages the share and reuse of such objects. An indexing and query tool uses a set of metadata elements to describe and enable search functionality on learning objects.

48.4   Mobile Computing Prototype

We have built a prototype system in the domain of mobile computing to demonstrate the ideas of the proposed model. Hybrid fragments of annotated resources in the domain are used in this prototype. Fragments are encoded in the four representations described above: Dublin cores, IEEE LOM, learning material markup language (LMML), and ARIANE. Figure 48.1 shows how the underlying ontologisms are used in the prototype to build virtual courses.

Figures 48.1, 48.2, 48.3, 48.4, and 48.5 show snapshots of the implemented prototype. Figure 48.1 shows part of the domain concept ontology of "mobile computing" in acronyms terminology with semantic feedback (by clicking on an Akron you get the balloon feedback with colored links to further explanation). Figure 48.2 shows part of the domain structural ontology of "mobile computing" in framework building blocks with semantic feedback (by clicking a framework block you get the star feedback). Figure 48.3 shows part of the domain structure ontology of the "mobile computing" in layering architecture with semantic feedback (by clicking on a layering element you get the balloon). Figure 48.4 shows part of the domain design ontology of "mobile computing" in hybrid wired/wireless networking with semantic feedback (the lamb feedback is always on during hybrid networking design). Figure 48.5 shows part of the domain task ontology of the "mobile computing" in sensor networking with semantic feedback (by clicking on any element you will get the scroll feedback).

48.5 Conclusion

In this paper, we have presented a flexible course/service generation environment to take advantage of reconceptualization and reutilization of learning/ service materials. Virtual courses/services are dynamically generated on demand from fragments' metadata entries stored in the repositories along with semantically powerful feedbacks.

References

1. http://www.mccombs.utexas.edu/kman/kmprin.html#hybrid
2. http://www.w3.org/2001/sw
3 http://www.indstate.edu/styles/tstyle.html
4. http://citeseer.nj.nec.com/context/1958738/0
5. Tendy SM, Geiser WF (1998) The search for style: it all depends on where you look. Natl Forum Teach Educ J 9(1)

Part VI
Information Communication Technologies

Information and communication technology (ICT) has become, within a very short time, one of the basic building blocks of modern society. Many countries now regard understanding ICT and mastering the basic skills and concepts of ICT as part of the core of education. The growth and development of information and communication technologies has led to their wide diffusion and application, thus increasing their economic and social impact. The papers in this part aim at improving our understanding of how ICTs contribute to sustainable economic growth and social prosperity and their role in the shift toward knowledge-based societies. The papers discuss the potentiality of information technology systems, a review of techniques to counter spam and spit, sequential algorithms for max-min fair bandwidth allocation, a temporal variation in indoor environment, workflow management for cross media publishing, a Bayesian approach to improve the performance of P2P networks, deploying BitTorrent in mobile environments, numerical simulation for the Hopf bifurcation in TCP systems, a computation study in mesh networks by scheduling problems, serially concatenated RS codes with ST turbo codes over ring, network performance monitoring and utilization measurement, blind adaptive multi-user detection based on the affine projection algorithm, multiple base station positioning for sensor applications, considerations for the design of a general purpose wireless sensor node, performance comparison and improvement of wireless network protocols, modeling the erroneous behavior of a sequential memory component with streams, FPGA implementation of PPM I-UWB base band transceiver, WSN-based audio surveillance systems, and smart antenna design using multi-objective genetic algorithms.

Chapter 49
Potentiality of Information Technology Systems

**Emmanouil Zoulias, Stavroula Kourtesi, Lambros Ekonomou,
Angelos Nakulas, and Georgios P. Fotis**

Abstract In this paper we would like to examine the potentiality of information technology systems. The modern business environment is characterized by dynamic products and services. A business environment is either expanding or contracting. Those changes have a relative flyer pace. In this paper we introduce the relationship within an IT system between the enabling communication technology on which it rests and the information needed by an organization. We first examine the cyclical nature of information and communication to describe the similarity between them. In the next part we examine the issues raised by the characteristics of information and how they affect, interact with, and relate to the communication system's configuration. Various examples illustrate all the issues in a practical way. Possible strategies for the management of this relationship are described in the second part of the paper, with examples. Finally, we conclude by summarizing some considerations and thoughts.

49.1 Introduction

In the past decades, manufacturing was the dominate occupation. Organizations could easily and profitably define no more than five-year manufacturing cycles, planning scheduled product enhancements and signing long-term contracts with suppliers. Companies could spend two or even three years building stand-alone applications to accomplish specific business needs without having to worry about integration, business requirements, customer needs, or new regulatory changes.

Nowadays, new dynamic products and services are needed for satisfying customer needs. Moreover, rapid economic and political changes and an increased focus on international commerce, global markets, and the new

E. Zoulias (✉)
National Technical University of Athens, 9 Iroon Politechniou Str., 157 80 Athens, Greece
e-mail: ezoulias@teemail.gr

N. Mastorakis et al. (eds.), *Proceedings of the European Computing Conference*, Lecture Notes in Electrical Engineering 27, DOI 10.1007/978-0-387-84814-3_49, © Springer Science+Business Media, LLC 2009

economy are shaping a demanding environment for every organization, regard-less of its size. This sets up an inherent difficulty for almost any organization to realistically define or set definite plans, even for a period of months. Good, solid planning and strong planning processes are necessary for modern organisa-tions, but today it's more important than ever to account for a wide variety of defined (or potential) dynamic elements in business processes.

All these aspects configure the modern market and its needs, which undoubt-edly results in a great need for information exchange. In this never-ending flow of information, IT technologies appear to support this demand in an efficient way. The holistic view of enterprise information and information technology assets as a vehicle for aligning business and IT in a structured and therefore more efficient and sustainable way, has attracted significant attention over the last years.

49.2 Cyclical Nature of Information and Communication Systems

The information subsystem of almost every organization and of every informa-tion type has a cyclical nature. Someone produces an information call; this is received by another person or machine and produces a reply or an action from the other person or persons. We can easily approach this with an everyday example using the very well known way of producing information which we call speech. Someone can say something by using his voice (*the source*), in a known language (*the transmitter*); the voice travels through the air (*the transmission system*) and can be heard by someone else (*the receiver*), who may or may not answer (Fig. 49.1). This might be followed by many iterations, which show the cyclical nature of information. Likewise, within a communication system, we start with a *source,* which generates data to be transmitted; a *transmitter,* which converts data into transmittable signals; a *transmission system,* which carries data; a *receiver,* which converts the received signal into data; and finally a destination, which takes the incoming data. The way of "traveling" can move backward and forward again and again, making a cycle, until we reach the end of the communication session.

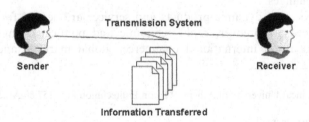

Fig. 49.1 Communication system

49.3 Nature of Information in a Modern Organization and Its Interaction with Communication Systems

The successful running of a modern organization requires the use of various modern technologies, delivering information about a wide variety of defined (or potential) dynamic elements of all business processes. These technologies include fax, analog or digital or voice over IP telephony, mail, web pages, pneumatic mail, etc.

Various other elements are important for every information system, regardless of the information type. These include transfer rate, availability, security of information travel, organizations plants, and potential of communication channels and receivers. In the next part we analyze how all those elements interact with the technology of communication systems and how they interact deeply with the information system of any organization.

49.4 Range of Information System and Communication System Aspects

The first issue that we will examine is the range over which this information system should work. This has to do with the organization's plants and the potential connections. The information system of ambulances and the central has completely different needs from the information system of a company that has its plants in one building. This highly affects the technology of the communications that will be used. This parameter should be taken seriously into account mainly during the design time, but also during the running time of the system. In the first example we need a wireless communication system, in the second example a well constructed LAN is more than enough.

Viewing this aspect from the information theory side the range of the information system will affect the configuration of the communication system with regard to many factors of coding functions. We can see step by step with some examples.

The system will affect the source codes and especially compression strongly. Suppose that we have a widespread network and its data has to travel via other extranets. In order to be travel swiftly, the data should have high compression. On top of that, suppose that the information is of high importance and we can not afford any loss; in this situation the proper compression algorithm is of high importance.

Furthermore, the security policy that should be applied on any system is affected by the range of the system. Assume that we have a company with many buildings in other countries and all of them have to transfer information to the central office using the Internet. This carries a high risk, and a serious security policy should be applied, taking into account attacks, lexicographical attacks, and stealing of accounts. On the other hand, for a small company on one floor, with a restricted Internet connection only for surfing and without a mail or web server, the security policy is completely different.

One more issue relevant to coding theory is bandwidth utilization. This issue is very important for every network, since bandwidth is like a road. If we build it small we will have heavy traffic; if we build it large we will have unused resources. In either case we have to utilize the available bandwidth in an effective way. The utilization of bandwidth is the most important issue for telecommunication companies and ISPs.

49.5 Types of Information Systems and Communication System Aspects

A second issue is the type of information that the system has to deliver. An example of that is medical information. Medical records are very sensitive to changes, since missing or wrong information might result even in the death of a person. On the other hand, information on products in an e-shop should be public, and preferably each customer can visually change their characteristics to make a customized product. The type of information is very important and affects the technology of a communication system on various levels.

From the information theory point of view, it is obvious that security aspects are predominant in these systems, and the security policy applied is important for the system's life. The new dynamic products and services require a dynamic security system. As a result, depending on the information type, we have to think of the proper security level; a medical or bank system should be very secure, for example.

Furthermore, information type affects compression issues. Again, in our example, we cannot afford any loss to medical data, since this could be dangerous. As a result we have to use the proper lossless compression algorithms and make the sacrifices to costs to keep the speed to affordable levels.

In addition to that, error control changes according to the type of information we have to deliver. When we have to transfer a medical image, we cannot afford any loss because this can reduce the image quality. But during the broadcast of a TV program, we can afford the loss of one frame; this will not be obvious to the human eye. The technology gives various solutions, according to the level of protocols.

The next issue is bandwidth utilization and its relation to the type of information. Let us again use the example of medical data. In order to achieve a high transfer rate and minimize losses, it is not enough to have the best compression and error checking algorithms; we also have to ensure the channel bandwidth needed. In other words, we have to make the road as broad as the necessary traffic and speed demand. This should be a main concern at design time.

49.6   Availability Needed by the Information System and Communication System Elements

Another element that is related to the nature of the information is the required availability of the data: in other words, when we need a piece of information, and with what time delay we can have it. As an example, medical information should be available any time, since lives are at risk. On the other hand, the web site of a well known supermarket, with no electronic interactions, can afford a short delay or even brief unavailability of the information. The level of needed availability in a system can change the system's structure completely.

During the establishment of a security policy in every system we have to take into account the availability needed. This can expressed in various technological aspects like UPS presence, fire insurance, backup presence, and many more. An information system can increase its availability by means of higher security. This is again a decision relevant to the system we have. The more sensitive the information and services the system accommodates, the higher availability we need.

One more issue that should be mentioned is bandwidth utilization. Any communications system is characterized by its bandwidth. The higher the bandwidth, the more data can be sent. This is in relation to information theory. The availability of a system is also affected by the bandwidth, and this can be made clear with an example. We can examine the system of a mobile company during an emergency situation, like an earthquake; the high number of calls can cause system collapse and block any new phone call. This could be dangerous, since some people might be in great danger. So mobile companies use a special number and a special part of the bandwidth for those emergency calls.

49.7   Sources of Information Systems and Communication Systems

In addition to this, the sources of the information system are important. Take, as an example, the police telephone center that accepts emergency calls, in contrast to the telephone center of a university. In the police telephone center every call should be answered without any delay, through a secure channel with full availability, because the callers are probably in jeopardy. The university call center can afford a short delay in answers during peak time, because the students probably have time to wait.

Moreover, information systems probably wish to communicate with other information systems. This happens only by following commonly agreed rules. This is the case with any language. People can communicate because they follow the rules of the language they speak. The same happens in every information system. As a result, the communication systems that support information systems should adopt protocols that are commonly agreed upon.

From the information theory point of view, the type of information source seriously affects the type of analog to digital (A/D) conversion. The proper conversion is very important for any data. This is obvious, because with incorrect conversion of the information, data might be lost, and in many cases this is not acceptable. We cannot use the same A/D conversion type for a high frequency system as for a low frequency one. We should carefully choose the proper A/D conversion type for any source which we know from the design time, in any potential information system.

In addition to that, compression is related to the source. A source can produce files of large size as well as those of small size. It is reasonable to use a high compression algorithm for the large files to achieve smaller files and eventually save bandwidth and time for their transfer. As a result, the communication system should adopt various compression algorithms and techniques to satisfy all types of information.

Moreover, security aspects differ relative to the source. Police information should be highly secured and no one, except the authorized personnel, should be able to interfere with the policy information system. This can be achieved by various technical means and described mainly by the security policy of the information system.

49.8 Strategies for Management

In this part we examine possible strategies to confront all possible issues that arise from the information and communication systems' relationship.

49.8.1 Design and Polices

The first and most important management tool is the good design of the system. We have to admit that design time is important for every technological system. Design is the process of finding the optimized configuration of a network and is based on the performance, bit rate, use of network parts, availability, and possible increase in the demand for network resources. The design of a network could be initial or corrective. The design phases are the following: collection of data for the network load, estimation of future loads, placement of network components, and optional solutions. Next comes an adequate assessment of needs, alternative technology solutions, quality of service, running costs, security issues, and fair allocation of resources. The third phase is the design, with some improvements and with consideration of the possible restrictions. Finally comes the design itself, with proper methods and confronting the weaknesses.

Bandwidth is the backbone of every system, and as a result it is very important to ensure the bandwidth needed, given the quality of service that is required. One very important aspect of this is the selection of proper and good communication channels. This has to do not only with the technology, but also with the construction if we have an internal network, or with the provider if we have a network that has to use public networks. There are many different solutions, and a deep market survey is essential. The term capacity planning is also used for this issue. This term has to do with the initial design for bandwidth needs and with the upgrade of the bandwidth when the needs are changing.

Security aspects are very important in modern life. Technology is on the front line in this area. The design, establishment, operation, and upgrade of a security system are very important for a communications system. Security management is a serious and stand-alone occupation. Specialists should take care of the system security and apply a policy to ensure data integrity, privacy, and availability. Although security is very important, the application of a strict security system can add difficulties to a company's operation. As a result, we have to apply the proper level of security for a given organization.

Applying the proper policies is a crucial aspect, and we have to use the proper tools to do so. The selection of good network devices is very important. High quality devices ensure regular operation of the communications system. The availability of the system is proportionally related to the quality of the devices, not only for proper operation but also for the proper level of support. We have to point out that the ability of selecting network devices is limited when the organization leases the networks of others, like the Internet. In this situation, the selection of a reliable provider is all one can do.

49.8.2 Equipment Selection and Repairing

At the same level the selection of proper terminal equipment is very important. The personal computers or terminals should have a configuration that satisfies the needs of the system. The selection of terminal equipment with lower capabilities reduces the efficiency of the system—for example, if we have the proper network to transfer images on time, but the terminal PC needs a long time to get those images on the screen.

From another point of view, the proper selection of a system is based on its protocols. The selection of equipment and devices should be based on the software protocols that they use. As an example, the selection of TCP for transferring videoconferencing over satellite is not proper, since TCP has high level error checking and resends any lost packet, although in a videoconferencing application the loss of one or even two frames is not important. As an alternative, UDP can be used, as this is the proper protocol. We conclude that selection of protocols is very important and has to be related to the selection of all equipment and software in the communications and information systems.

One aspect that is very important for the efficient operation of a communications system is problem spotting and solving. Those aspects have mainly to do with the system's efficient operation and availability. The ability to spot problems in a short time saves down time and/or operational problems. The technical department should have sufficient training and tools to spot the problems. Both help to spot a possible problem and act as soon as possible, or solve the problem in a better way, thus avoiding further problems resulting from a wrong repair action. In this part of the system management, a preventive action plan is necessary. This will reduce the possibility of problems appearing, money and time wasted on repairs will be saved, and problems for the final users will be minimized. A good example of this is the maneuvers in the communications system of the army. This ensures that a minimum number of problems will appear during battle time.

49.8.3 System Monitoring and Improvement

A very important part of management, which is omitted in many networks, is the monitoring and improvement of the system. This includes attention to the system by means of network administration software. With this software, we can see the traffic, any problems that might arise in the network, and spot them easily. Most of this software is user friendly and does not need a lot of effort in learning. As a result, using such software, we can spot any problem or potential problem and act proactively. In the example of a network that is attacked by hackers, network monitoring can spot the increased traffic to the network from outside. In this situation, we can see what type of traffic it is and clear up whatever is not normal or acceptable.

Any given company is either contracting or expanding, and the relevant communications system follows this alternation. In order for a system to keep the designed level of quality and performance, we have to continuously optimize the system. The tools for optimization comprise the monitoring system. This will guide the administrator through the useful upgrades, expansions, modifications, or any change to the system configuration. The next step is to apply those options that will bring the system to the quality expected.

Finally, any system has to be benchmarked. In order to evaluate the communication system's job and to make decisions for what should be done, we have to define what we measure and what the acceptable level is. This is done by a benchmarking system. Making correct decisions for changes without a benchmarking system is totally impossible.

49.9 Conclusions

In conclusion, we would like to point out the strong relationship between information technology and telecommunications technology. In addition, the very important issue of the dynamic nature of the modern business environment leads the new demand for efficient information systems. The evolution of telecommunications systems is a logical consequence of the evolution of information systems and, vice versa, the telecommunications evolution propels the information evolution. Common people are confused and believe that telecommunications and not information systems define this modern evolution and lead work processes. Only engineers and process developers are capable of discriminating the difference. The discrimination of the two systems, along with their relations and interactions, can help us to create a safe basis for more efficient work. Information theory is the basis, and a deep knowledge of this can help us to have the highest results.

References

1. Burgess S (2002) Managing information technology in small business: challenges and solutions. Idea Group Publishing, London, UK
2. Morgan T (2002) Business rules and information systems: aligning IT with business goals. Addison-Wesley, Reading, MA
3. Abbes S (2006) The information rate of asynchronous sources. 2nd IEEE international conference on information and communication technologies: from theory to applications. ICTTA'06, Damascus, Syria
4. Li SYR, Yeung RW, Cai N (2003) Linear network coding. IEEE T Inform Theory 49(2): 37–381
5. Bertsekas D, Gallager R (1992) Data networks. 2nd edn. Prentice Hall, Upper Sadle River, NJ
6. Cover T (1998) Comments on broadcast channels. IEEE T Inform Theory 44(6): 2524–2530
7. Ephremides A, Hajek B (1998) Information theory and communication networks: an unconsummated union. IEEE T Inform Theory 44(6):2416–2434
8. Shannon CE (1948) A mathematical theory of communication bell system. Tech J 27: 379–423, 623–656
9. Aziz S, Obitz T, Modi R, Sarkar S (2006) Making enterprise architecture work within the organization enterprise architecture. A governance framework, Part II
10. Addie RG (2002) Algorithms and models for network analysis and design. Online at http://waitaki.otago.ac.nz
11. Ahmad A (2003) Data communication principles for fixed and wireless networks. Kluwer, Boston
12. Benedetto S (1999) Principles of digital transmission with wireless applications. Kluwer (rich in error detection and correction (Ch10), includes Hamming Code), Boston
13. Prasad KV (2004) Principles of digital communication systems and computer networks. Charles River Media, Boston

Chapter 50
A Review of Techniques to Counter Spam and Spit

Angelos Nakulas, Lambros Ekonomou, Stavroula Kourtesi, Georgios P. Fotis, Emmanouil Zoulias

Abstract This paper studies the most important techniques with which to challenge the problem of unsolicited e-mails (spam) and unsolicited messages in Internet telephony (spit). First an introduction to the problem of spam demonstrates the importance (economic and technological) of finding a solution. Then we analyze the most important techniques that exist to counter the problem. After that we concentrate on a new problem: spam using new internet telephony technologies (spit). This problem, even if existing only theoretically until now, very soon will be one of the main factors affecting the broad use of VoIP. We analyze the most important methods and techniques of countering spit. Finally, we mentione differences between spam and spit and state some useful conclusions.

50.1 Introduction

In this paper we review the most important techniques with which to counter two similar problems: the existence of unsolicited messages in electronic mail (spam), and something that is until now only a hypothesis, the appearance of unsolicited calls in Internet telephony.

The spam problem is a very well known problem, and for years many techniques have been proposed to counter it; nevertheless the problem remains. Some of these techniques address part of the problem, but none address them all. We categorize all these techniques into three categories: those which can prevent spam, those which can detect it, and those which are mainly used to react to spam.

A new technology which is expected to have big growth in the next few years is telephony using IP networks. Many applications already exist that provide

A. Nakulas (✉)
National & Kapodistrian University of Athens, 11 Asklipiu Str.,
153 54 Athens, Greece
e-mail: aaatos@gmail.com

N. Mastorakis et al. (eds.), *Proceedings of the European
Computing Conference*, Lecture Notes in Electrical Engineering 27,
DOI 10.1007/978-0-387-84814-3_50, © Springer Science+Business Media, LLC 2009

the possibility of telephone calls via the Internet. Some of them use well-defined protocols like H.323 or SIP (session initiation protocol), some others use proprietary and undefined protocols like Skype. In either case, the trend towards the growth and extension of telephony via the Internet will lead to the appearance of spam calls using this technology. Although there are several similarities between spam and spit, not all of the techniques that counter spam are applied with the same success in the area of telephony via the Internet. Therefore we will first locate antispam techniques applicable to fighting spit, and then we will summarize new techniques applicable only to the Internet telephony domain.

50.2 Unsolicited Messages in Electronic Mail (Spam)

50.2.1 Electronic Mail

The mechanism of operation of electronic mail [1] appears in Fig. 50.1 and it can be described as follows:

a. The user (mail user agent—MUA) of the service writes the message and then is initially connected with his mail server (mail transfer agent—MTA), making use of the SMTP protocol.
b. The MTA of the sender locates the address of the MTA of the recipient, making use of the DNS service.
c. After the address of the recipient is located, the MTA of the sender, using the SMTP protocol, dispatches the message to the MTA of the recipient, where it is stored.

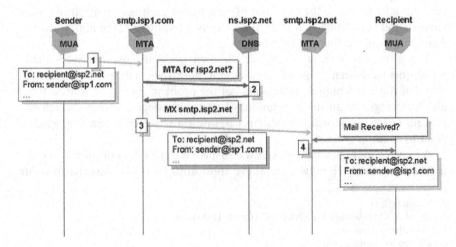

Fig. 50.1 How e-mail works

d. The recipient of the message recovers it when connected with the MTA, or via the POP or the IMAP protocol.

It is obvious that the operation of electronic mail involves various protocols, each one of which carries out a different task. These protocols and their roles are:

50.2.1.1 SMTP (Simple Mail Transfer Protocol)

This protocol is responsible for the exchange of messages between users. Each message of e-mail comprises two distinguishable parts: the headers and the body. Both headers and bodies are written in plain text. The purpose of headers is to determine important parameters of the message, like the address of the sender, the address of the recipient, etc. The existence of headers in simple text creates important problems concerning the security of e-mail. In its simplest form, this problem concerns the ability for malicious users to modify the content not only of the headers but also of the bodies.

50.2.1.2 MIME (Multipurpose Internet Mail Extensions)

The purpose of this protocol is to provide the ability to transport e-mail messages with content beyond simple text (like pictures, audio, etc.). This protocol does not solve the problems of security that were mentioned before.

50.2.1.3 S/MIME (Secure/Multipurpose Internet Mail Extensions)

The purpose of this protocol is to make the transfer of e-mail messages secure. In order to accomplish this task it uses cryptographic techniques (like public keys and digital signatures) to protect the header and the body of an e-mail.

50.2.1.4 POP (Post Office Protocol) and IMAP (Internet Message Access Protocol)

These protocols are used by the recipient of an e-mail for the retrieval of his messages from the MTA. The POP provides simple and basic abilities with which to retrieve messages, while the IMAP provides, beyond retrieval, further capabilities for message management which are stored in the MTA.

50.2.2 SPAM

By "spam" is meant the unsolicited sending of mass e-mail messages with varying content.

The word spam in this sense derives from a TV comedy sketch, "The Monty Python Spam Sketch," in which all the meals offered in a restaurant contain the

Table 50.1 Techniques for sending spam messages

Technique	Description
Massive e-mails	Sending of massive e-mails from malicious user, using his e-mail address
Joe job	Sending of massive e-mails from malicious user, using e-mail addresses of other users
Relaying through third parties	Sending of massive e-mails from malicious user, using open relays
Break-ins/Zombies	Invasion of system(s) of unsuspecting user(s) and using those system(s) for sending of spam messages
Open proxies	Use of open web proxies to send unsolicited messages using the HTTP CONNECT command
Web harvesting and dictionary attacks	Spammers use software that automatically visits millions of web pages, searching for anything that looks like an e-mail address, and adding it to a list of addresses to spam

processed meat called spam as a component. In this episode the word "spam" is heard 136 times. The frequency of repetition of this particular word was adopted as a description of the similar phenomenon in e-mail. Although historically the first spam e-mail message (to 600 recipients) turned up in 1978, nevertheless the term "spam" was used in the decade of the 1980s to describe the mass sending of messages generally (for example, in newsgroups). Eventually however this word ended up being used mainly in the area of e-mail. Thus today when we use the word "spam" we describe a phenomenon in the field of electronic correspondence. Despite the fact that the results of different researches [2, 3] indicate different growth rates, there is no doubt that the number of spam messages is continuously increasing and that the spam phenomenon is continuously expanding geographically.

Many researches [4–6] describe the techniques that are widely used to send spam messages. These techniques are shown in Table 50.1.

50.3 Techniques to Counter Spam

50.3.1 Techniques to Prevent Spam

The most important techniques for preventing spam [4, 7, 8] are:

Whitelists. An e-mail whitelist is a list of contacts that the user deems are acceptable to receive email from and that should not automatically be sent to the trash folder. More advanced whitelist techniques can also use verification. If a whitelist is exclusive, only e-mail from senders on the whitelist will get through. If it is not exclusive, it prevents e-mail from being deleted or sent to the junk mail folder by the spam filter. Usually, only end-users would set a spam filter to delete all e-mails from sources which are not on the white list, not Internet service providers or e-mail services.

Blacklists. A blacklist is an access control mechanism that means, allow everybody through except members of the blacklist. It is the opposite technique of a whitelist. An e-mail spam filter may keep a blacklist of addresses, any mail from which would be prevented from reaching its intended destination.

Challenge/Response. This is a technique that needs two parties: one party presents a question ("challenge") and the other party must provide a valid answer ("response") in order to be authenticated. The simplest example of a challenge-response protocol is password authentication, where the challenge is asking for the password and the valid response is the correct password. The main purpose of this technique is to ensure a human source for the message and to deter any automatically produced mass e-mails. One special case of this technique is the Turing test.

Consent token. In the consent framework, users and organizations express their wishes as consent and nonconsent, which may be implicit or explicit. Users and organizations also define their wishes with regard to e-mail messages without prior consent or lack of consent in place—for example, how to deal with e-mail strangers. Various antispam tools and systems store and enforce these decisions. Additional internal and external systems provide information that is used for enforcement—such as DNS-based block lists, reputation systems, e-postage systems, etc. The sender may also provide information within the body or headers of the email messages that is used during the enforcement process. Examples of such information are challenge/response messages, e-postage tokens, trusted sender digital certificates or tokens, etc., all of which are collectively called "consent tokens." There are also components which monitor the origin of the message in order to deter spammers.

Authentication. Ensuring a valid identity on an e-mail has become a vital first step in stopping spam, forgery, fraud, and even more serious crimes. E-mail authentication greatly simplifies and automates the process of identifying senders. After identifying and verifying a claimed domain name, it is possible to treat suspected forgeries with suspicion, reject known forgeries, and block e-mail from known spamming domains. It is also possible to "whitelist" e-mail from known reputable domains, and bypass content-based filtering, which always loses some valid e-mails in the flood of spam.

50.3.2 Techniques to Detect Spam

These techniques' function is to distinguish spam messages from regular e-mails. This operation is demanding because the system is called to take decisions on behalf of its users. Also, this operation should be performed with precision and speed. The whole process is made even more difficult by the fact that the system has limited information about the examined e-mail. The most important techniques with which to detect spam [4, 7, 9–11] are:

Reputation systems. This technique is a type of collaborative filtering algorithm which attempts to determine ratings for a collection of entities, given a collection of opinions that those entities hold about each other. This is similar to a recommendation system, but its intended purpose is for entities to recommend each other, rather than some external set of entities (such as books, movies, or music).

Static content filtering lists. These techniques require the spam blocking software and/or hardware to scan the entire contents of each e-mail message and figure out what's inside it. These simplistic but effective ways to block spam identify e-mail that contains certain words. Its defect is that the rate of false positives can be high, which would prevent someone using such a filter from receiving legitimate e-mails.

Learning content filtering systems. These techniques are based on Bayesian filters and on mechanisms of decision-making. Bayesian e-mail filters take advantage of Bayes's theorem. Bayes's theorem, in the context of spam, says that the probability of an e-mail being spam, given that it contains certain words in it, is equal to the probability of finding those certain words in spam e-mail, times the probability that any e-mail is spam, divided by the probability of finding those words in any e-mail. The filter doesn't know these probabilities at first, and needs to be trained to do so in advance so it can build on them. To train the filter, the user should manually indicate whether or not a new e-mail is spam. After training, the word's probabilities (also known as likelihood functions) are used to compute the probability that an e-mail with a particular set of words belongs to either category.

Quantity. These techniques try to find spam messages by examining the number of e-mail messages send by a particular user in a giver time period. As the number increases, the possibility that the sender is a spammer increases also.

50.3.3 Techniques to React to Spam

The purpose of these techniques is to handle a spam message in such a way as to minimize the effect of spamming. The main techniques are:

Quarantine. This technique functions by placing a spam message in provisional isolation until a suitable person (administrator, user, etc.) examines it for final characterization.

Limit rate. This technique acts mainly preventively and aims to restrict the rate of receipt of messages from a user, independently of whether or not this user has been characterized as a spammer. It is mainly used at the ISP (Internet Service Provider) level as an indirect way of restricting spam.

Reject. This technique works by rejecting an e-mail that has been characterized as spam.

Charge. This technique charges a spammer a small amount of money for each spam e-mail he sends. The purpose is mainly to raise the cost of sending spam messages.

50.4 Techniques to Counter Unsolicited Calls in Internet Telephony (Spit)

50.4.1 Spit Definition

Spit is an as yet nonexistent problem which has nonetheless received a great deal of attention from marketers and the trade press. Some people used to refer to it as SPIT (for "Spam over Internet Telephony"); however, this neologism is not used by VoIP technicians. Voice over IP systems, like e-mail and other Internet applications, are vulnerable to abuse by malicious parties who initiate unsolicited and unwanted communications. Telemarketers, prank callers, and other telephone system abusers are likely to target VoIP systems increasingly, particularly if VoIP tends to displace conventional telephony.

While few companies currently receive spit, this situation will soon change. According to a recent survey, enterprise spending on IP telephone systems is increasing in the United States, while spending on data networking products is decreasing. More than 31 percent of respondents already have IP-based private branch exchanges (PBXs) installed, and 12 percent have hosted VoIP (IP Centrex) services, IDC said.

VoIP provides companies the ability to consolidate telephony and networking infrastructures within the enterprise, integrate voice with e-mail and contact-management applications, and banish traditional office phones in favor of softphones that are integrated into desktop or laptop computers.

With the brand-new features of VoIP come new challenges, including the eventual threat of spit, which few enterprise networks are equipped to handle today.

50.4.2 Techniques to Counter Spit

Figure 50.2 shows the most important techniques with which to counter spit. Many of the antispam techniques can also be used to fight spit. This is because there are many similarities between e-mail and VoIP. Conclusively, antispam techniques which can be applied to spit are: blacklists and whitelists, Turing tests, reputation systems, and static and dynamic content filtering.

There are also new techniques that can be used against spit. These are:

Address obfuscation. The purpose of this technique is to make addresses of protocols (like e-mail addresses, SIP addresses, etc.) not readable by software programs that try to collect such addresses automatically. One typical example of such obfuscation is the transformation of the e-mail address "alice@example.com" to "Alice at example dot com."

SIP providers. This technique is a combination of the techniques called *payment* and *circle of trust*. It works by establishing mutual trust between VoIP providers in order to exchange information between them. If a user in

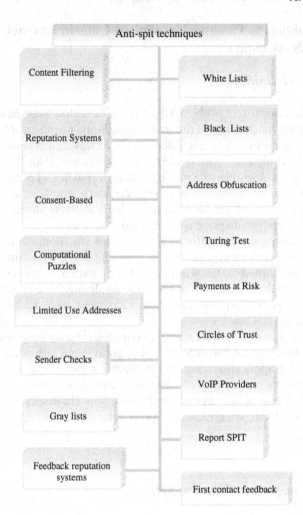

Fig. 50.2 Techniques to counter spit

one domain is characterized as a spitter, then this information spreads across the providers.

Graylists. This technique is a combination of the techniques *whitelists* and *blacklists.* Every user who is on neither a whitelist nor a blacklist is characterized as a gray user with a specific initial gray level. As long as this user initiates calls, this gray level changes continuously. When the level touches a threshold, the user is characterized as a spitter.

Report spit. The purpose of this technique is to immediately inform a spit identification system when a spitter is recognized.

First contact feedback. In this technique, when a user makes a call for the first time, his call is considered to be in an undefined state until the receiver of the call characterizes it. Only after this characterization has taken place is it possible for the caller to make more calls.

Legal actions. Finally, legal regulations at the state and global levels are necessary in order to achieve common legal reactions to acts like spam and spit.

50.5 Similarities and Differences Between Spam and Spit

50.5.1 Similarities

The main similarity between spam and spit is that nearly all the methods that are used to send spam messages can also be used to send spit messages. This is because both spam and spit have the same purpose: to send unsolicited messages.

A second similarity comes from the fact that both can be done with automation. This means that it's easy to write programs for sending such messages in big quantities and for a long time. This automation is the main problem with both spam and spit.

Another similarity is the fact that most of the techniques that are used to counter spam (like whitelists, blacklists, reputation systems, etc.) can also be used to fight spit. In general, we can say that except for the techniques which use content filtering or content examination (which means text manipulation), antispam techniques may be also applied in the antispit domain.

The final similarity is the cost of sending spam or spit messages compared to the cost of a phone call. It costs more [12] for the spammer to make a phone call than it does to send a spam or spit message. This includes both the cost for a system which can perform telemarketer calls, and in the cost per call. Considering the system cost, either spam or spit can run on a low-end PC and requires no special expertise to execute. The cost per call is also substantially reduced. Spam or spit is roughly three orders of magnitude cheaper to send than traditional circuit-based telemarketer calls. This low cost is certainly going to be very attractive to spammers. Indeed, many spammers utilize computational and bandwidth resources provided by others, by infecting their machines with viruses that turn them into "zombies" which can be used to generate spam. This can reduce the cost of call spam to nearly zero.

50.5.2 Differences

The main difference between spam and spit is their content. Spam messages contain mainly text (even if recently some spam messages carry images as well), but spit messages contain audio. This difference is the main problem if we want to apply some antispam techniques to the antispit domain. For example, we cannot apply content filtering antispam techniques to counter spit.

Finally, we should be point out the difference between spam and spit with respect to the nuisance that is caused by their existence. Obviously the nuisance of spit calls for the recipient is considered much greater than that of spam e-mail messages. The nuisance from continual audio notices of VoIP calls is enormous compared to the corresponding nuisance from the accumulation of undesirable correspondence in one's e-mail box.

50.6 Conclusions

This paper has presented the main techniques used to counter spam and spit. All the techniques which are described above solve the problem only partially. Some of them are effective against certain categories of attacks. Some others are successful against the special features of spam and spit messages. The radical difference between spam and spit with respect to their content renders imperative the necessity of discovering new techniques to counter spit. Up to now only a combination of the techniques and methods which we have described can deal with the problem of unsolicited messages.

References

1. Telecommunications Engineering and Certification Industry Canada (2005) Anti-spam technology overview. http://www.e-com.ic.gc.ca
2. Messaging Anti-Abuse Working Group (MAAWG) (2006) Email metrics program. http://www.maawg.org/about/publishedDocuments/
3. http://www.ironport.com/company/ironport_pr_2006–06–28.html
4. Bishop M (2005) Spam and the CAN-SPAM act expert report. Federal Trade Commission, Washington, DC 20580
5. Goodman J (2003) Spam: technologies and policies. White paper. Microsoft Research
6. Ramachandran A, Feamster N (2006) Understanding the network-level behavior of spammers. ACM SIGCOMM (Special Interest Group on Data Communications), Pisa, Italy
7. Judge QP, An analysis of technological and marketplace developments from 2003 to present and their impact on the effectiveness of CAN-SPAM. http://www.ftc.gov/reports/canspam05/judgerpt.pdf
8. Seltzer L (2003) Challenge-response spam blocking challenges patience. eWEEK
9. O'Donnell A (2006) Applying collaborative anti-spam techniques to the anti-virus problem. In: Virus bulletin 2006 (VB2006) conference, Montreal
10. Liu D, Camp J (2006) Proof of work can work. In: The fifth workshop on the economics of information security (WEIS 2006), Robinson College, University of Cambridge, England
11. Prakash V, O'Donnell A (2005) Fighting spam with reputation systems. ACM Queu 3(9)
12. Jennings C, Peterson J, Rosenberg J (2006) The session initiation protocol (SIP) and spam. Internet draft

Chapter 51
Sequential Algorithms for Max-Min Fair Bandwidth Allocation

Włodzimierz Ogryczak and Tomasz Śliwiński

Abstract Telecommunications networks are facing an increasing demand for Internet services. Therefore, a problem of network dimensioning with elastic traffic arises, which requires the allocation of bandwidth to maximize service flows with fair treatment for all the services. In such applications, the so-called Max-Min Fairness (MMF) solution is widely used to formulate the resource allocation scheme. It assumes that the worst service performance is maximized, and the solution is additionally regularized with the lexicographic maximization of the second worst performance, the third worst, etc. Because of lexicographic maximization of ordered quantities, the MMF solution cannot be tackled by the standard optimization model. It can be formulated as a sequential lexicographic optimization procedure. Unfortunately, the basic sequential procedure is applicable only for convex models; thus it allows dealing with basic design problems, but fails if practical discrete restrictions commonly arising in telecommunications network design are to be taken into account. In this paper we analyze alternative sequential approaches allowing the solving of nonconvex MMF network dimensioning problems. Both of our approaches are based on sequential optimization of directly defined artificial criteria. The criteria can be introduced into the original model with some auxiliary variables and linear inequalities; thus the methods are easily implementable

51.1 Introduction

A fair method of bandwidth distribution among competing demands becomes a key issue in computer networks [3], and telecommunications network design in general [7, 8, 17, 19]. Due to the increasing demand for Internet services, the problem of network dimensioning with elastic traffic arises, which requires the

W. Ogryczak (✉)
Institute of Control & Computation Engineering, Warsaw University of Technology,
00–665 Warsaw, Poland
e-mail: wogrycza@ia.pw.edu.pl

N. Mastorakis et al. (eds.), *Proceedings of the European*
Computing Conference, Lecture Notes in Electrical Engineering 27,
DOI 10.1007/978-0-387-84814-3_51, © Springer Science+Business Media, LLC 2009

allocation of bandwidth to maximize service flows with fair treatment for all the services [17]. The problem of network dimensioning with elastic traffic can be formulated as follows [16]. Given a network topology $G = <V,E>$, consider a set of pairs of nodes as the set $J = \{1, 2, ..., m\}$ of services representing the elastic flow from source v_j^s to destination v_j^d. For each service, we have given the set P_j of possible routing paths in the network from the source to the destination. This can be represented in the form of binary matrices $\Delta_e = (\delta_{ejp})_{j\in J; p\in P_j}$ assigned to each link $e \in E$, where $\delta_{ejp} = 1$ if link e belongs to the routing path $p \in P_j$ (connecting v_j^s with v_j^d) and $\delta_{ejp} = 0$ otherwise. For each service $j \in J$, the elastic flow from source v_j^s to destination v_j^d is a variable representing the model outcome and it will be denoted by x_j. This flow may be realized along various paths $p \in P_j$, and it is modeled as $x_j = \sum_{p\in P_j} x_{jp}$, where x_{jp} are nonnegative variables representing the elastic flow from source v_j^s to destination v_j^d along the routing path $p \in P_j$. The single-path model requires additional multiple choice constraints to enforce nonbifurcated flows.

The network dimensioning problem depends on allocating the bandwidth to several links in order to maximize the flows of all the services (demands). For each link $e \in E$, decision variables $\xi_e \geq 0$ represent the bandwidth allocated to link $e \in E$. Certainly, there are usually some bounds (upper limits) on possible expansion of the links capacities: $\xi_e \leq \bar{a}_e$ for all $e \in E$. Finally, the following constraints must be fulfilled:

$$0 \leq x_{jp} \leq Mu_{jp} \quad \forall j \in J;\ p \in P_j \tag{51.1a}$$

$$u_{jp} \in \{0, 1\} \quad \forall j \in J;\ p \in P_j \tag{51.1b}$$

$$\sum_{p\in P_j} u_{jp} = 1 \quad \forall j \in J \tag{51.1c}$$

$$\sum_{p\in P_j} x_{jp} = x_j \quad \forall j \in J \tag{51.1d}$$

$$\sum_{j\in J}\sum_{p\in P_j} \delta_{ejp}x_{jp} \leq \xi_e \quad \forall e \in E \tag{51.1e}$$

$$0 \leq \xi_e \leq \bar{a}_e \quad \forall e \in E \tag{51.1f}$$

$$\sum_{e\in E} c_e\xi_e \leq B \tag{51.1g}$$

where (51.1a), (51.1b), (51.1c), and (51.1d) represent single-path flow requirements using additional binary (flow assignment) variables u_{jp} and define the total service flows. Next, (51.1e) establishes the relation between service flows and links bandwidth. The quantity $y_e = \sum_{j\in J}\sum_{p\in P_j} \delta_{ejp}x_{jp}$ is the load of link e

and it cannot exceed the available link capacity. Further, while allocating the bandwidth to several links, the decisions must keep the cost within the available budget B (51.1 g), where for each link $e \in E$ the cost of allocated bandwidth is c_e.

The network dimensioning model can be considered with various objective functions, depending on the chosen goal. Typically, the fairness requirement is formalized with the lexicographic maxi-minimization (lexicographic Max-Min approach). Within telecommunications or network applications the lexicographic Max-Min approach has already appeared in [4] and now, under the name Max-Min Fair (MMF), is treated as one of the standard fairness concepts [2, 6, 10, 17, 20]. Indeed, the MMF approach generalizes equal sharing at a single link bandwidth to any network, also allowing the maximization of the second smallest flows provided that the smallest remains optimal, the third smallest, etc.

The lexicographic maxi-minimization can be seen as searching for a vector lexicographically maximal in the space of the feasible vectors with components rearranged in nondecreasing order. This can be mathematically formalized as follows. Let $\langle \mathbf{a} \rangle = (a_{(1)}, a_{(2)}, ..., a_{(m)})$ denote the vector obtained from $\mathbf{a}$ by rearranging its components in nondecreasing order. That means $a_{(1)} \leq a_{(2)} \leq ... \leq a_{(m)}$, and there exists a permutation π of set J such that $a_{(j)} = a_{\pi(j)}$ for $j = 1, 2, ..., m$. Lexicographically comparing such ordered vectors $\langle y \rangle$ one gets the so-called lex-min order. The MMF problem can be then represented in the following way:

$$\text{lex} \max \left\{ \left(y_{(1)}, y_{(2)}, ..., y_{(m)} \right) : y \in A \right\} \tag{51.2}$$

where A where depicts the set of attainable outcomes defined with constraints (51.1). Actually, we focus our analysis on the MMF bandwidth allocation problem, but the approaches developed can be applied to various lexicographic Max-Min optimization problems, i.e., to problem (51.2) with various attainable sets A.

The (point-wise) ordering of outcomes means that the lexicographic Max-Min problem (51.2) is, in general, hard to implement. Note that the quantity $y_{(1)}$, representing the worst outcome, can easily be computed directly by the maximization:

$$y_{(1)} = \max \left\{ r_1 : r_1 \leq y_j \quad \forall j \in J \right\}.$$

A similar simple formula does not exist for the further ordered outcomes $y_{(i)}$. Nevertheless, for convex problems it is possible to build sequential algorithms for finding the consecutive values of the (unknown) MMF optimal outcome vector. While solving Max-Min problems for convex models, there exists at least one blocked outcome which is constant for the entire set of optimal solutions to the Max-Min problem. Hence, the MMF solution can be found by solving a sequence of properly defined Max-Min problems with fixed outcomes (flows) that have been blocked by some critical constraints (link capacities) [12]. Indeed, in the case of LP models this leads to efficient algorithms taking advantage of the duality theory for simple identification of blocked

Fig. 51.1 Sample network without any critical link and blocked flow for max-min solution

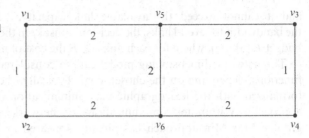

outcomes [1, 5, 18]. Unfortunately, in our network dimensioning model it applies only to the basic LP constraints (51.1d), (51.1e), (51.1f) and (51.1 g). In the case of a nonconvex feasible set, such a blocked quantity may not exist [11], which makes the approach not applicable to our case of nonbifurcated flows enforced by discrete constraints (51.1a), (51.1b), and (51.1c). This can be illustrated with the simplified network depicted in Fig. 51.1 with link capacity limits given in the figure, cost coefficients 4 for links (v_1, v_5), 3 for (v_3, v_5), and all others equal to 1, and the budget B = 11. We consider two demands: one connecting v_1 with v_2 along two possible paths (v_1, v_2) or (v_1, v_5, v_6, v_2); the second connecting v_3 with v_4 along two possible paths (v_3, v_4) or (v_3, v_5, v_6, v_4). The MMF solution is unique; it allocates flow 1 to path (v_1, v_2) (first demand) and flow 2 to path (v_3, v_5, v_6, v_4) (second demand). The Max-Min (single-path) problem leads us to the conclusion that one of two flows cannot exceed 1, but does not allow us to identify which one must be blocked. Note that the same difficulty arises also for the single path problem without any budget constraint, though the optimal solution is then not unique.

In this paper we analyze alternative sequential approaches to solving non-convex MMF network dimensioning problems. Both approaches are based on the lexicographic optimization of directly defined artificial criteria. The criteria can be introduced into the original model with some auxiliary variables and linear inequalities independently from the problem structure.

51.2 Cumulated Ordered Outcomes

The point-wise ordering of outcomes for lexicographic optimization within the MMF problem (51.2) is, in general, hard to implement. Following Yager [21], a direct formula, although requiring the use of integer variables, can be given for any $y_{\langle i \rangle}$. Namely, for any $k = 1, 2, ..., m$, the following formula is valid:

$$y_{\langle i \rangle} = \max r_i$$
$$\text{s.t.}$$
$$r_i - y_j \leq C z_{ij} \quad \forall j \in J \tag{51.3}$$
$$z_{ij} \in \{0, 1\} \quad \forall j \in J$$
$$\sum_{j \in J} z_{ij} \leq i - 1$$

where C is a sufficiently large constant (larger than any possible difference between various individual outcomes y_j) which allows us to enforce inequality $r_i \leq y_j$ for $z_{ij} = 0$, while ignoring it for $z_{ij} = 1$. Note that for $i = 1$ all binary variables z_{1j} are forced to 0, thus reducing the optimization in this case to the standard LP model. However, for any other $i > 1$, all m binary variables z_{ij} are an important part of the model. Nevertheless, with the use of auxiliary integer variables, any MMF problem (either convex or nonconvex) can be formulated as the standard lexicographic maximization with directly defined objective functions:

$$\text{lex} \max(r_1, r_2, ..., r_m)$$

s.t.

$$y \in A$$
$$r_i - y_j \leq C z_{ij} \quad \forall i, j \in J \tag{51.4}$$
$$z_{ij} \in \{0, 1\} \quad \forall i, j \in J$$
$$\sum_{j \in J} z_{ij} \leq i - 1 \quad \forall i \in J$$

We will refer to the above model as the Direct Ordered Outcomes (DOO) approach. Unfortunately, binary variables z_{ij} in the auxiliary constraints contribute to implementation difficulties of the DOO approach.

There is, however, a way to reformulate the MMF problem (51.2) so that only linear variables are used. Let us consider the cumulated criteria $\bar{\theta}_i(y) = \sum_{k=1}^{i} y_{\langle k \rangle}$, expressing, respectively, the worst (smallest) outcome, the total of the two worst outcomes, the total of the three worst outcomes, etc. Within the lexicographic optimization, a cumulation of criteria does not affect the optimal solution. Hence, the MMF problem (51.2) can be formulated as the standard lexicographic maximization with cumulated ordered outcomes:

$$\text{lex} \max\{(\bar{\theta}_1(y), \bar{\theta}_2(y), ..., \bar{\theta}_m(y)) : y \in A\}$$

Note that for any given vector $y \in R^m$ the cumulated ordered value $\bar{\theta}_i(y)$ can be found as the optimal value of the following LP problem:

$$\bar{\theta}_i(y) = \min \sum_{j \in J} y_j u_{ij}$$

s.t.

$$\sum_{j \in J} u_{ij} = k \tag{51.5}$$
$$0 \leq u_{ij} \leq 1 \quad \forall j \in J$$

The above problem is an LP for a given outcome vector y, while it becomes nonlinear for y being a variable. This difficulty can be overcome by taking advantage of the LP duality. Note that the LP duality of problem (51.5), with variable r_i corresponding to the equation $\sum_{j \in J} u_{ij} = k$ and variables d_{ij} corresponding to upper bounds on u_{ij}, leads us to the following formula:

$$\bar{\theta}_i(y) = \max i r_i - \sum_{j \in J} d_{ij}$$

s.t. (51.6)

$$r_i - y_j \leq d_{ij}, \ d_{ij} \geq 0 \quad \forall j \in J$$

It follows from (51.6) that $\bar{\theta}_k(y) = \max\left\{ k r_k - \sum_{j \in J} (y_j - r_k)_+ : y \in A \right\}$, where $(.)_+$ denotes the nonnegative part of a number and r_k is an auxiliary (unbounded) variable. The latter, with the necessary adaptation to the minimized outcomes in the location problems, is equivalent to the computational formulation of the k–centrum model introduced in [15]. Hence, the LP dual transformation provides an alternative proof of that formulation.

Following (51.6), we may express the MMF problem (51.2) as a standard lexicographic optimization problem with predefined linear criteria:

$$\text{lex max} \left(r_1 - \sum_{j \in J} d_{1j}, 2r_2 - \sum_{j \in J} d_{2j}, ..., m r_m - \sum_{j \in J} d_{mj} \right)$$

s.t.

$$y \in A$$ (51.7)

$$d_{ij} \geq r_i - y_j \quad \forall i, j \in J$$

$$d_{ij} \geq 0 \quad \forall i, j \in J$$

We will refer to the above model as the Cumulated Ordered Outcomes (COO) approach.

Theorem 1. *An attainable outcome vector $\in A$ is an optimal solution of the MMF problem (51.2), if and only if it is an optimal solution of the COO model (51.7).*

Note that this direct lexicographic formulation of the COO model remains valid for nonconvex (e.g. discrete) models, where the standard sequential approaches [9] are not applicable. Model COO preserves the problem's convexity when the original problem is defined with convex feasible set A. In particular, for an LP original problem, it remains within the LP class while introducing $m^2 + m$ auxiliary variables and m^2 constraints. Thus, for many problems with not too large a number of services (demands) m, problem (51.7) can easily be solved directly.

51.3 Shortfalls to Ordered Targets

For some specific classes of discrete, or rather combinatorial, optimization problems, one may take advantage of the finiteness of the set of all possible outcome values. The ordered outcome vectors may be treated as describing a distribution of outcomes y. In the case where there exists a finite set of all possible outcomes, we can directly describe the distribution of outcomes with frequencies of outcomes. Let $V = \{v_1, v_2, ..., v_r\}$ (where $v_1 < v_2 < ... < v_r$) denote the set of all attainable outcomes. We introduce integer functions $h_k(y)$ ($k = 1, 2, ..., r$) expressing the number of values v_k in the outcome vector y. Having defined functions h_k, we can introduce cumulative distribution functions:

$$\bar{h}_k(y) = \sum_{l=1}^{k} h_l(y), \quad k = 1, ..., r \tag{51.8}$$

Function $\bar{h}_k$ expresses the number of outcomes smaller than or equal to v_k. Since we want to maximize all the outcomes, we are interested in the minimization of all functions $\bar{h}_k$. Indeed, the following assertion is valid [11]: For outcome vectors $y', y'' \in V^m$, $\langle y' \rangle \geq \langle y'' \rangle$ if and only if $\bar{h}_k(y') \leq \bar{h}_k(y'')$ for all $k = 1, 2, ..., r$. This equivalence allows us to express the MMF problem (51.2) in terms of the standard lexicographic minimization problem with objectives $2\bar{h}(y)$:

$$\text{lex min} \{ (\bar{h}_1(y), ..., \bar{h}_r(y)) : y \in A \} \tag{51.9}$$

Theorem 2. *An attainable outcome vector $y \in A$ is an optimal solution of the MMF problem (51.2), if and only if it is an optimal solution of the lexicographic problem (51.9).*

The quantity $\bar{h}_k(y)$ can be computed directly by the minimization:

$$\bar{h}_k(y) = \min \sum_{j \in J} z_{kj}$$

$$\text{s.t. } v_{k+1} - y_j \leq C z_{kj}, \ z_{kj} \in \{0, 1\} \quad \forall j \in J$$

where C is a sufficiently large constant. Note that $\bar{h}_r(y) = m$ for any y, which means that the r-th criterion is always constant and therefore redundant in (51.9). Hence, the lexicographic problem (51.9) can be formulated as the following mixed integer problem:

$$\text{lex min} \left(\sum_{j \in J} z_{1j}, \sum_{j \in J} z_{2j}, ..., \sum_{j \in J} z_{r-1,j} \right)$$

s.t.

$$v_{k+1} - y_j \leq C z_{kj} \quad j \in J, \ k < r$$

$$z_{kj} \in \{0, 1\} \quad j \in J, \ k < r$$

$$y \in A$$

$\tag{51.10}$

Taking advantage of possible weighting and cumulating achievements in lexicographic optimization, one may eliminate auxiliary integer variables from the achievement functions. For this purpose we weight and cumulate vector $2\bar{h}(y)$ to get $\hat{h}_1(y) = 0$ and:

$$\hat{h}_k(y) = \sum_{l=1}^{k-1} (v_{l+1} - v_l)\bar{h}_l(y), \quad k = 2, ..., r \tag{51.11}$$

Due to the positive differences $v_{l+1} - v_l > 0$, the lexicographic minimization problem (51.9) is equivalent to the lexicographic problem with objectives $2\hat{h}(y)$:

$$\text{lex min}\left\{ \left(\hat{h}_1(y), ..., \hat{h}_r(y) \right) : y \in A \right\} \tag{51.12}$$

which leads us to the following assertion.

Theorem 3. *An attainable outcome vector $y \in A$ is an optimal solution of the MMF problem (51.2), if and only if it is an optimal solution of the lexicographic problem (51.12).*

Actually, vector function $\hat{\mathbf{h}}(y)$ provides a unique description of the distribution of coefficients of vector y, i.e., for any $y', y'' \in V^m$ one gets: $\hat{\mathbf{h}}(y') = \hat{\mathbf{h}}(y) \Leftrightarrow \langle y' \rangle = \langle y \rangle$. Moreover,s $\hat{\mathbf{h}}(y') \leq \hat{\mathbf{h}}(y'')$ if and only if $\bar{\Theta}(y') \geq \bar{\Theta}(y)$ [11].

Note that $\hat{h}_1(y) = 0$ for any y, which means that the first criterion is constant and redundant in problem (51.12). Moreover, putting (51.8) into (51.11) allows us to express all achievement functions $\hat{h}_k(y)$ as piecewise linear functions of y:

$$\hat{h}_k(y) = \sum_{j \in J} \max\{v_k - y_j, 0\} \quad k = 1, ..., r \tag{51.13}$$

Hence, the quantity $\hat{h}_k(y)$ can be computed directly by the following minimization:

$$\hat{h}_k(y) = \min \sum_{j \in J} t_{kj}$$

$$\text{s.t.} \tag{51.14}$$

$$v_k - y_j \leq t_{kj}, \ t_{kj} \geq 0 \quad \forall j \in J$$

Therefore, the entire lexicographic model (51.12) can be formulated as follows:

$$\text{lex min}\left(\sum_{j \in J} t_{2j}, \sum_{j \in J} t_{3j}, ..., \sum_{j \in J} t_{rj} \right)$$

$$\text{s.t.} \tag{51.15}$$

$$v_k - y_j \leq t_{kj} \quad j \in J, \ k = 2, ..., r$$

$$t_{kj} \geq 0 \quad j \in J, \ k = 2, ..., r$$

$$y \in A$$

We will refer to the above model as the Shortfalls to Ordered Targets (SOT) approach.

Note that the above formulation, unlike the problem (51.10), does not use integer variables and can be considered as an LP modification of the original constraints (51.1). Thus, this model preserves the problem's convexity when the original problem is defined with a convex set A. The size of problem (51.15) depends on the number of different outcome values. Thus, for many problems with not too large a number of outcome values, the problem can easily be solved directly. Note that in many problems of telecommunications network design, the objective functions express the quality of service, and one can easily consider a limited finite scale (grid) of the corresponding outcome values. One may also notice that model (51.15) opens a way for the fuzzy representation of quality measures within the MMF problems.

51.4 Computational Experiments

We have performed some initial tests of the sequential approaches to the MMF network dimensioning problem (51.1). We have not assumed any bandwidth granulation and thereby no grid of possible bandwidth values that can be allocated. Therefore, in the case of the Shortfalls to Ordered Targets approach, the resulting bandwidth allocation is only an approximation to the exact MMF solution.

For the experiments we used a set of 10 randomly generated problems for each tested size. The problems were generated as follows. First, we created a random but consistent network structure. Then we chose random node pairs to define services. For each service three different possible flow routes between the two end nodes were generated. Two of them were fully random and one was the shortest path between the nodes (with the smallest number of links). We decided to use the integer grid of the v_k values in the ordered values approach, that is, to check each integer value from the feasible set of objective values. In this case the number of targets depends on the range of the feasible objective values. We managed to restrict the number of targets to the range of 5–10, applying different link capacities for different problem sizes. We set the large budget limit to B, thus relaxing the budget constraints (51.1 g).

We analyzed the performance of the three sequential approaches: the Direct Ordered Outcomes (DOO) model (51.4), the Cumulated Ordered Outcomes (COO) model (51.7), and the Shortfalls to Ordered Targets (SOT) model (51.15), with the condition $y \in A$ representing the bandwidth allocation problem defined with constraints (51.1). Each model was computed using the standard sequential algorithm for lexicographic optimization with predefined objective functions. For the lexicographic maximization problem $\text{lex} \max\{(g_1(y), ..., g_m(y)) : y \in Y\}$, the algorithm reads as follows:

Step 0: Put $k := 1$.

Step 1: Solve problem P_k:
$$\max_{y \in Y}\{\tau_k : \tau_k \leq g_k(y), \ \tau_j^0 \leq g_j(y) \ \forall j < k\}$$
denote the optimal solution by (y^0, τ_k^0).

Step 2: If $k = m$, **STOP** (y^0 is MMF optimal).
Otherwise, put $k := k + 1$ and
go to **Step 1**.

For example, the algorithm for the COO model worked according to the above scheme with functions g_k defined as $kt_k - \sum_{j \in J} d_{kj}$. Let $k = 1$. Following (51.7), we built the initial problem P_1 with the objective $\tau_1 = t_1 - \sum_{j \in J} d_{1j}$ being maximized and m constraints of the form $t_1 - d_{1j} \leq y_j$, $j = 1...m$. The expression $y \in A$ of (51.7) was replaced by (51.1). Each new problem P_k in subsequent iterations ($k > 1$) was built by adding new constraints $\tau_{k-1}^0 \leq t_{k-1} - \sum_{j \in J} d_{k-1,j}$ and $t_k - d_{kj} \leq y_j$, $j = 1...m$ to problem P_{k-1}, where τ_{k-1}^0 was the optimal objective value of P_{k-1}. A similar algorithm was performed for the DOO and the SOT approaches. The difference was in the objectives and auxiliary constraints, as defined in (51.4) and (51.15), respectively. All the tests were performed on a Pentium IV 1.7 GHz computer employing the CPLEX 9.1 package.

Table 51.1 presents solution times for the three approaches being analyzed. The times are averages of 10 randomly generated problems. The upper index denotes the number of tests out of 10 for which the timeout of 120 s occurred. The minus sign '−' shows that the timeout occurred for all 10 test problems. One can notice that, while for smaller problems with the number of services equal to five, all three approaches perform very well, for bigger problems only the SOT approach gives acceptable results (in the sense of solving a majority of the problems within the 120 s time limit).

To examine how the number of targets in the SOT approach influences the test results, we also performed similar experiments increasing the capacities of the links and considering 15 to 25 targets. This did not significantly affect the DOO and COO approaches. For the SOT approach the computing times

Table 51.1 Computation times (in s) for different solution approaches

	# of nodes	# of links	5	10	20	30	45
			Number of services				
	5	10	0.0	1.2			
DOO	10	20	0.0	6.8	−	−	−
(4)	15	30	0.0	3.9	−	−	−
	5	10	0.0	0.2			
COO	10	20	0.0	1.3	[3]62.9	−	−
(7)	15	30	0.1	1.0	[5]78.0	−	−
	5	10	0.1	0.1			
SOT	10	20	0.0	0.3	4.1	[2]35.0	[7]101
(15)	15	30	0.1	0.3	7.1	[4]72.4	[8]106

Table 51. 2 Computation times (in s) for problems with increased link capacities

	# of nodes	# of links	Number of services				
			5	10	20	30	45
	5	10	0.1	0.1			
SOT	10	20	0.1	1.3	23.8	[4]74.3	–
(15)	15	30	0.1	1.2	33.9	[8]108.0	–

increased (Table 51.2), but it still outperformed both the DOO and COO approaches.

51.5 Conclusion

As lexicographic maximization in the Max-Min Fair optimization is not applied to any specific order of the original outcomes, the MMF optimization can be very hard to implement in general nonconvex (possibly discrete) problems. We have shown that the introduction of some artificial criteria with auxiliary variables and linear inequalities allows one to model and to solve the MMF problems in a very efficient way. We have performed initial tests of computational performance of the presented models for the MMF network dimensioning problem. It turns out that both the models outperform the Direct Ordered Outcomes model. The Shortfall to Ordered Targets model enables to solve within 2 min a majority of the MMF single-path dimensioning problems for networks with 15 nodes and 30 links. Such performance is enough for the efficient analysis of a country's backbone network of ISPs (12 nodes and 18 links in the case of Poland [14]). Nevertheless, further research is necessary on the models and corresponding algorithms tailored to specific MMF network optimization problems. The models may also be applied to various MMF resource allocation problems, not necessarily related to networks.

Acknowledgments This research was supported by the Ministry of Science and Information Society Technologies under grant 3T11C 005 27, "Models and Algorithms for Efficient and Fair Resource Allocation in Complex Systems."

References

1. Behringer FA (1981) A simplex based algorithm for the lexicographically extended linear maxmin problem. Eur J Oper Res 7:274–283
2. Bertsekas D, Gallager R (1987) Data networks. Prentice-Hall, Englewood Cliffs, NJ
3. Denda R, Banchs A, Effelsberg W (2000) The fairness challenge in computer networks. LNCS 1922:208–220
4. Jaffe J (1980) Bottleneck flow control. IEEE T Commun 7:207–237
5. Klein RS, Luss H, Rothblum UG (1993) Minimax resource allocation problems with resource-substitutions represented by graphs. Oper Res 41:959–971

 6. Kleinberg J, Rabani Y, Tardos E (2001) Fairness in routing and load balancing. J Comput Syst Sci 63:2–21
 7. Lee CY, Choo HK (2007) Discrete bandwidth allocation considering fairness and transmission load in multicast networks. Comput Oper Res 34:884–899
 8. Linardakis Ch, Leligou HC, Angelopoulos JD (2004) Performance evaluation of a distributed credit-based fairness mechanism for slotted WDM rings. In: 8th WSEAS international conference on communications, paper 487–455
 9. Luss H (1999) On equitable resource allocation problems: a lexicographic minimax approach. Oper Res 47:361–378
10. Nace D, Doan LN, Klopfenstein O, Bashllari A (2008) Max-min fairness in multicommodity flows, Comput Oper Res 35:557–573
11. Ogryczak W (1997) Linear and discrete optimization with multiple criteria: preference models and applications to decision support. Warsaw University Press (in Polish), Warsaw
12. Ogryczak W, Pióro M, Tomaszewski A (2005) Telecommunication network design and maxmin optimization problem. J Telecomm Info Tech 3:43–56
13. Ogryczak W, Śliwiński T (2006) On direct methods for lexicographic min-max optimization. LNCS 3982:802–811
14. Ogryczak W, Śliwiński T, Wierzbicki A (2003) Fair resource allocation schemes and network dimensioning problems. J Telecomm Info Tech 3:34–42
15. Ogryczak W, Tamir A (2003) Minimizing the sum of the k largest functions in linear time. Info Proc Let 85:117–122
16. Ogryczak W, Wierzbicki A, Milewski M (2008) A multi-criteria approach to fair and efficient bandwidth allocation. Omega 36:451–463
17. Pióro M, Medhi D (2004) Routing, flow and capacity design in communication and computer networks. Morgan Kaufmann, San Francisco
18. Pióro M, Nilsson P, Kubilinskas E, Fodor G (2003) On efficient max-min fair routing algorithms. In: Proceedings of the 8th IEEE ISCC'03, pp 365–372
19. Salles RM, Barria JA (2005) Fair and efficient dynamic bandwidth allocation for multiapplication networks. Comp Netw 49:856–877
20. de Silva R (2004) A simple approach to congestion control of ABR traffic with weighted max-min fairness. WSEAS T Comput 3:75–78
21. Yager RR (1997) On the analytic representation of the Leximin ordering and its application to flexible constraint propagation. Eur J Oper Res 102:176–192

Chapter 52
A Temporal Variation in Indoor Environment

A. Jraifi, E.H. Saidi, A. El Khafaji, J. El Abbadi

Abstract In order to model the propagation of radio waves and develop a real mobile communications system, a good knowledge of environmental propagation is crucial. Usually, the temporal variations of signal amplitude between transmitter (Tx) and receiver (Rx) are neglected in modeling radio propagation. According to Hashemi (IEEE J Sel Areas Commun, 2003), the effects of temporal variations in the indoor environment caused mostly by human body shadowing cannot be neglected. In this paper we demonstrate the effects of temporal variations due to the movement of persons between Tx and Rx. The method used to predict wave propagation in the indoor environment is a deterministic technique called ray tracing, mainly used for simulating a static environment; and the human body is modeled as a parallelepiped form full of salty water. Finally, we have proposed an algorithm to mitigate the unavailability of the channel when the period of obstruction caused by human bodies becomes significant.

52.1 Introduction

To model radio propagation channels, designers require detailed understanding of environmental propagation. In an indoor radio channel (when the Tx and Rx are within a closed room or building) the temporal variation is very significant, especially in the high frequency bands (millimeter waves). This temporal variation within an indoor environment is usually caused by the movement of people between Tx and Rx. Increased attenuation can be generated when a person passes through the LOS (line of sight) between transmitter and receiver. Such attenuation can negatively affect the QoS (quality of service) of the radio channel. A deterministic method called ray tracing is used for modeling radio waves in an indoor environment. Usually, this method is used for a static

A. Jraifi (✉)
Groupe CPR, Canal, Radio & Propagation, Lab/UFR-PHE , Faculté des Sciences de Rabat, Morocco

N. Mastorakis et al. (eds.), *Proceedings of the European*
Computing Conference, Lecture Notes in Electrical Engineering 27,
DOI 10.1007/978-0-387-84814-3_52, © Springer Science+Business Media, LLC 2009

environment. However, an effect of the ray tracing technique is used to predict the impulse response, local mean power, and rms delay spread of an arbitrary indoor environment. The model is based on geometrical optics and on the uniform geometrical theory of diffraction (UTD), in which the geometry of the environment is user definable. The model makes full use of reflection, transmission, and diffraction. The mobile radio channel with variable characteristics in time is modeled in the following way: For each point of space, the resulting channel is a linear time-varying filter [1] with the impulse response given by:

$$h(t, \tau_k) = \sum_{k=0}^{N(\tau)-1} a_k(t)\delta(t - \tau_k)\exp(j\varphi_k(t)) \qquad (52.1)$$

Where t is the observation time and τ is the application time of the emitted impulsion. $N(\tau)$ the number of paths components, a_k, and φ_k describe respectively the random time-varying amplitude, arrival time, and phase sequences. The channel is completely characterized by these path variables. This general model is often simplified and given in time-invariable form. This form was introduced first by Turin [2] and was used successfully to model a mobile radio channel [3, 4]. For a stationary channel, the above equation is reduced to:This modeling of the radio environment as described in eq. (52.1) does not take into account the effect of the human body shadowing caused by the movement of persons between Tx and Rx. These temporal variations cannot be obtained directly by the classical ray tracing technique. However, it is possible to benefit from the advantages of ray tracing modeling by creating a dynamic model of the propagation environment. In this paper we present a technique for determining the temporal variations of indoor radio channels due to body shadowing. The paper is organised as follows: In Section 52.2, we give a brief presentation of the previous work of Hashemi [5]. In Section 52.3, we present the human body model proposed by [6] used in this work. In Section 52.4, we give some results of the simulation. In Section 52.5, we propose an algorithm to mitigate the unavailability of the channel. In Section 52.6, we present our conclusion.

52.2 Effect of Body Shadowing

52.2.1 Statistical Distribution

A typical study of the temporal variations of an indoor radio channel was reported by Hashemi [5]. Measurements were taken in four different buildings, with a variable number of people moving, each recording lasting 60 s. The results showed (in certain cases) very deep fading, the fading dependent on the nature of the environment and the number of people moving around the base station or around the mobile station. The fading reached as much as −30 dB, compared to the average value of the signal. By using the mean square error, Hashemi [1] showed that the temporal variations of the signal were described by the densities of probability of Weibull, Nakagami, Rice, lognormal, and Rayleigh, with percentages of 37, 31.8, 22.3, 7.8, and 1%, respectively. The raised results showed a good fit to the Nakagami and Weibull distributions, but on the other hand the Rayleigh distribution was the worst to represent them. Several measurements of propagation in the narrow band at 1 GHz [6,7] showed that in the line of sight path, the temporal variations can be modeled by the Rice distribution. Another study of propagation around 60 GHz showed that the temporal variations of the signal envelope followed the law of Rayleigh [7].

52.2.2 Amplitude Correlation

In research by [5], the coefficient of amplitude correlation ρ was recorded with a periodicity equal to t and a transmitter/receiver distance equal to 5 m. The typical values raised were: for t = 0.2 s, $\rho(t)$ varies from 0.48 to 0.6; for t = 0.5 s, $\rho(t)$ varies from 0.21 to 0.34; generally $\rho(t)$ remains lower than 0.2 with t durations higher than 1 s. This result shows we can have several versions of the signal with a weak correlation, and therefore a great statistical independence by a simple temporal shift of the signal. This property is exploited to produce a temporal micro- diversity given by the delay lines with surface acoustic waves (SAW). This technique is used particularly in the structure of receiver RAKE of the systems with spread out spectrum with direct sequence [8,9].

52.2.3 Level Crossing Rate and Duration of Fading

Other parameters which characterize the temporal variations are LCR (level crossing rate) and the average fade duration (AFD). These two parameters directly influence the performances of the communication systems. The measurements derived by Hashemi [5] showed that the LCR and the durations of fading depend on the transmitter/receiver distance, the number of people moving near the transmitter, and the number of people moving near the receiver.

52.3 Adopted Model of the Human Body

There are two models [10] to simulate human body shadowing in an indoor environment. The first is known as "Phantom Human," and the second is called "Salty-Lite;" and for both, the human body is modeled as a cylinder containing salty water. In this paper, we have adopted the model proposed by [6]. In this model, the human body is represented as a parallelepiped circumscribed with the salty cylinder model (see Fig. 52.1). The length of the sides of the basic

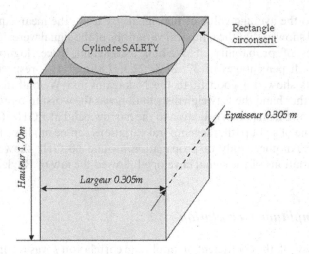

Fig. 52.1 Model of human body

rectangle will be 0.305 m. The height is 1.7 m, the thickness is 0.305 m. We model the persons moving near the mobile radio link by objects having finished dimensions in the form of parallelepipeds, characterized by a permittivity r and conductivity. The model assigns to each object a position which will be modified according to its speed and direction. After having evaluated the impulse response of the channel, the amplitude, and the mean square delay, the technique calculates the position of the objects. The results are stored in a file to be analysed and interpreted. This model has the advantage of being suitable for the deterministic technique (ray tracing) used mainly for plane surfaces.

52.4 Simulation of Temporal Variations of Signal

52.4.1 Simulating Setting

To implement the model described above, we consider a room of dimensions 10×20 m in which we simulate several people moving around near the radio link at 2.4 GHz. The starting positions of the people are random, the speeds are supposed to be constant at 0.5 m/s, the positions of the transmitter and receiver are indicated in Fig. 52.2 and remain fixed during the simulation. The model representing the human body is like a wall with two identical facets: these two facets can reflect waves emitted by the transmitter just as they can refract them and transmit them. A series of measurements is obtained by an automatic change in the positions of the objects modeling the people moving, with respect to their speed of movement and their direction. Measurements are taken at regular time intervals, which makes it possible to compute the positions of the

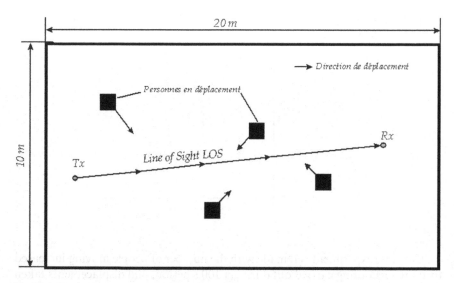

Fig. 52.2 Environment propagation

Table 52.1 The input data for our simulation tool

Height of walls	2.8 m
Thickness of walls	0.1 m
Relative permittivity of wall	1.6
Conductivity of walls	0.00105 S/m
Frequency	2.4 GHz
Polarization	Vertical
Power	0 dBm
Gain antenna	0 dB
Height transmitter/receiver	1 m
Number of reflections for each path	5
Transmission	Activated
Diffraction	Deactivated
Minimal level	−150 dBm
Interval of time computing	0.25 s

objects again and to make a new calculation of the parameters of the channel. The input data for our simulation tool are given in Table 52.1.

52.4.2   Analysis of Narrow Band Channel

52.4.2.1 The Temporal Variation of the Signal Envelope

The results recorded during 90 s of simulation and a transmitter/receiver distance of 15 m is presented in Fig. 52.3. This figure shows fast fading and variations around an average value of −62 dBm. The maximum depth of fading is −12 dB.

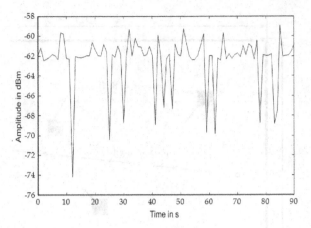

Fig. 52.3 Temporal
variations for four persons

The same experiment is remade with the number of people moving increased to 10, the recording carried out during a 100 s period, and displacements taken to be random within the room. The resulting measurements are presented in Fig. 52.4. Analysis of these data shows that the fading in this case is deeper and can reach a value of −30 dB; the frequency of fading is more significant. The average value of the amplitude has not changed because the transmitter/receiver distance is always 15 m.

52.4.2.2 Level Crossing Rate

The level crossing rate is the number of times the signal crosses a level in the positive direction. This measurement of the frequency of fading in the fixed level represents the sensitivity of the receiver. The LCR also allows estimation of the average duration of fading in order to determine the code detecting and correct channel error most suitably. To evaluate the LCR, we carried out four

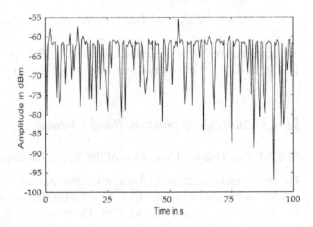

Fig. 52.4 Temporal
variations for 10 persons

Fig. 52.5 LCR of signal

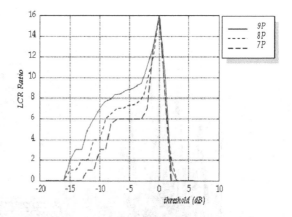

recordings of the amplitude of the signal with 6, 7, 8, and 9 people moving in the simulated propagation environment. Figure 52.5 illustrates the obtained results. The LCR is calculated for thresholds varying from -20dB to + 10dB compared to the average value of the amplitude of the signal and a transmitter/ receiver distance of 15 m. Figure 52.5 shows the variations of the LCR according to the threshold for the four simulated situations. For increasingly negative thresholds, the LCR growth reaches its maximum at 0dB compared to the average, then decreases to reach 0 with a few dB compared to the average value. Another obvious result illustrated by Fig. 52.5 is that the LCR is as large as the number of people moving in the environment. This conclusion was also deduced by Hashemi in his study presented previously [5], i.e., the fading becomes increasingly frequent.

52.4.3 Analysis of Wide Band Channel

52.4.3.1 Impulse Response of the Channel

The temporal variations of the channel also result in a temporal variation of the multipath components of the impulse response. The model of ray tracing makes it possible to predict the impulse response of the channel for a given transmitter/ receiver link. We can record the evolution of the impulse response over time in the case of the radio link and the simulated environment. Figure 52.6 shows an example of impulse response determined by the model. The same preceding parameters are used with 10 people moving, a transmitter/receiver distance of 15 m, and a duration of simulation of 60 s. The impulse response shows that although the simulated situation presents a LOS situation, the amplitude of the first path (the path of weaker time) is not always the largest; this indicates that the moving people mask the direct path from time to time and create a temporary obstruction. The temporal variations of the multipath components of the impulse response give place to temporal variations of the rms delay spread. The

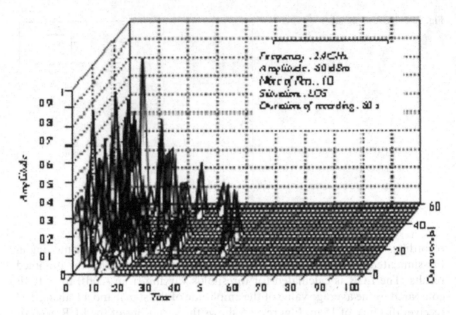

Fig. 52.6 Impulse response

previous run makes it possible to calculate and trace the variations of this factor in the form of cumulative distribution. The analysis of the results shows a weak variation of the rms delay spread, which remains lower than 25 nS.

52.5 Solution to the Unavailability of the Channel

An uncontested fact in wireless transmission is the vulnerability of the radio channel in contrast to other methods (coaxial, FO). This vulnerability is due mainly to the various mechanisms of propagation (reflections, refractions, etc.) and to the temporal variations caused by the movement of the people in the channel, especially in an indoor environment. In some cases, in an indoor environment, when the duration of the obstruction caused by people between the line of sight of Tx and Rx increases, the radio link goes out of service and prevents communication between Tx and Rx. Usually, the obstruction caused by temporal variation results in the unavailability of the channel. This unavailability can be defined as the probability that the channel is out of service at any given time. It is expressed as:

The duration of the unacceptable response time is generally very long, and more than ten seconds can cause the unavailability of the channel. Very often, we have to distinguish two situations: if the duration of the period of inaccessibility is relatively short, this period will not be counted as unavailability, only the longer periods will be counted in the calculations of unavailability and average accessibility. This unavailability of the radio channel affects the QoS negatively. It causes a reduction in revenue for the operators. For this reason, the search for another palliative solution to mitigate the problem of unavailability is essential. In this context, we propose an algorithm which aims the diversity of the radio link in the event of failure of the principal link between Tx and Rx.

Proposed algorithm:

1. Observe the duration of the obstruction.
2. Compare to the threshold duration.
3. If Tobs > Tshoold————' Un handover (HO) between TXs must be activated.
4. Return to 1.

52.6 Conclusion

In this paper, we have elaborated a new method with which to study and characterize the temporal variations of an indoor channel which are caused mainly by the random movements of people. We have made a model of the human body as a parallelepiped form full of salty water; we have then applied ray tracing techniques to determine the temporal variation of the indoor radio channel. Comparison of our simulation results with various statistical distributions shows a good fit with the Nakagami distribution. We have also analyzed the case of the wideband channel. This method permits us to find very significant results, similar to the empirical results already published [5].

References

1. Hashemi H (1993) The indoor radio propagation channel. In: Proceedings of the IEEE 81(7):941–968
2. Turin GL et al (1972) A statistical model of urban multipath propagation. IEEE T Veh Technol VT-21:1–9
3. Suzuki H (1977) A statistical model for urban radio propagation. IEEE T Commun COM 28(7):673–680
4. Hashemi H (1979) Simulation of the urban radio propagation channel. IEEE T Veh Technol VT-28:213–224
5. Hashemi H (1993) Impulse response modeling of indoor radio propagation channels. IEEE J Sel Area Commun
6. El Abbadi J, Khafaji A, Belkasmi M, Benuna A (2003) A human body model for ray tracing indoor simulation.
7. Siwiak K (1995) Radiwave propagation and antennas for personal communications. Artech House Publishers, London
8. Howard SJ, Pahlavan K (1991) Fading results from narrowband measurements of the indoor radio channel. In: Proceedings of the second IEEE international symposium on personal indoor and mobile radio communications, London, pp 92–97

9. Rappaport TS (1989) Indoor radio communications for factories of the future. IEEE Commun Mag 117–126
10. Howard SJ, Pahlavan K (1990) Measurements of the indoor radio channels. Electron Lett 26(2):107–109
11. Bultitude RJC, Mahmoud SA (1987) Estimation of indoor 800/900 MHz digital radio channels performance characteristics using results from radio propagation measurements. In: Proceedings of the international communication conference ICC'87, pp 70–75

Chapter 53
Workflow Management for Cross-Media Publishing

Andreas Veglis and Andreas Pomportsis

Abstract During the last two decades of the 20th century, ICTs (Information Communication Technologies) have revolutionized newspaper organizations. As technology for the distribution of journalistic information in various forms has become more easily available, and with the Internet and the World Wide Web's introduction into companies and households, the tendency has been for the larger media organizations and companies to have several publication channels at their disposal. This paper addresses the issue of workflow management in newspaper organizations that implement cross-media publishing.

53.1 Introduction

Workflow is the specification and the execution of a set of coordinated activities representing a business process within a company or an organization [1]. A workflow is an organization-wide task-sharing process typically involving a large number of people and software systems. A process includes work activities which are separated into well-defined tasks, roles, and rules [3].

Workflow management is considered to be a crucial factor because by knowing the underlying business process and its structure, a workflow management system can provide support in assigning activities to people, in allocating resources, and in monitoring the status of activity execution. It is also worth noting that workflow management is often used synonymously with administration and automation of business processes. A workflow management system is a software system consisting of several tools supporting the tasks and functions of workflow management [1, 3].

A published newspaper is the net result of the integrated accumulative work of a group of people [5, 11]. Digital technologies have been a part of the daily newspaper world for many decades. Newspapers began setting type using computers in the early 1960s. By the early 1980s, most newspapers were using

A. Veglis (✉)
Department of Journalism and MC, Aristotle University of Thessaloniki, 54006,
Thessaloniki, Greece

N. Mastorakis et al. (eds.), *Proceedings of the European*　　　　　　　533
Computing Conference, Lecture Notes in Electrical Engineering 27,
DOI 10.1007/978-0-387-84814-3_53, © Springer Science+Business Media, LLC 2009

digital systems to set type in galleys, which were cut and pasted into pages, and then imaged. Today the world's leading-edge newspapers are moving to 100 percent digital page assembly and distribution, streamlining workflow processes while ensuring higher quality.

Usually, newspaper organizations are occupied only with the distribution of print newspapers. But as technology for the distribution of journalistic information in various forms has become more easily available, there has been a trend for the media organizations and companies to publish news in various publication channels [9]. Cross-media is defined as any content (news, music, text, and images) published in multiple media (for example print, web, and TV). The content is posted once and it is available on other media. All the previously mentioned changes lead to the conclusion that newspapers have to alter their document workflow in order to adapt to cross-media publishing. That means that workflow management systems for cross-media publishing is an important issue for modern newspaper organizations.

This paper addresses the issue of workflow management in newspaper organizations. More precisely, it investigates the workflow management that facilitates cross-media publishing.

The rest of the paper is organized as follows: Section 53.2 briefly discusses cross-media publishing. The workflow in a paper edition is presented in Section 53.3. The following section discusses the workflow in a cross-media publishing organization. Publishing speed and rhythm are examined in Section 53.5. Concluding remarks can found in the last section.

53.2 Cross-Media Publishing

During the previous decade an increasing number of newspaper companies began publishing electronic editions in addition to the printed editions of their newspapers. Today, newspaper publishers are increasingly distributing their editorial and advertising material over several delivery channels, primarily a combination of print and the WWW [7]. More specifically, they are re-purposing content from the printed editions in various secondary electronic editions, notably on the WWW [8]. A list of the publishing channels used by newspapers nowadays is included in Table 53.1.

Table 53.1 List of publishing channels used by newspapers nowadays

Publishing channels	
Print	CD/DVD-ROM
WWW	Webcasting
PDA	Tablet PC
E-mail	SMS
PDF	WAP
RSS	Wi-Fi

The problem is that although the news and information sources for the printed and the alternative editions are often the same, because the cooperation between the different editorial departments is poor, the information is processed independently for nearly every channel. The above work scheme results in more workload and moderate outcome. There are examples from newspaper companies showing photographers and reporters from different channels within the same company covering the same story or event as if they were competitors [4].

Studies indicate that some Swedish newspaper organizations have separated their editorial departments, thus creating complicated editorial workflows. Because the departments work very much in parallel, the research, creative, and composition work is duplicated or even triplicated [10].

In order to solve this problem, newspaper and media companies worldwide are moving towards integrating their editorial departments for the different output channels. The underlying idea behind these kinds of integration projects most often is to have a common news desk for all publishing channels, where incoming news information is evaluated regarding, for example, news value and suitable channel for pub-lishing.

Until recently, the aim of the newspaper or media company management often was to require that most reporters work with all possible output channels in mind at all times. In order that no limit as to the number of possible output channels be set in advance, the single multimedia news reporter should be able to collect content in the form of, for example, images, video clips, written and recorded interviews—all at the same time [10].

53.3 Paper Edition

The workflow in a traditional editorial department for printed newspapers has been set for decades, or even centuries. There are a number of people (depending on the size of the newspaper organization) that are involved in the production workflow, namely, reporters, editors, etc. The editor-in-chief assigns a reporter to a certain task. The reporter researches and gathers background information about the news. Images are shot by the photographer, and, if needed, graphics/illustrations are drawn. The assembled information is evaluated and a decision is made. If the decision is positive, the story is written and another evaluation occurs. If the article is not adequate, the previous steps are repeated. The completed pages are composed and sent to the printing plant for printing, bundling, and distribution. The whole process is depicted in Fig. 53.1. We must note that in order to keep the workflow as simple as possible, we have not included the possibility of permanently rejecting a story, or of holding it in order to publish it in the near future.

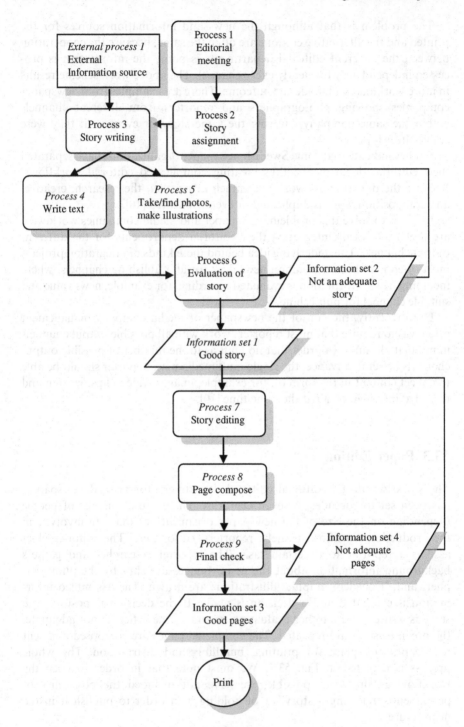

Fig. 53.1 Document workflow in a newspaper

53.4   Cross-Media Publishing

In the case of a cross-media newspaper organization, the news information is collected and evaluated collectively by one or more news editors at an input desk. Assignments are given to reporters, photographers, illustrators, and other people on the editorial staff. The content is edited and composed for different output channels, then published and distributed. In this model, there is a so-called publishing channel selection process involved. The information workflow in a newspaper organization that employs alternative publishing channels has been proposed by Sabelström [7].

We have extended this model in order to include more publishing channels, and we have also included the possible interactions that may take place between the different information streams (text, voice, video, etc.).

The full version of the information workflow is included in Fig. 53.2. The grey processes indicate the output of a publishing channel.

We must note that in WWW, tablet PC, PDF, and printed editions, we have included an evaluation of the story and final check processes. These processes are not included in the other more immediate publication channels. There is of course some kind of evaluation and final check, but because in these channels the time element is very crucial, these processes are more informal.

It is worth mentioning the feed from video webcasting to the WWW and CD/DVD editions that enhances those channels with multimedia features. This means that multimedia files are used multiple times, probably with different output formats (for example, video in 640×480 format can be included in a CD/DVD-ROM, but a lower resolution is required for webcasting).

53.5   Publishing Speed and Rhythm

One other parameter that must been taken into account is time. In Fig. 53.3 we plot the publishing channels versus time. The first channels, that relay the headline news to the readers, are SMS, e-mail, RSS, and WWW. Next, short story descriptions are available via voice or video webcasting. The full story is available first on web pages (WWW for PCs and tablet PCs, WAP) and PDF files sent via e-mail. Finally, we get the printed version of the story, which may be accompanied by a CD/DVD supplement perhaps including videos from the webcasting or other interactive features of the WWW version of the story. Except for the printed version, all other editions of the news can be updated several times during the day.

The rhythms of the publishing channels differ, depending on the type of information they include [6]. The lead time from the collection of news by a reporter until the point when the audience is able to receive the information

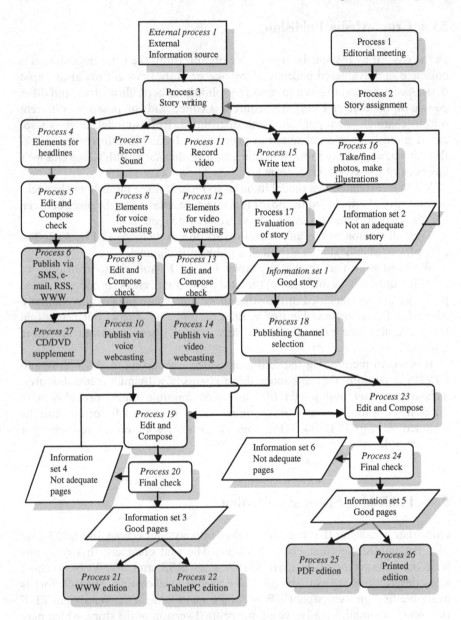

Fig. 53.2 Information workflow model for cross-media publishing

varies among the channels. A reporter writing for the printed edition of a newspaper often has several hours to do research and write the report, whereas the same reporter has less than half an hour to write a news item for the Web. On the other hand, webcasting publishes updated news bulletins on a schedule

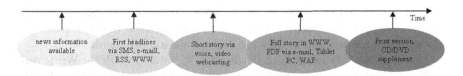

Fig. 53.3 Publishing channels versus time

several times a day; that is to say, the time for research and content creation is shorter than for the printed daily newspaper, which is published only once a day. The same thing holds for all other publishing channels, always depending on the publication schedule of each channel.

We propose the classification of publishing channels into four categories, depending on the frequency of updates in one day (Fig. 53.4). Printed newspapers are usually rather slow, with a publishing rhythm of 24 h (except very important events that may force the publication of a second or even third edition). In the same category we must include the PDF version of the printed edition (an exact duplicate), as well as CD/DVD supplements to the printed edition.

The next category with more than a single publication per day is webcasting. Webcasting resembles radio and television bulletins. The production and publishing rhythm is somewhat faster. The created content is published

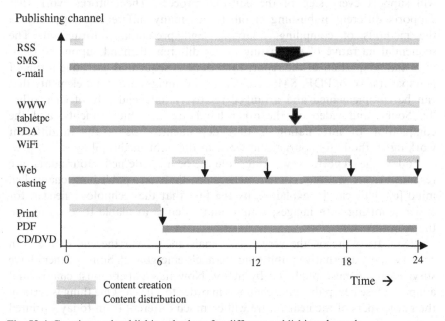

Fig. 53.4 Creation and publishing rhythms for different publishing channels

(broadcast) at least two or three times, depending on the newspaper's policy. Unlike newspaper production, there is no real difference between broadcast editing and production and they are carried out almost simultaneously.

The next category includes channels (WWW, tablet PC, PDA, WiFi) with more frequent publication periods. Content is published continuously day and night, with no deadlines. The distribution phase is integrated with the production phase. The final category includes RSS, SMS, and e-mail. These are the channels that usually broadcast breaking news faster than the other channels. One might argue that they must belong to the same category as WWW, but we chose to include them in a separate category in order to stress the speed of their publication and the fact that they are delivered (pushed) to the readers. When the news is received, they can be instantly published.

53.6 Conclusions

In this paper we have investigated the workflow management for cross-media publishing. We have proposed a detailed workflow that supports multiple publishing channels. The study has also focused on the publishing speed and rhythm of the publishing channels, which are considered to be crucial factors in workflow management. The above scheme can be implemented by a computer-supported collaboration work system (CSCWS) that will support every step of the editorial process. The editorial work that supports different publishing channels has many differences, especially in the methods of compiling, adapting, and presenting information. The sequential narrative of webcasting places different demands upon the planning and adaptation of the report than the less sequential narrative forms of newspapers, Web, PDF, SMS, etc. Texts and images are static elements that can be complemented and arranged relatively independently of each other [8]. Sound and video on the other hand are dynamic elements, and the compilation of information is time dependent. This results in different working methods for journalistic work in different media [6].

Today the reporter's working role is being rede?ned. Journalists are responsible for creating content with more than one publishing channel in mind [6]. This can be explained by the fact that the technology needed for easier reporting with images, sound, and video is available to everyone on the editorial staff.

Newspapers are on the verge of a renaissance. Near the end of the 20th century many alternative publishing channels emerged [2]. Some of them have survived the demise predicted by many. Now newspapers must change and adapt to these new publishing channels in order to survive. One thing is certain: the newspapers of the near future will be much different from today's printed editions.

References

1. Borghoff UM, Schlichter JH (2000) Computer-supported cooperative work. Springer, New York
2. Chisholm J (2002) The distribution revolution, shaping the future of the newspaper. Strategy Report, vol 2, Report No. 1, WAN
3. Mentzas G, Halaris C (1999) Workflow on the web: integrating e-commerce and business process management. Int J E-Bus Strat Manag 1(2):147–157
4. Northrup K (2001) Finding a happy medium for handling multiple media. Newspaper Techniques, English ed, pp 12–16
5. Reavy M (2001) Computer-assisted reporting: a journalist's guide. Mayfield Publishing Company, Mountain View, CA, USA
6. Sabelström K (2000) The multimedia news reporter: technology and work processes. In: TAGA's 52nd annual technical conference, Colorado Springs, Colorado, USA. Published in TAGA Proceedings, TAGA Office, Rochester, New York, USA, pp 53–66
7. Sabelström K, Enlund N (1999) Newspaper premedia work?ows for parallel publishing. In: TICGC—Taipei international conference on graphic communications, Taipei, Taiwan
8. Sabelström K, Nordqvist S, Enlund N (1997) Synergy in integrated publishing of printed and electronic newspapers: advances in printing science and technology. In: Bristow A (ed) Pira International, Surrey, UK vol 24
9. Veglis A, Pomportsis A (2004) New production models for newspaper organizations. WSEAS Trans Commun 3(1):218–222.
10. Veglis A, Pomportsis A, Avraam E (2004) Computer supported cooperative work in newspaper organizations. WSEAS Trans Inf Sci Appl 1(1):127–132
11. Zack MH (1996) Electronic publishing: a product architecture perspective. Inf Manage 31:75–86

Chapter 54
A Bayesian Approach to Improve the Performance of P2P Networks

Bertalan Forstner and Hassan Charaf

Abstract The task of efficient peer-to-peer information retrieval has been one of the most serious challenges in the history of information technology. Since the computing resources and increased usability of smartphones make these devices with their high proliferation a suitable platform for presenting different kinds of information, it becomes highly important to integrate them into the peer-to-peer (P2P) information retrieval world. There is an increasing need for mobile software that can locate and retrieve well-defined documents (texts, music, or video files) that are in the fields of interest of the mobile user. There are different proposals to make P2P information retrieval more efficient using semantic data; however, the application of these protocols in the mobile environment is not straightforward, because they generate higher network traffic and require longer online time for the participants. However, because of various considerations, such as the limited connectivity of these devices, they do not spend much time connected to the network. The Bayesian approach presented in this paper targets this strong transient character of mobile P2P clients. We will show how the network topology can be quickly improved to increase the hit rate with an appropriate protocol and algorithm using a Bayesian process based on local decisions derived from the fields of interest of the nodes.

54.1 Introduction

With the spread of broadband wireless communication, network applications have moved from desktop computers to intelligent mobile devices. One of the most important applications is the fully distributed information sharing client, the peer-to-peer application, because of its versatile usability. However, the relatively high cost of wireless communication requires the use of advanced

B. Forstner (✉)
Department of Automation and Applied Informatics, Budapest University
of Technology and Economics, H-1111 Budapest, Goldmann György tér 3., Hungary
e-mail: bertalan.forstner@aut.bme.hu

N. Mastorakis et al. (eds.), *Proceedings of the European*
Computing Conference, Lecture Notes in Electrical Engineering 27,
DOI 10.1007/978-0-387-84814-3_54, © Springer Science+Business Media, LLC 2009

protocols with small generated network traffic. These protocols should also tolerate the strong transient property of these networks, which derives from economic considerations and the limited connectivity of such mobile devices. The connection of a smartphone can be interrupted by various unexpected situations, such as low battery conditions or loss of network coverage. An average smartphone device can operate with constant network traffic less than a day without recharging its batteries.

Consequently, we should use an efficient protocol that fulfills the following requirements: (1) It reaches a higher hit rate than unstructured protocols in a short period of online time, while (2) it generates low network traffic and (3) also tolerates the strong transient character of the network.

54.2 Related Work

Most P2P protocols suffer from scalability issues: with the growth of the number of nodes, the amount of requested network traffic (or other resources) also increases notably in order to reach a reasonable hit rate. In the case of the basic protocols, the network is established in a random manner, therefore, the queries reach random nodes in the network. It is reasonable to reorganize the connections of the nodes in such way that new neighbors can answer the query with a higher probability.

The efforts dealing with the scalability issue can be divided into two significantly different groups. The first group consists of the structured P2P networks (for example [1–3]). These protocols specify strict rules for the location of stored documents, or define which other peers a node can connect to. Although these networks usually have good scalability properties, and their performance can be estimated quite accurately, they become disadvantageous in networks with a strong transient character: they handle the frequent changes in the network population only with difficulty, and with great resource expense.

The second approach tries to extend a basic protocol (for example, Gnutella [4]) in an unstructured manner. In this case there is no rule for the location of stored documents, and the connections of the nodes are controlled by a few simple rules. For that reason, these systems have limited protocol overhead and can tolerate it when nodes frequently enter and leave the network.

The advantages of the two approaches can be combined using semantic overlay networks. In our research, we have examined different classes of protocols with some kind of semantic layers. These are built on the fact that the fields of interest belonging to the nodes can be determined, or nodes with probable greater hit rates can be found. The first group of these algorithms tries to achieve better hit rates based on run-time statistics [5, 6]; however, they require too much time to collect enough data, or they achieve only slight performance gains.

The second group of the content-aware peer-to-peer algorithms uses metadata provided for the documents in the system. However, as our aim is to locate well-defined documents (for example, music files with a known title), some of these algorithms cannot be applied to achieve our goals because they assume the kind of information that one would not expect in a real system. For example, [7] assumes that the user is aware of the keywords of the documents being searched for. Since these keywords are produced by some algorithmic method [8, 9]) based on the document itself, we lose accuracy right at the beginning of the search, because we cannot expect the user to produce these keywords in the absence of the requested document. However, keyword-based document discovery in a mobile environment is also a topic which requires further research.

Another shortcoming of the existing structured content location algorithms is that they cannot generalize the collected semantic information. References [8] and [7] store and use metadata for selecting the neighbors for semantic routing. However, they do not utilize deeper information, such as semantic relationships in the concept hierarchy, that can be extracted from the available data.

The most important drawback which restricts the rather efficient existing algorithms is that they either cannot operate at all in highly transient networks, or the overhead for keeping the overlay network up-to-date in such an environment overwhelms the benefit of the intelligent protocol. Therefore, their performance falls back to the level of the pure unstructured networks, or they generate even larger communication costs.

There are solutions that require the presence of superpeers or cache peers in the network to solve the issue of transient peers. These can be quite effective; however, it is hard to use them when the need for semantic overlays arises. Therefore, we have been searching for a fully distributed solution with no peers with a special role.

54.3 Solution Overview

Based on our experiences with the different protocol extensions, we developed a new solution, the SemPeer protocol. The main idea of the protocol is that the individual nodes establish semantic connections to groups of nodes with similar fields of interest. Such suitable nodes are found with an efficient Bayesian process.

The nodes in the network set up hypotheses on the effectiveness of whole query propagation paths; more precisely, they approximate the probability that the set of nodes reachable through another node by a message with a given lifespan can answer queries on a given topic. The nodes should be agreed on a common document-analyzing algorithm that can extract the topics (concepts) from the shared documents which characterize the given file. This algorithm depends on the aim of the peer-to-peer network. It could be, for example, a method using human languages that determines the keyword(s) of the documents, or even a simple function that retrieves the genre byte from MP3 files.

The nodes of the SemPeer network calculate probabilities as hypotheses on the effectiveness of whole query propagation paths reachable through different nodes. Because the aim of the protocol is to find the nodes which can answer the queries in the fields of interest of the user, these probabilities should be stored for each concept (or topic) that are recognized by the document-analyzing algorithm. Most of these algorithms construct a hierarchy from the concepts (called a taxonomy), in which the concepts are organized in a tree according to the "is a" relation. This means that the concepts closer to the root of the taxonomy are more general than these that are close to the leaves. Our protocol extension constructs representative descriptions (profiles) for each node based on its stored documents and behavior; and by the similarity of the profiles, the nodes can establish the connections assumed to be useful. The comparison of the profiles is always restricted to a specified level of the taxonomy, which means that the nodes can select the concept generalization level in accordance with the quantity of the reachable nodes with a given field of interest. There have been various methods elaborated to compare the similarities of concepts and taxonomies, and we found that [10] describes algorithms adequate to serve our goals.

Each node should maintain its hypotheses with the help of local decisions based on its observations. An observation means finding and downloading a given document on a topic through one or more connections. This is a standard operation in peer-to-peer networks, therefore there is no need for additional network traffic for maintaining the hypotheses.

The prior hypotheses are set up based on the hypotheses of the immediate neighbors and the locally stored documents. It is important that no explicit information about distant (non-neighbor) nodes are stored, therefore the changes in the networks should not be communicated in any way to the nodes. The nodes become aware of these changes by their observations, which they build into their hypotheses.

Because the SemPeer protocol acts as a new, semantic layer on Gnutella, it can be implemented in such a way that the advanced nodes supporting this protocol extension can collaborate seamlessly with the standard clients that know only the basic Gnutella protocol. In the following sections we will present this new protocol in detail. Four kinds of profiles (that store the hypotheses) are maintained by the nodes in order to find the most appropriate connections for the user. First, we introduce the semantic profile, which characterizes the stored documents of a node. This profile can be a good prior hypothesis until the client learns the fields of interest of the user from the dynamic behavior of the node, since supposedly these stored documents in some degree reflect the fields of interest of the user. Second, the connection profile will be introduced. Connection profiles are maintained by each node for all of their connections. The purpose is to compare the expected answer ratios of the different propagation paths reachable through different node connections. Third, the reply profile is described, which is the profile that is sent by a node in its answers to help other nodes decide whether they want to initiate a connection with it or not. The reply profile acts as the prior hypothesis when initiating the connection. The fourth taxonomy, the search profile, characterizes the queries transmitted by the node. The search profile is supposed to store the fields of interest of the user; therefore, it can be used to determine with high probability whether a propagation path described by a connection profile can deliver documents in the fields of interest of the user. This is followed by Section 54.5, which describes the protocol extension that utilizes all these profiles.

54.4 The Profiles

54.4.1 Constructing the Semantic Profile for Each Node

Because the effectiveness of the information retrieval system will be increased with the addition of semantic information, we needed a tool with which the nodes can categorize the stored documents. To be precisely defined, the profiles are weighted taxonomies, in which the nodes represent the different synsets (a set of one or more concepts that are synonyms) describing the shared documents, along with the count of occurrences of the given keyword. During our research we used the generally accepted and up-to-date WordNet taxonomy to reveal the relationships of the stored keywords. WordNet is an online lexical reference system whose design is inspired by current psycholinguistic theories of human lexical memory [11]. Other research projects in this field use various versions of WordNet, therefore we also utilized this taxonomy during our work.

The stored documents contain enough information for the peers to compose their own semantic profile. As already mentioned, there are several methods of acquiring even quite complex ontologies from documents, therefore the user does not need to produce the keywords manually for each document. When a new document is stored in a node, it updates its semantic profile by increasing a counter for each keyword's synset and all of its hypernyms. This is quite different from the systems discussed in the related work section, because the semantic profile will contain information on concepts that are not directly gathered from the metadata of the documents, but are deduced from them.

With this approach, a probability can be calculated from the semantic profile that approximates the chance of finding a document in the document store of the node whose topic is about the selected synset. We denote this probability with P_s.

54.4.2 Constructing and Maintaining the Profiles for the Connections of the Nodes

As described in Section 54.3, the nodes with the advanced protocol select their neighbors in such a way that their on-topic queries will be answered with high probability. Therefore, they need the connection profiles of the neighboring nodes. The connection profile is a weighted taxonomy, where, together with the synsets, a probability is stored whose value represents the chance that a query for a document that can be described with the given concept s is found through the given connection C. This probability is denoted with $P_s{}^C$.

This probability depends on the profiles of the nodes being accessible through the whole query propagation path. Because of the frequent joining and leaving of the network, we are unable to gather these profiles from all the reachable nodes. As they are constantly changing, it might require huge network traffic to update these data. We also suppose that there is only moderate computing capability available for the devices. Therefore we had to find a solution that is resource-aware. However, we can suppose that the nodes are using the same algorithm to extract metadata from the documents.

We developed a solution that can be applied in these circumstances. It follows a Bayesian process to estimate the probability of finding a document in a specific topic with a given connection, using the prior information of the semantic profile propagated by the newly connected node.

With the Bayesian process we want to approximate the probability $P_s{}^C$ that a direct connection C to the candidate node c can deliver a positive answer to a query for a document on the topic s. The candidate node will be denoted by a lower case c and the connection to that node by a capital C. We can regard the different queries of a node as independent, therefore the probability of getting exactly α successes and β negative answers out of $\alpha + \beta$ queries through a connection in a context s is given by the probability mass function of the binomial distribution:

$$f(\alpha; \alpha + \beta, P_s^C) = \binom{\alpha + \beta}{\alpha} \left(P_s^C\right)^\alpha \left(1 - p_s^C\right)^\beta \qquad (54.1)$$

Moreover, it is well known that the beta distribution is a good choice for representing the prior belief in the Bayesian estimation because it is conjugate prior for binomial likelihood [12]. Returning to our problem, let $P_s{}^C$ be a probability with beta distribution that represents the *belief* of a node that connection C gives positive answers in context s. Its parameters are α and β, where α is the number of observations when connection C did give results to a query for a document on the topic s, and β is the number of observations when it did not.

Initially, the prior is *Beta(1,1)*, the uniform distribution on *[0, 1]*; this represents the absence of information. However, as nodes spend little time on the network, we should use a more precise prior approximation, the reply profile, sent by the candidate node c as a weighted initial observation. The *reply profile* for a node c is a taxonomy, where, together with each synset s, the maximum number of known documents in that topic and also the estimation of the answering ratio from node c are represented. These probabilities are denoted with R_s^c. Let the connecting node regard the profile sent by the candidate node as an *n-fold* observation. In that case, the prior distribution can be stated as:

$$P_s^C = Beta\left(nR_s^C, n(1 - R_s^C)\right) \qquad (54.2)$$

The actual value of n can be set by the connecting node individually according to its past experience as to the truthfulness of the propagated data.

After each observation (query), the connecting node can update the hypothesis (the connection profile) for a given synset for each connection. We assume that the node with prior distribution Beta(α_0, β_0) transmits queries for documents which have been found and downloaded. Suppose that the synset s extracted from the documents by the agreed algorithm characterizes the files. If α' pieces of documents among all the downloaded files were found via connection C, and this connection gave no results β' times, the profile for connection C should be updated according to the following equation:

$$\left(P_s^C\right)' = Beta(\alpha_0 + \alpha', \beta_0 + \beta') \qquad (54.3)$$

With the growth of the number of queries, the probability distribution function becomes close to a Dirac at P_s^C. Although the probability distribution function for the beta distribution is quite complex, the expected value of a beta random variable X can be calculated as easily as $E(X)$ equals; $\alpha/(\alpha + \beta)$. We use squared-error loss for the deviation from the true P_s^C as commonly done, that is, we consider $E(\alpha,' \beta')$ for P_s^C, which calculation does not need many computational resources. Therefore, this theory is very suitable for mobile environments.

As we expect that the neighboring node does not change its fields of interest radically, we might suppose that it will replace its connections with some other nodes with similar fields of interest, when one of its existing connected nodes leaves the network. This principle can be applied in an inductive manner to all the nodes in the network. Therefore, the precise value of P_s^C can be approximated by a constant, and the hypothesis of the nodes tends to that constant.

The nodes can cache connection profiles of known hosts for faster bootstrapping when joining the network. Moreover, they can use connection profiles sent by other peers as prior knowledge, because they can be more precise than reply profiles. However, learning from other peers may require the use of an appropriate reputation algorithm [13]. Designers of the network might decide whether this overhead is allowable in a given mobile P2P system, based on the average computing and communication capabilities of the participating devices.

54.4.3 The Reply Profile

The aim of the reply profile is to represent the knowledge of a node about the number of documents on a topic, along with their availability in the set of reachable documents, as a probable answer ratio. The reply profile serves as prior knowledge for the connecting nodes that is later refined with the observations of these nodes. Therefore, it is more important for the probabilities of the profile to be quickly computable than to be absolutely precise.

The reply profile contains the maximum number of known documents in each synset s: $D_{s,max}$. It is the maximum number of documents known either by the node itself or by the connected nodes. It can be seen that the maximum number of reached documents on the topic might be greater, because the stored documents are not necessarily overlapping. The maximum known number of documents for the synsets is important because, based on that data, nodes can approximate their own answering ratio.

The reply profile also stores the answering probability for each synset s. This is approximated by (54.4).

$$R_s = \max\left\{ P_S^{C_i}; \frac{D_s}{D_{s,max}} \right\}$$ (54.4)

54.4.4 The Query Profile

Our protocol extension transforms the connections of a node in such a way that the query propagation path contains nodes with similar interests as the given node. Therefore the fields of interest of the nodes should be represented in some way. As mentioned earlier, the semantic profile of a node can be regarded as prior information for this reason, assuming that the stored documents more or less reflect the fields of interest of a user. However, the real interests can be extracted from the documents that are sent as query hits from other nodes. Therefore, each node will set up a structure in which these interests can be stored.

The query profile of a node is a weighted taxonomy, where, along with the synsets, a probability is stored whose value represents the chance that a query issued by the given node is pointed to a document that can be described by the synset s. This probability will be denoted as Q_s. As with the connection profile, a probability value in the query profile for a synset s will be a hypothesis, which means that the node *believes* that a question sent out by the node will result in a reply consisting of a document which can be described with the synset s. For the same considerations, we regard this belief as a beta distribution probability which is updated with the observations of the query hits. After each observation (query hit) the node extracts the keywords from the sent documents, then updates the hypothesis of the given synset in the query profile in the same way as described at the connections profile. With the growth of the number of queries, the probability distribution function becomes close to a Dirac at Q_s.

Let the node regard its semantic profile as an *n-fold* observation. In that case, the prior distribution for the query profile can be formulated as:

$$Q_s = Beta\big(nP_{s,n}(1 - P_s)\big)$$ (54.5)

54.5 Protocol Extension

In order to let the nodes transform their connections in the way described above, we should extend the basic unstructured protocol. We demonstrate our results based on the message types of the well-known Gnutella protocol, because it was straightforward to implement our results as a standard *GGEP* extension [4]. However, we believe that our approach can be applied to most existing semantic solutions to make them more efficient in a mobile environment. Our message extension consists of two modifications.

First, the connected node sends its reply profile to the connecting side in its *Pong* message. The connecting node may request only one level from the profile according to its actual comparison level, as described in Section 54.3, in order to decrease the amount of protocol overhead to a few bytes. Second, when a non-neighbor node finds a matching document to a query, it attaches (a level of) its semantic profile to the answer *Queryhit* message.

The operation of SemPeer nodes differs only in a few properties from that of the basic clients. (1) They should construct their semantic profiles from the stored documents. (2) In case they receive a *Queryhit* from one or more of its connections, they should update their query profile and all of the connection profiles as described in Section 54.4. (3) When a node receives a reply profile from a non-neighbor node as part of a *Queryhit* message, it decides whether to initiate a connection to it based on the comparison of the query and connection profiles [10]. The nodes maintain a specified number of connections for the few topics that have increased probability in their query profile (denoted with $C_{s,\max}$). To be able to discover new similar nodes and to find documents not belonging to the fields of interest of the user, the nodes also maintain a specified number of "random" (standard Gnutella) connections, as proposed by the researches made on the "small world" phenomenon [6]. (4) Finally, the candidate node may permit the initiated connection in its answer message. In case the number of connections of the connecting node reaches the predefined maximum value for the given topic ($C_{s,\max}$), then the semantic connection with the lowest similarity should be replaced with the new one.

54.6 Results

We have validated in a real mobile environment that the extension fulfills our requirements. First, we used Symbian-based smartphones to run the modified version of a Gnutella client with our algorithm. We found that our algorithm does not allocate noticeable computing and memory resources on widespread smartphone device types (Nokia 3650, 6630, N80, N91, and N95), because the number of necessary operations and amount of memory are very low and depend only on the taxonomy used. In the case of the mentioned music genre taxonomy, the extra amount of necessary memory was around 6 Kbytes, and the surplus in used computing power was less than 8%.

We have also conducted simulations with the GXS simulation environment [14] to measure the performance of the protocol extension with different ontologies. We found that our extension reaches the hit rate of the known metadata-based extensions of unstructured protocols. In addition, its performance does not fall down when nodes massively join or leave the network.

Because of the protocol design, our process does not require extra messages to discover nodes with similar fields of interests. What is more, the simulations showed us that nodes can find stable semantic connections after only queries are sent out.

In Fig. 54.1, we show the graphical results of a simulation of a typical situation with 24,000 nodes and 15 main fields of interest. Each node contained 40 documents from the set of 2600 documents per topic. Each node was connected to exactly 5 other nodes randomly. We set the TTL parameter of the protocol to 4. At the first simulation steps the hit rate was around 0.48, which was equal to the performance of the standard Gnutella client. However, as the SemPeer protocol started to transform the network, it has grown to around 0.98 when the nodes send out queries for files in their fields of interest.

However, the protocol did not give promising results when nodes did not have recognizable fields of interest (e.g., entirely random queries).

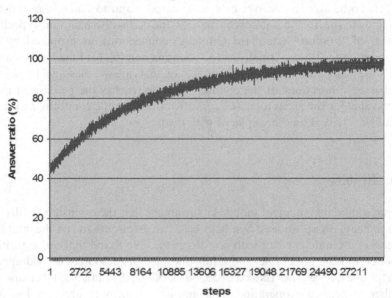

Fig. 54.1 Simulation results

54.7 Conclusion

In this paper, we have described a proposal for a semantic protocol extension in a transient P2P network. Our extension transforms the connections of the peers in a very transient mobile environment in order to increase the hit rate. The connections are transformed based on the fields of interest of the nodes.

We designed the extension in a manner that takes the characteristics of a mobile P2P environment into account; therefore the peers do not need to store a significant amount of data (because the profiles are represented by small-sized integer arrays), and also the extension does not imply huge protocol overhead. The key advantage of our proposal is that it reaches the raised hit rate in a very transient environment.

However, reaching a high hit rate in the case of nodes without recognizable fields of interest requires further research. We should also investigate how the stored metadata can be used in query routing in a transient mobile peer-to-peer network.

References

1. Ratnasamy S, Francis P, Handley M, Karp R, Shenker S (2001) A scalable content-addressable network. In: Proceedings of SIGCOMM'2001
2. Stoica I, Morris R, Karger D, Kaashoek F, Balakrishnan H (2001) Chord: a scalable peer-to-peer lookup service for internet applications. In: Proceedings of SIGCOMM'2001
3. Zhao BY, Kubiatowicz J, Joseph A (2001) Tapestry: an infrastructure for fault-tolerant wide-area location and routing. Technical report UCB/CSD-01-1141, University of California at Berkeley, Computer Science
4. The gnutella project homepage, http://www.the-gdf.org
5. Sripanidkulchai K, Maggs B, Zhang H (2003) Efficient content location using interest-based locality in peer-to-peer systems. Infocom
6. Jovanovic MA, Annexstein FS, Berman KA (2001) Modeling peer-to-peer network topologies through "small-world" models and power laws. IX Telecommunications Forum, TELFOR
7. Joseph S (2003) P2P metadata search layers. In: Second international workshop on agents and peer-to-peer computing (AP2PC)
8. Kietz JU, Maedche A, Volz R (2000) Semi-automatic ontology acquisition from a corporate intranet. In: Proceedings of the learning language in logic workshop (LLL-2000), ACL, New Brunswick, NJ
9. IBM UIMA Framework, http://www.research.ibm.com/UIMA/
10. Resnik P (1999) Semantic similarity in taxonomy: an information-based measure and its application in problems of ambiguity in natural language. J Artif Intell Res 11:95–130
11. The WordNet project homepage, http://www.cogsci.princeton.edu/~wn/
12. Berger JO (1985) Statistical decision theory and Bayesian analysis, 2nd ed. Springer, New York
13. Esfandiari B, Chandrasekharan S (2001) On how agents make friends: mechanisms for trust acquisition. In: Proceedings of the 4th workshop on deception, fraud and trust in agent societies, pp 27–34
14. Forstner B, Csúcs G, Marossy K (2005) Evaluating performance of peer-to-peer protocols with an advanced simulator. Parallel and Distributed Computing and Networks, Innsbruck, Austria

Chapter 55
Deploying BitTorrent in Mobile Environments

Imre Kelényi and Bertalan Forstner

Abstract BitTorrent is a peer-to-peer protocol focused on efficient content delivery. It is used extensively for distributing large files between users of the Internet. However, deploying the protocol over mobile networks has not been analyzed in detail yet. The transient characteristics of mobile networks and the limited resources of mobile devices form an environment which requires a special approach. In this paper, we propose an architecture and present a complete application framework capable of sharing files between mobile phones by using the BitTorrent protocol. Since BitTorrent relies on other components for file search, we primarily focus on the effective distribution of resources and only briefly discuss the issues of actually finding the data on the network. Our solution is not limited for use only in homogeneous mobile networks; any kind of system which implements the BitTorrent protocol can participate in the process of sharing files. In addition, we discuss different issues that we faced during the design and implementation of a file-sharing system based on the proposed architecture.

55.1 Introduction

The use of peer-to-peer (P2P) architecture for file-sharing in mobile networks is becoming very common. There is extensive ongoing research in this area, but the majority of publications focus on the theoretical aspects of the field without addressing real-life scenarios.

Mobile phones with advanced connectivity features and considerable processing power are becoming widespread. Furthermore, due to the increased multimedia capabilities and larger storage capacity of these devices, the amount of information handled and generated by applications running on mobile phones increases every year. Transferring this large amount of content over a

I. Kelényi (✉)
Department of Automation and Applied Informatics, Budapest University of
Technology and Economics, H-1111 Budapest, Goldmann György tér 3., Hungary

N. Mastorakis et al. (eds.), *Proceedings of the European*   555
Computing Conference, Lecture Notes in Electrical Engineering 27,
DOI 10.1007/978-0-387-84814-3_55, © Springer Science+Business Media, LLC 2009

network of nodes can be done in several different ways. However, the key characteristics of mobile networks, such as the ad hoc nature of network topology, the limited bandwidth of the participating nodes, and the typically unreliable connections, raise some concern.

Peer-to-peer systems aim to share information among a large number of users with minimal assistance of explicit servers, or no servers at all; thus the network relies primarily on the bandwidth and computing power of the participating nodes, rather than on the scarce resources of a limited number of central servers [1]. Furthermore, the technology supports easy one-to-one communication between the devices and the extensible distribution of resources. These characteristics make P2P architecture one of the most suitable networking technologies for mobile systems and ubiquitous communication.

Another important factor that should be taken into account is that 3rd-generation mobile networks are improving and extending rapidly. Recent investigations in the field have shown that these networks have sufficient infrastructure to deploy P2P services.

Our principal goal with the research presented in this paper was to propose an architecture based on the BitTorrent protocol [2] that can be deployed over currently available cell phone networks, and to fully implement an application suite that supports the fast and reliable transfer of resources among the participants of the network.

The rest of this paper is organized as follows. After this introduction, a brief description of some P2P file-sharing technologies and their applicability in mobile communication is given, including a summary of related work in the field. Section 55.3 gives an overview of our BitTorrent-based mobile file-sharing architecture, and describes its key elements. Furthermore, we discuss arising issues and the proposed solutions. Finally, we show how the architecture performs in a real-life environment with a particular mobile platform, by introducing a file-sharing application based on the proposed architecture.

55.2 P2P File-Sharing Over Mobile Networks

To enable file-sharing, we had to consider whether to choose an already available protocol that could be adapted to mobile networks, or to develop a new technology. However, our previous experiments with the Gnutella protocol for Symbian-based smartphones [3] have shown that supporting a commonly used, standardized protocol is a great advantage, because it can join with existing P2P communities from the PC world.

Peer-to-peer protocols can be applied in several layers of a file-sharing network. It is crucial to differentiate between protocols designed for searching and protocols which participate in the actual transfer of resources between the peers [4]. In the following part, we investigate two candidates which take different approaches.

55.2.1 Gnutella

Gnutella [3] is an unstructured, pure P2P system, which mainly aims at managing a network of nodes and searching among the shared resources (in most cases files). The actual file transfers are handled using HTTP. Gnutella has already been successfully deployed over cell-phone networks, but the results were not significant enough to propose it as a standard file-sharing solution for the mobile environment. Our mobile Gnutella implementation, Symella [3], is currently the only available Gnutella-based file-sharing client for mobile phones running Symbian OS.

There are several problems with applying Gnutella to a mobile environment. First of all, the lack of control over starting and updating shares is a serious issue, because there is no built-in method for announcing newly shared resources, or even for providing some kind of reference for a shared file. The users must depend on the queries, which does not guarantee that the searched resources will be found. The success rate is proportional to the popularity of the content; the probability of finding a rare document is extremely low. It should also be noted that the search service of the protocol is basically based on a query-flooding algorithm, and even if there are several extensions available which aim to improve its performance [5–7], it still results in relatively high bandwidth consumption. In addition, legal issues are also a concern, since there is no way to track back illegally shared, copyrighted material to its original sharer.

55.2.2 BitTorrent

BitTorrent takes a different approach than Gnutella by concentrating only on distributed file transfers [8]. It is designed to distribute large amounts of data without incurring the corresponding consumption in costly server and bandwidth resources; hence, it can be adequate for mobile file-sharing. With BitTorrent, when several peers are downloading the same file at the same time, they upload pieces of the file to each other. This redistributes the cost of upload to downloads.

Sharing files over BitTorrent needs at least one dedicated peer in the network, which is called the *tracker*. The tracker coordinates the file distribution and can be queried for the shared resources which are under its supervision. A network can contain several trackers. A particular shared file, however, can be associated with only one tracker. This means that when a peer starts sharing a file (or group of files), a specific tracker must be chosen, and cannot be changed afterwards. This mechanism could cause some issues in a transient network environment. We will address this issue in more detail in Section 55.3.

The process of sharing a file or a group of files begins with the creation of a *torrent file,* which contains metadata about the chosen tracker and the files to be shared. The torrent file must be registered with the tracker; afterwards, any client which obtains the torrent file can connect to the swarm and download or upload the shared files. The torrent file itself is relatively small in size; transferring it over the network does not consume significant amount of bandwidth. The torrent file is usually hosted on a web server. Peers are required to periodically check in with the tracker (this process is called *announcing*); thus, the tracker can maintain an up-to-date list of the participating clients. The tracker can also offer several additional features such as *leech resistance*, which encourages clients to not just download, but upload file fragments as well.

Concerning legal issues, BitTorrent, similarly to any other file transfer protocol, can be used to distribute files without the permission of the copyright holder. However, a person who wishes to make a file available must run a tracker on a specific host and distribute the tracker's address in the torrent file. This feature of the protocol does imply some degree of vulnerability that other protocols lack. It is far easier to request that the server's ISP shut the site down than to find and identify every user sharing a file in a traditional peer-to-peer network.

Currently, BitTorrent is very seldom used with mobile devices and especially with cell phones. A client implementation is available for Windows Mobile Pocket PCs [9], but there are no options for cell phones or for devices based on other operating systems. Furthermore, no attempts have been made to deploy a complete, independent BitTorrent system (including clients, tracker, and web server hosting the torrent file) over a mobile network. The already mentioned Pocket PC software consists of a client only, and without a tracker it is not capable of sharing new files on its own.

55.3 The Proposed Architecture

In a traditional BitTorrent set-up, clients mainly focus on the download and upload of some previously announced files. Sharing new information is a fairly complex procedure which involves creating a torrent file, registering with a dedicated tracker, and finally uploading to a web server where others can access the information. In a mobile network, clients rarely have access to dedicated web servers and trackers. Furthermore, there is currently no standardized mechanism available to register the torrent file with the tracker; the users must do it by themselves, and the actual process is also implementation-dependent.

Our solution enables mobile clients to share resources in a partly-automatic manner, even in a constantly changing network environment that consists of only mobile devices. No assistance of dedicated servers hosted on fixed computers is necessary. The only help we make use of is a proxy running on a server with a fixed IP address, which is necessary to give the phones a virtually public

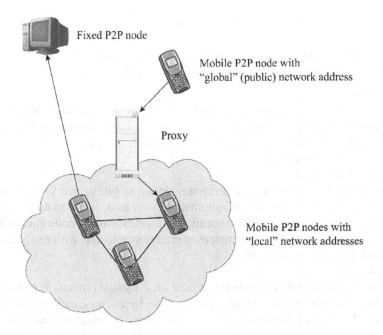

Fixed P2P node

Mobile P2P node with
"global" (public) network address

Proxy

Mobile P2P nodes with
"local" network addresses

Fig. 55.1 Basic architecture of the mobile peer-to-peer file-sharing network

IP address. It should be noted that this component is needed only if not all the mobile devices have a public network address.

Figure 55.1 shows an architectural overview of the BitTorrent-based mobile file-sharing network. The basic components of the architecture are as follows:

1. Mobile P2P node
   A mobile device that implements the BitTorrent protocol. It can be further categorized into two subclasses:

   - nodes possessing a "public" IP address (can be accessed directly by all other nodes in the network);
   - nodes behind NAT-enabled routers (can only be accessed directly by nodes in the same subnet).

2. Fixed P2P node
   This is a desktop computer, server, or workstation which implements the BitTorrent protocol. Its presence in the architecture emphasizes that not only mobile phones, but any device which implements the common protocol can participate in the network.

3. Proxy
   This is a server that allows nodes behind NAT-enabled routers to participate in the network by providing public IP addresses. We discuss the implementation details of the proxy in a separate section.

55.3.1 The Mobile Peer-to-Peer Node

The mobile P2P node is the most significant component of the architecture. A node integrates four separate services, as shown in Fig. 55.2.

1. BitTorrent client
   This is the client application which implements the standard BitTorrent protocol. It can communicate with a tracker and exchange file fragments with different BitTorrent clients. It is capable of downloading multiple torrents simultaneously.
2. BitTorrent tracker
   This is a fully functional tracker running on the mobile device. It coordinates the distribution of the files initially shared by the node itself, so no assistance of external nodes (e.g., a PC) is required. However, when a node goes offline, its tracker drops as well. Our proposed mechanism to deal with this is issue is called the *multi-tracker extension*.
3. Web server
   This is a simple web server that can host files. Its sole purpose is to provide access to the torrent files created by the node. In this way, users only need to share the URL of the torrent file to allow others to download the files.
   This is not a mandatory component because the sharing party can send the torrent file, which is presumably small in file size, to the participants via other channels (like Bluetooth or MMS).
4. Torrent maker
   This service is responsible for creating torrent files on demand.

55.3.2 The Multi-Tracker Extension

If a tracker goes offline, all of its hosted torrents become invalid. This is a major criticism of the BitTorrent protocol, because clients are required to communicate with the tracker to initiate downloads. In a transient network environment, it is crucial to find a mechanism that deals with this issue. There is already a trackerless version of the BitTorrent protocol [10], although it relies heavily on DHTs (Distributed Hash Tables). The maintenance of DHTs needs considerable computing power, especially in a very transient environment, and also results in a great increase in network traffic, which is highly undesirable [11].

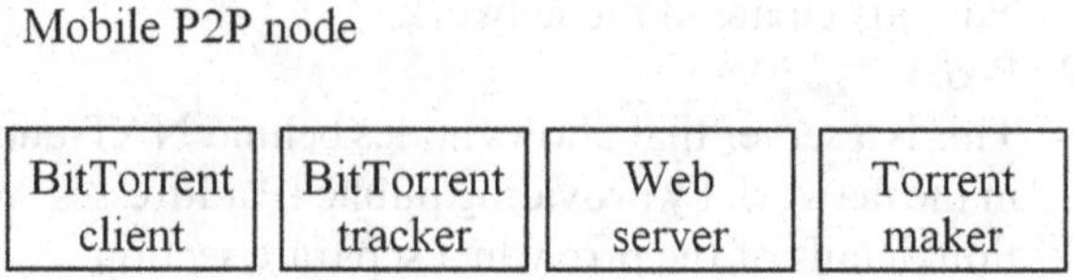

Mobile P2P node

Fig. 55.2 Services of the mobile P2P node

BitTorrent client	BitTorrent tracker	Web server	Torrent maker

Our solution is based on the fact that although each torrent is associated with only one tracker through the torrent file, it is not prohibited to register the torrent itself with multiple trackers.

To support the multi-tracker extension, the nodes must take the following steps:

1. Each time the client starts a new download, it registers the torrent with its own tracker.
2. Each time the client connects to another node, it also announces itself to the connected node's tracker.

If the client cannot access the tracker referred by the torrent file, it can still check in with another node's tracker. However, finding a node with a tracker that serves a particular torrent is another problem. We tested the most basic approach, when the client randomly tries nodes that it has recently seen online. In a network that consists of a couple of hundred nodes, the results were acceptable. The basic approach works especially well for those nodes which had already been connected when the original tracker was still up. A possible-use case: somebody shares a video clip with four friends. If this user sees that the other clients have downloaded at least one "distributed copy," he can switch off the device that runs the tracker and hosts the "initial copy," but the four friends can still finish the transfer between each other.

If the number of nodes increases, a more sophisticated mechanism must be found. Further analysis can be performed, but it is an extensive field, which is beyond the scope of this paper.

55.3.3 Proxy

One of the main issues in cell phone networks is that mobile operators usually use network access translation (NAT) and provide only "local" IP addresses for network users. Furthermore, hosts behind a NAT-enabled router do not have true end-to-end connectivity, which is one of the key principles of peer-to-peer file sharing. Although there are no technical obstacles that would prevent operators from providing "public" IP addresses for the users, this is not a common practice in the current cellular phone networks. And even if a public address is provided, we still have to deal with the lack of fixed IP addresses, which is another problem.

Our solution is based on a dedicated proxy, referred to as a "connector server," developed to run on a separate desktop computer. The main task of the proxy is to provide public and semi-permanent (or long-term) IP addresses for the connecting clients. Without going into details, we now discuss the principal features that the proxy should implement to support the concept.

The proxy maintains two dedicated ports for supporting the process. The one responsible for service messages is referred to as the *control port*. Clients connect to this port to apply for a public IP address. This *control connection* remains active during the whole session.

Each client obtains the (same) IP address of the proxy as a public address but with a different port number. The proxy stores the IMEI number (International Mobile Equipment Identity, a number unique to every GSM and UMTS mobile phone) of the connecting device; thus, the next time a client connects to the proxy, the same port can be supplied. Generally speaking, the address and port uniquely identify the client. The steps of this process are shown in Fig. 55.3.

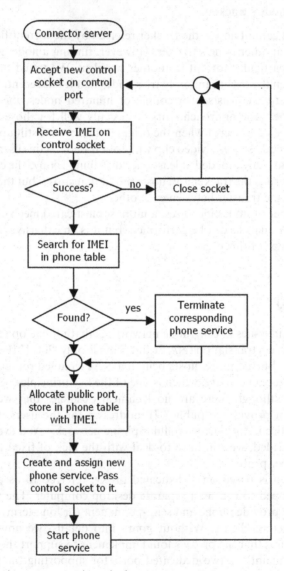

Fig. 55.3 Accepting new connections via the connector server

When the proxy receives an incoming connection to one of the assigned public addresses, it notifies the corresponding client, which then opens a new connection to the proxy's other dedicated port, called the *forward port*. As the name suggests, this port is responsible for forwarding incoming connections towards the mobile clients. The *forwarder connection* is initialized by the client, and acts as the transmitter between the host connecting to the public address, which is handled by the proxy, and the mobile client. Fig. 55.4 shows the details.

This architecture provides full transparency for the connecting party. Connecting to the public address through the standard TCP seems as if the connection were made directly to the phone. The architecture is semi-transparent from the phone point of view as well. After the forwarder connection is initialized, the socket can be used in a normal way as if it were connected directly to the connecting party.

If the proxy is implemented in a platform-independent language, such as Java, then it can be used under different operating systems. Actually, in an extreme case, even a mobile device can run the proxy if it has a permanent IP address and enough bandwidth resources.

It should also be noted that connecting through the proxy is optional; participating devices can still listen on their own IP address if it is accessible to other hosts.

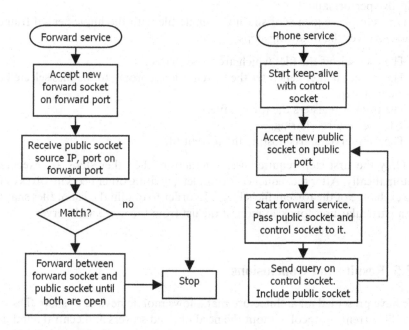

Fig. 55.4 Details of the forward and phone service processes

55.4 BitTorrent for Symbian-Based Smartphones: A Case Study

In order to analyze the architecture in a real-life environment, we decided to implement an application suite based on the proposed architecture. Since such a complex system implies the use of several advanced services, including multi-tasking, receiving incoming TCP/IP connections, and communicating between threads, we could not use mobile Java because it is restricted in features and its performance is not sufficient.

We selected Symbian OS [12] because it is the global industry standard operating system for smartphones; it has a real-time, multithreaded kernel; and it has broad support for communications protocols. Furthermore, it is an open system, and most APIs for advanced features are also available for the developers. Applications can be developed in native C++ language.

We implemented all four components of the presented P2P node: the client, the tracker, a simple web server, as well as the torrent maker. The client on its own can have a major impact on the peer-to-peer world, because it enables millions of mobile users worldwide to join the BitTorrent swarms. The client application itself is called SymTorrent.

For demonstrating the whole system, a possible-use case could be distributing files in a relatively small mobile community which consists of a couple of hundred cell phones. The phones connect to each other through an IP-based network. The users can share their documents, such as photographs, notes, or timetables. Optionally, we can also include fixed servers in the network to improve performance.

An example sequence of sharing a single file with the implemented framework involves the following steps:

1. The user selects the files which are to be shared.
2. The torrent maker generates the torrent file according to the files selected in the first step.
3. The tracker registers the torrent file.
4. The client starts sharing.
5. The web server starts hosting the torrent file.

Only the first step requires user interaction, the other steps are executed automatically. After the sequence is complete, inviting other peers involves only the exchange of the torrent file's URL. In order to obtain the list of files shared by a particular user, one can request the list from the user's web server.

55.5 Results and Conclusions

We have presented an architecture that allows mobile devices to share files via the BitTorrent protocol without the need of fixed servers and central administration. Based on this architecture, we have implemented a file-sharing system for Symbian-based smartphones. While we have proposed solutions for the issues raised by the special environment, we have also shown that BitTorrent can be deployed over transient mobile networks and used as a general file-sharing mechanism.

There is much room for improvement, though. While BitTorrent performs reasonably well as a file distribution protocol, its lack of a built-in mechanism for searching implies that it must rely on other components. Currently, we are developing solutions to make it possible to integrate P2P protocols in mobile environments in order to enable both effective searching and resource distribution

References

1. Oram A (2001) Peer-to-peer: harnessing the benefits of a distributed technology. O'Reilly, Sebastopol, CA
2. Cohen B (2003) Incentives build robustness in BitTorrent. In: Proceedings of the 1st workshop on economics of peer-to-peer systems
3. Symella homepage, http://symella.aut.bme.hu
4. Arthorne N, Esfandiari B, Mukherjee A (2003) U-P2P: a peer-to-peer framework for universal resource sharing and discovery. In: Proceedings of USENIX 2003 annual technical conference, FREENIX track, pp 29–38
5. Merugu SS, Zegura E (2005) Adding structure to unstructured peer-to-peer networks: the use of small-world graphs. J Parallel Distr Comput 65(2):142–153
6. Yang B, Garcia-Molina H (2002) Efficient search in peer-to-peer networks. In: Proceedings of the 22nd international conference on distributed computing systems
7. Haase P, Broekstra J, Ehrig M, Menken M, Mika P, Plechawski M, Pyszlak P, Schnizler B, Siebes R, Staab S, Tempich C (2004) Bibster—a semantics-based bibliographic peer-to-peer system. In: Proceedings of the international semantic web conference (ISWC2004), Hiroshima, Japan
8. BitTorrent specification, http://wiki.theory.org/BitTorrentSpecification
9. WinMobil Torrent homepage, http://www.adisasta.com/wmTorrent.html
10. Cohen BBitTorrent trackerless DHT protocol specifications v1.0. Experimental draft, http://www.bittorrent.org/Draft_DHT_protocol.html
11. Ding G, Bhargava B (2004) Peer-to-peer file-sharing over mobile ad hoc networks. In: Proceedings of PERCOMW '04, Orlando, FL, USA, pp 104–108
12. Symbian OS homepage, http://www.symbian.com/symbianos/
13. The Gnutella homepage, http://gnutella.wego.com

Chapter 56
Numerical Simulation for the Hopf Bifurcation in TCP Systems

Gabriela Mircea and Mihaela Muntean

Abstract In this paper, an Internet model with one route, one resource, and a nonadaptive user and round-trip continuous distributed delay has been studied. We have shown that a Hopf bifurcation occurs in such a model, yielding a family of periodic orbits bifurcating out from the network equilibrium. A simulation is given to verify the theoretical analysis.

56.1 Introduction

Congestion control mechanisms and active queue management schemes for the Internet have been extensively studied in earlier works [1–3].

In this paper, we consider an Internet model with one route, one resource, a nonadaptive user, and round-trip continuous distributed delay. The model is described by:

$$\dot{x}(t) = k\left[w - \int_0^\infty \rho(r)x(t-r)p(x(t-r))dr\right]$$

$$\dot{y}(t) = h\left[m - y(t)\int_0^\infty \rho(r)p(y(t)+x(t-r))dr\right] \tag{56.1}$$

where $x(t)$ is the sending rate of the source at the time t, $y(t)$ represents the flow along a nonadaptive user [1], k,h are the positive gain parameters, and the congestion indication function $p : R \to R_+$ is increasing, nonnegative, and not identical to zero, which can be viewed as the probability that a packet at the source receives a "mark"—a feedback congestion indication signal. The weight function $\rho : R_+ \to R$ is a nonnegative bounded function that describes the

G. Mircea (✉)
Faculty of Economic Sciences, West University of Timisoara, Romania
e-mail: gabriela.mircea@fse.uvt.ro

N. Mastorakis et al. (eds.), *Proceedings of the European Computing Conference*, Lecture Notes in Electrical Engineering 27, DOI 10.1007/978-0-387-84814-3_56, © Springer Science+Business Media, LLC 2009

influence of past states on current dynamics. We consider that $\rho(r)$ is normalized, that is:

$$\int_0^\infty \rho(r)dr = 1 \tag{56.2}$$

If $\rho(r)$ is the Dirac function for $\tau > 0$, the model (56.1) is given by [1]:

$$\rho(r) = \delta_\tau(r - \tau) \tag{56.3}$$

If $\rho(r)$ is given by:

$$\rho(r) = \begin{cases} \frac{1}{\tau}, & r \in [0, \tau] \\ 0, & r > \tau, \end{cases} \tag{56.4}$$

for $\tau > 0$, the model (56.1) is given by:

$$\begin{aligned} \dot{x}(t) &= k[w - \frac{1}{\tau} \int_0^\tau f(x(t-r)p(x(t-r))dr] \\ \dot{y}(t) &= h[m - \frac{1}{\tau}y(t) \int_0^\tau p(y(t) + x(t-r))dr] \end{aligned} \tag{56.5}$$

Using a reflected Brownian motion approximation [2], the function p can be:

$$p(x) = \frac{\theta\sigma^2 x}{\theta\sigma^2 x + 2(C - x)} \tag{56.6}$$

where σ^2 denotes the variability of the traffic at the packet level, C is the capacity of the virtual queue, and θ is a positive constant.

According to the existing theoretical results [4] on the stability analysis of the system (56.1), we will show in this paper that a Hopf bifurcation will occur when the gain parameter k passes through a critical value;, i.e., a family of periodic solutions will bifurcate out of the equilibrium. Therefore, a careful study of bifurcations in the Internet congestion control system model is quite important.

The rest of this paper is organized as follows. In Section 56.2, we study the asymptotical stability of the equilibrium point of (56.5) and the existence of the Hopf bifurcation. Based on the normal form and the center manifold theorem in [4], the formulas for determining the properties of the Hopf bifurcating periodic solutions are derived in Section 56.3. In Section 56.4, a numerical example is given to demonstrate the theoretical analysis, and in Section 56.5, we formulate the conclusions.

56.2 Local Stability and Existence of Hopf Bifurcation

For system (56.5) the following affirmations hold:

Proposition 2.1
a. *If the function p is given by* (56.6), *the equilibrium point of* (56.5) *is given by:*

$$x_0 = \frac{\beta w + \sqrt{\beta^2 w^2 + 4\gamma w \alpha}}{2\alpha}$$

$$y_0 = \frac{(m\beta - \alpha)x_0 + \sqrt{(mp - \alpha)^2 x_0^2 + 4\alpha m(\gamma + \beta x_0)}}{2\alpha},$$

where

$$\alpha = \theta\sigma^2, \beta = \alpha - 2, \gamma = 2C. \tag{56.8}$$

b. *If* $\alpha_1 = px_0 + x_0 p'(x_0)$,

$$b_{10} = y_0 p'(x_0 + y_0), \; b_{01} = p(x_0 + y_0) + y_0 p'(x_0 + y_0),$$
and $u(t) = x(t) - x_0$, $v(t) = y(t) - y_0$, *the linearized version of* (56.5) *is:*

$$\dot{u}(t) = -\frac{ka_1}{\tau} \int_0^\tau u(t - r)dr$$

$$\dot{v}(t) = -\frac{kb_{10}}{\tau} \int_0^\tau u(t - r)dr - hb_{01}v(t). \tag{56.9}$$

The characteristic equation of (56.9) *is:*

$$\left(\lambda + \frac{ka_1}{\tau} \int_0^\tau e^{-\lambda r} dr\right)(\lambda + hb_{01}) = 0. \tag{56.10}$$

One method of studying the stability is to find conditions for which the eigenvalue equation (or characteristic equation) has no roots with the positive real part.

Proposition 2.2 *If* $\alpha < \tau k < \frac{2\pi}{ca_1}$, *where* $c = \sup\{d | \cos x = 1 - \frac{dx}{\pi}, x > 0\}$ $\cong 2.2754$, *the solution* (x_0, y_0) *is asymptotically stable.*
From the condition that equation (56.10) *accepts the roots* $\lambda_1 = i\omega$, $\lambda_2 = \bar{\lambda}_1$, *results:*

Proposition 2.3
a. *If* $\tau k > \frac{2\pi}{ca_0}$, *then the characteristic equation* (56.10) *has pure imaginary roots* $\lambda_{1,2} = \pm i\omega_0$, *for given* $\tau > 0$, *where*

$$\omega_0 = \frac{\pi}{\tau}, \quad k_0 = \frac{\pi^2}{2a_1\tau} \tag{56.11}$$

b. If $\lambda(k)$ are the roots of equation:

$$\lambda + \frac{ka_1}{\tau} \int_0^\tau e^{-\lambda r} dr = 0 \qquad (56.12)$$

then $\dfrac{d\,\mathrm{Re}(\lambda(k))}{dk}\bigg|_{k=k_0} \neq 0.$

The above analysis can be summarized as follows:

Theorem 2.1 When the positive gain parameter k passes through the value k_0, then there is a Hopf bifurcation of the system (56.5) at its equilibrium (x_0, y_0).

56.3 Direction and Stability of Bifurcating Periodic Solutions

In this section, we study the direction, stability, and period of the bifurcating periodic solutions in equation (56.5). The method we use is based on the normal form theory and the center manifold theorem from [4].

For notational convenience, let $k = k_0 + \mu$, where k_0 is given by (56.11). Then, $\mu = 0$ is the Hopf bifurcation value for system (56.5).

Then system (56.5) can be rewritten as:

$$\dot{x}_1 = -\frac{ka_1}{\tau} \int_0^\tau x_1(t-r)dr + F^1(\mu, x(t-r)) + O_1(|x_1(t-\tau)|^4)$$

$$\dot{x}_2 = -\frac{kb_{10}}{\tau} \int_0^\tau x_1(t-r)dr - hb_{01}y_2(t) + F^2(\mu, x_1(t-r), y_2(t)) \qquad (56.13)$$

$$+ O_2(|x_1(t-r)|^4, y_2(t)^2)$$

where

$$x_1(t) = x(t) - x_0, \quad x_2(t) = y(t) - y_0,$$

$$F^1(\mu, x_1(t-r)) = -\frac{1}{2}\frac{ka_2}{\tau}\int_0^\tau x_1(t-\tau)^2 dr - \frac{1}{6}\frac{ka_3}{\tau}\int_0^\tau x_1(t-\tau)^3 dr$$

$$F^2(\mu, x_1(t-r), x_2(t)) = -\frac{1}{2}\left(\frac{hb_{20}}{\tau}\int_0^\tau x_1(t-r)^2 dr + 2\frac{hb_{11}}{\tau}y_2(t)\right.$$

$$\left.\int_0^\tau x_1(t-r)dr + hb_{02}x_2(t)^2\right) + \qquad (56.14)$$

$$+ \frac{1}{6}\left(\frac{hb_{30}}{\tau}\int_0^\tau x_1(t-r)^3 dr + 3\frac{hb_{21}}{\tau}x_2(t)\int_0^\tau x_1(t-r)^2 dr + \right.$$

$$\left. + 3\frac{hb_{12}}{\tau}x_2(t)^2\right)\int_0^\tau x_1(t-r)^2 dr + hb_{03}x_2(t)^3\right)$$

and

$$a_2 = 2p'(x_0) + x_0 p''(x_0) \qquad\qquad a_3 = 3p''(x_0) + x_0 p'''(x_0)$$
$$b_{20} = y_0 p''(x_0 + y_0), \qquad\qquad b_{30} = y_0 p'''(x_0 + y_0),$$
$$b_{11} = p'(x_0 + y_0) + y_0 p''(x_0 + y_0), \quad b_{21} = p''(x_0 + y_0) + y_0 p'''(x_0 + y_0), \quad (56.15)$$
$$b_{02} = 2p''(x_0 + y_0) + y_0 p'''(x_0 + y_0), \quad b_{12} = 2p''(x_0 + y_0) + y_0 p'''(x_0 + y_0),$$
$$b_{03} = 3p''(x_0 + y_0) + y_0 p'''(x_0 + y_0).$$

Theorem 3.1 *In the parameter formulas, μ_2 determines if the Hopf bifurcation is supercritical (subcritical) and the bifurcating periodic solutions exist for $k > k_0 (< k_0)$; β_2 determines the stability of the bifurcating periodic solutions: the solutions are orbitally stable (unstable) if $\beta_2 < 0 (> 0)$; and T determines the period of the bifurcating periodic solutions; the period increases (decreases) if $T > 0 (< 0)$.*

56.4 A Numerical Example Illustrated by Computer Simulations

For function $p(x)$ given by (56.6), we assume that $\theta\sigma^2 = 0.5$; let the capacity of virtual queue be 1Mbps and the round trip delay be 40 ms. Let one round trip time be the unit of time. If the packet sizes are 1000 bytes each, then the virtual queue capacity can be equivalently expressed as $C = 5$ packets per time unit. Let the parameters $w = 1$, $m = 2$, and $h = 2$ increase, so that the controllers are designed as:

$$p(x) = \frac{x}{20 - 3x}$$

The equilibrium (x_0, y_0) given by (56.7) is $x_0 = 3.2169$. With the soft Maple 11, for $\tau = 0.2$, we can determine that: $k_0 = 49.58735$, $\omega_0 = 15.707963$. Because $\mu_2 > 0$, the Hopf bifurcation is supercritical, $\beta_2 < 0$, the period of the bifurcating solutions is orbitally stable, and $T > 0$, the period of bifurcating solutions, is increasing. For $\tau = 2$, it follows that $\omega_0 = 1.57079$, $k_0 = 4.9587$. Because $\mu_2 < 0$ the Hopf bifurcation is subcritical; $\beta_2 > 0$ the solutions are orbitally unstable; and $T > 0$ the period of bifurcating solutions increases.

For $\tau = 2$, Fig. 56.1 is the wave form plot for $(t, x(t))$, Fig. 56.2 is the phase plot for the wave form plot for $(x(t), y(t))$, and Fig. 56.3 is the wave form plot for $(t, y(t))$.

From Figs. 56.4, 56.5, and 56.6, for $\tau \in (0, 0.8)$ result $\beta_2 < 0$ (orbitaly stable), $\mu_2 > 0$ (supercritical), $T > 0$.

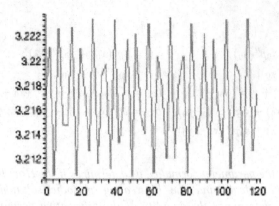

Fig. 56.1 The orbit $(t, x(t))$, $\tau = 2$

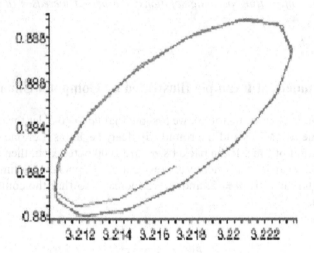

Fig. 56.2 The orbit $(x(t), y(t))$, $\tau = 2$

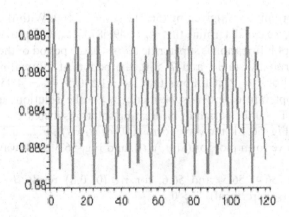

Fig. 56.3 The orbit $(t, y(t))$, $\tau = 2$

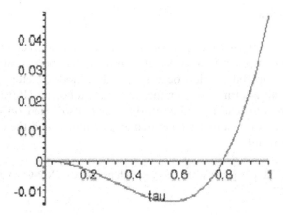

Fig. 56.4 The orbit (τ, β_2)

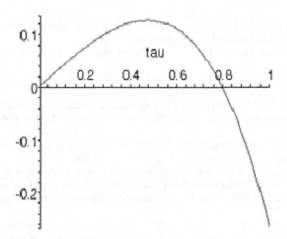

Fig. 56.5 The orbit (τ, μ_2)

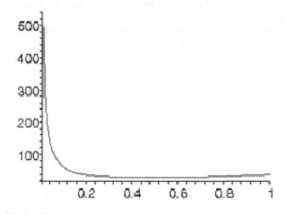

Fig. 56.6 The orbit (τ, T)

56.5 Conclusion

In this paper, an Internet model with one route, one resource, a non-adaptive user, and round-trip continuous distributed delay has been studied. We have shown that a Hopf bifurcation occurs in such a model, yielding a family of periodic orbits bifurcating out from the network equilibrium. Both stability and direction of the bifurcating periodic orbits have also relied on the manifold theorem. Simulation results have verified and demonstrated the correctness of the theoretical results.

Acknowledgment Results paper within the research project CEEX P-CD, code CEEX-M1-C2–505/2006–2008.

References

1. Hassard BD, Kazarinoff ND, Wan YH (1981) Theory and applications of Hopf bifurcation. Cambridge University Press, Cambridge
2. Kelly FP (2000) Models for a self-managed internet. Philos Trans Roy Soc A 358:2335–2348
3. Li C, Chen G, Liao X, Yu J (2004) Hopf bifurcation in an internet congestion control model. Chaos Soliton Fract 19:853–862
4. Mircea G, Neamtu M, Opris D (2003) Dynamical systems in economics, mechanics, biology described by differential equations with time delay. Mirton Publishing House, Timisoara
5. Mircea G, Neamtu M, Opris D (2004) Hopf bifurcations for dynamical systems with time delay and application. Mirton Publishing House, Timisoara
6. Mircea G, Neamtu M, Opris D (2004) Systems of harmonic oscillators with time delay and applications. In: 5th international conference on nonlinear problems in aviation and aerospace ICNPAA 2004. Cambridge Scientific Publishers Ltd., proceedings of ICNPAA 2004, pp 475–485
7. Mircea G, Opris D (2004) Nonlinear effects in a discrete-delay dynamic model for a stock market. In: 8th international congress on insurance: mathematics and economics, Rome, pp 506–517
8. Mircea G, Muntean M (2004) Nonlinear analysis of network traffic. In: The Central and East European conference on business information systems 2004, Cluj-Napoca, Risoprint Publishing House, pp 587–602
9. Voicu M, Mircea G (2005) Algorithms for exploiting multidimensional databases. WSEAS Trans Inf Sci Appl 2(12):2176–2183

Chapter 57
Communications by Vector Manifolds

Guennadi Kouzaev

Abstract The topological structure of the electromagnetic fields of signals is studied. It is shown that the topology of field-force line maps is a natural carrier of digital information. In this paper, a new type of topologically modulated signals is proposed. They are impulses the vectorial content of which varies with the time and spatial coordinates. Impulses can have topologically different spatiotemporal shapes of fields described by a combination of 3-D vector manifolds, and they carry logical information by this spatiotemporal content. The noise immunity of these signals is estimated, and hardware design principles are proposed based on the geometrical interpretation of the energy conservation law. The derived results are interesting for communications through dispersive and noisy media and for advanced computing.

57.1 Introduction

Recently, an increased amount of scientific attention has been paid to topology. Topology studies the global spatial characteristics of figures, as opposed to point-oriented differential equations. In contrast to geometry, topology focuses on the "soft" shape properties that are preserved by bicontinuous transformations. The main goal of topology is to detect the shapes that cannot be changed into each other by such transformations.

Some objects under intensive topological study are the manifolds. They are defined as "topological spaces equipped with a family of local coordinate systems that are related to each other by coordinate transformations belonging to a specified class" [1]. More complicated topological objects are composed of manifolds. They have different shapes, and the goal of topology is to study them.

G. Kouzaev (✉)
Department of Electronics and Telecommunications, Norwegian University of Science and Technology, Trondheim, Norway

N. Mastorakis et al. (eds.), *Proceedings of the European*
Computing Conference, Lecture Notes in Electrical Engineering 27,
DOI 10.1007/978-0-387-84814-3_57, © Springer Science+Business Media, LLC 2009

Our world has four dimensions: three of them are assigned to space, and the fourth is time. Unfortunately, to find a way to distinguish topological shapes in a four-dimensional world was the most difficult task during the last 100 years. Only recently, G. Perelman solved this "problem of the millennium" and derived the proof for the Poincaré conjecture on the characterization of the three-dimensional sphere amongst three-dimensional manifolds [2]. Other papers related to this subject are from a highly abstract field of mathematics [1, 3, 4]. These results can be applied to the theory of curved space-time, quantum mechanics, nonautonomous systems and the qualitative theory of partial differential equations.

One study started in 1988 on topological solutions of boundary problems of electromagnetism [5–10]. Traditionally, since Faraday's time, electric and magnetic fields have been described by their force lines. Each line is a trajectory of a hypothetical electric or magnetic charge moving in the electric or magnetic field, respectively. These trajectories are the solutions of ordinary differential equations/dynamical systems. A set of trajectories composes a field-force line map qualitatively described by its skeleton—topological scheme T that is a set of the field separatrices and equilibrium field positions [5]. The first ones are the vector manifolds in the phase space of dynamical systems. The topological solution of an electromagnetic problem consists of analytical or semi-analytical composing of topological schemes according to the given boundary conditions [5–7].

Later, this developed theory allowed proposing the topological modulation of the field [7–8]. Logical variables are assigned to topological field schemes composed of manifolds of different dimensions, and the signal is a series of field impulses with their discretely modulated spatiotemporal forms. The developed hardware recognizes these impulses and compares them according to Boolean logic [8, 10]. A more generalized view of these studies and signaling is shown here.

57.2 Topological Description of the Electromagnetic Field

A geometrical description of the field by field-force lines traditionally supposes a solution of the nonstationary wave equations with the boundary and initial conditions for the electric $\mathbf{E}$ and magnetic $\mathbf{H}$ fields [11]:

$$\Delta\mathbf{E}(\mathbf{r}, t) - \frac{\varepsilon_r\mu_r}{c^2}\frac{\partial^2\mathbf{E}(\mathbf{r}, t)}{\partial t^2} = \frac{1}{\varepsilon_0\varepsilon_r}\nabla\rho(\mathbf{r}, t) + \\ +\mu_0\mu_r\frac{\partial\mathbf{J}(\mathbf{r}, t)}{\partial t}, \tag{57.1}$$

$$\Delta\mathbf{H}(\mathbf{r}, t) - \frac{\varepsilon_r\mu_r}{c^2}\frac{\partial^2\mathbf{H}(\mathbf{r}, t)}{\partial t^2} = -\nabla \times \mathbf{J}(\mathbf{r}, t) \tag{57.2}$$

where $\rho(\mathbf{r}, t)$ is the time-dependent electric charge, $\mathbf{J}(\mathbf{r}, t)$ is the time-dependent electric current, c is the velocity of light, ε_0 and ε_r are the vacuum absolute and relative media permittivities, respectively, and μ_0 and μ_r are the vacuum absolute and relative media permeabilities, respectively.

There are two types of the above mentioned nonstationary phenomena. The first of them is described by the separable spatiotemporal functions $\mathbf{E}(\mathbf{r}, t) = \mathbf{E}(x, y)e(z)e(t)$ and $\mathbf{H}(\mathbf{r}, t) = \mathbf{H}(x, y)h(z)h(t)$. Typically, it is the effect in regular TEM and quasi-TEM lines. Others require more complicated solutions with nonseparable spatiotemporal functions $\mathbf{E}(\mathbf{r}, t)$ and $\mathbf{H}(\mathbf{r}, t)$.

Mathematically, field-force lines are calculated with a system of differential equations [12], if the solutions of (57.1) and (57.2) have been derived analytically or numerically:

$$\frac{d\mathbf{r}_{e,h}(t)}{ds_{e,h}} = \mathbf{E}, \mathbf{H}(\mathbf{r}_{e,h}, t) \tag{57.3}$$

where $\mathbf{r}_{e,h}$ are the radius-vectors of the field-force lines of the electric and magnetic fields, respectively, and $s_{e,h}$ are the parametrical variables.

During a long period of time the geometrical representation of the fields only helps in the understanding of the electromagnetic phenomena which is the secondary mean in the boundary-problem study. In [5], topological schemes were proposed as a convenient and fast tool to analyze the field inside complicated components of 3-D microwave integrated circuits. Then, in [6, 7], a new approach was developed for harmonic and quasi-static fields, allowing direct analytical composition of topological schemes in accordance with boundary conditions and excitation frequency. The method analyzes the magnitudes of the excited modes with respect to the criteria of frequency and spatial resonances. The modes with increased magnitudes form the topological schemes, and the modal series is reduced when a topological scheme is stabilized. Due to the coarseness of the schemes, the reduced number is not high. This simplified and visually oriented model describes the most important effects in microwave components, and it can be an initial approximation for more accurate calculations.

The nonstationary electromagnetic problems (57.1) and (57.2) are more complicated in their solutions with regard to the harmonic fields, and any qualitative approach that allows fast calculations of spatiotemporal field structures is very attractive. One idea consists of presenting the solutions of (57.1) and (57.2) by integrals [11] on the surface and volume currents and charge densities, similarly to [5]. Then these unknown functions are substituted by approximate functions derived from simplified physics. These qualitatively derived fields are substituted into (57.3), giving the field geometry. Unfortunately, this system (57.3) is a nonautonomous one, and any qualitative study is a difficult task [12]. To avoid the unsolved theoretical problems with the nonautonomous systems, (57.3) is transformed into an autonomous one by introducing a new pseudo-spatial variable τ [8,12–14]:

$$\frac{d\mathbf{r}_{e,h}}{d\tau} = \frac{\mathbf{E}, \mathbf{H}}{|\mathbf{E}|, |\mathbf{H}|} \left(\frac{ds_{e,h}}{d\tau}\right)^{-1},$$

$$\frac{d\tau}{dt} = 1. \tag{57.4}$$

The phase space of this autonomous dynamical system now has four dimensions, and the qualitative study of (57.4) is the composition of topological schemes from manifolds/separatrices, including the three-dimensional ones. These qualitative techniques for nonstationary phenomena need further study and development, and potentially could be a powerful tool for spatiotemporal electromagnetics.

One idea for the visualization of 4-D phase space is the use of 3-D hyperplanes for projecting the 4-D objects [9]. Such projections are easily handled by contemporary software like Matlab. Visualization starts with the definition of the hyperplane and its position. Then, according to standard formulas from linear algebra, the integral lines of (57.4) are projected on the 3-D plane and visualized with a 3-D software function.

In addition to this analysis, topological descriptions of the field can help to generate new ideas on spatiotemporal field signaling. One of them is the topological modulation of the field proposed in [7, 8] and generalized here as signaling by vector manifolds.

57.3 Signaling by Vector Manifolds

A field map is described with its topological scheme $T_{e,h}$ composed of the oriented manifolds/separatrices and the equilibrium states of (57.3) of the electric or magnetic field. A topological scheme is defined at the 4-D phase space if the extended autonomous system (57.4) is considered. The schemes of the i-th and j-th fields can be nonhomeomorphic to each other $T_{e,h}^{(i)} \not\leftrightarrow T_{e,h}^{(j)}$, and then a natural number i or j can be assigned to each scheme. A set of two nonhomeomorphic topological schemes of the propagating modes can correspond to the binary system $\{i = 0, j = 1\}$. Besides the field signal topology, these impulses can carry information by their amplitudes, and they are two-place signals.

Manipulation of topological schemes according to a digital signal is topological field modulation [8]. Operations with these signals are topological computing [12]. Mathematically, these operations refer to the main goal of topology, which is the recognition of topological shapes in spaces of different dimensions. The attractive feature of these techniques is the increased noise immunity of the introduced signals. Intuitively, this follows from the coarseness of the topological schemes of dynamical systems [12]. An accurate estimation of the noise immunity is derived from the general theory of space-time modulated signals [8].

Let $u_0(\mathbf{r}, t)$ and $u_1(\mathbf{r}, t)$ be the binary space-time modulated impulses defined for the spatial volume V and the time interval T. The lower limit of the conditional probability error P_{err} is estimated according to a formula from the theory of general space-time signals:

$$P_{err} \geq \frac{1}{2}\left(1 - \Phi\sqrt{\frac{1}{N_0}\int_0^T\int_0^V [u_0(\mathbf{r}, t) - u_1(\mathbf{r}, t)]^2 dvdt}\right) \qquad (57.5)$$

where N_0 is the signal norm and Φ is the Krampf function. It follows that P_{err} is minimal if the impulses are orthogonal to each other in space-time, i.e., they are the impulses of the waveguide eigenmodes having different field topological schemes.

57.3.1 Quasi-TEM Modal Signaling

One of the first forms of topological signaling is the transmitting of modal impulses along the TEM or quasi-TEM lines [8, 10]. One employed waveguide consists of two conducting strips placed at the dielectric substrate, covered by grounded conducting plates. This coupled strip line supports the even and odd modes of the TEM type, which have different topological schemes (Fig. 57.1).

In general, the field maps can be derived from the study of nonautonomous systems (57.3), but the quasi-TEM modal nature allows representing the

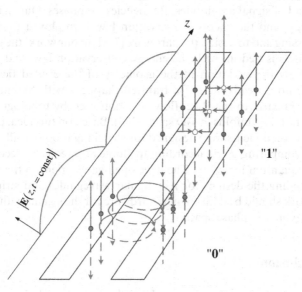

Fig. 57.1 Coupled strip line and modal electric field force-lines. The odd and even modal fields are marked by logical "0" and "1," correspondingly. The upper and bottom conducting plates are not shown

impulses by separable spatial and temporal functions. A more complicated situation arises when the signals are transferred through highly dispersive media. In this case, the quasi-TEM models are not applicable, and the signal requires a full topological description.

57.3.2 Signaling by Nonseparable Field Impulses

In this case, the signal is described by the system (57.4) for nonseparable fields and with the topological schemes $T_{e,h}(\mathbf{r}, \tau)$ defined at the extended 4-D phase space$(\mathbf{r}, \tau)$. Topological communications means signaling by electromagnetic impulses with topologically different schemes in the 4-D phase space and the Boolean units assigned to them. Such extension of topological signaling allows developing a new approach to communications along highly dispersive lines and wireless traces.

57.4 Hardware: Theory and Circuitry

Digital hardware for topologically modulated signals detects the impulses with different time-varying 3-D shapes or compares them according to a certain logic [10, 14]. Formally, this hardware needs an established theory of nonautonomous 3-D dynamical systems and a topological theory of 3-D manifolds and oriented graphs in 4-D phase space (see Introduction), because it is dealing with shapes instead of signal amplitudes, frequencies, or phases. Our signals carry a certain energy, and the energy conservation law is employed to describe the signal processing and to design the hardware [7, 8]. In our work, the geometry of the field signals is tied up with the energy conservation law, and the derived equations show the evolution of the geometry of the excited fields inside a component into a steady state if a transient happens with the incident field. This works similarly to the Ricci flow equation used by topologists to prove their theorems on manifolds in 4-D space [1–4]. Based on this idea, published in 1993, several components were designed for switching topologically modulated signals and comparing them according to Boolean, predicate, reconfigurable, and pseudo-quantum logic [10]. The developed gates orient to the quasi-TEM signals, excluding the designs for rectangular waveguides and strip-slot lines; and more work should be done on new hardware for the signals with modulated field topology in 4-D phase space.

57.5 Conclusions

A new type of electromagnetic signal has been proposed. It carries digital information by its topological scheme of the field-force line maps defined at the extended 3 + 1-D phase space and composed of 3-D manifolds. Analysis of the signals and design of the components have been performed using the energy conservation law in geometrical interpretation. These proposed signals, hardware, and design principles are for electromagnetic and optical signaling and advanced computing.

References

1. Milnor J (2003) Towards the Poincaré conjecture and the classification of 3-manifolds. Not Am Math Soc 50:126–1233
2. Perelman G (2002) The entropy formula for the Ricci flow and its geometric applications. http://arxiv.org/abs/math.DG/0211159
3. Hamilton RS (1982) Three-manifolds with positive Ricci curvature. J Differ Geom 17:255–306
4. Thurston WP (1982) Three-dimensional manifolds. Kleinian groups and hyperbolic geometry. Bull Am Math Soc 6:357–381
5. Gvozdev VI, Kouzaev GA (1988) A field approach to the CAD of microwave three-dimensional integrated circuits. In: Proceedings of the microwave 3-D integrated circuits. Tbilisy, Georgia, pp 67–73
6. Gvozdev VI, Kouzaev GA (1991) Physics and the field topology of 3-D microwave circuits. Russ Microelectron 21:1–17
7. Kouzaev GA (1991) Mathematical fundamentals for topological electrodynamics of 3-D microwave IC. In: Electrodynamics and techniques of micro- and millimeter waves. MSIEM, Moscow, pp 37–48
8. Bykov DV, Gvozdev VI, Kouzaev GA (1993) Contribution to the theory of topological modulation of the electromagnetic field. Russ Phys Doklady 38:512–514
9. Kouzaev GA (1996) Theoretical aspects of measurements of the topology of the electromagnetic field. Meas Tech 39:186–191
10. Kouzaev GA (2006) Topological computing. WSEAS Trans Comput 6:1247–1250
11. Fabrizio M, Morro A (2003) Electromagnetism of continuous media. Oxford University Press, Oxford
12. Andronov AA, Leontovich EA (1973) Qualitative theory of second-order dynamical systems. Transl. from Russian. Halsted Press, New York
13. Peikert R, Sadlo F (2005) Topology guided visualization of constrained vector fields. In: Proceedings of the TopolnVis 2005, Bumerize, Slovakia
14. Shi K, Theisel H, Weinkauf T, Hauser H, Hege H-C, Seidel H-P (2006) Path line oriented topology for periodic 2D time-dependent vector fields. In: Proceedings of the Eurographics/ IEEE-VGTC Symp. Visualization.
15. Gvozdev VI, KouzaevGA (1992) Microwave flip-flop. Russian Federation Patent, No 2054794, dated 26 Feb 1992

Chapter 58
A Computation Study in Mesh Networks by Scheduling Problems

G.E. Rizos, D.C. Vasiliadis, E. Stergiou, and E. Glavas

Abstract The problem of scheduling n dependent tasks, with arbitrary processing times, on m identical machines so as to minimize the range criterion is considered. Since this problem is NP-complete in the strong sense, it can be solved only suboptimally using heuristic approaches. Four existing heuristic algorithms (dispatching rules), namely the CP/MISF, DHLF/MISF, MVT/MISF, and DMVT/MISF algorithms, are proposed for this problem. These algorithms are then used together as the dispatching rule base of a new combinatorial positional value dispatching (CPVD) rule. This combinatorial dispatching rule has a superior behaviour compared to simple dispatching rules. Extended experimentation with these algorithms supports this argument. In addition, some empirical rules are derived and proposed for the selection of a simple dispatching rule (heuristic), if such a selection is required, for each particular input data set.

58.1 Introduction

Scheduling problems deal with the assignment of tasks to machines (processors) over time so that a certain number of constraints are satisfied, while a performance measure (criterion) is optimized [1]. These problems have received great attention because of their theoretical interest. On the other hand, they are faced in almost all real-time environments, namely computing and manufacturing environments.

In this paper we are concerned with a set of n dependent tasks to be scheduled on a set of m identical processors to minimize the positional value. Since this problem is known to be NP-complete in the strong sense [1, 2], it can be solved only suboptimally according to some heuristics (dispatching rules). Applying

G.E. Rizos (✉)
Department of Computer Science and Technology, University of Peloponnese,
GR-22100 Tripolis, Greece
e-mail: georizos@uop.gr

N. Mastorakis et al. (eds.), *Proceedings of the European
Computing Conference*, Lecture Notes in Electrical Engineering 27,
DOI 10.1007/978-0-387-84814-3_58, © Springer Science+Business Media, LLC 2009

such a rule to a scheduling problem, the nonassigned tasks are sorted into a priority list according to some objective function. Then the first task in the list is dispatched to the schedule. If the list remains unchanged during the schedule construction, then the respective algorithm is called "static," while if the list may vary from step to step, then the algorithm is called "dynamic."

Some very important results have been derived in this direction. We mention here two classical heuristic scheduling algorithms, namely CP/MISF and DHLF/ MISF [3], that are based on the idea of the "critical time" of a task. These two algorithms provide very efficient solutions, ensuring a maximum deviation from the optimal solution. These two algorithms ensure a maximum error ratio equal to 2–1/m. The other two known heuristic scheduling algorithms we present are MVT/MISF and DMVT/MISF. These algorithms are based on the idea of the "task value" instead of the critical time. They seem to be very efficient ensuring the same maximum deviation from the optimal solution (2–1/m).

It is well known that the performance of a scheduling algorithm depends on the specific example under solution in an unknown way. That is why there are no heuristic scheduling algorithms always providing better solutions than any other one [4]. Taking into account this important feature of scheduling problems, we developed a new combinatorial dispatching rule consisting of these four algorithms, where their scaling factors are adaptively estimated in every specific example. This scaling estimation is performed taking into account how efficient these algorithms were found to be in every specific example. The new dispatching rule seems to have a very much improved behaviour compared to each single simple heuristic algorithm. This superiority derives from a large scale experimental analysis carried out.

The structure of this paper is as follows: In Section 58.2 the problem is precisely formulated, while in Section 58.3 the existing scheduling algorithms, as well as the combinatorial positional value dispatching (CPVD) rule are presented. In Section 58.4 we present some important results through an experimental analysis, and also a typical example with a graph and the results. Finally we present some conclusions, along with suggestions for future research such as decision support systems or robotic intelligence systems.

58.2 Problem Formulation

We are given a set $T = \{ T_1, T_2, \ldots, T_n \}$ of n tasks and a set $P = \{ P_1, \ldots, P_m \}$ of m identical processors. The tasks have processing times p_j ($j = 1 \ldots n$) and are assumed to be non pre-emptive, i.e., if a task started its execution on a particular processor, it has to be completed on that processor without any interruption.

The tasks are interdependent in the sense that before a task starts its execution on any processor, it is required that a certain set of other tasks (predecessors) must have already been executed.

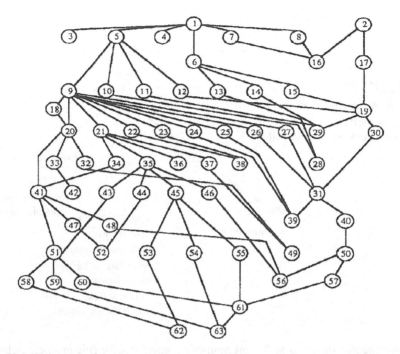

Fig. 58.1 Task graph of group

These dependencies are represented by a directed acyclic graph G. The nodes of G stand for the tasks, and the arcs of G represent the task dependencies.

The problem is to find an allocation of tasks on processors over time such that the positional value criterion is minimized [1, 2], while the following additional constraints are satisfied (Fig. 58.1):

- Each task is executed only once and by only one processor.
- Each processor is capable of executing at most one task at every time instant.
- There is no setup time between tasks on any processor.
- All processors are available for task execution from the time instant $t = 0$.

58.3 Problem Solution

Since this problem is well known to be NP-complete in the strong sense [2], it can be treated only suboptimally using heuristic approaches.

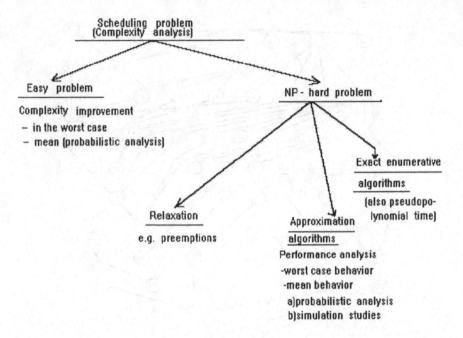

We present the most sufficient heuristic algorithms for this problem, which are the MVT/MISF (most valuable task/most immediate successors first) and DMVT/MISF (dynamic most valuable task/most immediate successors first) algorithms. These algorithms belong to a broader class of scheduling algorithms known as *list scheduling algorithms* [5]. When such an algorithm is applied to a scheduling problem, a priority list of tasks is constructed according to some dispatching rule. Then the nonassigned task of the list that has the highest priority is assigned to the processor on which it will be executed as early as possible.

We define the *task value* parameter (V_i) of a task T_i as follows: Successors of the task T_i are all tasks to which one can go from T_i via some path(s) of the task graph. The task T_i, together with all its successors, constitutes a set S_i [4]. The sum of the processing times of all tasks that belong to S_i is defined as the "value" V_i of T_i, i.e.,

$$V_i = \sum_{k:T_k \in S_i} p_k \qquad (58.1)$$

The MVT/MISF algorithm constructs the priority list by sorting the tasks in nonincreasing order of their values. If two or more tasks have the same value, then these tasks are sorted at a second level in nonincreasing order of the number of their immediate successors. Having constructed the priority list, the list scheduling is applied.

The DMVT/MISF algorithm is similar to the MVT/MISF algorithm, but has a dynamic nature. This means that the priority list is not constructed once and for all before the application of list scheduling, but it changes dynamically every time a new task is dispatched to the schedule. More specifically, the next task to be dispatched to the schedule is selected from the tasks that can be assigned as early as possible to the processor which has been available earlier.

For the same problem, two other list-scheduling algorithms were developed, namely CP/MISF [1] and DHLF/MISF [3], before the others. These algorithms adopt the idea of the priority list, and their construction is based on the *critical time* of a task in the graph. The first algorithm is static, and the second, dynamic.

These four algorithms (dispatching rules) are used as the dispatching rule base of a new combinatorial positional value dispatching rule, namely, CPVD.

This CPVD rule is a dynamic rule, since two of the involved rules (DMVT/MISF and DHLF/MISF) are dynamic as well. The behaviour of the CPVD rule was shown to be more satisfactory on average, compared to the simple rules.

Step 1. Apply the MVT/MISF, DMVT/MISF, CP/MISF, and DHLF/MISF algorithms and get the four positional values V^1_{max}, V^2_{max}, V^3_{max}, and V^4_{max}, respectively.

Step 2. Calculate the rule weights W_i ($i = 1, 2, 3, 4$), using the following formula:

$$W_i = 1 + \frac{\min_k\{V^k_{min}\} - V^i_{min}}{\min_k\{V^k_{min}\}}, \quad i = 1, ..., 4 \tag{58.2}$$

and construct the rule weight vector $W = [W_1, W_2, W_3, W_4]$.

Step 3. In every phase of dispatching a task to the schedule, construct the $4 \times n$ proposal matrix PM with elements PM_{ij} ($i = 1, ..., 4$, $j = 1, ..., n$), where:

- 1 if the task n is proposed by the rule i,
- 0 otherwise.

Then the task priority vector PR with elements PR_j ($j = 1, ..., n$) is computed as:

$$PR = W \times PM \tag{58.3}$$

where PR_j is the priority of the task T_j ($j = 1, ..., n$).

Step 4. The task of higher priority is dispatched to the schedule, to be executed as early as possible. If more than one task is of the same higher priority, then the one that has more immediate successors is dispatched.

Step 5. When all tasks have been assigned to the processors, compute the resulting positional value V^5_{max} and finally select the minimum among V^1_{max}, V^2_{max}, V^3_{max}, V^4_{max}, and V^5_{max}.

58.4 Experimental Analysis

An experimental analysis was carried out to evaluate the average performance of the four simple rules, and the new combinatorial rule. We used randomly generated examples ($10 \leq n \leq 50, 3 \leq m \leq 6$). The results from this analysis are the following:

1. If the task processing times involve a considerable deviation, then the rules CP/MISF and DHLF/MISF provide better solutions than the other two simple rules (this was known from the beginning of paper [6]).
2. If the task graph is of large depth and small width, then the dynamic rules provide better solutions than the static ones (this was known from the beginning of paper [7]).
3. The new combinatorial rule has provided solutions better than, or at least equivalent to, the best of the simple rules in 71% of the tested cases.
4. The new combinatorial rule improves its behaviour when applied to examples of large size.

58.4.1 A Typical Example

The processing times of all tasks were measured. They are given in μsec. These tasks are dependent, and the respective task graph is shown in Fig. 58.2.

In the graph we assume that the arcs are directed from lower-order tasks to higher-order tasks. Applying the five algorithms mentioned above, we obtained the results shown in Table 58.1.

As we can see, the CPVD rule results in a better schedule than any other scheduling algorithm.

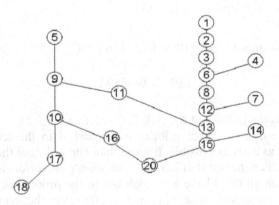

Fig. 58.2 Task graph of example

Table 58.1 Results of the five algorithms applied to the example

Algorithm	Results
CP/MISF	2480
DHLF/MISF	2480
MVT/MISF	2450
DMVT/MISF	2450
CPVD	2425

58.5 Conclusions

In this paper, four simple dispatching rules (algorithms), namely, CP/MISF, DHLF/MISF, MVT/MISF, and DMVT/MISF, were presented. These algorithms were combined in an adaptively combinatorial dispatching rule with very improved behaviour on an average basis. This is because each simple rule contributes to the CPVD rule in every specific example as much efficient as it were found to be.

This idea can be applied to other scheduling problems with multiple constraints, including many different dispatching rules that seem to be efficient, taking into account these constraints. In addition, the same methodology can be used for solving other combinatorial optimization problems besides scheduling.

References

1. Blazewicz J (1987) Selected topic on scheduling theory. Ann Discr Math 31:1–60
2. Garey MR, Johnson DS (1979) Computers and intractability: a guide to the theory of NP-completeness. Freeman, San Francisco
3. Chen CL, Lee CSG, Hou ESH (1988) Efficient scheduling algorithms for robot inverse computation on a multiprocessor system. IEEE T Syst Man Cybern 18(5):729–742
4. Pinedo M (2002) Scheduling: theory, algorithms and systems. Prentice Hall, Englewood Cliffs, NJ
5. Blum A, Furst M (1995) Fast planning through plan-graph analysis. In: 14th international joint conference on artificial intelligence. Morgan-Kaufmann Editions, pp 1636–1642
6. Tzafestas SG, Triantafyllakis A (1993) Deterministic scheduling in computing and manufacturing systems: a survey of models and algorithms. Math Comput Simulat 35(5):397–434
7. Brucker P (1998) Scheduling algorithms. Springer, Berlin
8. Luh JYS, Lin CS (1982) Scheduling of parallel computation for a computer controlled mechanical manipulator. IEEE T Syst Man Cybern 12(2):214–234

Chapter 59
Serially Concatenated RS Codes with ST Turbo Codes Over Ring

P. Remlein and R. Orzechowski

Abstract In this paper, we investigate serially concatenated Reed-Solomon codes (RS) with the space-time turbo codes over ring Z_M. The encoded QPSK signals are transmitted through the quasi-static and fast-fading multiple-input-multiple-output (MIMO) channel. In the receiver, we use the iterative soft input soft output (SOVA) algorithm. Simulation results show that the system performance is similar to the performance of the previously published space-time trellis code system (STTC) [7] for a fast-fading channel and is advantageous for a quasi-static channel.

59.1 Introduction

The insatiable demand for high data rates and high link quality has caused broad research into new transmitting technologies. Nowadays multiple input/multiple output (MIMO) systems are commonly used in wireless transmission [1]. Thanks to the MIMO systems it is possible to increase the rate of transmission or improve the quality in comparison to one receive-transmit antenna systems [2]. Recently the space-time turbo codes (STTC) [2, 1, 3] have been proposed as an alternative that integrates space-time code and turbo codes.

In this paper, we propose a serial concatenation of Reed-Solomon encoder (RS), interleaver, and space-time (ST) turbo coder over ring Z_M. The turbo encoder consists of two recursive systematic convolutional encoders (RSC) over ring of integers modulo-M.

The performance of the proposed system with QPSK modulation and transmission through a MIMO quasi-static and fast-fading channel is evaluated by simulation.

This paper is organised as follows. Section 59.2 describes the analyzed system. Section 59.3 introduces the simulation system model. Section 59.4 presents the results, and finally, the conclusions are drawn in Section 59.5.

P. Remlein (✉)
Chair of Wireless Communication, Poznan University of Technology, Piotrowo 3a, 60–965 Poznan, Poland

N. Mastorakis et al. (eds.), *Proceedings of the European Computing Conference*, Lecture Notes in Electrical Engineering 27, DOI 10.1007/978-0-387-84814-3_59, © Springer Science+Business Media, LLC 2009

59.2 System Description

In this paper a single point-to-point system consisting of n_T transmit antennas and n_R receive antennas is taken into consideration. Data are transmitted over the Rayleigh fading channel. The RS code is used in this system as an outer code. Reed-Solomon codes are often used where burst errors occur, for example, in satellite communication systems or xDSL modems. The RS codes achieve the greatest possible minimum Hamming distance for every linear code of the same code rate [3]. Reed-Solomon codes (RS) are nonbinary cyclic codes [3]. The symbols in the RS codeword are usually written as m-bit binary blocks, where m is a positive integer greater than 2. All operations are done in $GF(2^m)$.

RS(n,k) codes exist for such m-bit symbols when the following equation is fulfilled:

$$0 < k < n < 2^m + 2, \tag{59.1}$$

where k is the number of information symbols, n is the number of all symbols in the codeword, and $(n-k)$ is the number of parity check symbols. The most popular RS(n,k) codes exist for n and k defined by:

$$(n, k) = (2^m - 1, 2^m - 1 - 2t), \tag{59.2}$$

where t is the number of correctable symbols in a codeword, and it can be expressed as:

$$t = \left\lfloor \frac{n-k}{2} \right\rfloor. \tag{59.3}$$

Every cyclic code has its own generator polynomial. A Reed-Solomon codeword is achieved by multiplication of the information polynomial $a(x)$ by the generator polynomial $g(x)$. The result of this multiplication is the polynomial codeword $c(x)$:

$$c(x) = a(x)g(x). \tag{59.4}$$

The information polynomial has rank k, and the achieved codeword polynomial has rank n, so the rank of generator polynomial is $n-k$. Generally the generator polynomial for the RS code, which has the ability to correct t erroneous symbol errors, can be written as [3]:

$$g(x) = (x-)(x-^2)\dots(x-^2). \tag{59.5}$$

In this paper the RS(63,59) code is used as the outer code. This Reed-Solomon code operates on 6-bit symbols and is able to correct up to 2 erroneous symbols.

As the inner encoder, the space-time turbo trellis encoder over ring Z_M is used. At time t the encoder maps a group of $m = \log 2M$ information bits of the input sequence, denoted by:

$$c_t = (c_1^t, c_2^t, \ldots c_m^t) \tag{59.6}$$

into the M-PSK signal.

Let us write the signal transmitted at time t by n_T antennas as:

$$x_t = (x_t^1, x_t^2, \ldots, x_t^{n_T}). \tag{59.7}$$

In Fig. 59.1 is shown the block diagram of the space-time turbo encoder, consisting of two recursive space-time trellis encoders (STTC).

Information bits input the recursive STTC encoder in the upper branch and the interleaver, which is connected to the STTC encoder, in the lower branch. The interleaver is a UMTS type interleaver [4]. Each of the STTC encoders operates on a block consisting of groups of m information bits. The block diagram of the recursive STTC encoder for M-ary modulation is shown in Fig. 59.2 [2].

The input sequence of binary vectors $[c^1, \ldots, c^m]t$ is transformed by the convolutional encoders shown in Figs. 59.1 and 59.2 into a coded sequence of the symbols which belong to the ring $Z_M = \{0, 1, 2, \ldots, M-1\}$, ($\Re = Z_M$; M = 4).

The coefficients in the encoder structure (Fig. 59.2) are taken from the set $\{0, \ldots, M-1\}$ (M = 4), and belong to the same ring Z_M. The memory cells are capable of storing binary symbols. Multipliers and adders perform multiplication and addition, respectively, in the ring of integers modulo-M [5].

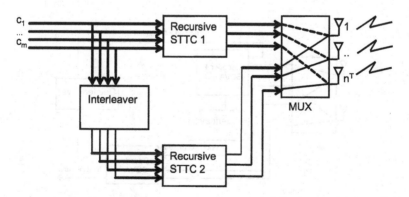

Fig. 59.1 Space-time turbo encoder

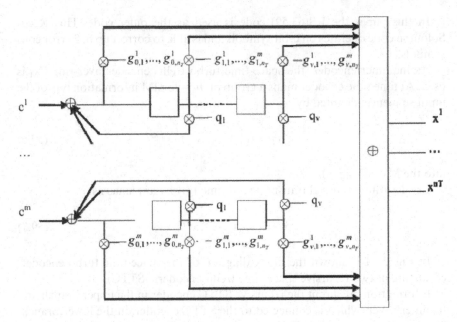

Fig. 59.2 Recursive STTC encoder for M-ary modulation

In the system under consideration a space-time turbo trellis decoder is used as a receiver. It is shown in Fig. 59.3 [2].

In the receiver, decoder 1 and decoder 2 (Fig. 59.3) use the iterative soft output Viterbi algorithm (SOVA) [6]. It selects the maximum likelihood path on the trellis diagram with a priori received symbols probability taken into consideration.

At the receiver side there are n_R antennas. The signal received by each antenna is demultiplexed into two vectors, denoted by r_1^j and r_2^j, contributed

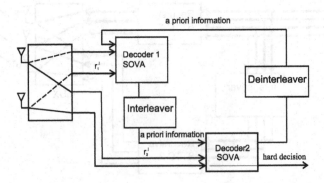

Fig. 59.3 Space-time turbo trellis decoder

by the recursive STTC 1 (upper) and recursive STTC 2 (lower) encoder, respectively. The received signal r_t can be described as [1]:

$$r_t = H_t x_t + n_t, \tag{59.8}$$

where H is the channel state matrix, n_t is an additive white Gaussian noise at time t. $H \in C^{n_R \times n_T}$ is the channel matrix, which describes the connections between the transmitter and receiver and can be expressed as:

$$H = \begin{bmatrix} \alpha_{11} & \alpha_{12} & \cdots & \alpha_{1n} \\ \alpha_{21} & \alpha_{22} & \cdots & \alpha_{2n} \\ \vdots & \vdots & \ddots & \vdots \\ \alpha_{m1} & \alpha_{m2} & \cdots & \alpha_{mn} \end{bmatrix}, \tag{59.9}$$

where α_{mn} is a complex transmission coefficient between the mth element of the transmitter (TX) and the nth element of the receiver (RX). To generate channel matrix H, the narrowband Kronecker model presented in [1] is used. The principle of this model lies in the assumption that the receive correlation matrix is independent of the corresponding transmit matrix, and vice versa. The model presented in the paper was further simplified according to the assumptions given in [7]. To calculate the correlation coefficients between the antennas in the transmitting as well as in the receiving array, we followed the approach in [7].

The receiver has a perfect channel state information (CSI). The receiver algorithm searches for the maximum likelihood codeword by computing the squared Euclidean distance between each of the pattern symbols at time t and the received signal at time t [2]:

$$\sum_t \sum_{j=1}^{n_T} \left| r_t^j - \sum_{i=1}^{n_T} h_{j,i}^t x_t^i \right|^2. \tag{59.10}$$

The iterative SOVA decoders (Fig. 59.3) compute the symbols' probabilities. The soft output of the SOVA decoder 1 is interleaved (in the same way as in the encoder) and used to produce an improved estimate of the a priori probabilities of the received symbols for SOVA decoder 2. The soft decision from SOVA decoder 2 is deinterleaved and passed to SOVA decoder 1 as the a priori information about the received symbols. After some iterations, the system produces a hard-decision.

59.3 Simulation Model

The block diagram of the simulated system is shown in Fig. 59.4. (63,59) RS encode over field $GF(2^6)$ is used as an outer encoder. The turbo space-time encoder over the ring Z_4 is used as an inner encoder. The interleaver between the encoders used in the analyzed system is a 63×6 block bit interleaver.

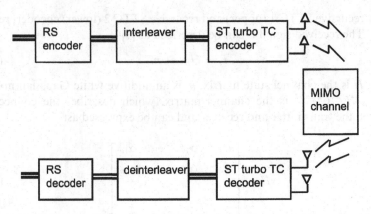

Fig. 59.4 Block diagram of an analyzed system

The random information source generates 59 6-bit symbols. Those symbols are encoded by the RS encoder and binary interleaved. Next the symbols are encoded by the turbo space-time encoder over ring Z_4 and sent through a MIMO fading channel to the receiver. We used QPSK modulation, and two transmit and two receive antennas. The recursive encoders of the turbo encoder (Fig. 59.2) are described by the feedforward coefficients:

$$g^1 = [(0,2) \quad (1,2)], \quad g^2 = [(2,3) \quad (2,0)] \tag{59.11}$$

and feedback coefficients $q^1 = 3$ i $q^2 = 3$.

In the simulation, as a channel model, we used the narrowband Kronecker model presented in [7] and described in the previous section. In this paper, both quasi-static and fast-fading channels were considered. For quasi-static fading channels, it is assumed that the fading coefficients are constant during a frame (80 space-time symbols) and vary from one frame to another independently. For fast-fading channels, the fading coefficients are constant within each symbol period and vary from one symbol to another independently. In both fading types, the fading coefficients are modeled as independent zero-mean complex Gaussian random variables with variance 1/2 per dimension. The simulation model used in this paper assumes that the channel is narrowband. A narrowband channel means that the spectral characteristics of the transmitted signal are preserved at the receiver. It is called a flat or nonselective channel. Considering the frequency domain, this means that the signal has narrower bandwidth than the channel.

59.4 Results

In this section, we present the simulation results. The simulations were performed in the MATLAB environment. The (63, 59) RS code is the outer code. The inner code is a turbo ST code with convolutional encoder over ring Z_4 as a recursive encoder. It is assumed that each transmitted frame consists of 354 bits. The system performance was investigated over a quasi-static Rayleigh fading channel and a fast-fading channel with the parameter $f_d T_s = 0.01$. The time interval of a single fade (in the quasi-static channel) was equal to 80 space-time symbols. It was also assumed that we have perfect knowledge of the channel. The system has 2 transmit and 2 receive antennas (2Tx/2Rx). The modulation scheme was QPSK. The frame error rate (FER) of the described system was evaluated using the Monte Carlo simulation.

Figures 59.5 and 59.6 show the frame error rate of our system over Rayleigh for different numbers of iterations. Aside from the channel type, when the number of iterations increases, the decoder performs significantly better. However, after four iterations there is only nonsignificant improvement possible for further iterations. Specifically, the gain between the fist and second iteration is significant. For transmission over the fast-fading channel this gain is about 2 dB at FER of 10^{-2}, and for transmission over the quasi-static channel the gain is about 1.5 dB at FER of 10^{-2}.

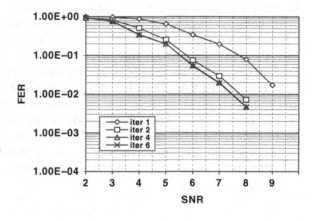

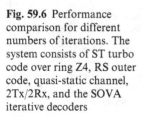

Fig. 59.5 Performance comparison for different numbers of iterations. The system consists of: ST turbo code over ring Z_4, RS outer code, fast-fading channel, 2Tx/2Rx, and the SOVA iterative decoders

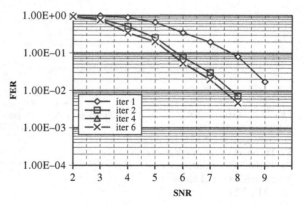

Fig. 59.6 Performance comparison for different numbers of iterations. The system consists of ST turbo code over ring Z4, RS outer code, quasi-static channel, 2Tx/2Rx, and the SOVA iterative decoders

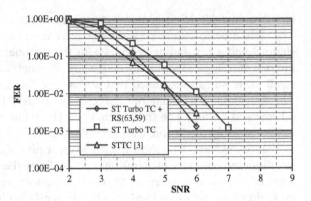

Fig. 59.7 Performance comparison between different systems: STTC [2], ST Turbo code (TC) over ring Z4, ST Turbo TC with RS outer code (fast-fading channel with 2Tx/2Rx)

Figure 59.7 shows FER results obtained for the transmission over the fast-fading channel (six iterations of the SOVA decoder). We can observe that the outer RS code improves the performance about 0.7 dB for FER $= 10^{-2}$.

We compare our results with the results shown in [2] for binary STTC with the same number of turbo encoder states (equal to 4).

In the fast-fading channel, our system with outer RS encoder performs better by almost 0.4 dB at the FER of 10^{-3} than the binary STTC system in [2]. In this case, the system proposed in this paper, ST turbo coded over ring Z_4 (without RS code), performs worse by 0.5 dB at the FER of 10^{-3} than the system in [2].

In Fig. 59.8 FER is shown for an ST turbo coded over ring Z_4 system with and without outer code (RS). The transmission over the quasi-static fading channel (six iterations of the SOVA decoder) is investigated. In this case, we can see that the outer RS code improves the performance of the system proposed in this paper with ST turbo code over ring Z_4 by about 1 dB for FER $= 10^{-2}$. Fig. 59.8 also illustrates the performance comparison for the system proposed

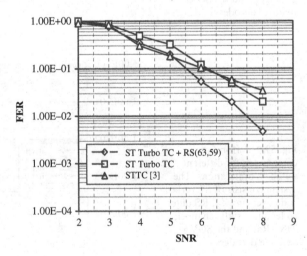

Fig. 59.8 Performance comparison between different systems: STTC [2], ST Turbo TC over ring Z_4, ST Turbo TC with RS outer code (quasi-static channel with 2Tx/2Rx)

in this paper and STTC [5] for the quasi-static channel. It is shown that the proposed system with and without outer RS code and space-time turbo code over ring Z_4 performs better than the binary system proposed in [5] by 1.5 dB and 0.5 dB, respectively, at FER of 10^{-2}.

59.5 Conclusions

In this paper we investigated the system which consists of serially concatenated RS code, interleaver, and ST turbo code over ring Z_M. A new structure of the turbo encoder was proposed, which consists of two convolutional encoders over ring. In the iterative turbo decoder, two SOVA blocks exchange soft informa- tion. We showed how iterative decoding improves the performance (FER) of the system. The computer simulations were performed for the system with 2 transmit and 2 receive antennas. The quasi-static and the fast-fading channels were considered. We found in different simulation scenarios how much the outer RS improved system performance.

We compared our results with results presented in [2] for binary STTC systems. The system proposed by us performs as well or better than the system in [2].

Further research is necessary for the optimum selection of convolutional encoder over the ring used in the space-time encoder.

References

1. Jafarkhani H (2005) Space-time coding: theory and practice. Cambridge University Press, Cambridge
2. Hong Y, Yuan J, Chen Z, Vucetic B (2004) Space-time turbo trellis codes for two, three, and four transmit antennas. IEEE T Veh Technol 53(2):318–328
3. Wesolowski K (2002) Mobile communication systems. Wiley, New York
4. 3GPP TS 25.212 version 5.2.0 Release 5
5. Remlein P (2003) The encoders with the feedback for the packed transmission without tail symbols. In: VIIIth Poznan workshop on telecommunication, PWT '03, Poznan 11–12 Dec. 2003, pp 165–169 (in Polish)
6. Cong L, Xiaofu W, Xiaoxin Y (1999) On SOVA for nonbinary codes. IEEE Commun Lett 3(12):335–337.
7. Kermoal JP, Schumacher L, Pedersen JP, Mogensen PE, Fredriksen F (2002) A stochastic MIMO radio channel model with experimental validation. IEEE J Sel Area Comm 20(6)

Chapter 60
Network Performance Monitoring and Utilization Measurement

Goran Martinovic, Drago Zagar, and Drazen Tomic

Abstract Computer intensive applications require efficient monitoring of computer networks. This can be enabled by operational analysis. In addition to other parameters, utilization measurement of a local network was carried out for a period of 24 hours by an artificially generated continuous workload, and at varying workloads caused by normal and increased use of network applications. Regardless of the rather high network workload, the commutated network proved its resistance to collisions and errors. It also showed a tendency to a change of utilization depending on the load itself. Utilization diagrams show the true behavior pattern of particular network segments. Utilization changes show a certain regularity by means of which future changes of performance might be foreseen.

60.1 Introduction

The efficiency of any institution cannot be envisaged without a computer network. In case of computer intensive applications, special attention should be paid to efficient network monitoring and management. The requests mentioned were additionally made harder by the great number of users, as well as by the heterogeneity of the users, the service providers, and the infrastructure [1]. The most complete overview of real performances and predictions of performance patterns can be provided by operational analysis [2]. The prerequisites for its use are the availability of a real system and the necessary software and hardware for performance monitoring. Experimental results obtained by operational analysis should enable performance analysis at the observed interval. One of the important performance indicators is definitely network utilization [3, 4]. Network communication is, according to [5], limited by various factors, such as available bandwidth, network congestion, delay, server performance, and

G. Martinovic (✉)
Faculty of Electrical Engineering, Josip Juraj Strossmayer University of Osijek, Kneza Trpimira 2b, Croatia
e-mail: goran.martinovic@etfos.hr

N. Mastorakis et al. (eds.), *Proceedings of the European Computing Conference*, Lecture Notes in Electrical Engineering 27, DOI 10.1007/978-0-387-84814-3_60, © Springer Science+Business Media, LLC 2009

complexity of the protocols. The ethernet is a broadcast technology [6], and mutual communication between network devices is established through common network media which can be negatively affected by many factors [2, 7]. Network congestion is caused by an increase in the number of people using the network for sharing big files, access to file servers, and connection to the Internet. The consequences are increased response times, slower transmissions of files, and thereby less productive users [8]. In order to reduce network congestion, what is necessary is either larger bandwidth or a more effective use of bandwidth. Network efficiency is improved by monitoring and management of its performance.

Section 60.2 describes the need for performance evaluation. Section 60.3 shows software tools which enable traffic generation, performance control, and monitoring, as well as the commutator as a hardware condition for designing a network with increased redundancy and availability. Section 60.4 gives an overview of an experiment conducted in which network utilization was measured. Broadcast errors and collisions are also taken into consideration [3].

60.2 Network Performance Evaluation

The proper selection of a computer system for a specific application requires a certain knowledge and a preliminary analysis. Specific computer architecture, operating system, database, and LAN represent elements for improvement of efficiency [2]. Today's computer systems enable resource sharing at a global level, and fulfill requests such as flexibility, extensibility, availability, and reliability [7]. Therefore, optimal design and system selection are of great importance.

Modeling enables an efficient analysis of complex systems. According to [2], there are four types of modeling tools: analytic, simulation, testbed, and operational analysis. For utilization measurement of a local network an operational analysis of the existing system was used. Compression and simplification of the system are not important in the operational analysis, but rather getting information from the real system. Analysis of this information provides a good projection of future system behavior. Operational analysis covers measurement and estimation of the real system. Measurements are conducted by means of software and hardware monitoring devices. The choice of a corresponding evaluation method depends on the following criteria: level of system development, time necessary for analysis, tool availability, and costs of necessary hardware or software. Network Inspector [9] is selected for utilization measurement of the observed network.

60.3 Testing Platform

Network Inspector, in combination with Fluke Networks equipment, enables network monitoring and planning, as well as error detection on a LAN segment. The standard version enables network "visibility" by showing devices and local subnets. It also monitors the state of devices and generates errors, warnings, and messages in order to prevent, detect, and solve problems. Network Inspector consists of three main components. Agent Manager detects and analyzes information about devices, domains, and subnets. A MS Access database provides a location where the Agent stores the information. The console shows the information from the selected database. On the console, reports can be made, and auxiliary tools can be run, which altogether provide information about problems, devices, or subnets detected on the network. Network Inspector reports about collision rate, broadcast rate, error rate, average utilization, switch performance, switch details, and problem log. Utilization is a measure of effective activity of the resources observed [4]. For ports under monitoring, a diagram of utilization (in %) shows available bandwidth used on that port. For ports with rates greater than 4 Mbps, there is a warning if utilization crosses the warning threshold (50%); on the other hand, if the error threshold is crossed (80%), an error is reported.

Cisco Catalyst 2950SX-24 [10], is an independent and controlled 10/100 commutator with a fixed configuration, which enables the connection of users on networks. It represents an extremely suitable solution for delivery of Gb speeds by optical fiber. Dual ports enable redundancy, increased availability, and adequate commutator connection in series, with the possibility of controlling them as a cluster.

60.4 Experimental Results and Analysis

60.4.1 Experimental Setup

Measurements were conducted on a local network equipped with 14 personal computers (PIV) and two servers networked through a Cisco commutator [10]. The computers ran MS Windows XP and the servers ran the MS W2K Server OS. The first server was used for antivirus program updates, whereas the second server was used for network monitoring and a series of educational activities. All computers were networked by means of the previously described commutator. Network monitoring was carried out by using the Network Inspector tool. Since measurements of the commutated LAN were executed in conditions of increased workload, the traffic generator TfGen was also used [10]. Measurement was carried out for 24 hours. The behavior of the network and computers on the network was observed in conditions of a continuous workload of 80%, with additional workload from the usual usage. Utilization was measured on every port on the commutator.

60.4.2 Experimental Results

A continuous workload of 80% was achieved by means of a traffic generator, TfGen. Additional workload was achieved by different data transfers, as in [1].

Artificially generated traffic was carried out in a closed loop. Through the commutator each computer simultaneously received the data from the preceding computer and transmitted it to the next computer. After 24 hours Network Inspector recorded 10 ports with the greatest network utilization (u), as shown in Fig. 60.1. As examples, Fig. 60.2 shows only three diagrams of u, at ports 8, 15, and 24. Other ports indicated similar utilization patterns. According to Fig. 60.2a, the greatest u was reached at port 8. Average utilization ($\bar{u}$) was 82.58%, while peak utilization (u_{preak}) was 89.96 % at 6:21 p.m. At port 15 (Fig. 60.2b) at 9:00 p.m. ($\bar{u}$) and u_{preak} were 64.24% and 84.04%, respectively. A significant change in values of u could be noticed during the day, whereas a utilization decrease to about 40% was noticeable between midnight and 9:00 a.m. This was caused by computer overload, and it caused failures of certain computers. Within that failure interval, the computer at port 15 was only receiving traffic from the neighboring computer connected to port 16, but it was not able to generate and transmit traffic to the computer connected to port 14, which failed. At port 24 (Fig. 60.2c) at 9:12 a.m. ($\bar{u}$) and u_{preak} usage was 51.09% and 84.56%, respectively. A significant change of u was noticeable by 3:00 p.m., and after that by 9:00 a.m. was at about 40%. Then the computer stopped receiving traffic from the computer at port 17, but it was only transmitting traffic to the computer at port 22.

Table 60.1 shows average utilization ($\bar{u}$) and peak utilization (u_{preak}) at every port [11]. The total average network utilization ($\bar{u}_{total}$) at these ten ports was 69.44%, whereas the standard deviation (σ_u) was 13.82%. Figure 60.3 shows a diagram of average and peak utilizations at the top ten ports. Computers at ports at which individually measured values $\bar{u}$ were greater than $\bar{u}_{total}$ had successfully completed their tasks, achieving $\bar{u}$ greater than 80%. Computers connected to ports at which individual measured values of $\bar{u}$ were less than $\bar{u}_{total}$, did not meet the expected $\bar{u}$ of 80%, the reason being the failures of computers with which they communicated and exchanged traffic, because of a rather high workload and insufficient processor power. Peak utilizations u_{peak} were somewhat equal at all ports, excluding port 8, to which the server was connected. During measurements, no network congestion was recorded.

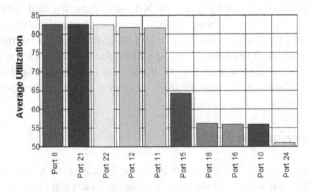

Fig. 60.1 Top 10 ports for network utilization

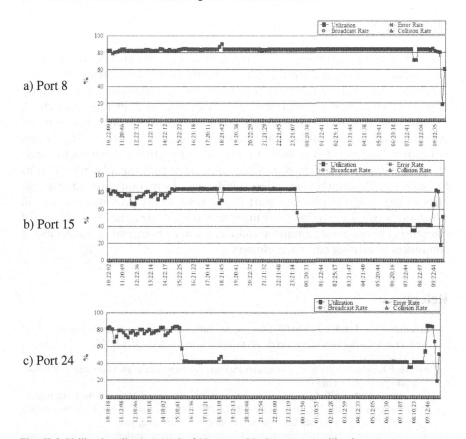

a) Port 8

b) Port 15

c) Port 24

Fig. 60.2 Utilization diagrams at 3 of 10 ports with the greatest utilization

Table 60.1 Average and peak utilizations

Port	8	21	22	12	11	15	18	16	10	24
$(\bar{u})$ (%)	82.58	82.52	82.43	81.79	81.68	64.24	56.22	55.95	55.88	51.09
u_{preak} (%)	89.96	84.56	84.62	84.93	84.56	84.04	84.56	87.08	87.43	84.55

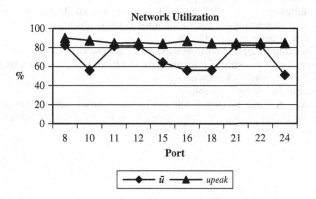

Fig. 60.3 Network utilization diagram

60.5 Conclusion

Regardless of the continuous and rapid development of network technologies, demanding applications require careful performance monitoring. The best insight into network performance is provided by operational analysis. In addition to other indicators, utilization was also measured in this research. A commutated network is not prone to errors in transmission, congestion, or collision. The average utilization was 69.44% (deviation of 13.82%). From utilization diagrams it is possible to inspect a utilization pattern on the basis of which periodicity in behavior of the observed part of the network might be foreseen. A utilization pattern offers significant information about the possibilities of engaging the observed part of the network. It proposes intervals in which the network is overloaded. Future research will use a similar procedure of monitoring performance aiming at more efficient engagement of nondedicated computers in existing computer clusters.

References

1. Guoqiang M, Habibi D (2002) Loss performance analysis for heterogeneous on-off sources with application to connection admission control. IEEE ACM T Network 10(1):125–138
2. Fortier PJ, Michel HE (2003) Computer systems performance evaluation and prediction. Digital Press, New York
3. Hanemann A et al. (2006) Complementary visualization of perfSONAR network performance measurements. In: Proceedings of the 2006 international conference on internet surveillance and protection. Cap Esterel, Côte d'Azur, France, p 6
4. Shiomoto K et al. (1998) A simple bandwidth management strategy based on measurements of instantaneous virtual path utilization in ATM networks. IEEE ACM T Network 6(5):625–634
5. Parthasarathy P (2006) Analysis of network management of remote network elements. In: Proceedings of the international conference network, systems and mobile. Mauritius, p 57
6. Breitbart Y et al. (2001) Efficiently monitoring bandwidth and latency in IP networks. In: Proceedings of the 20th annual joint conference IEEE computer and communications societies. Anchorage, AK, USA, pp 933–942
7. Li M, Sandrasegaran K (2005) Network management challenges for next generation networks. In: Proceedings of the 30th annual IEEE conference on local computer networks. Sidney, Australia, pp 593–598
8. Jain M, Dovrolis C (2003) End-to-end available bandwidth: measurement methodology, dynamics and relation with TCP throughput. IEEE ACM T Network 11(4):537–549
9. Fluke Networks, www.flukenetworks.com
10. Cisco Systems Inc, www.cisco.com
11. Papagiannaki K, Cruz R, Diot C (2003) Characterization: network performance monitoring at small time scales. In: Proceedings of the 3rd ACM SIGCOMM conference on internet measurement. Miami, FL, USA, pp 295–300

Chapter 61
Blind Adaptive Multiuser Detection Based on Affine Projection Algorithm

Amin Khansefid and Ali Olfat

Abstract We introduce an algorithm for blind adaptive multiuser detection of synchronous direct sequence code division multiple access (DS-CDMA) signals. This algorithm is based on the affine projection algorithm (APA). We first derive the algorithm for a single antenna at the receiver and then generalize it to the multiple antenna case at the receiver. Simulation results show that the proposed algorithm has lower misadjustment, an improved convergence rate, and superior bit error rate performance compared to existing conventional methods.

61.1 Introduction

For efficient use of available bandwidth, different multiple access techniques are used in communication systems. One of the multiple access techniques that is widely used in wireless mobile systems is code division multiple access (CDMA), which is a candidate for most of the next generation of cellular communication systems. The performance of conventional receivers for CDMA systems is limited by interference and multipath distortion. An optimum receiver for DS-CDMA was introduced in [1], but its high computational complexity is a prohibiting factor for practical implementation. Several suboptimal receivers are proposed that achieve acceptable performance with lower computational complexity [2]. There has been much interest in the design of blind adaptive multiuser detectors that operate without any training data and with no assumption about the signature vectors of interfering users. Most of these blind adaptive multiuser detectors employ linear MMSE and decorrelating detectors adaptively [3]. Two adaptive methods based on LMS and RLS algorithms are also discussed in [4, 5].

A. Khansefid (✉)
Electrical and Computer Engineering Department, University of Tehran, Campus #2,
North Kargar Ave., Tehran, Iran
e-mail: a.khansefid@ece.ut.ac.ir

N. Mastorakis et al. (eds.), *Proceedings of the European*
Computing Conference, Lecture Notes in Electrical Engineering 27,
DOI 10.1007/978-0-387-84814-3_61, © Springer Science+Business Media, LLC 2009

In [6], a novel blind adaptive multiuser detector based on the affine projection algorithm is introduced for synchronous DS-CDMA with a single antenna at the receiver end. That algorithm tries to minimize changes of the coefficients of linear detector vectors at each step in such a way that new coefficients satisfy some constraints. It is also shown in [6] by simulations that the proposed method with reduced computational complexity achieves convergence speed and misadjustment performance comparable to the RLS algorithm.

This paper is organized as follows. In Section 61.2 we will formulate the proposed multiuser detector and its adaptive implementation based on the affine projection algorithm. Simulation results show that the proposed algorithm has significant misadjustment reduction compared to the algorithm in [6] while achieving a higher convergence rate. In Section 61.3 we extend the algorithm to the case of a receiver with multiple antennas. In Section 61.4 simulation results are presented for single and multiple antenna cases. Simulation results show that the use of antenna arrays at the receiver will significantly improve the performance of the receiver in interference reduction and convergence.

61.2 Proposed Blind Adaptive Multiuser Detector Using Single Antenna

In synchronous DS-CDMA systems with additive white Gaussian noise (AWGN), by passing the received signal through a chip matched filter and then sampling at chip rate, the discrete-time output of the receiver can be modeled as:

$$r[n] = \sum_{k=1}^{K} A_k b_k[n] s_k + v[n] \qquad (61.1)$$

where v[n] is the ambient noise vector that is assumed to be a white multivariate Gaussian random vector with zero mean and covariance matrix $\sigma^2 I_N$ (I_N is $N \times N$ identity matrix). N, K, A_k are processing gain, number of active users, and received amplitude of kth user, respectively. $b_k[n] \in \{\pm 1\}$ is the information bit of k_{th} user and s_k is $N \times 1$ signature vector of kth user that is assumed to have unit norm, i.e., $s_k^T s_k = 1$. The linear multiuser detector uses coefficients vector, w, to combine elements of $r[n]$ and detect information bits:

$$\hat{b}_k[n] = sign(\Re(A_k^* w^H r[n])) \qquad (61.2)$$

where $\Re(.)$ denotes real part. We also assume that our desired user is user 1 ($k = 1$).

The so called affine projection algorithm (APA) [7] is the solution of the following constrained optimization problem:

$$w[n+1] = \arg\min_w \|w - w[n]\|^2$$

$$\text{s.t. } w^H u[n-l] = d[n-l] \qquad 1 = 0,...,L-1 \tag{61.3}$$

where $u[n]$ is input data vector at time instant n, $d[n]$ is desired response, and L is number of constraints.

The proposed algorithm of this paper is formulated as the solution of the following constrained optimization problem that has the general form of (61.3):

$$w[n+1] = \arg\min_w \|w - w[n]\|^2$$

$$\text{s.t. } w^H s_1 = 1$$

$$w^H r[n-l] = sign(\Re(A_1^* w[n-l]r[n-l])) \tag{61.4}$$

$$= \hat{b}_1[n-l] \qquad 1 = 0,...,L-1$$

In this optimization problem, the first constraint tries to receive the signal of the desired user, while other constraints try to eliminate interference.

We define:

$$X[n] \triangleq [s_1 \quad r[n] \quad \cdots \quad r[n-L+1]] \tag{61.5}$$

$$d[n] \triangleq \left[1 \quad \hat{b}_1[n] \quad \cdots \quad \hat{b}_1[n-L+1]\right]^T \tag{61.6}$$

where $\triangleq$ denotes by definition and $[\ ]^T$ denotes matrix transpose.

So the constraints of (61.4) can be written as:

$$X[n]^H w[n+1] = d^* \tag{61.7}$$

where $()^*$ denotes complex conjugate. The solution of the optimization problem of (61.4) can be found by the Lagrange multiplier method as:

$$w[n+1] = w[n] + X[n][X[n]^H X[n]]^{-1}[d^* - X[n]^H w[n]] \tag{61.8}$$

So the algorithm for blind adaptive multiuser detection is given by Table 61.1.

The main difference between the proposed algorithm and that of [6] is in the constraints of the optimization problem of (61.4), where in [6] the d vector in (61.6) is defined by $d = [1,0,...,0]^T$. If we consider the steady state performance of the detector in [6], there is no difference between $w[n+1]$ and $w[n-l]$ for $l = 0,...,L-1$ and we will get $d = [1,0,...,0]^T$, which is not reasonable, because we don't want to have $w[n-l]^H r[n-l] \approx w[n+1]^H r[n-l] = 0$. Simulation results show that the proposed algorithm has a much lower misadjustment and better convergence rate.

Table 61.1 Proposed blind adaptive multiuser detection algorithm

Initializing: $w[0] = s_1$ (start from match filter)

$n = 0, 1, 2, \ldots$

$$e[n] \overset{\Delta}{=} d[n] - X[n]^H w[n] \qquad (61.9)$$

$$p[n] \overset{\Delta}{=} [X[n]^H X[n]]^{-1} e[n] \qquad (61.10)$$

$$w[n+1] = w[n] + \mu X[n] p[n] \qquad (61.11)$$

61.3 Proposed Blind Adaptive Multiuser Detector Using Antenna Array

In this part we assume the receiver has M antennas. The received signal at the mth antenna is passed through a chip match filter and then is sampled at chip rate. Discrete time output of the mth receiver can be modeled as:

$$r_m[n] = \sum_{k=1}^{K} A_k a_m(\theta_k) b_k[n] s_k + v_m[n] \qquad (61.12)$$

θ_k is direction of arrival (DOA) of signal of kth user. $a_m(\theta_k)$ is mth array element response in direction of θ_k, the other parameters are the same as in the single antenna case of the previous section. We define $z[n]$ and $\zeta[n]$, $MN \times 1$ vectors, as the concatenation of discrete time output vectors of M antennas and the concatenation of noise vectors, respectively.

$$z[n] \overset{\Delta}{=} \begin{bmatrix} r_1[n]^T & \cdots & r_M[n]^T \end{bmatrix}^T \qquad (61.13)$$

$$\zeta[n] \overset{\Delta}{=} \begin{bmatrix} v_1[n]^T & \cdots & v_M[n]^T \end{bmatrix}^T \qquad (61.14)$$

We define $d(\theta_k)$ as the spatial-temporal signature vector of kth user in (61.15):

$$d(\theta_k) \overset{\Delta}{=} a(\theta_k) \otimes s_k \qquad (61.15)$$

where $a(\theta_k)$ denotes the steering vector of the receiving array antenna and $\otimes$ denotes the Kronecker matrix product. We can write $z[n]$ in the form of (61.1), which is given by:

$$z[n] = \sum_{k=1}^{K} A_k b_k[n] d(\theta_k) + \zeta[n] \tag{61.16}$$

We need to know the DOA of the desired user and we will use the method proposed in [8] to estimate the DOA of users in the DS-CDMA system. The estimation is obtained as [8]:

$$\hat{\theta}_k = \arg \max_{0 \le \theta \le 2\pi} \frac{1}{d(\theta)^H R_z^{-1} d(\theta)} \tag{61.17}$$

$$R_z \triangleq E\{z[n]z[n]^H\} \tag{61.18}$$

where $E\{\}$ denotes expectation.

In case we don't have the complete information on R_z we can use an estimation of it using Q observed samples:

$$\hat{R}_z \triangleq \frac{1}{Q} \sum_{n=1}^{Q} z[n]z[n]^H \tag{61.19}$$

The linear receiver uses $w[n]$, the coefficients vector at step n, to combine elements of $z[n]$. The estimated bit is then given by (61.20)

$$\hat{b}_1[n] = sign(\Re(A_1^* w[n]^H z[n])) \tag{61.20}$$

Here by redefining $d[n]$ and $X[n]$ as (61.21) and (61.22), we can use (61.9), (61.10), and (61.11) to compute $w[n]$ for a receiver with multiple antennas. The initial value for $w[0]$ is set to $d(\hat{\theta}_1)$.

$$X[n] \triangleq [d(\theta_1) \quad z[n] \quad \cdots \quad z[n-L+1]] \tag{61.21}$$

$$d[n] \triangleq \begin{bmatrix} 1 & \hat{b}_1[n] & \cdots & \hat{b}_1[n-L+1] \end{bmatrix}^T \tag{61.22}$$

61.4 Simulation Results

In this part we show the performance of the proposed blind adaptive multiuser detection method via simulation. To assess the multiple access interference (MAI) suppression capability of linear blind adaptive multiuser detectors, it is usual to plot the time average of SINR. Also we plot the bit error rate versus the SNR for the proposed method. The SNR of the k_{th} user is defined by $SNR \triangleq 10 \log_{10} \frac{|A_k|^2}{\sigma^2}$ (dB). As before, we assume user 1 is the desired user.

The choice of step size parameter in (61.11) is a trade-off between speed of convergence and misadjustment, as with the LMS algorithm. For our simulation we use variable step size $\mu[n] \triangleq \max\{\mu_0 \times 0.992^n, 0.003\}$.

61.4.1 Case 1

In this simulation we compare our algorithm for a single antenna receiver, with the LMS-based receiver of [4], and the RLS-based receiver of [5], and with a blind multiuser detector based on the APA [6].

There are $K = 10$ active users in the system. The SNR of the desired user is 20 dB. There are nine interfering users, the SNR of five of them is 30 dB, that of three of them is 40 dB, and that of one is 50 dB (so near-far is considered). Gold sequences of length $N = 31$ are used as spreading codes.

For LMS, $\mu = 3 \times 10^{-4}$ is considered. For RLS, an initial value of $R^{-1} = 100I$ and a forgetting factor $\lambda = 0.997$ is considered. For an APA-based blind adaptive multiuser detector [6] and our proposed method, $L = 4$ and $\mu = \max\{0.2 \times 0.992^n, 0.003\}$ is considered. We compute time-average SINR according to (61.23), which is given by [4]:

$$SINR_{av}[n] = 10\log_{10} \frac{\sum\limits_{i=1}^{N_{run}} |A_1 w_i[n]^H s_1|^2}{\sum\limits_{i=1}^{N_{run}} |w_i[n]^H (r_i[n] - A_1 b_{1,i}[n]s_1)|^2} \tag{61.23}$$

where N_{run} is the number of independent runs.

As Fig. 61.1 shows, the proposed method has better misadjustment and speed of convergence. The final SINR of the proposed algorithm is approximately 19.8 dB, which is close to 20 dB, the SNR of the desired user. This means that the proposed algorithm has eliminated the interference from other users very effectively.

61.4.2 Case 2

In this simulation we investigate the effect of the number of constraints (L) for the single antenna case. The simulation scenario is the same as in the first case.

Figure 61.2 shows that changing $L = 1$ to $L = 2$ gives significant improvement in SINR, but that increasing L more is not so significant.

61.4.3 Case 3

In this simulation we inspect the performance of the proposed method with multiple antennas at the receiver. We have repeated the scenario of simulation 1. We plot the time-average SINR using (61.24) and estimate θ_1 with $Q = 200$ samples.

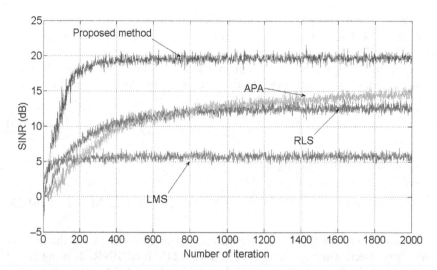

Fig. 61.1 Time-average SINR comparison of four algorithms over 200 independent runs in synchronous CDMA system with processing gain $N = 31$, $K = 10$ active users, and single antenna at receiver

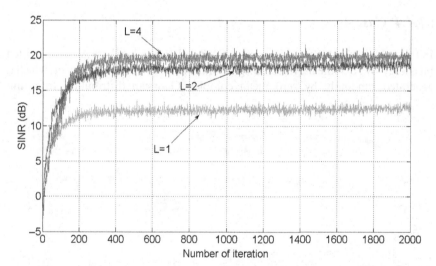

Fig. 61.2 Comparison of proposed blind adaptive multiuser detector with different numbers of constraints (L). Time-average SINR comparison of proposed algorithm over 200 independent runs in synchronous CDMA system with processing gain $N = 31$, $K = 10$ active users, and single antenna at receiver

$$SINR_{av}[n] = 10\log_{10}\frac{\sum\limits_{i=1}^{N_{run}}|A_1 w_i[n]^H d(\hat{\theta}_1)|^2}{\sum\limits_{i=1}^{N_{run}}|w_i[n]^H (z_i[n] - A_1 b_{1,i}[n]d(\hat{\theta}_1)|^2} \qquad (61.24)$$

We use variable step size parameter in (61.11) $\mu = \max\{0.2 \times 0.992^n, 0.003\}$ and set $L = 4$. The directions of arrival of the users are (in degrees): 50, 264, 193, 80, 132, 4, 320, 215, 160, and 20, corresponding to the users in increasing SNR. We use a circular symmetric array antenna with half wavelength spacing between adjacent antennas, of which the steering vector is given by (61.25) and (61.26)

$$d(i) = \frac{\lambda}{2 \sin(\frac{\pi}{M})} \sin\left((i-1)\frac{\pi}{M}\right) \tag{61.25}$$

λ is the wavelength of the signal.

$$a(\theta) = \left[c\, 1 \cdots e^{j\frac{2\pi}{\lambda}d(i)\cos\left(\frac{\pi}{2}+(i-1)\frac{\pi}{M}-\theta\right)} \cdots \right]^T, i = 0, 1,..., M-1 \tag{61.26}$$

Figure 61.3 shows that increasing the number of antennas (M) at the receiver has a very effective impact on the improvement of the final SINR. By using $M = 3$ antennas at the receiver, the proposed algorithm has a 5 dB gain compared to the single antenna case.

61.4.4 Case 4

In this part we plot bit error rate versus $SNR = 10 \log_{10} \frac{A^2}{\sigma^2}$, where A is transmitted bit energy. We assume the received signal of the k_{th} user is multiplied by g_k, which is generated randomly by zero mean complex Gaussian distribution with unit

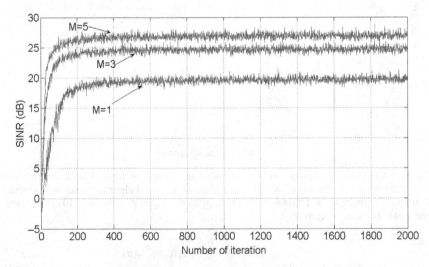

Fig. 61.3 Time-average SINR of blind adaptive proposed method with multiple antennas at receiver, over 200 independent runs, in synchronous CDMA system with processing gain $N = 31$ and $K = 10$ active users

variance. Received signals have different powers, so the near-far situation is considered. We assume $K = 10$ active users in the system, and their spreading codes are generated randomly with processing gain $N = 15$. The desired user is user 1.

Simulation parameters are given in Table 61.1. For ease of representation, in signature vectors, -1 is mapped to 0.

We specifically compare the performance of the proposed method with a conventional space-time matched filter receiver and a space-time MMSE receiver [9]. We apply exact values of DOA and parameters for these two algorithms (Table 61.2).

Figure 61.4 shows the bit error rate of the proposed blind adaptive multiuser receiver compared to some conventional space-time receivers. These conventional receivers are not adaptive. They use M = 3 antenna elements. We can see from

Table 61. 2 Parameters of synchronous CDMA system

K	Signature vector	DOA (deg)	Channel gains g_k	$\|g_k\|$
1	110101011001110	50	$-0.53 + 0.94i$	1.082
	100111111001011			0.942
2		264	$0.553 + 0.763i$	
3	101001001011111	193	$-0.605 - 0.566i$	0.828
4	101110001101001	80	$-0.758 - 0.852i$	1.140
5	011001001000001	132	$0.554 - 1.27i$	1.386
6	000110110001011	4	$0.933 - 0.705i$	1.169
7	110101111111001	320	$0.885 + 0.338i$	0.947
8	100000111001110	215	$-0.092 + 0.756i$	0.762
9	011010010001110	160	$0.141 - 0.491i$	0.511
10	010010011101110	20	$-0.409 + 0.045i$	0.411

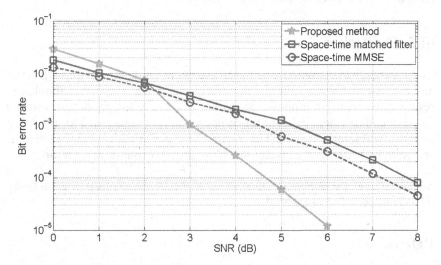

Fig. 61.4 Comparing bit error rate of proposed algorithm for blind adaptive multiuser detection and some conventional space-time receivers in synchronous CDMA, with processing gain $N = 15$, $K = 10$ active users, and a uniform circular array with $M = 3$ antenna elements

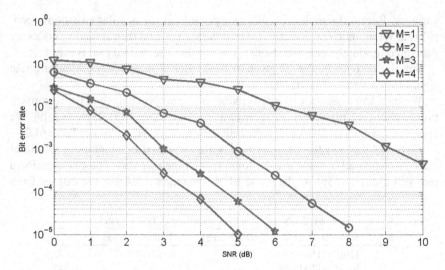

Fig. 61.5 Bit error rate of proposed algorithm for blind adaptive multiuser detection in synchronous CDMA, with processing gain $N = 15$, $K = 10$ active users, and a uniform circular array with different numbers of elements (M)

Figure 61.4 that the proposed method achieves 2.5 dB gain at bit error rate 1e-4, versus the space-time MMSE receiver.

Figure 61.5 shows the bit error rate performance of the proposed receiver with different numbers of antennas (M) at the receiver. It is clear that using 3 antennas at the receiver, 6 dB gain is achieved at a bit error rate of .001 compared to the receiver with a single antenna.

61.5 Conclusions

In this paper, we proposed a blind adaptive multiuser detector with both single and multiple antennas at the receiver end for DS-CDMA systems. Simulation results show that our proposed algorithm has a very low misadjustment. Its speed of convergence is also faster compared to some other well known blind adaptive multiuser detectors. The bit error rate performance of our receiver is shown to be better than conventional blind adaptive multiuser detectors.

Acknowledgment The authors wish to thank the Iran Telecommunication Research Centre (ITRC) for helpful financial support.

References

1. Verdu S (1986) Minimum probability of error for asynchronous Gaussian multiple-access channels. IEEE T Inform Theory 32:85–96
2. Verdu S (1998) Multiuser detection. Cambridge University Press, Cambridge

3. Wang X, Poor HV (2002) Wireless communication systems, advanced techniques for signal receptions. Prentice Hall, Englewood Cliffs, NJ
4. Honig M, Madhow U, Verdu S (1995) Blind adaptive multiuser detection. IEEE T Inform Theory 41(7):944–960
5. Poor HV, Wang X (1997) Code-aided interference suppression for DS-CDMA communications. IEEE T Commun 45(9):1112–1122
6. Li J, Zhang X (2005) Blind adaptive multiuser detection based on affine projection algorithm. IEEE Signal Proc Lett 12(10):673–676
7. Haykin S (2001) Adaptive filter theory. Prentice Hall, Englewood Cliffs, NJ
8. Olfat A, Nader-Esfahani S (April 2004) New receiver for multiuser detection of CDMA signals with antenna arrays. IEE P-Commun 151(2):143–151
9. Wang X, Poor VH (1999) Space-time multiuser detection in multipath CDMA channels. IEEE T Signal Process 47(9):2356–2374

Chapter 62
Multiple Base Station Positioning for Sensor Applications

M. Amac Guvensan, Z. Cihan Taysi, and A. Gokhan Yavuz

Abstract Sensor applications have many different constraints, such as the network lifetime and the number of sensor nodes and base stations (BS). The main goal of the researchers is to maximize the network lifetime at the minimum cost. BS positioning and the choice of routing algorithm are two important criteria for maximizing the lifetime. Many wireless sensor network (WSN) applications, like intelligent agriculture, habitat, and weather monitoring, are based on a continuous data delivery model. In these applications, each sensor node generally collects data of the same size and periodically transfers this data to BS by multi-hop communication. To maximize the network lifetime of such applications, we propose a new BS placement algorithm for deploying multiple base stations, called K-Means Local+, which provides up to 45% longer network lifetime than the single k-means algorithm [8].

62.1 Introduction

The importance of recent technological advancement is a result of the networking of sensors that led the way to integrated information dissemination. Sensor systems can be active in a wide area and involved in tasks such as localization, detection, and preventive access. Sensor nodes, generally embedded with a microprocessor, wireless transceivers, storage media, and a battery unit, communicate with their counterparts deployed in various dispositions [1].

As with many other technologies, the first examples of wireless sensor networks were designed and implemented for military applications. Nowadays, wireless sensor networks are being used in habitat monitoring, intelligent agriculture, industrial monitoring and control, home automation, and security applications.

M. Amac Guvensan (✉)
Department of Computer Engineering, Yildiz Teknik University, Besiktas,
Istanbul, Turkey
e-mail: amac@ce.yildiz.edu.tr

N. Mastorakis et al. (eds.), *Proceedings of the European*
Computing Conference, Lecture Notes in Electrical Engineering 27,
DOI 10.1007/978-0-387-84814-3_62, © Springer Science+Business Media, LLC 2009

Sensor networks are different from other networks because of the limitations on battery power, node densities, and the significant amount of desired data information. Sensor nodes tend to use energy-constrained small batteries for their energy supply. Therefore, power consumption is a vital concern in prolonging the lifetime of a network operation.

Different methods for reducing power consumption in wireless sensor networks have been explored in the literature. Various approaches were suggested, such as increasing the density of the sensor nodes to reduce transmission range, reducing standby power consumption via suitable protocol design, and advanced hardware implementation methodology. Algorithms for finding the minimum energy disjoint paths in an all-wireless network were developed [2]. SEAD [3] was proposed to minimize energy consumption in both building the dissemination tree and disseminating data to sink nodes. Few researches, however, have studied how the placement of sensor nodes/base stations can affect the performance of wireless sensor networks.

In some applications, large scale sensor networks would be necessary. Thus, we have to deploy thousands of sensor nodes over the monitored region. In this case, the scalability of the network is a very important design issue. In order to deal with this issue, it might be necessary to use more than one BS within the monitored region. In such cases, placement of the base stations becomes the main problem we have to deal with.

In this research, we examine the locations of multiple base stations to increase the network lifetime. In Section 62.2, we refer to research about the base station location problem. In Section 62.3, we describe the detailed specifications, assumptions, and constraints of our network topology. In Section 62.4, we give detailed information about our proposed algorithm. In Section 62.5, we present our test parameters and results.

62.2 Base Station Placement Approaches

Considerable research was done on optimal initialization, that is, at network set-up time, positioning of single or multiple BSs in the WSN. Published works [4–6] generally differ, based on the assumptions made, the network model considered, the network state information available, and the metrics to be optimized [7].

Due to the limited battery capacity of sensor nodes, base station placement directly affects the lifetime of a sensor network. Moreover, the expensive price of a BS makes the designer focus on accurate placement. Some approaches for the BS problem are listed below [8]:

- finding the best sink locations,
- minimizing the number of sinks for a predefined minimum operation period,
- finding the minimum number of sinks while maximizing the network life.

62.2.1 Placement of N-Base Stations

Depending on the assumptions and network models, most of the optimal BS positioning problem formulations are NP-complete. A common way to counter such complexity is to employ approximation. For example, in [9], the search space is restricted to the sensor node locations, and the best position *s* among them in terms of the network lifetime is selected. On the other hand, there are methods which use integer linear programming to determine new locations for the base stations.

In some applications, the monitoring duration is a predefined value. In such cases, the main goal is to find the minimum number of sinks to satisfy this predefined lifetime for the sensor network. On the other hand, in other applications, there may not be any such constraints like number of sink nodes, network lifetime, etc. From this point of view, placement of multiple BSs is an optimization problem which requires us to research the correlation between the number of BSs and the lifetime of the wireless sensor network.

On the other hand, the number of BSs is closely related to project budgets, and one should estimate the number of BSs before deployment. In this case, this problem can be considered to be finding the location for each BS which can best be handled using a classic clustering approach [8].

Many different clustering algorithms have been proposed to classify various data groups, which fall into two categories [10]:

- hierarchical,
- non-hierarchical.

The k-means clustering algorithm can be given as an example of a non-hierarchical method. Both the hierarchical and non-hierarchical methods need an initial parameter. That leads us to design a system which finds the optimum point in the relation between the number of BSs and the lifetime of the network.

One can use clustering for finding the exact locations of *n* BSs, by calculating the center of mass of the sensor nodes within a cluster. Euclidean distance or power aware distance could be used as a clustering metric [8].

In WSN applications, the sensor nodes one hop away from a BS with more subnodes die more quickly and result in many unreachable sensor nodes. To solve this problem, we decided to place the base stations near these critical sensor nodes with respect to the number of their subnodes.

62.3 System Model

In the previous section, we discussed the design issues of sensor networks with multiple base stations. In this section we will delve into the details of our design. Figure 62.1 shows the block diagram of our proposed system.

Fig. 62.1 Block diagram of
the proposed system model

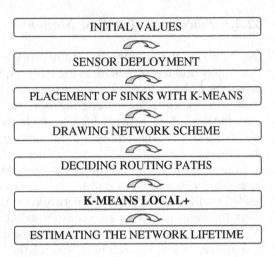

62.3.1 *Deployment of the Sensor Nodes*

With regard to the limitations and needs of applications, different deployment
techniques can be used in wireless sensor networks. In military applications, the
network might be deployed beyond enemy lines with the help of an aircraft or
artillery. On the other hand, in agriculture monitoring applications, sensor
nodes can be placed by humans one by one.

The deployment is either deterministic or self-organizing. In the deterministic case, the sensor nodes are manually placed and data is routed through
predetermined paths. However, in self-organizing systems, the sensor nodes
are scattered randomly. This kind of deployment increases the importance of
the BS location in terms of energy efficiency and performance.

In our simulation, we chose to deploy nodes randomly over a rectangular area.

62.3.2 *Finding the Location Information*

In order to calculate BS locations we need to know the location information of
each node. A node can obtain its location information by using specialized
hardware (GPS, ultrasound, acoustic, etc.). The location information can also
be derived using centralized or distributed methods [8]. We assume that the
information is calculated with a centralized method.

62.3.3 *Collecting Location Information*

After the deployment of the sensor nodes, we can gather the location information of each sensor node via a mobile BS. If we don't have a mobile BS, we can
deploy a BS temporarily. This BS can be removed from the field during the final

BS deployment phase. If it is not possible to remove this BS, we can use it as a final BS and consider it as a fixed BS in the clustering algorithm.

62.3.4 Determination of the Best Locations for K Base Stations

Since the sensing range of a sensor node is very short, we must deploy sensor nodes densely in order to provide maximum coverage over the monitored field. Sensor nodes may communicate directly with the base station, but this kind of communication is very energy-consuming. It will be more efficient to perform multi-hop communications (Fig. 62.2).

However, since all sensor nodes are equipped with the same limited power sources, sensor nodes that are closer to the BS will deplete their energy supply at a much quicker pace because of their additional communication burden. Thus, a large number of nodes will be disconnected from the network.

In order to prolong the network lifetime, we should share this communication burden between as many nodes as possible by placing base stations where the sensor node density is high. To find these dense locations, a clustering algorithm can be used. K-means is one of the popular clustering algorithms to find the best locations for k base stations. To maximize the network lifetime, we created our own heuristic algorithm, called K-Means Local +, and compared its results with the results of the k-means algorithm.

62.3.5 Routing

In our research, the main problem is the placement of the base stations. Therefore, we decided to implement one of the classical routing algorithms, called the shortest path routing algorithm. After assigning all sensor nodes to any of the n-BSs, a shortest path to the BS is calculated for each node of each group headed by one BS. Since the distance between nodes directly affects the power consumption, and therefore the network lifetime, we draw the shortest route for each node to the BS. Thus, all sensor nodes learn their up and down sensor nodes to transceive packets.

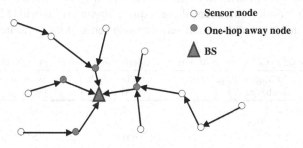

Fig. 62.2 Multi-hop communication

62.3.6 Sensor Node Energy Model

Usually sensor nodes are powered with two AA batteries. The energy capacity of a AA battery varies between 9072 and 11,050 J, as shown in Table 62.1 [11]. In our simulation we assumed each sensor node has two AA batteries, which provide 22,100 J total energy.

Basically, sensor nodes perform three operations: receiving data from other nodes, sensing environmental data, and transferring data to the BS over other nodes. Thus, the total energy consumption of a sensor node can be formalized as in (62.1):

$$E = E_S + E_R = E_T \tag{62.1}$$

E_S: consumed energy for the sensing operation,
E_R: consumed energy for the receive operation,
E_T: consumed energy for the transmit operation.

Sensor nodes are usually capable of adjusting their transmit power level. The energy consumption for the transmit operation (E_T) is proportional to the distance between the transmitter and the receiver, where k is a real number:

$$E_T(d) = kd \tag{62.2}$$

The energy consumption for the receive operation is independent of the distance (d) and can be expressed with a real constant τ. Since the energy required for sensing operations is negligible compared to the energy required for transmission and reception, we did not include this energy in our design decisions.

$$E_R = \tau \tag{62.3}$$

62.3.7 Lifetime Estimation

Depending on the application of the sensor network, the data delivery model to the BS can be continuous, event-driven, query-driven, or hybrid [12, 13]. In the continuous delivery model, each sensor node sends data periodically. In the

Table 62.1 Comparison of battery properties according to their types [11]

Battery type	Average voltage during discharge (V)	milli-Amp hours (mAh)	Watt-hours (Wh)	Joules (J)
Alkaline long-life	1.225	2122	2.60	9360
NiMH	1.2	2100	2.52	9072
Lithium Ion	3.6	853	3.1	11,050

event-driven and query-driven models, the transmission of data is triggered when an event occurs or a query is generated by the BS. Some sensor networks apply a hybrid model using a combination of continuous, event-driven, and/or query-driven data delivery. The routing protocol is highly influenced by the data delivery model, especially with regard to the minimization of energy consumption and route stability. Our K-Means Local+ algorithm gives better results, especially for the continuous data delivery model.

62.4 K-Means Local +

As described in Section 62.3.6, sensor nodes use their major energy for communication operations. The network lifetime can be prolonged by minimizing the power consumption for those operations. The power consumption for communication operations is proportional to the number of bits transferred and to the distance to the receiver. It can be reduced by decreasing the distance between the BS and its neighbors. In K-Means Local+, we first calculate the BS locations using k-means, and then we simply move these base stations closer to their one-hop neighbor nodes with more subnodes and obviously with heavier traffic loads [14], as shown in Fig. 62.3. With this approach, we prolong the lifetime of the network.

The K-Means Local + algorithm has three main steps:

- k-means clustering,
- establishment of the routing,
- positioning of the base station.

62.4.1 K-Means Clustering

In the first step, the initial positions of the base stations are calculated using the k-means clustering algorithm. The implementation details of this algorithm are given in Section 62.3.4.

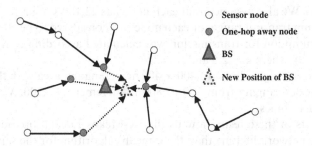

Fig. 62.3 Positioning of BS with K-Means Local+

62.4.2 Establishment of the Routing

Our K-Means Local+ algorithm is independent of the routing algorithm. Thus, any routing algorithm suitable for WSN can be used in this step. However, the resulting routing scheme directly affects the placement of the base station.

62.4.3 Positioning of the Base Station

In this step, the length of the data packets relayed by each sensor node is calculated according to the routing scheme. The new position of the BS is based on the position of its one-hop neighbors and the length of the data packets relayed by those neighbors. Equation (62.4) gives the formula:

$$X_{bs} = \frac{\sum\limits_{i=1}^{m}(ds_i)^2 \cdot x_i}{\sum\limits_{i=1}^{m}(ds_i)^2} \qquad Y_{bs} = \frac{\sum\limits_{i=1}^{m}(ds_i)^2 \cdot y_i}{\sum\limits_{i=1}^{m}(ds_i)^2} \qquad (62.4)$$

X_{bs}, Y_{bs}: final position of base station,
ds_i : relayed data packet size by i-*th* one-hop neighbor of BS,
x_i, y_i: position of i-*th* one-hop neighbor of BS,
m: number of one-hop neighbors of BS.

62.5 Computational Results

Using our sink placement simulator (SPS), several tests were run to observe the performance of the K-Means Local+ algorithm.

In our simulations, we implemented the continuous packet delivery model. In this model, each node generates 64-byte data packets and can communicate at 250 Kbps. We also assumed that each node has 22,100 J of battery capacity. The maximum transmit range for each node was considered to be 50 m, and the power consumption for transmission was calculated according to the formula given in (62.2), where k = 2.5 and $\alpha = 2$.

We tested our algorithm on a 400 × 400 m rectangular area with the number of sensor nodes ranging from 100 to 800 in increments of 100. We ran each deployment 10 times.

The results of these tests show us that K-Means Local+ provides up to a 45% longer network lifetime than the k-means algorithm for the same deployment. Figures 62.4 and 62.5 present the results of the tests.

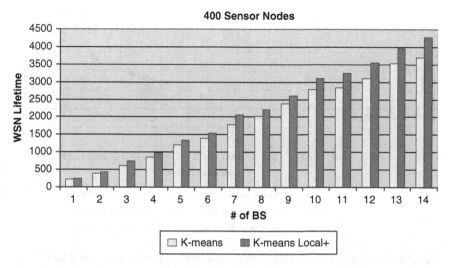

Fig. 62.4 K-means vs. K-Means Local+ with 400 sensor nodes

Analyzing Fig. 62.6 clearly shows that in WSN applications with multiple base stations placed by K-Means Local+, the number of unreachable sensor nodes is always less than those placed by the k-means algorithm, at any given time.

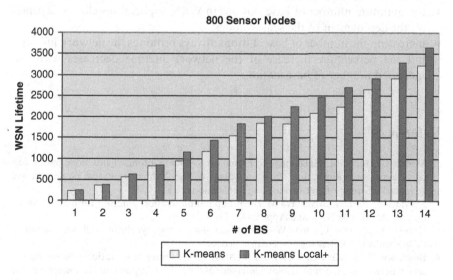

Fig. 62.5 K-means vs. K-Means Local+ with 800 sensor nodes

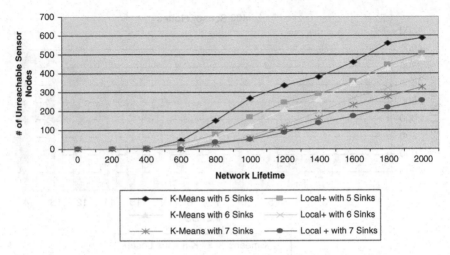

Fig. 62.6 Distribution of unreachable sensor nodes with different numbers of BSs

62.6 Conclusion

In this research, we have focused on the BS placement problem in wireless sensor networks. We have proposed a new algorithm which produces better results than k-means, the classical clustering algorithm.

The test results presented in Section 62.5 clearly show that:

- our heuristic algorithm, K-Means Local+, provides up to 45% longer network lifetime than the standard k-means algorithm;
- the optimum number of base stations in WSN applications closely depends on the deployment of the sensor nodes;
- increasing the number of base stations always prolongs the network lifetime, but the percentage increase of the network lifetime decreases with the increased number of base stations.

References

1. Mendis C, Guru SM, Halgamuge S, Fernando S (2006) Optimized sink node path using particle swarm optimization. In: Proceedings of the 20th international conference on advanced information networking and applications, 2:5
2. Savvides A, Park H, Srivastava M The n-hop multilateration primitive for node localization. ACM Mobile Netw and Appl 8:443–451
3. Kim H, Abdelhazer T, Kwon W (2003) Minimum-energy asynchronous dissemination to mobile sinks in wireless sensor network. Sensys
4. Bogdanov A, Maneva E, Riesenfeld S (2004) Power-aware base station positioning for sensor networks. 23rd annual joint conference of the IEEE computer and communications, Vol 1

5. Pan J, Cai L, Hou YT, Shi Y, Shen SX (2005) Optimal base-station locations in two-tiered wireless sensor networks. IEEE T Mobile Comput 4(5):458–473
6. Hou YT, Shi Y, Sherali HD (2006) Optimal base-station selection for anycast routing in wireless sensor networks. IEEE T Veh Technol 55(3):813–821
7. Akkaya K, Younis M, Youssef W (2007) Positioning of base stations in wireless sensor networks. IEEE Commun Mag 45(4):96–102
8. Oyman I, Ersoy C (2004) Multiple sink network design problem in large scale wireless sensor networks. IEEE Int Conf Commun 6:3663–3667
9. Efrat A, Har-Peled S, Mitchell JSB (2005) Approximation algorithms for two optimal location problems in sensor networks. Proceedings of the 3rd international conference on broadband communications, networks and systems
10. Alpaydin E (2004) Machine learning. MIT Press, Cambridge, MA
11. http://www.allaboutbatteries.com/Energy/tables.html
12. Akkaya K, Younis M (2005) A survey on routing protocols for wireless sensor networks. Ad Hoc Netw 3:325–349
13. Tilak S, Heinzelman W(2002) A taxonomy of wireless microsensor network models. ACM Mobile Comput Commun Rev (MC2R)
14. Cheng P, Chuah C, Liu X (2004) Energy-aware node placement in wireless sensor networks. IEEE GLOBECOM 5:3210–3214

Chapter 63
Considerations for the Design of a General Purpose Wireless Sensor Node

Z. Cihan Taysi and A. Gokhan Yavuz

Abstract Wireless sensor networks (WSNs) consist of small, autonomous nodes which are capable of communicating with each other over a wireless link. Minimizing the size and the power consumption of these nodes is an important research topic. A WSN node consists of a microcontroller, a wireless communication controller, a secondary storage module, and different types and combinations of sensors. We propose a wireless sensor node design that aims to be minimal in size and power consumption without significantly degrading processing power.

63.1 Introduction

Wireless sensor networks provide unique opportunities in industrial, health care, military, and environmental monitoring applications. WSNs are networks of several smart devices that have multiple onboard sensors, networked through wireless links and deployed in large numbers [1].

Minimizing energy consumption and size of sensor nodes is an important research topic aimed at making WSNs more deployable. Many uses for energy-aware wireless sensor networks are rapidly emerging from industry and academia. The numerous military and civilian applications of this technology have the potential to make a considerable impact on society. As a result, the sensor network research community has grown steadily over the past few years [2].

As most WSN nodes are battery powered, their lifetime is highly dependent on their energy consumption [3]. In cases with hundreds of sensor nodes, changing batteries can be an almost an unachievable task. It is perhaps more cost effective to replace the entire node than to locate the sensor node and replace or recharge its battery supply, because of the low cost of an individual

Z. Cihan Taysi (✉)
Department of Computer Engineering, Yildiz Technical University, 34349, Besiktas, Istanbul, Turkey
e-mail: cihan@ce.yildiz.edu.tr

N. Mastorakis et al. (eds.), *Proceedings of the European*
Computing Conference, Lecture Notes in Electrical Engineering 27,
DOI 10.1007/978-0-387-84814-3_63, © Springer Science+Business Media, LLC 2009

sensor node. In other scenarios the location of the node might make battery changes infeasible, because the node might be physically inaccessible (e.g., embedded into the hosting equipment), or the node might be located in an environment where human intervention is undesirable, such as in a bird nest [4], or in a dangerous chemical plant, or in a rugged, inaccessible terrain.

63.2 Previous Work

Typically, sensor nodes are comprised of sensors to collect environmental data, a low-power microcontroller to process the collected data and to control the system, an RF transceiver with limited range to transfer data from node to node, and on-chip memory to store the collected data. The sensor nodes are characterized by their small size and significant energy/resource constraints. In addition, once deployed, the sensor nodes are difficult to retrieve, so they must function for months on a battery pack. Also, the sensor nodes need to be extremely durable, because they will be released into potentially harsh environments. These issues affect all areas of hardware design of a sensor node, from the selection of the microcontroller to the design of the power supplies [2].

Due to these fundamental constraints, hardware and software design guidelines for sensor nodes are different from those for other applications. While theoretical research on sensor networks is still ongoing, researchers from academia and industry have increasingly deployed working prototype sensor networks, in which theories could become reality. Great Duck Island [4], ZebraNet [3], and Hogthrob [5] are just some of these projects.

There are several wireless sensor node designs used for either academic research or for industrial usage. WeC, Rene, Rene2, Dot, Mica, Mica2Dot, Mica2, and Telos were designed by the University of California at Berkeley during the TinyOS project [6] and used by other researchers and also companies like Crossbow and Mote IV, with some minor changes in the original design. Eyes is a prototype developed within a three-year European research project (IST-2001-34734), on self-organizing and collaborative energy-efficient sensor networks. MULLE [7] is a small embedded Internet system (EIS) platform designed and implemented by EISLAB [8]. Current research issues on the MULLE platform include power partitioning techniques and alternate power sources, with the aim of creating very long lifetimes with minimal battery capacity

63.3 System Design Constraints

Figure 63.1 shows the design of a sensor node consisting of five modules common to most wireless sensor node designs. These modules are the processor module, the RF transceiver module, the memory module, the sensor board

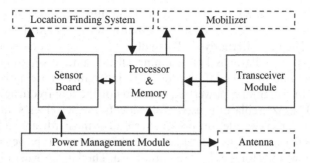

Fig. 63.1 Block diagram of a sensor node

module, and the power management module. In this section, we will discuss the design constraints for each module

63.3.1 Processor Module

Since the control of the sensor node is performed by the processor module, selecting the appropriate processor module is a very important issue. While making the decision, candidate processor modules must be analysed by means of power saving, clock scaling, instruction set complexity, and support for different types of I/O operations.

63.3.1.1 Power Saving

For most sensor applications, an average duty cycle of 1% will provide optimum operating conditions in regard to power consumption. To achieve this goal, a processor module must schedule operations for some time in the future and then enter a power saving mode, in which the processor core and the peripherals are hibernated, whereas an asynchronous timer independent of the core is kept running to cause a wake-up interrupt when necessary [2].

Once the processor core is awakened from the hibernated state, it will execute instructions to power on the peripherals for interacting with them. Two of the major caveats of this scenario are how fast the processor core can be awakened, and how fast the peripherals will be ready for operation after power-up.

63.3.1.2 Clock Scaling

There are different techniques employed to dynamically switch the operating frequency of the processor cores. These techniques are generally termed "clock scaling." One method is reducing the output of the primary system oscillator down to the operating frequency actually required for the processor core. Another is to use a low frequency oscillator with a controllable hardware multiplier. The first alternative will usually not save much power, because the high frequency oscillator is kept running all the time, dissipating maximum power. The multiplier-based approach is inherently more power friendly because of the minimum power requirements of the low frequency oscillator. Usually a 32 kHz oscillator is used as this low frequency clock source [2].

One should also consider the power overhead of the hardware multiplier. For both alternatives there may exist a latency, where the core is powered on after a wake-up but waiting for the clock to settle down. This latency will result in a dead time, causing loss of precious power while the processor core is executing nothing useful. Therefore, the clock startup latency must be minimized.

63.3.1.3 Instruction Set

Current microcontroller architectures are not very well justified for complex data processing tasks, especially ones involving floating-point calculations. Most microcontrollers are built around an eight-bit architecture and they do not tend to have a large number of general purpose registers, which is common in microprocessor architectures.

Eight-bit architectures can be enhanced either by adding special instructions to ease 16 or 32 bit arithmetic, or by including a hardware multiplier to enable fast floating-point operations. Nevertheless, the more complex the instruction set becomes, the more power will be required by the core, thus adversely affecting the suitability of a given microcontroller for sensor node designs [2].

63.3.1.4 I/O Ports/Operation

The processor module is required to interact with a number of external devices. This requirement imposes a large number of digital I/O pins on the processor module. For example, the Nordic nRF2401 [9], a radio transceiver operating at 2.4 GHz, needs at least six I/O pins when only one data channel is to be used. Otherwise, three additional I/O pins are required.

Most of the time interaction with peripherals can be carried out via serial communication; but, most of the time as well, up to four or five pins per sensor (considering that the sensors will be powered on or off individually for power saving) will be required. This means that processor modules must have a really large number of configurable I/O pins in order to accommodate a large number of sensors.

Another aspect of the serial interfacing of peripherals is that the serial communication should be carried out mostly by hardware, so the processor module is relieved of shifting in or shifting out individual bits and can be put into a power-saving mode as soon as possible.

63.3.2 RF Transceiver Module

Since WSNs are used in many different types of applications, the RF transceiver module should satisfy different metrics like low-power consumption, variable communication range, high communication rate, and security.

63.3.2.1 Power Consumption

As most WSN nodes are battery powered, their lifetime is highly dependent on their energy consumption. Thus, all modules in WSN nodes must have low power consumption.

In WSN applications, nodes transmit data to other nodes at intervals with very large periods. Thus, the RF transceiver module must be capable of being shut down completely and being awakened from the shut-down state very quickly to minimize energy consumption. Also, the RF transceiver module must be capable of adjusting its transmission range in order to minimize power consumption while communicating with closer sensor nodes.

63.3.2.2 Ease of Use

Data that will be sent by the RF transceiver module must be modulated before transmission and demodulated during reception. These modulation and demodulation operations can be performed either by the RF transceiver module or the processor module. Doing these operations with the processor module will add an additional processing burden to the sensor node, which will increase both the complexity of the software running on the sensor node and its overall power consumption.

In WSN applications, sensor nodes are generally densely deployed, which will result in multiple nodes trying to communicate with their neighbors almost at the same time. To avoid unwanted collisions on the communication medium, the RF transceiver module must support the use of multiple channels dynamically.

To adjust the transmission range, the RF transceiver module needs a feedback mechanism to estimate the approximate distance to its neighboring sensor nodes. While custom solutions to this problem can be developed in software, RF transceiver modules equipped with received signal strength indication (RSSI) capability will obtain more accurate results about the distance of their neighbors. Thus, the sensor node will be able to adjust its transmit power efficiently.

The RF transceiver module's methods of setting the communication parameters, like transmit power and communication channel, affect the complexity of the sensor node software. Thus, they determine how many cycles must be run to set up the RF transceiver module with the desired parameters.

63.3.3 Memory Module

Like all the other modules conprising a sensor node, the memory module must also have low power consumption for writing, reading, and storing data.

Another important issue for selecting a memory module for sensor nodes is the storage capacity. In some WSN applications, mobile sensor nodes might be used. Thus, some sensor nodes might not find any sink to which to transfer their data for long periods of time. In this case, depending on the application type, the sensor node may be required to store its data in the memory module.

Also, block writing to and block reading from the memory module is very desirable, because once the memory module is powered on and dissipating power, an adequate amount of data will be transferred so that the power consumed for the power-up of the memory module will become negligible.

63.3.4 Sensor Board Module

Usually sensor nodes have an expansion slot for connecting different types of sensors that will satisfy the requirements of the application. In WSN applications, various types of sensors can be used, including, but not limited to, pressure, temperature, light, humidity, and acceleration sensors. Also, some special sensors like gyroscopes, global positioning systems (GPSs), solar cells, and oximeters can be used to satisfy the needs of the application.

63.3.5 Power Management Module

Sensor nodes are generally battery powered; thus power must be consumed very effectively. Each sensor node must perform simple tasks, including sensing, sending, and receiving data. During these operations, modules which are not needed should be shut down. For example, for sensing environmental data, only the sensors and the processor module have to be running; for communication operations, only the RF transceiver and the processor modules are needed. Also, advanced operations like charging the batteries or moving a sensor node equipped with a mobilizer are sometimes performed by the sensor nodes.

These simple and advanced power control operations can either be done by a power supervisor module, or this power management can be achieved by multiple modules like the processor module and a real-time clock with an alarm function.

63.4 Our Design

In this section we will outline the design, as based on the constraints given in Section 63.3, of our sensor node, called VF-1S. Our two primary design goals were:

- to create a sensor node that has a very compact design with most of the required sensors built-in, which makes it small in size, and suitable for many WSN applications; and
- to minimize the power consumption by use of low-power sensors and external components and by selecting an optimized processor module and an RF transceiver module.

63.4.1 Processor Module

The existing platforms that are mentioned in Section 63.2 use the ATMEL AT90LS8535, AtMega128L, AtMega163, Mitsubishi M16C/62 M, and Texas Instruments MSP430F149 as their processor modules. After some research, we decided to consider the Analog Devices ADuC845 [10], which contains an 8051 core; the ATMEL ATMega128L [11]; the Microchip PIC16F877 [12]; the Microchip PIC18F4525 [13]; the Mitsubishi M16C/62 M [14]; and the TI MSP430F149 [15] for our sensor node design. Below, we give a detailed comparison of these processor modules under the topics that were discussed in Section 63.3.

63.4.1.1 Power Saving

As mentioned earlier, the processor module must have several power saving states, which will allow the use of only the needed functions from the entire set of functions, in order to reduce power consumption. Also, both the so-called clock recovery time, i.e., the startup latency of a stopped clock, and the transition between different power saving states, which changes the clock frequency, must be at a minimum in order not to waste precious power. Table 63.1 clearly shows that the MSP430F149 has the shortest wake-up time compared to other processor modules. With the exception of the AduC845, the MSP430F149 has, on average, a wake-up time 200 times better than the other processor modules.

The power consumption figures for different running speeds of the candidate processor modules are given in Table 63.2. Not all of them support the idle and/ or 32 kHz operating modes. Also, the PIC16F877 may seem to have the best power consumption values for both the 1 MHZ and the 8 MHZ modes; but the MSP430F149 has the lowest power consumption for shutdown, idle, and

Table 63.1 Wake-up times for candidate processor modules [10–15]

Microcontroller	Wake-up time
AduC845	30 µs
AT90LS8535	1.2 ms + 4clk
AtMega163L	4.2 ms + 6clk
AtMega128L	2 ms + 4 clk
ML67Q500x	10 ms
MSP430F149	< 6 µs
PIC16F877	~ 32 ms (1024t_{osc})
PIC18F4320	~ 32 ms (1024t_{osc})

low-speed working states, which makes it more suitable for our sensor node, considering the 1% average duty cycle goal.

63.4.1.2 Clock Scaling

As mentioned in Section 63.4.1.1, the MSP430F149 processor module has a remarkable wake-up time, which is less than 6 µs. This high performance is achieved by the use of a frequency locked loop (FLL), which is a method of clock generation by multiplication [15], as stated in Section 63.3. Other processor modules under consideration use Phase Locked Loop (PLL) for clock generation. In PLL clock generation, a low-pass filter, which has loop delays, is used [15].

63.4.1.3 Instruction Set

All candidate processor modules, except the MSP430F149 and the ML67Q500x, have 8-bit word sizes (Table 63.3). Since all the processor modules have ADC with more than 8-bits, it will take two cycles to complete an ADC reading. Also the implications (refer to Section 63.3) of using an 8-bit architecture need to be compensated for.

Moreover, the MSP430F149 has a hardware multiplier which enables performing multiple operations in one cycle and makes the MSP430F149 a better choice for our sensor node design.

Table 63.2 Current consumptions of candidate processor modules [10–15]

	Power down	Idle (1 MHz)	Active (32 KHz)	Active (1 MHz)	Active (8 MHz)
AduC845	20 µA	–	2.78 mA	4.05 mA	13.3 mA
AT90LS8535	15 µA	0.6 mA	–	2.5 mA	9 mA
AtMega128L	25 µA	0.4 mA	0.7 mA	2 mA	8 mA
MSP430F149	0.5 µA	55 µA	19.2 µA	420 µA	1.9 mA
PIC16F877	30 µA	220 µA	35 µA	220 µA	1.5 mA
PIC18F4320	7.2 µA	250 µA	70 µA	650 µA	3 mA

Table 63.3 Architectures of candidate processors [10–15]

	Architecture	Word size (bit)	Number of instructions	Number of registers
AduC845	CISC	8	255	–
AT90LS8535	RISC	8	118	32
AtMega163L	RISC	8	130	32
AtMega128L	RISC	8	133	32
ML67Q500x	RISC	32	–	31
MSP430F149	RISC	16	51	16
PIC16F877	RISC	8	35	–

Table 63.4 I/O peripherals of candidate processor modules [10–15]

	ADC	I2C	SPI	UART	USART	HW multiplier	General I/O
AduC845	24 bit	1	1	1	–	–	24
AT90LS8535	10 bit	–	1	1	–	–	32
AtMega163L	10 bit	–	1	1	–	–	32
AtMega128L	10 bit	–	1	2	2	–	53
ML67Q500x	10 bit	1	–	1	–	1	
MSP430F149	12 bit	–	2	1	2	1	48
PIC16F877	10 bit	1	1	–	1	–	24
PIC18F4320	10	1	1	–	1	–	24

63.4.1.4 I/O Ports/Operation

Since the processor module is required to interface with a large number of both sensors and external peripherals, a processor module with more I/O pins is highly desirable. A quick look at Table 63.4 shows that the AtMega128L and the MSP430F149 are the two best candidates. We have chosen the MSP430F149 because this processor module has two hardware SPI channels, which will ease the interfacing of serial peripherals and an ADC with a resolution of 12 bits, which in turn will enable more accurate readings from analog sensors like an accelerometer.

63.4.2 RF Transceiver Module

The RF transceiver modules that we have considered for our design were the RFM TR1000 [16], the RFM TR1001 [17], the Chipcon CC1000 [18], and the Chipcon CC2420 [19]. The characteristics of these RF transceiver modules are given in Table 63.5.

The reasons for choosing the CC2420 as the RF transceiver module are threefold. First, working at 2.4 GHz, longer ranges will be attained with the same power consumption, thus the connectivity in a WSN will be increased. Second, we will be able to use more communication channels, which will help to

Table 63.5 Comparison of candidate RF transceiver modules [16–19]

	Frequency (MHz)	Bitrate (kbps)	Output power (dBm)	Receiver sensivity (dBm)	Passive (µA)	Transmit (mA)	Receive (mA)
TR1000	916.50	115.2	1.5	−106	700	12	3.8
TR1001	868.35	115.2	1.5	−106	–	12	3.8
CC1000	868.35	76.8	5	−110	200	14.8	7.4
CC2420	2483.5	250	0	−95	90	17.4	18.8

avoid unwanted collisions and increase the number of simultaneous communications. Third, the CC2420 includes the Zigbee [20] standard, thereby providing a readily available data link layer (DLL) beyond the raw physical connectivity. The use of a standard DLL greatly simplifies communication tasks between sensor nodes and relives the software burden of a custom DLL developed from the processing module.

63.4.3 Memory Module

The characteristics of the memory modules from the aforementioned platforms are given below in Table 63.6.

Although the M25P80 has the maximum read and write current requirements, less power will be consumed compared to the other memory modules because of the M25P80's very high bit rate.

63.4.4 Sensor Board Module

After examining the WSN applications that are mentioned in Section 63.2, we decided to include some widely used sensors in our sensor node design. We also included an expansion slot in our design to enable integration of different sensors in order to satisfy the requirements of other applications. The four sensors we have included in our design are:

Table 63.6 Comparison of candidate memory modules [25–28]

	Capacity (Kbit)	Write (ms)	Bitrate (KHz)	Passive (µA)	Read (µA)	Write (mA)
24LC256	256	5	400	1	400	3
AT45DB041B	4096	7	5000	20	4000	15
M24M01S	1024	10	400	2	2000	–
M25P80	8192	1.5	25000	1	4000	15

- the SHT15[21] humidity and temperature sensor,
- the EL7900[22] light sensor,
- the ADXL321[23] 3-axis accelerometer, and
- the DS1629[24] real-time clock and temperature sensor.

In many WSN applications, the TI OPT101 light sensor is used, but this sensor has a very high stand-by current compared to the EL7900. This is why we replaced the OPT101 with the EL7900.

63.4.5 Power Management Module

Since we do not have any solar cells or a mobilizer, we did not include an advanced power control function in our design. Thus we decided to implement the power control functions with the processor module using a power supervisor and a real-time clock. Basically, the processor module enables the required modules for a given operation and then powers them down before putting itself into a hibernation state. The real-time clock (RTC) is used for accurate long-term sleeping and routine wake-up operations. The power supervisor module is used for monitoring residual battery power. The inclusion of an RTC provided us with the ability to wake up the sensor node at given date. For this purpose, the RTC is programmed with the desired date of wake-up and then the node is put into hibernation. Considering the time-keeping current of 0.3 μA of the RTC and the power-down current of 0.5 μA of the MSP430F149, pretty good values for future wake-up for our sensor node can be programmed.

63.5 Conclusion

In this paper we presented both the design criteria and a sample design for wireless sensor nodes minimal in size and power consumption, but equipped with adequate processing power.

To minimize the size of the sensor node, we included several mostly required types of sensors in our design, in contrast to designs where these sensors are attached to the processing module via an expansion slot, which further increases the overall footprint of the sensor node. To further decrease the power consumption of our sensor node, unneeded sensors may be left unpopulated on the sensor node. Our design also permits the connection of other sensors by the use of an expansion slot.

Our sensor has 129.5 μA current consumption in the idle state, and it uses less than 20 mA during receive operations. The current required for transmit operations varies from 13.4 to 21.5 mA, depending on the transmission range.

References

1. Chong C, Kumar S (2003) Sensor networks: evolution, opportunities, and challenges. P IEEE 91:1247–1256
2. Lynch C, O'Reilly F (2005) Processor choice for wireless sensor networks. In: Workshop on real-world wireless sensor networks, REALWSN'05
3. Sadler C, Zhang P, Martonosi M, Lyon S (2002) Hardware design experiences in Zebranet. In: Proceedings of the 2nd international conference on embedded networked sensor systems, pp 227–238.
4. Mainwaring A, Polastre J, Szewczyk R, Culler D, Anderson J (2002) Wireless sensor networks for habitat monitoring. In: WSNA'02, Atlanta
5. Hogthrob Homepage, www.hogthrob.dk/
6. TinyOS Homepage, www.tinyos.net/
7. MULLE Homepage, www.csee.ltu.se/~jench/mulle.html
8. EISLAB Homepage, www.ltu.se/csee/research/eislab
9. nRF2401 single chip 2.4 GHz transceiver datasheet. June 2004, Rev 1.1, www.sparkfun.com/datasheets/RF/nRF2401rev1_1.pdf
10. ADuC845 recision analog microcontroller datasheet. Feb. 2005, Rev B, www.analog.com/UploadedFiles/Data_Sheets/ADUC845_847_848.pdf
11. 8-bit AVR microcontroller ATMega128L datasheet. Oct. 2006, Rev O, www.atmel.com/dyn/resources/prod_documents/doc2467.pdf
12. PIC16F87X microcontroller datasheet. Rev C0, Jan. 2007, ww1.microchip.com/downloads/en/DeviceDoc/30292c.pdf
13. PIC18F4525 enhanced flash microcontroller datasheet. Rev B, Apr. 2007, ww1.microchip.com/downloads/en/DeviceDoc/80282b.pdf
14. M16C/62 M single chip 16-bit cmos microcomputer datasheet. Rev B1, Jun. 2001, documentation.renesas.com/eng/products/mpumcu/62meds.pdf
15. MSP430F149 mixed signal microcontroller datasheet. Rev F, Jun. 2004, www.ti.com/lit/gpn/msp430f149
16. TR1000 hybrid transceiver datasheet. www.rfm.com/products/data/tr1000.pdf
17. TR1001 hybrid transceiver datasheet. www.rfm.com/products/data/tr1001.pdf
18. CC1000 singlechip very low-power RF transceiver datasheet. Rev A, Feb.2007, www.chipcon.com/files/CC1000_Data_Sheet_2_2.pdf
19. CC2420 ZigBee ready RF transceiver datasheet. Rev B, Mar. 2007, focus.ti.com/lit/ds/symlink/cc2420.pdf
20. ZigBee Homepage, www.zigbee.org
21. SHT15 humidity sensor datasheet. Rev 3.0, Mar. 2007, www.sensirion.com/en/pdf/product_information/Data_Sheet_humidity_sensor_SHT1x_SHT7x_E.pdf
22. EL7900 ambient light photo detect IC datasheet. Rev FN7377.5, Feb. 2007, www.intersil.com/data/fn/fn7377.pdf
23. ADXL321 small and thin accelerometer datasheet. Rev 0, Jan. 2005, www.analog.com/UploadedFiles/Data_Sheets/ADXL321.pdf
24. DS1629 digital thermometer and real-time clock/calendar datasheet. Rev A4, Jul 2005, www.maxim-ic.com/reliability/dallas/DS1629.pdf
25. 24LC256 I2C CMOS serial EEPROM datasheet. Rev, Aug. 2005, ww1.microchip.com/downloads/en/DeviceDoc/21203 N.pdf
26. AT45D041B serial interface flash datasheet. Rev C, May 2005, www.atmel.com/dyn/resources/prod_documents/doc3443.pdf
27. M24M01S I2C bus EEPROM datasheet. Rev 1.2, Jan. 2003, www.ortodoxism.ro/datasheets/stmicroelectronics/8279.pdf
28. M25P80 serial flash memory datasheet. Rev 10, Jun 2006, www.st.com/stonline/products/literature/ds/8495/m25p80.pdf

Chapter 64
Performance Comparison and Improvement of Wireless Network Protocols

G.E. Rizos, D.C. Vasiliadis, and E. Stergiou

Abstract Ad hoc networks require efficient dynamic routing protocols to tackle the issues raised by their change-prone network topology and the fact that packets typically traverse multiple hops to reach their destinations. Ad hoc networks combine arbitrary mobile hosts with wireless communication capabilities, which come together into a temporary network formation consisting of self-contained, not centrally managed, resources. Nodes in an ad hoc network may be further apart than the range of the wireless devices used for communication; in such cases, intermediate nodes participating in the network can be used to transfer packets between the data source and sink. Two widespread routing protocols that handle such transfers are the dynamic source routing (DSR) and the ad hoc on-demand distance vector routing (AODV) protocols. In this work, we use simulation to compare the performances of these two protocols. Although the approaches taken by these protocols share a considerable number of common features, our simulation has shown that differences in performance figures exist, which may be attributed either to implementation issues or to interlayer operations. In our study, we have considered different network loads, the number of nodes within the network, and mobility parameters.

64.1 Introduction

With the advent of technology, the availability of wireless networking hardware and mobile hosts has dramatically increased. Nodes communicating through these technologies can form closed networks, or even be connected with the Internet. Wireless networks may be formed with the aid of *hot spots* (or *base stations*), where terminal nodes connect and exchange packets; hot spots

G.E. Rizos (✉)
Department of Computer Science and Technology, University of Peloponnese,
GR-221 00, Tripolis, Greece
e-mail: georizos@uop.gr

N. Mastorakis et al. (eds.), *Proceedings of the European Computing Conference*, Lecture Notes in Electrical Engineering 27, DOI 10.1007/978-0-387-84814-3_64, © Springer Science+Business Media, LLC 2009

undertake the task of delivering the packets to their ultimate destinations, either directly or via other, interconnected hot spots.

In some cases, though, no hot spots are available at specific premises; any form of communication, under these circumstances, must be performed directly between nodes. This formation is called an *ad hoc network*. In such a setup, each node may communicate with any other member of the ad hoc network using its wireless hardware capabilities. Besides being a packet source (sending packets to other nodes) and a packet sink (accepting packets addressed to it), each node in an ad hoc network assumes routing responsibilities, being an intermediary for forwarding data packets addressed to other nodes. Since node population in ad hoc networks is not predetermined, and since, moreover, a node's location is not fixed, a major issue in such a context is to determine viable routes between two nodes that need to communicate. This task must be performed efficiently, i.e., without performing tasks that are expensive in terms of network communications, processing power, or memory, taking into account that in mobile environments these resources are scarce. The routing protocol should include provisions to monitor and adapt to changes in the network population and topology, which is highly prone to changes due to node mobility. Defence-related research has published results discussing characteristics and mechanisms in similar networks (e.g. [1]).

In this work, we estimate and compare the performance of two routing protocols for ad hoc networks, namely the *dynamic source routing* protocol (DSR) [2, 3] and the ad hoc on-demand distance vector protocol (AODV) [4, 5]. Both protocols have an "on-demand" nature, i.e., nodes initiating routing-related activities only when packets need to be transmitted/forwarded, refraining from constant monitoring of network topology so as to be already prepared when the need for packet transmission arises.

Although DSR and AODV are similar in regards to their "on-demand" nature, the instrumentations of the routing mechanisms are considerably different in many respects. A major difference is that DSR employs source routing [6], while AODV employs an approach based on routing tables [7] complemented with sequence numbers. AODV uses timers to determine certain events, whereas DSR does not.

The rest of the paper is organized as follows. In Section 64.2 an overview of the DSR and AODV protocols is presented. Section 64.3 outlines the simulation setup and its parameters, while section 64.4 presents and discusses the simulation results. Finally, in Section 64.5 conclusions are drawn.

64.2 Description of Protocols

DSR makes very aggressive use of source routing and route caching. It does not require any mechanism for detecting routing loops. Additionally, forwarding nodes cache the source routes found in forwarded packets for possible future usage. The authors of the protocol have proposed additional optimizations, which they have evaluated and found to be effective. These optimizations are described in [8] and are, in brief:

1. *Salvaging:* An intermediate node can use an alternate route from its own cache, when a data packet meets a failed link on its source route.
2. *Gratuitous route repair:* A source node receiving an RERR packet piggy-backs the RERR in the following RREQ. This helps clean up the caches of other nodes in the network that may have the failed link in one of the cached source routes.
3. *Promiscuous listening:* When a node overhears a packet not addressed to itself, it checks whether the packet could be routed via itself to gain a shorter route. If so, the node sends a *gratuitous* RREP to the source of the route with this new, better route. Aside from this, promiscuous listening helps a node to learn different routes without directly participating in the routing process (Fig. 64.1).

AODV [4, 5] resembles DSR regarding its on-demand nature, since it too finds routes on an "as needed" basis. AODV, though, employs a radically different mechanism for routing information maintenance. AODV maintains routing tables, in which each destination node is mapped to a single route (another difference with DSR, where each destination may be mapped to multiple routes through cache entries). In the absence of source routing, AODV uses only the routing table for RREP propagation back to the source and subsequent data packet routing to the node they are addressed to. Routing loops in AODV are prevented through sequence numbers (present in all packets), which are also used to determine how "up-to-date" the routing information within tables is.

An important feature of AODV is maintenance of timer-based states in each node, regarding the utilization of individual routing table entries [9]. A routing table entry is "expired" if not used recently. A set of predecessor nodes is maintained per routing table entry, which denotes the set of neighbouring nodes that use this entry to route data packets. These nodes are notified with RERR packets when the next hop link breaks. Each predecessor node, in turn, forwards the RERR to its own set of predecessors, thus effectively erasing all routes using the broken link.

The recent specification of AODV [5] includes an optimization technique to control the RREQ flood in the route discovery process. It uses an

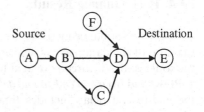

Fig. 64.1 DSR protocol

Fig. 64.2 AODV protocol

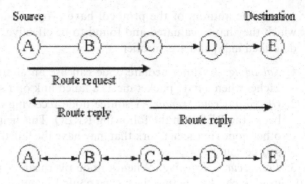

expanding ring search initially to discover routes to an unknown destination. In the expanding ring search, increasingly larger neighbourhoods are searched to find the destination. The search is controlled by the TTL field in the IP header of the RREQ packets. If the route to a previously known destination is needed, the prior hop distance is used to optimize the search (Fig. 64.2).

64.3 The Simulation Network Model

We use a detailed simulation model based on *ns-2* [10]. The distributed coordination function (DCF) of the new IEEE standard 802.11 [11] for wireless LANs is used as the MAC layer. Data packet transmission is followed by an ACK. "Broadcast" data packets and the RTS control packets are sent using physical carrier sensing. An unslotted CSMA technique with collision avoidance (CSMA/CA) is used to transmit these packets [11]. The radio model uses characteristics similar to a commercial radio interface, Lucent's WaveLAN [12]. WaveLAN is a shared-media radio with a nominal bit-rate of 2 Mb/sec and a nominal radio range of 250 meters. A detailed description of the simulation environment and the models is available in [13, 10].

64.4 Performance Results

Two key performance metrics are evaluated: (i) *packet delivery ratio (PDR)*— the ratio of the data packets delivered to the destination to those generated by the sources; this is directly related to the packet loss ratio (PLR), since PLR = 1–PDR; and (ii) *mean end-to-end delay (MEED)* of the data packets. Note that the end-to-end delay includes time spent in any phase of the packet trip, including actual propagation and transfer time in the physical layer, retransmissions due to collisions or packet losses, time waiting for the physical medium to become available, time spent in queues behind other packets, processing time, etc. In the absence of QoS criteria, PDR and MEED are the most important performance metrics.

It should be noted, however, that dependencies exist between these metrics: if a network drops a significant number of packets, then there exist fewer packets in the network to contend for the available bandwidth, thus the end-to-end delay metric is expected to drop (i.e., improve). Routing algorithms that choose lengthier routes are bound to have higher packet loss ratios, since, for instance, packet TTLs are more likely to expire along a lengthy route.

The simulation experiments were conducted using a varying number of sources with a moderate packet rate and changing packet interarrival times. When the network size was set to 50 nodes, we used 10, 20, 30, and 50 traffic sources and a transmission rate of 4 packets/sec (interarrival time [or pause time] was set to 0.25 sec). For 50 sources, in particular, a packet rate of 3 packets/sec was used, as the network congestion was too high otherwise for a meaningful comparison, in accordance with the results presented in (Perkins et al. (2001) Performance comparison of two on-demand routing protocols for ad hoc networks. IEEE Personal Communications). Pause times, reflecting the relative speed of the mobiles, were also varied.

Note that the packet delivery ratio for DSR and AODV are very similar when 10 and 20 sources are used (see Fig. 64.3(a) and (b)). With 30 and 50 sources, however, AODV performs better than DSR (Fig. 64.3(c) and (d)). Packet loss in DSR exceeds that of AODV by 30–50% when longer pause times are considered (higher mobility).

DSR outperforms AODV with respect to end-to-end delay with 10 and 20 sources (see Fig. 64.4). Performance differences for 10 sources are quantified to be large, especially for lower pause times. The gap reduces as pause times increase (and thus mobility drops). When the number of sources rises to 20, the performance difference is much smaller. With a source greater than 30, AODV has a lower delay than DSR for all pause times (Fig. 64.4(c) and (d)); especially for lower pause times (higher mobility), the performance difference is about 50%.

When the number of sources reaches 50 and pause times are high (low mobility), the delays for both protocols increase (see Fig. 64.4(d)). The root cause of this can be traced back to increased congestion at portions of the network. When pause times drop (and mobility increases), sources scatter more uniformly across the network area, thus congestion spots are more infrequent. The inclusion of an appropriate mechanism in the routing protocols that would try to avoid congestion would be bound to improve overall performance in similar situations.

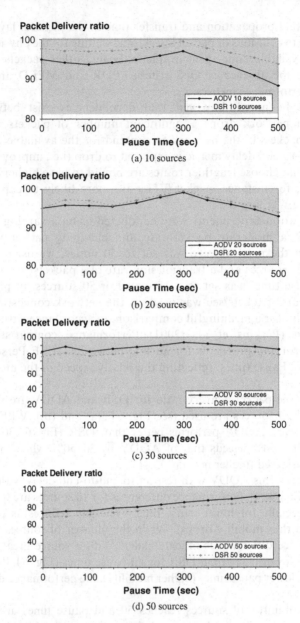

Fig. 64.3 Packet delivery fraction (50 nodes)

64.5 Conclusion

In this work, we compared performance characteristics of two widely used protocols for ad hoc network routing, namely DSR and AODV. Although we expected similar—or comparable—results for both protocols, having in mind

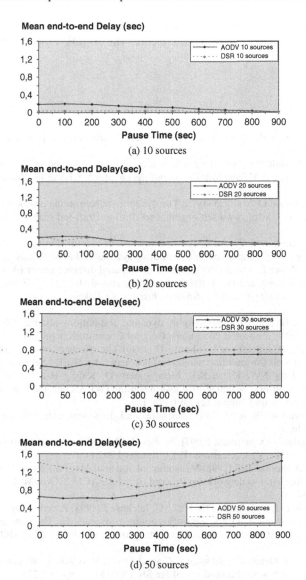

Fig. 64.4 Average data package delays (50 nodes)

that they both perform routing activities "on-demand" rather than proactively, considerable differences were identified in various cases, depending on the network population, the offered load, and the mobility of the nodes. The differences can be attributed to the different mechanisms employed by the two protocols to implement their routing activities. Summarizing our results, we conclude that delay-wise DSR behaves better than AODV when the number of nodes within the ad hoc network is small, the offered load is little, and node mobility is limited. When any of these parameters (nodes, load, or mobility) increases, AODV appears to have a performance edge over DSR; in fact, the more these parameters increase, the clearer the AODV advantage over DSR becomes.

References

1. Jubin J, Tornow JD (1987) The DARPA packet radio network protocols. In: P IEEE 75(1):21–32
2. Johnson D, Maltz D (1996) Dynamic source routing in ad hoc wireless networks. In: Imielinski T, Korth H (eds) Mobile computing, chapter 5. Kluwer Academic, Dordrecht, Netherlands
3. Broch J, Johnson D, Maltz D (1998) The dynamic source routing protocol for mobile ad hoc networks. http://www.ietf.org/internet-drafts/ draft-ietf-manet-dsr01.txt (IETF Internet Draft)
4. Perkins C, Royer E (1999) Ad hoc on-demand distance vector routing. In: Proceedings of the 2nd IEEE workshop on mobile computing systems and applications, pp 90–100
5. Perkins C, Royer E, Das S (1999) Ad hoc on-demand distance vector (AODV) routing. http://www.ietf.org/internet-drafts/draft-ietf-manet-aodv-03.txt (IETF Internet Draft)
6. Linktionary, source routing definition, http://www.linktionary.com/s/source_routing. html
7. Perkins CE, Bhagwat P (1994) Highly dynamic destination-sequenced distance-vector routing (DSDV) for mobile computers. Comput Commun Rev pp 234–244
8. Maltz D, Broch J, Jetcheva J, Johnson D (1999) The effects of on-demand behavior in routing protocols for multihop wireless ad hoc networks. IEEE J Sel Area Commun
9. Choi D-I, Jung J-W, Kwon KY, Montgomery D, Kahng H-K (2005) Design and simulation result of a weighted load aware routing (WLAR) protocol in mobile ad hoc network. Lect Notes Comput Sci 3391:178–187
10. Fall K, Varadhan K (eds) (1999) NS notes and documentation. http://www-mash.cs. berkeley.edu/ns/
11. IEEE Standards Department (1997) Wireless LAN medium access control (MAC) and physical layer (PHY) specications. IEEE standard 802.11
12. Eckhardt D, Steenkiste P (1996) Measurement and analysis of the error characteristics of an in-building wireless network. In: Proceedings of the ACM SIGCOMM '96 conference, pp 243–254
13. Broch J, Maltz DA, Johnson DB, Hu Y-C, Jetcheva J (1998) A performance comparison of multi-hop wireless ad hoc network routing protocols. In: Proceedings of the 4th international conference on mobile computing and networking (ACM MOBICOM'98), pp 85–97
14. Bharghavan V, Demers A, Shenker S, Zhang L (1994) MACAW: a media access protocol for wireless LAN's. In: Proceedings of the SIGCOMM'94, pp 212–225
15. Das SR, Castaneda R, Yan J, Sengupta R (1998) Comparative performance evaluation of routing protocols for mobile, ad hoc networks. In: 7th international conference on computer communications and networks (IC3N), pp 153–161
16. Haas ZJ, Pearlman MR (1998) The performance of query control schemes for the zone routing protocol. In: Proceedings of ACM SIGCOMM'98 conference, pp 167–177, Vancouver
17. Johansson P, Larsson T, Hedman N, Mielczarek B (1999) Routing protocols for mobile ad hoc networks—a comparative performance analysis. In: Proceedings of the 5th international conference on mobile computing and networking (ACM MOBICOM'99)
18. Macker J, Corson S (1997) Mobile ad hoc networks (MANET). http://www.ietf.org/ html.charters/manet-charter.html (IETF Working Group Charter)

Chapter 65
Modeling the Erroneous Behaviour of a Sequential Memory Component with Streams

Walter Dosch

Abstract A sequential memory component stores data in addressable locations. The component serves an input stream in a regular way iff all read commands retrieve data from locations with a previous assignment. We study the component's erroneous behaviour for input streams outside the service domain. We specify a fault sensitive memory component, a fault tolerant memory component, a robust memory component, and a fault correcting memory component in the setting of stream functions. We implement the different versions by state transition machines in a modular way. Beyond the case study, we express adequate notions for modeling the services of interactive components.

65.1 Introduction

Modern computer systems are composed of software and hardware components which store information and provide services [1] through interfaces. The overall system evolves through an ongoing interaction [2] between the components and the environment.

The behaviour of an interactive component [3] describes a function from input streams to output streams [4, 5]. An input history, for a short stream, records the sequence of messages passing through the component's interface.

In general, an interactive component provides the contracted service only for a subset of input histories, called the service domain. A sequential memory component cannot serve a read operation in a regular way when the requested location was not initialized before.

An interactive component should behave in a predictable way when processing unexpected input outside its service domain. The erroneous behaviour of a component may significantly influence the behaviour of the overall system for critical inputs.

W. Dosch (✉)
Institute of Software Technology and Programming Languages, University of Lübeck, Lübeck, Germany

N. Mastorakis et al. (eds.), *Proceedings of the European Computing Conference*, Lecture Notes in Electrical Engineering 27, DOI 10.1007/978-0-387-84814-3_65, © Springer Science+Business Media, LLC 2009

Against this general background, we model different erroneous behaviours of a sequential memory component in the setting of stream functions. The component allows write, read, and reset commands as the input interface. A write command enters a datum into the specified location, possibly overwriting a previously entered datum. A read command retrieves the datum stored under a specified location, provided it exists. A reset command clears the current contents of the memory at all locations.

A sequential memory component forms an abstraction of different types of devices that store and retrieve data, using addresses or keys for indexing the locations. The attribute "sequential" refers to the fact that the memory component does not support concurrent write and read commands, since it consumes a single stream of input commands.

We partition the set of all input histories into the classes of regular and erroneous input streams. As the starting point, we specify the input/output behaviour of the memory component for regular input histories (Section 65.2). Then we extend the regular behaviour in a systematic way to input streams outside the service domain (Section 65.3).

We identify four major types of irregular behaviours. A fault sensitive memory breaks upon the first offending read command (Section 65.4). A fault tolerant memory ignores an offending read command and continues its service afterwards (Section 65.5). A robust memory signals an offending read command to the environment (Section 65.6). A fault correcting memory suspends an erroneous read command until a write command assigns a datum to the requested location (Section 65.7).

The four versions of the sequential memory component are discussed, employing a black-box view on the behavioural level and a glass-box view on the implementation level. In the black-box view, we characterize the input/output behaviour by a stream function mapping input histories to output histories. In the glass-box view, we describe the implementation of the four memory components by state transition machines.

The specification techniques, the description methods, the transformations, and the underlying concepts used can be transferred from this case study to other interactive components with a restricted service domain [6]. In the long term, software engineering should be based on a sound engineering theory of services.

For the basic notions about (untimed) streams and state transition machines, we refer to [7]; for a comprehensive treatment, to [8].

65.2 Regular Behaviour

A *sequential memory* is a communications component with one input channel and one output channel. The component provides the services of an addressable memory, storing and retrieving data in specified locations.

65.2.1 Interface

The memory component receives a stream of write, read, and reset commands:

$$Input = write(Address, Data) \cup read(Address) \cup \{reset\} \qquad (65.1)$$

A write command stores a datum in the specified location and produces no output. A read command requests the datum stored most recently in the specified location, if it exists. A reset command clears the current contents of the memory component at all locations.

The type $Address \neq \emptyset$ of *addresses* (*locations*) and the type $Data \neq \emptyset$ of *data* to be stored will not be specified further.

65.2.2 Service Domain

The component's service domain comprises the set of all regular input histories whose processing leads to no read errors. Such errors occur when a read command addresses a location without a previous assignment.

An input stream represents a *regular* input history from the *service domain* $ServDom \subseteq Input^*$ iff each read command to a specified location is preceded by at least one write command to that location which was not cancelled afterwards by a reset command. We have $X \in \leftarrow ServDom$ iff

$$X = Y\&\langle read(a)\rangle\&Z \Rightarrow Y = L\&\langle write(a, d)\rangle\&R \wedge reset \notin R. \qquad (65.2)$$

The notation $reset \notin \leftarrow R\psi$ means that the stream $R\psi$ contains no reset command.

65.2.3 Regular Behaviour

We specify the *regular behaviour* of a sequential memory component as a stream function mapping regular input histories to output histories; compare Fig. 65.1.

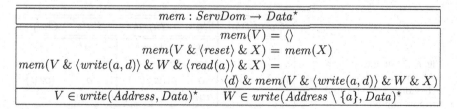

$mem : ServDom \rightharpoonup Data^\star$
$mem(V) = \langle\rangle$
$mem(V \,\&\, \langle reset\rangle \,\&\, X) = mem(X)$
$mem(V \,\&\, \langle write(a, d)\rangle \,\&\, W \,\&\, \langle read(a)\rangle \,\&\, X) =$
$\langle d\rangle \,\&\, mem(V \,\&\, \langle write(a, d)\rangle \,\&\, W \,\&\, X)$
$V \in write(Address, Data)^\star \qquad W \in write(Address \setminus \{a\}, Data)^\star$

Fig. 65.1 Regular behaviour of a sequential memory component

A (possibly empty) sequence of write command generates no output. A reset command clears all preceding write commands. A read command retrieves the datum stored most recently under the requested location.

For a regular input stream, the length of the output stream agrees with the number of read commands.

65.2.4 Implementation

We implement the regular behaviour by a partial *state transition machine* in which the transition functions are specified for regular input histories only. The internal state of a sequential memory maps locations to data values:

$$State = \{\sigma : Address \to Data \cup \{nodata\}$$
$$|\sigma(a) = nodata \text{ for almost all } a \in Address\}$$

(65.3)

A location with an assignment records the data value stored most recently. A location without a previous assignment carries the default value *nodata* $\notin$ *Data*. In the *initial state init* $\in$ *State* we have *init*(a) = *nodata* for all $a \in$ *Address*.

Figure 65.2 displays the state transition table for regular input histories. The three transition rules relate the current state and the next input to the successor state (denoted by prime) and the output sequence generated.

A reset command clears the memory component to the initial state. A write command updates the state at the specified location with the value entered, while leaving the contents of other locations unchanged. Here .[.←.] denotes the update operation on states. A read command retrieves the stored datum from the requested location. This location indeed witnessed a previous assignment, since we confined the input to regular input histories—the enabling condition $\sigma(a) \in$ *Data* is always true!

Condition	State	Input	State'	Output
	σ	reset	init	$\langle\rangle$
	σ	$write(a, d)$	$\sigma[a \leftarrow d]$	$\langle\rangle$
$\sigma(a) \in Data$	σ	$read(a)$	σ	$\langle\sigma(a)\rangle$
	init		initial state	

Fig. 65.2 State transition table for the regular behaviour of a sequential memory component

65.3 Irregular Behaviour

The irregular behaviour of a sequential memory component captures its reaction on input histories which contain at least one erroneous read command. The different erroneous behaviours surveyed in this section will be evaluated in greater detail in the subsequent sections.

65.3.1 Erroneous Input Histories

The set $Error = Input *\backslash ServDom$ of *erroneous input histories* comprises all input streams outside the service domain. Prolongations of erroneous input histories are erroneous as well. Every erroneous input history can be uniquely decomposed into a maximal regular prefix and a remaining input stream starting with an offending read command.

65.3.2 Classification

Figure 65.3 explicates different irregular behaviours of the sequential memory component using an erroneous input stream with two offending read commands.

A fault sensitive memory breaks upon the first read command to a non-initialized location. A fault tolerant memory ignores offending read commands in the input stream and continues its service. A robust memory signals an offending read command to the environment by sending an error message. A fault correcting memory suspends an offending read command until a datum is assigned to the requested location.

input stream	$write(1, a)$	$read(1)$	$read(2)$	$read(3)$	$write(3, b)$	$write(2, c)$	$read(2)$
fault-sensitive	$\langle\rangle$	$\langle a\rangle$	$\langle\rangle$	$\langle\rangle$	$\langle\rangle$	$\langle\rangle$	$\langle\rangle$
fault-tolerant	$\langle\rangle$	$\langle a\rangle$	$\langle\rangle$	$\langle\rangle$	$\langle\rangle$	$\langle\rangle$	$\langle c\rangle$
robust	$\langle\rangle$	$\langle a\rangle$	$\langle error(2)\rangle$	$\langle error(3)\rangle$	$\langle\rangle$	$\langle\rangle$	$\langle c\rangle$
fault-correcting	$\langle\rangle$	$\langle a\rangle$	$\langle\rangle$	$\langle\rangle$	$\langle b\rangle$	$\langle c\rangle$	$\langle c\rangle$

Fig. 65.3 Output generated by different versions of a sequential memory component in response to the erroneous input stream $\langle write(1, a), read(1), read(2), read(3), write(3, b), write(2, c), read(2)\rangle$. The output $error(a)$ signals a read error to location $a \in Address$

65.4 Fault Sensitive Memory Component

A *fault sensitive memory component* breaks upon the first read command to a non-initialized location. The service is resumed with the subsequent reset command, provided it exists.

65.4.1 Input/Output Behaviour

Figure 65.4 specifies the input/output behaviour of a fault sensitive memory component by two additional equations.

A fault sensitive memory produces no output, if an offending read command is followed by no reset command. Otherwise, the contracted service is resumed after the next reset command.

$mem : Input^\star \rightarrow Data^\star$
$mem(W \mathbin{\&} \langle read(a) \rangle \mathbin{\&} Z) = \langle \rangle$
$mem(W \mathbin{\&} \langle read(a) \rangle \mathbin{\&} Z \mathbin{\&} \langle reset \rangle \mathbin{\&} X) = mem(X)$
$W \in write(Address \setminus \{a\}, Data)^\star \qquad reset \notin Z$

Fig. 65.4 Extending the input/output behaviour for a fault sensitive memory component

65.4.2 Implementation

The fault sensitive memory component is implemented by a state transition machine. The extended state space $FsState = \{fail\} \cup State$ comprises the control state $fail \notin State$ recording a failure.

The transition rules can systematically be derived from the input/output behaviour using a history abstraction [9, 10]. Figure 65.5 displays the resulting transition rules which extend the state transition table of the regular behaviour from Fig. 65.2.

Fig. 65.5 Extending the state transition table for a fault sensitive memory component

Condition	FsState	Input	FsState'	Output
$\sigma(a) \notin Data$	σ	$read(a)$	$fail$	$\langle \rangle$
	$fail$	$reset$	$init$	$\langle \rangle$
	$fail$	$write(a, d)$	$fail$	$\langle \rangle$
	$fail$	$read(a)$	$fail$	$\langle \rangle$
$\sigma \in State$				

The additional transition rules center around the failure state, which is entered upon an offending read command, left upon a reset command, and retained for read and write commands.

65.5 Fault Tolerant Memory Component

A *fault tolerant memory component* ignores offending read commands in the input stream and provides the contracted service for the remaining input stream. The output stream concatenates the output from the regular segments of the input stream where offending read commands were removed.

65.5.1 Input/Output Behaviour

The input/output behaviour of a fault tolerant memory component is specified in Fig. 65.6 by one additional equation. A fault tolerant memory prolongs the output stream generated by a fault sensitive memory component.

65.5.1.1 Implementation

A fault tolerant memory component can be implemented by a state transition machine without enlarging the state space: $FtState = State$. The extension of the state transition table is shown in Fig. 65.7.

An offending read command neither changes the state nor produces output. A fault tolerant memory possesses a simpler implementation than a fault sensitive memory, since past errors are not recorded in the state.

Fig. 65.6 Extending the input/output behaviour for a fault tolerant memory component

$$mem : Input^\star \to Data^\star$$
$$mem(W \mathbin{\&} \langle read(a) \rangle \mathbin{\&} X) = mem(W \mathbin{\&} X)$$
$$W \in write(Address \setminus \{a\}, Data)^\star$$

Condition	FtState	Input	FtState'	Output
$\sigma(a) \notin Data$	σ	$read(a)$	σ	$\langle \rangle$

Fig. 65.7 Extending the state transition table for a fault tolerant memory component

65.6 Robust Memory Component

A *robust memory component* signals an offending read command to a non-initialized location with an error message to the environment. After an offending read command, the component continues with the regular service.

65.6.1 Input/Output Behaviour

We enlarge the output alphabet with error messages indicating a read error together with the requested location. Figure 65.8 specifies the input/output behaviour of a robust memory component with one additional equation.

The output of a robust memory agrees with the output of a fault tolerant memory, if we forget possible error messages. Moreover, the input stream and the output stream validate a simple length balance, since each read command generates exactly one output.

65.6.2 Implementation

Figure 65.9 shows the implementation of a robust memory component with the state space *RbState* = *State*.

The fault tolerant memory allows "silent" read errors, whereas the robust memory provides a feedback about read errors. When using a robust memory component, the environment must be prepared to handle error messages.

$$mem : Input^\star \to (Data \cup error(Address))^\star$$

$$mem(W \mathbin{\&} \langle read(a)\rangle \mathbin{\&} X) = \langle error(a)\rangle \mathbin{\&} mem(W \mathbin{\&} X)$$

$$W \in write(Address \setminus \{a\}, Data)^\star$$

Fig. 65.8 Extending the input/output behaviour for a robust memory component

Condition	RbState	Input	RbState'	Output
$\sigma(a) \notin Data$	σ	$read(a)$	σ	$\langle error(a)\rangle$

Fig. 65.9 Extending the state transition table for a robust memory component

65.7 Fault Correcting Memory Component

A fault correcting memory component serves regular read commands in the input history as expected, but postpones offending read commands until data becomes available at the requested location. The memory component continues the contracted service for other addresses while postponing read requests.

65.7.1 Input/Output Behaviour

The input/output behaviour of a fault correcting memory component is specified in Figure. 65.10.

The operation $R\#read(a)$ counts how often $read(a)$ occurs on the stream R. The operation $R\ominus read(a)$ removes all occurrences of $read(a)$ from the stream R. d^i denotes the constant stream with $i \geq 0$ elements d.

A fault correcting memory postpones offending read commands as long as necessary, but not longer. While postponing illegal read commands, legal read commands are served in a regular way. As soon as the requested memory location receives an assignment, the postponed read commands are served. A reset command clears the memory including pending read requests.

65.7.2 Implementation

We implement the fault correcting memory component extending the state space. Now the state stores for each address either the datum entered most recently or the number of postponed read commands:

$$State = \{\sigma : Address \rightarrow Data \stackrel{.}{\cup} \mathbb{N} \mid \sigma(a) \in \mathbb{N} \text{ for almost all } a \in Address\} \quad (65.4)$$

In the initial state we have $init(a) = 0$ for all $a \in Address$. Figure. 65.11 displays the state transition machine implementing the fault correcting memory component.

A reset command clears the memory, returning it to the initial state. A write command to a location without postponed read commands updates the state

$mem : Input^\star \rightarrow Data^\star$
$mem(W \,\&\, R) = \langle\rangle$
$mem(W \,\&\, R \,\&\, \langle write(a,d)\rangle \,\&\, X) =$
$d^{R\#read(a)} \,\&\, mem(W \,\&\, \langle write(a,d)\rangle \,\&\, (R \ominus read(a)) \,\&\, X)$
$mem(U \,\&\, \langle write(a,d)\rangle \,\&\, V \,\&\, R \,\&\, \langle read(a)\rangle \,\&\, X) =$
$\langle d\rangle \,\&\, mem(U \,\&\, \langle write(a,d)\rangle \,\&\, V \,\&\, R \,\&\, X)$
$mem(W \,\&\, R \,\&\, \langle reset\rangle \,\&\, X) = mem(X)$
$W \in write(A, Data)^\star \quad R \in read(B)^\star \quad A \cap B = \emptyset$
$U \,\&\, \langle write(a,d)\rangle \,\&\, V \in write(A, Data)^\star \quad V \in write(A \setminus \{a\}, Data)^\star$

Fig. 65.10 Input/output behaviour of a fault correcting memory component $(A, B \subseteq Address)$

Fig. 65.11 State transition
table of a fault correcting
memory component

Condition	FcState	Input	FcState'	Output
	σ	reset	init	$\langle\rangle$
$\sigma(a) \in Data \cup \{0\}$	σ	$write(a, d)$	$\sigma[a \leftarrow d]$	$\langle\rangle$
$\sigma(a) = k + 1$	σ	$write(a, d)$	$\sigma[a \leftarrow d]$	d^{k+1}
$\sigma(a) \in Data$	σ	$read(a)$	σ	$\langle\sigma(a)\rangle$
$\sigma(a) \in \mathbb{N}$	σ	$read(a)$	$\sigma[a \leftarrow \sigma(a) + 1]$	$\langle\rangle$
	init		initial state	

without generating output. A write command to a location with postponed
read commands additionally generates output. A read command to an
assigned location retrieves the requested datum. A read command to a non-
initialized location increases the number of postponed read commands to that
location.

65.8 Conclusion

A sequential memory component allows various design decisions on how to
handle unexpected read commands. The irregular behaviour influences the
overall behaviour of a composite system where the memory is embedded as a
subcomponent. Software engineering should provide sound methods and transparent guidelines for how to document, specify, and implement interactive
components with a restricted service domain. The specifications of the four
memory components document the design decisions for regular and erroneous
input streams. The implementations encode particular "situations" in an
extended state space. The local transitions of the state transition machine
have to be integrated into an overall behaviour [11] in order to understand the
memory component in a compositional way [12].

We advocate a clear separation of concerns as a specification and design
methodology. In the first step, the designer should concentrate on capturing the
required service of an interactive component. In the second step, the designer
should consider the irregular behaviour for unexpected input streams. The effects
of the (ir)regular behaviour should be traceable through the entire software life
cycle. This helps software engineers modify an implementation in a disciplined
way when adjusting the component's behaviour to changing requirements.

The distinction between a fault sensitive memory, a fault tolerant memory, a
robust memory, and a fault correcting memory represents a coarse classification which allows further refinements and various other combinations. Because
of page restrictions, we confined the presentation to four basic types of irregular
behaviours to demonstrate the specification and design methodology. A more
comprehensive classification would incorporate further dimensions like the
number of (subsequent) faults or the type of faults.

Acknowledgment I gratefully acknowledge valuable improvements suggested by A. Stümpel.

References

1. Broy M (2004) Service-oriented systems engineering: specification and design of services and layered architectures—the JANUS approach. Working material, international summer school 2004, Marktoberdorf, Germany
2. Wegner P (1997) Why interaction is more powerful than algorithms. Commun ACM 40(5):80–91
3. Broy M, Stølen K (2001) Specification and development of interactive systems: focus on streams, interfaces, and refinement. Monographs in computer science. Springer, Berlin
4. Kahn G (1974) The semantics of a simple language for parallel programming. In: Rosenfeld J (ed) Inf Process, North–Holland, 74:471–475
5. Kahn G, MacQueen D B (1977) Coroutines and networks of parallel processes. In: Gilchrist B (ed). Inform Process North-Holland 77:993–998
6. Dosch W, Hu G (2007) On irregular behaviours of interactive stacks. In: Latifi S (ed) Proceedings of the 4th international conference on information technology: new generations (ITNG 2007). IEEE Computer Society Press, Washington, DC, pp 693–700
7. Stephens R (1997) A survey of stream processing. Acta Informatica 34(7):491–541
8. Stümpel A (2003) Stream-based design of distributed systems through refinement. Logos Verlag, Berlin
9. Dosch W, Stümpel A (2005) Transforming stream processing functions into state transition machines. In: Dosch W, Lee R, Wu C (eds) Software engineering research and applications (SERA 2004). Lect Notes Comput Sci 3647:1–18
10. Dosch W, Stümpel A (2007) Deriving state-based implementations of interactive components with history abstractions. In: Virbitskaite I, Voronkov A (eds) Perspectives of systems informatics (PSI 2006). Lect Notes Comput Sci 4378:180–194
11. Breitling M, Philipps J (2000) Step by step to histories. In: Rus T (ed) Algebraic methodology and software technology (AMAST'2000), Lect Notes Comput Sci 1816:11–25
12. de Roever WP, Langmaack H, Pnueli A (eds) (1998) Compositionality: the significant difference. Lect Notes Comput Sci 1536

Chapter 66
FPGA Implementation of PPM I-UWB Baseband Transceiver

K.S. Mohammed

Abstract In this paper the FPGA implementation of an impulse ultra-wideband (UWB) baseband transceiver is discussed. UWB has two major advantages, capacity and simplicity. The modulation scheme used here is pulse position modulation. The transceiver design was simulated using Matlab and was targeted to a Xilinx Spartan3 starter kit. The hardware was tested using a Xilinx ChipScope logic analyzer.

66.1 Introduction

Ultra-wideband (UWB) is a candidate technology for short range and high data rate communications. Its principle is to transmit over a huge spectrum band at a very low power, avoiding interference with existing systems such as existing WLANs (wireless area networks) and ISM band devices [1].

There are two main types of UWB signals: multicarrier UWB (MC-UWB) and impulse UWB (I-UWB).; Multicarrier UWB uses multiple carrier frequencies concurrently, employing techniques like orthogonal frequency domain multiplexing (OFDM); whereas impulse UWB uses very short duration pulses [2, 3].

The modulation schemes available for an I-UWB system include pulse position modulation (PPM), on-off keying (OOK), and pulse amplitude modulation (PAM). This paper focuses on the PPM modulation.

The paper is organized as follows: First a brief description of ultra-wideband systems and the differences between them and narrowband systems are presented in Section 66.2. Section 66.3 displays PPM modulation/demodulation. Section 66.4 displays the Matlab and VHDL simulation results for the system. In Section 66.5, VHDL synthesis results and hardware testing are discussed.

K.S. Mohammed (✉)
Electronic Department, National Telecommunication Institute, Cairo, Egypt

N. Mastorakis et al. (eds.), *Proceedings of the European Computing Conference*, Lecture Notes in Electrical Engineering 27, DOI 10.1007/978-0-387-84814-3_66, © Springer Science+Business Media, LLC 2009

66.2 Ultra-Wideband Versus Narrowband

UWB differs substantially from conventional narrowband radio frequency
(RF) and spread spectrum technologies (SS), such as Bluetooth and 802.11a/g.
UWB uses an extremely wide band of RF spectrum to transmit data (Fig. 66.1).
In so doing, UWB is able to transmit more data in a given period of time than
the more traditional technologies.

The FCC has defined the −10 dB bandwidth (F_H–F_l) of an UWB signal to be
greater than 25% of the center frequency or greater than 0.5 GHz, as shown in
equation (66.1)

$$F_H - F_l > = 0.25(F_H + F_l/2) \quad \text{or} \quad F_H - F_l > 0.5 \text{GHz} \qquad (66.1)$$

where

- F_H is the upper frequency of the −10 dB bandwidth,
- F_l is the lower frequency of the −10 dB bandwidth.

The bandwidth is clearly much greater than the one used for today's narrowband communication systems; but the maximum permissible average power per
MHz is about three orders of magnitude smaller than those in the ISM bands
This results in a decrease of signal-to-noise ratio for UWB [4], as each pulse has
very little energy (low average power).

In addition, it has all the advantages of spread spectrum but to a much larger
extent, because it has high resistance to multipath fading as it isolates direct
paths from reflected paths, as shown in Fig. 66.2, where path (1–2) is the direct
path and path (1–3) is the reflected path.

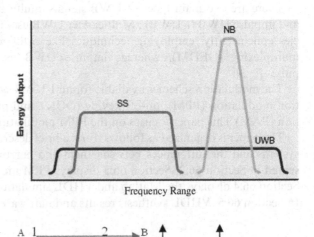

Fig. 66.1 Comparison of
narrowband (NB), spread
spectrum (SS), and ultra-
wideband (UWB) signal
concepts

Fig. 66.2 UWB multipath
fading isolation

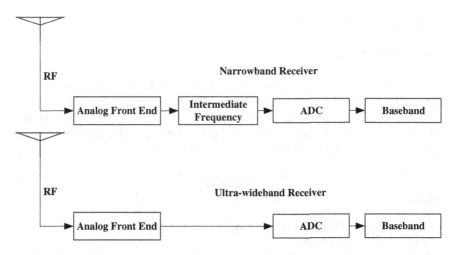

Fig. 66.3 Transceiver architecture comparison

UWB also has simplicity in hardware design compared to narrowband transceivers, as shown in Fig. 66.3; an UWB transceiver is much simpler than a narrowband one because there is no need for an intermediate frequency stage. Table 66.1 summarizes the major differences with respect to antenna, analog front end, intermediate frequency, baseband, and channel [5].

UWB also fuses the capability of low data rate communication with accurate ranging (on the order of centimeters), and is thus ideally suited to the design of a position location network.

The use of narrow pulses implies that a digital implementation is feasible, avoiding otherwise expensive and power-hungry analog components. The bandwidth available is from 3.1–10 GHz; so, even with other non-UWB signals present, it is still possible to achieve data rates on the order of hundreds of megabits per second [6].

Table 66. 1 Comparison of narrowband and UWB transceiver design

	Ultra-wideband	Narrowband
Antenna	Need high bandwidth antenna	Need high Q antenna
Analog front end	Wideband LNA is hard to match	Narrowband LNA is easy to match
Intermediate frequency	No need	Mixer, oscillators, PLL
baseband	Coherent demodulation	Coherent and non-coherent demodulation
channel	UWB channel characteristics are not completely known, but studies are underway	Narrow channels are well characterized (fading models)

66.3 Pulse Position Modem Architecture

66.3.1 PPM Signal Structure

In pulse position modulation the information is modulated on the position of
pulses. This means that if a bit with the value "0" is represented by a pulse
originating at a certain time t, a bit with a value "1" is shifted in time by the
amount of δ from t. Figure 66.4 describes the PPM signal [7, 8]. An analytical
expression of the PPM signal is given by equation (66.2)

$$X(t) = W_{PPM}(t - d_j) \tag{66.2}$$

where

- W_{PPM} is the UWB pulse waveform,
- J is the bit transmitted,
- d_j assumes the values "1" for j = 1 and "0" for j = 0.

The most popular I-UWB pulses are the Gaussian pulse and its derivatives
(the Gaussian monocycle and the Gaussian doublet).They are shown in Fig. 66.5.
The Gaussian pulse is given by equation (66.3).

$$p(t) = \frac{1}{\sqrt{2\pi\sigma^2}} \, e^{\dfrac{-(t-\mu)^2}{2\sigma^2}} \tag{66.3}$$

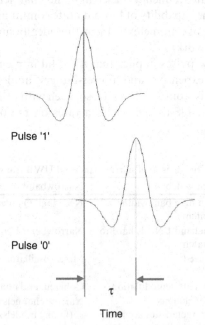

Fig. 66.4 Pulse position modulation

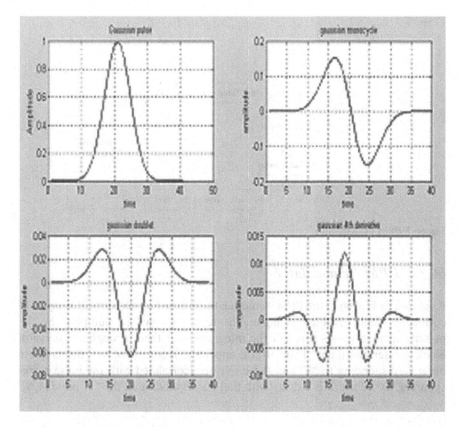

Fig. 66.5 Typical impulse UWB pulses

66.3.2 PPM Modulator

The pulse position modulator consists of three main building blocks: a timing delay, a pulse generator, and a shaping filter. The timing delay block is responsible for generating the timing offset between the "0" symbol and the "1" symbol, the pulse generator generates very narrow pulses, and the shaping filter sends the pulse as a Gaussian pulse or its derivative, as shown in Fig. 66.6 [4].

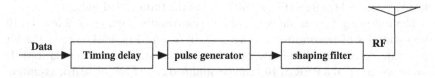

Fig. 66.6 PPM modulator

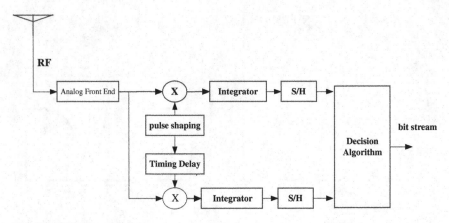

Fig. 66.7 PPM demodulator

66.3.3 PPM Demodulator

As shown in Fig. 66.7, the PPM demodulator is based on the symbol by symbol correlator demodulation technique. The received signal is multiplied by the pulser and then passes through the integrator and then through the decision block to determine if it is "1" or "0"; the upper arm is correlated with the template pulse corresponding to "1" and the lower arm correlated with the pulse corresponding to "0" which is δ delay with that corresponding to the "1" pulse [4, 9].

66.4 PPM I-UWB Software and Hardware Simulation

66.4.1 Matlab Algorithm and Simulation Results

At the transmitter, generate random bits and the pulse to be transmitted, which is a fourth derivative of a Gaussian pulse; define the delay between two possible positions in the PPM, transmit these bits over an AWGN channel, and observe the % bit error rates at different signal-to-noise ratios (SNRs).

At the receiver, look at the correlations at two different pulse positions corresponding to "0" and "1", compare the correlations, and choose the bit (as the estimated received bit) corresponding to the one that has the larger correlation, and then calculate the bit error rate. The system parameters are chosen to be 36 Mpulse/s. Figure 66.8 shows the transmitted pulse.

The following figures show the simulation results. Figures 66.9 and 66.10 show samples of transmitted and received bits, and Fig. 66.11 shows the bit error rate versus signal-to-noise ratio (theoretical and Matlab simulation). It can be noticed that a BER of 10^{-3} can be obtained at an SNR of 10 dB, assuming an additive white Gaussian noise (AWGN) channel [10–15].

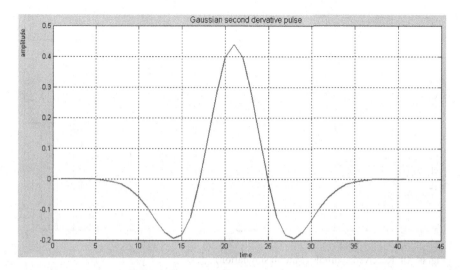

Fig. 66.8 PPM I-UWB pulse (Matlab simulation)

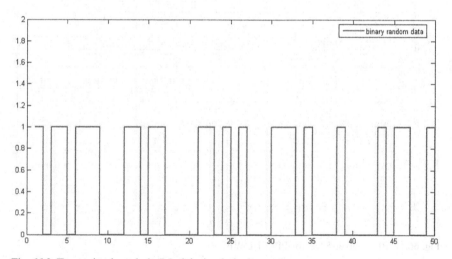

Fig. 66.9 Transmitted symbols (Matlab simulation)

66.4.2 VHDL Algorithm and Simulation Results

The VHDL model was built for the PPM I-UWB baseband transceiver using Mentor Graphics tools. The VHDL top level entity of the transceiver is shown in Fig. 66.12 and the simulation results are shown in Fig. 66.13, where

a = the system clock,
b = the reset signal for the entire system,

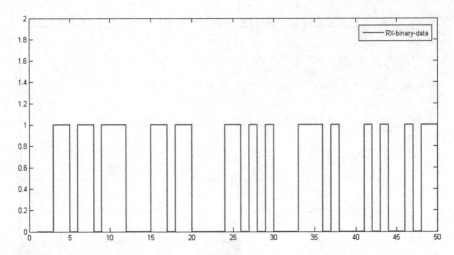

Fig. 66.10 Received symbols (Matlab simulation)

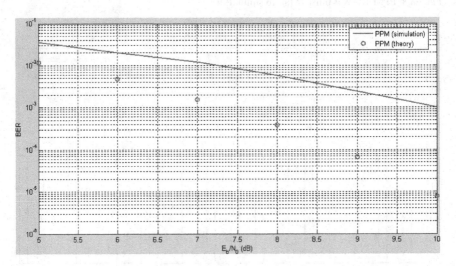

Fig. 66.11 BER versus SNR for PPM I-UWB

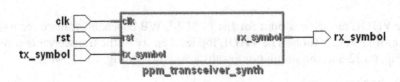

Fig. 66.12 VHDL top level entity for PPM I-UWB

Fig. 66.13 VHDL
simulation results for PPM
I-UWB

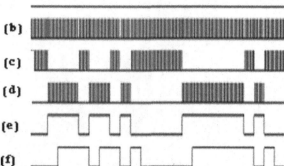

c = the PPM I-UWB transmitted pulses,
d = the correlation output with the pulse template,
e = the correlation output with the delayed pulse template,
f = the transmitted symbols,
g = the received symbols after the decision block.

66.5 PPM I-UWB Hardware Synthesis and Verification

66.5.1 VHDL Synthesis Results

The VHDL code was synthesized and targeted to the Spartan3 (400g208pq) kit.
Table 66.2 shows the mapping results for the PPM I-UWB baseband transceiver. As can be noticed, the transceiver has a small area.

The floorplanning, using the Xilinx floorplanner tool for the design, is shown
in Fig. 66.14; using the floorplanner data, paths can be placed at the desired
location on the die [16, 17].

Table 66. 2 Mapping results for PPM I-UWB

Resource type	Available	Used	Utilization (%)
Number of 4 input LUTS	7.168	112	1
Number of slice flip flops	7.168	112	1
Number of occupied slices	3.584	78	2
Number of GCLKS	8	1	12
Number of bonded I/O	4	141	2

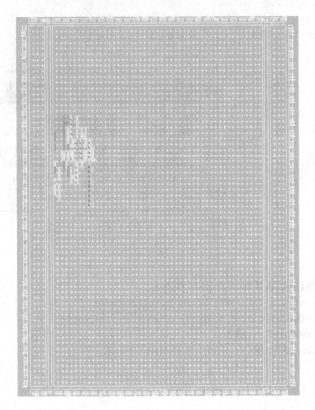

Fig. 66.14 A snapshot of the Xilinx FPGA editor for PPM I-UWB

66.5.2 Xilinx ChipScope Verification Results

The ChipScope[TM] Pro tools integrate key logic analyzer hardware components with the target design inside Xilinx; the ChipScope Pro tools communicate with these components and provide the designer with a complete logic analyzer.

Using the Xilinx ChipScope logic analyzer [18], the transmitted symbols were compared to the received ones using the PPM I-UWB baseband modem. Samples of the transmitted and the received symbols are shown in Fig. 66.15.

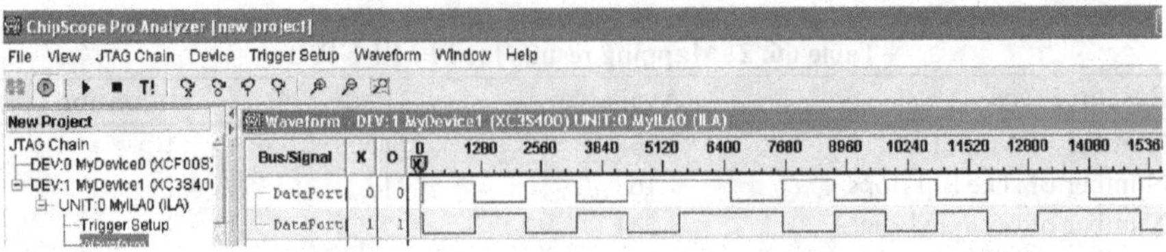

Fig. 66.15 PPM I-UWB ChipScope verification result

Fig. 66.16 PPM I-UWB
LCD verification result

66.5.3 LCD Verification Results

This test depends on using the liquid crystal display (LCD) of the Nu Horizons Electronics Spartan3 development board. A constant pattern was generated, and it passes through loop-back PPM I-UWB modulation-demodulation; the result is displayed on the LCD screen. The results are shown in Fig. 66.16.

66.6 Conclusions

This paper described a flexible FPGA design and the implementation of a Pulse position modulation impulse ultra-wideband baseband transceiver. Simulation results show that this transceiver is able to achieve a BER of 10^{-3} % at a SNR of 10 dB. The VHDL synthesis results showed how small the transceiver design is; hardware verification was done using the Xilinx ChipScope logic analyzer and liquid crystal display screen.

References

1. IEEE 802.15.3a Task Group Website (2005) Available online: http://www.ieee802.org/15/pub/TG3a.html
2. Federal Communications Commission (FCC) (2002) Revision part 15 of the commission's rules regarding ultra wideband transmission systems. First report and order, ET Docket 98–153, FCC 02–48, adopted: Feb 2002; released: Apr 2002
3. Aiello GR, Rogerson GD (2003) Ultra-wideband wireless systems. IEEE Microw Mag 4(2):36–47
4. Oppermann I, Hämäläinen M, Iinatti J (2004) UWB theory and applications. Wiley, New York
5. Barras D, Ellinger F, Jäckel H (2002) A comparison between ultra wideband and narrowband transceivers. In: Conference proceedings of TR-Labs wireless, pp 211–214

6. Blazquez R, Newaskar PP, Lee FS, Chandrakasan AP (2005) A baseband processor for impulse ultra-wideband communications. IEEE J Solid-State Circ 40(9)

7. Zhang H, Gulliver TA (2004) Pulse position amplitude modulation for time-hopping multiple access UWB communications. IEEE Wireless Commun Netw Conf, 2(1):895–900

8. Limbodal C, Meisal K, Lande TS, Wisland D (2005) A spatial RAKE receiver for real-time UWB-IR applications. IEEE international conference on ultra-wideband (ICU 2005), pp 65–69

9. Shuo M, Chen W (2005) Ultra wide-band baseband design and implementation. Research Project, Department of Electrical Engineering and Computer Sciences, University of California at Berkeley

10. www.opencores.org

11. Norimatsu T et al (2005) A novel UWB impulse-radio transmitter with all-digitally-controlled pulse generator. European solid-state circuits conference, pp 267–270

12. Nakache Y, Molisch A (2006) Spectral shaping of UWB signals for time hopping impulse radio. IEEE J Sel Area Commun 24(4):738–744

13. Barajas E, Cosculluela R, Coutinho D, Molina M, Mateo D, Gonzalez JL, Cairo I, Banda S, Ikeda M (2006) A low-power template generator for coherent impulse-radio ultra wideband receivers. IEEE international conference on ultra-wideband, pp 97–102

14. www.vhdl-online.de

15. Wentzloff DD, Chandrakasan AP (2007) Delay-based BPSK for pulsed-UWB communication. IEEE international conference on speech, acoustics, and signal processing

16. Wentzloff DD, Chandrakasan AP (2007) A 47pJ/pulse 3.1-to-5 GHz all-digital UWB transmitter in 90 nm CMOS. In: IEEE international solid-state circuits conference, pp 118–119

17. Welborn M (2001) System considerations for ultra-wideband wireless networks. In: Proceedings of the IEEE radio wireless conference, pp 5–8

18. Lee Fred S, Wentzloff David D, Chandrakasan Anantha P (2004) An ultra-wideband baseband front-end. In: IEEE radio frequency IC symposium, pp 493–496

19. http://www.xilinx.com/

Chapter 67
WSN-Based Audio Surveillance Systems

R. Alesii, G. Gargano, F. Graziosi, L. Pomante, and C. Rinaldi

Abstract This work focuses on the analysis of fundamental issues which have to be considered in order to obtain voice from a wireless sensor network (WSN). The latter is assumed to be constituted by MicaZ wireless sensor nodes. Problems related to the transport of an audio signal through a wireless channel and sensor nodes are thus analyzed, and a project for an audio surveillance system is presented.

67.1 Introduction

With the main goal of providing secure living environments for people in the world, the demand for surveillance systems has seen an incredible growth during recent years. One approach to this goal involves the use of wireless sensor networks (WSN [1]), since these systems are designed to be low-cost, high–technology, pervasive, and easy to use.

In general, the main application domains of WSN can be grouped into three classes [2]: *environmental data collection, security monitoring,* and *sensor node tracking.* However, when considering *audio surveillance systems,* it is worth noting that these represent an atypical WSN application class, quite different from the previous ones and imposing contrasting requirements on the WSN itself (e.g., sample rate vs. power consumption).

Nevertheless, the benefits of a *Voice over WSN (VoWSN)* approach could be very valuable. In particular, a VoWSN compared with the classic *COTS* analog audio surveillance system could guarantee:

R. Alesii (✉)
Università degli Studi dell'Aquila—Centro di Eccellenza DEWS, 67100 Poggio di Roio (AQ), Italy
e-mail: alesii@ing.univaq.it

N. Mastorakis et al. (eds.), *Proceedings of the European Computing Conference*, Lecture Notes in Electrical Engineering 27, DOI 10.1007/978-0-387-84814-3_67, © Springer Science+Business Media, LLC 2009

- more pervasiveness;
- reduced power consumption;
- capability to perform local signal processing to improve the communication quality and/or to reduce the bandwidth.

In order to achieve all these potential benefits, research efforts have been addressed toward joining the WSN world and toward audio processing issues. In [3], the authors focused their attention on the use of WSN to provide a sort of *walkie-talkie* and *voice-mail* system. The audio source localization problem has been studied in [4]. A real-time acoustic monitoring system, with a simple gender classifier, is proposed in [5]. In [6], the authors focused on an audio coding algorithm in order to use the correlation of the signal in both space and frequency; while an interesting approach to solve synchronization problems has been proposed in [7].

This work explicitly focuses on *surveillance systems* (i.e., continuous or event-driven audio monitoring tailored to the voice signal) and presents the fundamental issues to be considered in order to realize an innovative *VoWSN Audio Surveillance System*.

The remainder of this work is organized as follows. The next section describes the proposed system architecture. Then, main design and implementation issues are discussed in Sections 67.3 and 67.4; while in Section 67.5, the validation results are presented. Finally, conclusions are drawn in Section 67.6.

67.2 System Architecture

The communication platform assumed to support the *VoWSN Audio Surveillance System* is based on the *IEEE 802.15.4* standard [8] (in particular at the 2.4 GHz ISM band implementation). This standard has been developed for *low rate-wireless personal area networks* (LR-WPAN) and has all the prerequisites needed to match the stringent constraints imposed by the surveillance system of our interest (particularly in terms of low power consumption).

The devices constituting the network belong to two different categories: *full function devices* (FFD) and *reduced function devices* (RFD). FFDs can be used for every topology, they are able to work as network coordinators, and can communicate with every device belonging to the network. On the other hand, RFDs can be used only as end-nodes, and can communicate only with a coordinator.

In order to create a *personal operating space* [8], it is possible to organize the devices discussed into three different network topologies named *star, peer-to–peer,* and *cluster tree*. In order to choose the proper network topology, it is necessary to cope with problems related to the use of an audio channel in an energy-aware context. The limit of a point-to-point link for the standard is considered theoretically equal to 250 Kbps (various overheads actually limit the bit rate to some tens of Kbps in ideal radio propagation conditions), while the bandwidth of an audio channel is directly related to the source coding approach. For instance, for a typical *PCM* approach, with a bit rate of

64 Kbps, almost all the available bandwidth has to be used. In this limited radio resource scenario, the role of the source coding techniques is of paramount importance. Therefore, every effort to reduce the rate of source coded signals compatible with the available processing resources needs to be considered.

Another important aspect to be considered is that, when a certain number of nodes are placed around the audio source to be observed, each node will detect the same signal at different distances and with different background noises. Therefore, in order to use signals correlation with the aim of obtaining a single, stronger audio signal, it is possible to utilize the concept of *distributed source coding* [9], which aims to wisely exploit the redundancy introduced by multiple measures of the same signal.

Since the LR-WPAN protocol allows for the use of end nodes which are not designed for routing purposes, it is possible to use them, implemented with RFDs, as terminals for audio capture, thus reducing energy consumption. Afterwards, RFDs have to communicate with their coordinator node (FFD) in order to transmit their information. Moreover, since the channel capacity is limited, we have to establish the maximum number of RFDs which can be connected to a single FFD. It is obvious at this point that we are considering a cluster tree topology.

The task of computing the maximum number of leaves connected to a single parent is strictly dependent on the modulation used for data transmission (the modulation of interest for us is *offset quadrature phase shift keying,* with half-sine pulse shaping, O-QPSK), as well as on the use of source coding algorithms. Indeed, the number of possible RFDs connected to a single parent increases as the source coding rate increases.

Figure 67.1 shows the proposed architecture for the *VoWSN audio surveillance system,* where the RFDs are embedded in the environment to be monitored (so they should be very few and energy-aware), while the FFD are deployed surrounding the environment and so are easily replaceable/rechargeable. Finally, one (or more) gateway nodes will have the task of effectively forwarding the acquired data.

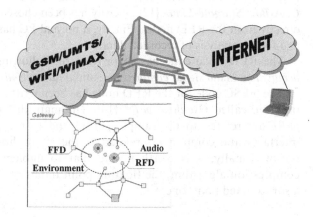

Fig. 67.1 The proposed
system architecture

67.3 Design Issues: Audio Source Coding Algorithms

Once having defined the system architecture, a fundamental design choice should be made with regard to the proper source coding technique suitable for the developed system, taking into account the limited bandwidth and the low computational capabilities of end nodes.

There are three main audio source coding methods: *waveform, vocoder,* and *hybrid codec* [10]. Waveform coding treats speech signals as normal signal waveforms, and it has relatively simpler algorithms, good adaptive capabilities, and good speech quality. It has been widely used at the 16–64 kbit/s range also, thanks to its low complexity. However, the coding rate cannot decrease further; otherwise the distortion in the waveform may be too high to achieve an acceptable performance in audio perception.

Unlike waveform coding, and based on a speech synthetic model, parametric and hybrid coding attempt to obtain high quality speech by extracting and encoding the feature parameters of the speech signals. They can achieve a lower bit rate than waveform coding, but the quality of the synthesized speech is not as good, either in clearness or naturalness.

In this case, it has to be taken into account that the computational complexity of both parametric and hybrid coding techniques is higher than *10 MIPS* (too much for a typical WSN node) in the absence of dedicated hardware [11], so the forced solution for the compression algorithm is represented by the waveform coding. Then the final choice has been to use a proper implementation of the *interactive multimedia association-adaptive pulse code modulation (Ima-Adpcm,* [10]).

67.4 Implementation Issues: RFD, FFD, and GW Software

To develop a prototype of the system, the *CrossBow MicaZ-TinyOS* [12, 13] platform has been used for both the FFD and RFD nodes, and a C-programmed *CrossBow Stargate-Linux* [12] gateway has been chosen in order to collect data directly from the RFD nodes. Finally, a normal PC has been used for listening and for the analysis of recorded data.

The RFD node code has to provide the following main functions: *data acquisition, audio signal compression,* and *data transmission.*

The NESC code for the RFD node has been written based on a preexisting program called *OscilloscopeRF* [13]. The main differences with *OscilloscopeRF* are related to the sample period and to the fact that we are using interfaces and components related to the microphone instead to the light sensor. Finally, it is worth noting that to implement the *IMA-ADPCM* compression algorithm, the floating point data type has been replaced with a sort of fixed point one.

In the system prototype, the FFD node is directly connected with the *Stargate-Linux* gateway. This implies that the only functionality of an FFD node is to collect data coming from all the RFD nodes referring to it and forwarding them to the gateway by means of a serial communication. The program accomplishing such *packet forwarding* between the radio interface and the UART controller is called *TOSBase* [13], and it is already present on existing TinyOS distributions. The only needed changes are related to the packet dimension in the *OscopeMsg.h* library.

The network gateway has to provide two main services: packet *decoding* and packet *forwarding* in order for the decoded packets to be heard. This implies that the software has to:

- capture the data flow coming from onboard MicaZ nodes (FFD);
- decode the *ADPCM* flow;
- write the decoded audio to a file.

The implementation has been based on software furnished with the Stargate gateway, called *motetest* [12]. It has to be emphasized that a lot of changes have been required in this case. In particular, a bug encountered with the transmission of a *bit stuffing* by the FFD was fixed through a proper *bit destuffing* procedure.

67.5 Validation

In order to analyze results coming from the decoded audio flow, which has been written to a file by the *Stargate,* we have used a multimedia PC with two kinds of software, one for data numerical elaboration and the other for audio editing. In particular, *OpenOffice Calc* has been used for the first operation. Then the correctness of the captured data has been verified by using known tones as input signals for the RFD nodes.

For spectral analysis of and listening to recorded data *OpenOffice Audacity* has been used. In Fig. 67.2 it is possible to observe the vocal frequency peak around 500 Hz. Moreover, effects such as *ADPCM drift* and *microphone saturation* are indicated by red arrows. The latter effect is due to a vocal source with excessive amplification, or which is too close to the microphone. In order to avoid this effect, the amplification gain on the sensor board *MTS300CA* should be properly adapted [12].

Regarding the drift phenomenon, it is caused by the loss of a packet, after which the coder and decoder are no longer aligned on the last predicted value and on the step-size table (*offset* and *amplification* [10]). This effect has been dealt with by the transmission of some additional values for every packet.

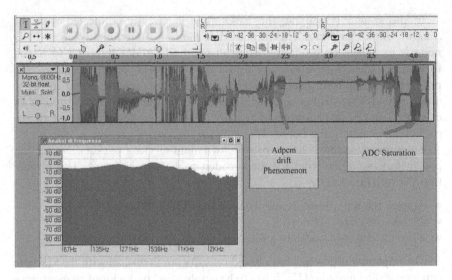

Fig. 67.2 Screenshot of a vocal recording performed by Audacity

67.6 Conclusion and Future Work

This work has analyzed the main problems that arise while designing a *VoWSN audio surveillance system*. System architecture, design, and implementation issues have been dealt with, and a possible prototypical implementation for a MicaZ node has been described and validated for the problem of audio signal acquisition, without taking into account the remaining functionalities that a node has to implement while part of a network. In particular, it is important to allow the wake-up of the node only when a source is active, in order to save energy; and to adjust the compression rate based on the current traffic conditions of the network. Moreover, it would be of interest to allow the gateway to control the amplification of the signal captured by the microphone, thus realizing an AGC (*automatic gain control*) function and improving the quality of the acquired signal. Finally, one of the major problems to be further considered is related to audio flows routing. Indeed, it is very difficult to forward high data rate packets on a MicaZ platform, which is characterized by scarce computational resources such as radio bandwidth.

References

1. Romer K, Mattern F (2004) The design space of wireless sensor networks. IEEE Wireless Commu, Dec.
2. Hill JL (2003) System architecture for wireless sensor networks. PhD Thesis, University of California-Berkeley, AAT 3105239. ACM Press, New York

3. Mangharm R, Rowe A, Rajkumar R, Suzuki R (2006) Voice over sensor networks. In: Real-time system symposium, Rio de Janeiro, Brazil
4. Masson F, Puschini D, Julian P, Crocce P, Arenghi L, Mandolesi PS, Andreou AG (2005) Hybrid sensor network and fusion algorithm for sound source localization. In: IEEE International symposiumon circuits and systems
5. Berisha V, Kwon H, Spanias A (2006) Real time acoustic monitoring using wireless sensor motes. IEEE international symposium on circuits and systems
6. Dong H, Lu J, Sun Y (2006) Distributed audio coding in wireless sensor networks. In: IEEE international conference on computational intelligence and security
7. Zhang J, Zhou G, Son AH, Stankovic JA (2006) Ears on the ground: an acoustic streaming service in wireless sensor networks. In: IEEE/ACM international conference on information processing in sensor networks, Nashville, TN
8. IEEE Standard 802.15.4, www.ieee.org
9. Mayumdar A, Ramchandran K, Kozintsev I (2003) Distributed coding for wireless audio sensors. IEEE workshop on applications of signal processing to audio and acoustics
10. Hanzo L (2007) Voice and audio compression for wireless communications. Wiley, New York
11. Kiviluoto A (2003) Speech coding standards. Helsinki University, Helsinki, Finland
12. www.xbow.com
13. www.tinyos.net

Chapter 68
Smart Antenna Design Using Multi-Objective Genetic Algorithms

I.O. Vardiambasis, N. Tzioumakis, and T. Melesanaki

Abstract Optimizing antenna arrays to approximate desired far-field radiation patterns is of exceptional interest in smart antenna technology. This paper shows how to apply evolution and natural selection, in the form of genetic algorithms (GAs), to achieve specific beam-forming with linear antenna arrays. A multi-objective GA is used to maximize multiple main beams' radiation and a simple GA to minimize the side lobe level of a linear antenna array. Dealing with these two design criteria separately, we manage to achieve better results than other methods based on GAs. In particular, a triple beam radiation pattern with low side lobes is presented in order to demonstrate the effectiveness and the reliability of the proposed approach. The results show that multi-objective GAs are robust and can solve complex antenna problems.

68.1 Introduction

Smart antennas have been widely used in mobile and wireless communication systems to increase signal quality, improve system capacity, enhance spectral efficiency, and upgrade system performance. The performance of these systems depends strongly on the antenna array design [1]. In this paper we consider two design criteria for the evaluation of smart antenna arrays' performance: multiple main beams and minimum side lobe level.

The synthesis of an antenna array with a specific radiation pattern is a nonlinear optimization problem, which cannot be effectively treated by traditional optimization techniques using gradients or random guesses [2, 3]. Especially in complex cases of radiation shapes with multiple main beams and nulls

I.O. Vardiambasis (✉)
Division of Telecommunications, Microwave Communications and Electromagnetic Applications Laboratory, Department of Electronics, Technological Educational Institute (T.E.I.) of Crete –Chania Branch, Romanou 3, Chalepa, 73133, Chania, Crete, Greece
e-mail: ivardia@chania.teicrete.gr

N. Mastorakis et al. (eds.), *Proceedings of the European Computing Conference*, Lecture Notes in Electrical Engineering 27, DOI 10.1007/978-0-387-84814-3_68, © Springer Science+Business Media, LLC 2009

at given directions, there are too many possible phase-amplitude excitations, and exhaustive checking of the best solution is very difficult. However, genetic algorithms are capable of solving this kind of complicated and nonlinear search problem, providing robust searches in complex spaces [4]. They are computationally simple, easily improved, applied to continuous and discrete search spaces, not limited by continuity and differentiability requirements of the objective functions, and they also have an inherent parallel nature. Therefore, many researchers have developed GAs for antenna array synthesis and design. A plethora of studies have investigated GA-based methods for reducing array side lobes by thinning, perturbing element positions, tapering excitation amplitude or phase, or designing arrays with specified null locations or specific pattern shapes [5–9]. In most cases these works consider antenna array design to be a problem of optimizing a single objective. However, the proposed antenna array design aims to find a set of excitations that make the radiation pattern satisfy two probably conflicting objectives at the same time, which is a natural multi-objective problem. The application of multi-objective methods for designing antenna arrays has been rather rare in the literature [1]. This paper shows that antenna array design can be dealt with as a multi-objective optimization problem, using a multi-objective GA in order to maximize main beams to various directions, and a simple GA in order to reduce the side lobe level. Thus the radiation pattern of a linear antenna array with 30 elements and relative side lobe level about -20 dB is computed efficiently, resulting in better performance compared to other published results [3].

68.2 Formulation of the Antenna Array Pattern

In this paper, we will concentrate on finding the current excitations of all antenna array elements, which is the standard technique for designing antenna arrays. If the elements in the linear array are taken to be isotropic sources, the pattern of this array can then be described by its array factor.

The array factor for the linear array of Fig. 68.1 is given by

$$S(\theta, \varphi, \bar{I}) = \sum_{n=1}^{M} I_n \cdot \exp[j\,nkd\,\psi(\theta, \varphi)] \tag{68.1}$$

where $\bar{I} = [I_1, I_2,..., I_M]$, $I_n = A_n \exp[j\delta_n]$ represents the excitation of the nth element of the array (with A_n and δ_n being the amplitude and the phase of the current excitation), $k = 2\pi/\lambda$ is the phase constant, λ is the signal wavelength, d is the uniform distance between elements, (θ, φ) is the direction of interest, $\psi(\theta, \varphi) = \cos\theta \cos\theta_a + \sin\theta \sin\theta_a \cos(\varphi - \varphi_a)$, and (θ_a, φ_a) is the direction of the array axis.

To synthesize a radiation pattern of multiple main beams and low side lobe level (SLL) for the linear array of Fig. 68.1, we develop: (a) a multi-objective

Fig. 68.1 The linear array geometry

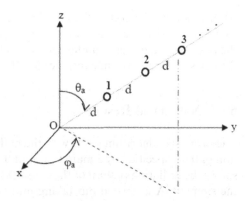

GA in order to calculate the phase excitation coefficients, and (b) a simple GA in order to calculate the amplitude excitation coefficients. The basic difference between multi-objective and simple GAs is that the multi-objective GA aims to find all the non-dominated solutions of the problem and present the set of the non-dominated solutions to the decision maker, who selects the best solution.

The fitness function for the multi-objective GA is given by:

$$\text{fitness} = w_1 S_1 + w_2 S_2 + \ldots + w_n S_n \tag{68.2}$$

where $S_1, S_2, \ldots, S_n$ are the amplitudes of the array factor given in (68.1) for every desired direction, and $w_1, w_2, \ldots, w_n$ are the non-negative weights for the n objectives. These weights satisfy the following relations:

$$w_1 + \ldots + w_n = 1; \quad w_i \geq 0 \ (i = 1, 2, \ldots, n) \tag{68.3}$$

The fitness function for the simple GA is given by:

$$\text{fitness} = \sum_{i=1}^{n} S_i \bigg/ \text{SLL} \tag{68.4}$$

where n is the number of main beams in the radiation pattern and SLL the side lobe level.

Our approach avoids binary encoding and uses real encoding to simplify computer programming and speed up computation. So, each solution (chromosome) is represented directly by a real weighting vector:

$$C = [c_1, \ c_2, \ \ldots, \ c_n, \ldots, \ c_M] \tag{68.5}$$

where c_n (gene) represents the excitation (phase or amplitude) of the nth radiator of the M-element linear array.

Both GAs use populations of 500 array excitations. Each array has 30 elements equally spaced at $d = \lambda/2$ and positioned along the z-axis $\theta_a = 0°$.

After random selection of the initial populations, the search and optimization technique works on a similar evolutionary principle as chromosomes, reflecting the actions of crossover, mutation, and natural selection based on survival of the fittest. The algorithm continues for 500 iterations.

68.3 Numerical Results

The proposed algorithm starts with the multi-objective GA for a desired radiation pattern specified by 3 main beams at θ = 40, 100, and 135°. Figure 68.2(a) shows the radiation pattern of the chosen solution with SLL = -6.26 dB. Next the simple GA is carried out taking into account the phase excitation coefficients given in Table 68.1. Figure 68.2(b) shows the radiation pattern of the

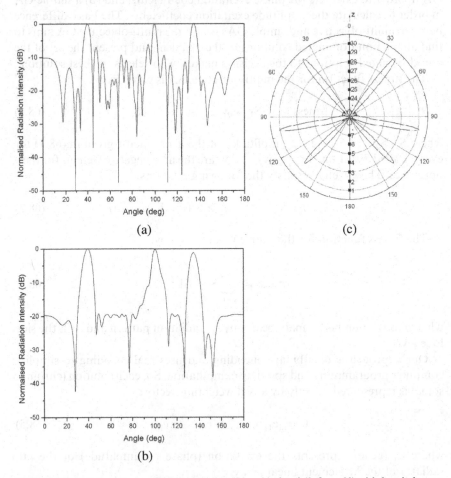

Fig. 68.2 Radiation patterns of the linear array (M = 30, d = $\lambda/2$, θ_a = 0°) with 3 main beams at θ = 40, 100, and 135°, (**a**) after completion of just the multi-objective GA, (**b**) after completion of both GAs, (**c**) polar form

Table 68.1 Amplitude A_n and phase δ_n of all calculated excitation coefficients

Element	Excitation's A_n	Excitation's δ_n	Element	Excitation's A_n	Excitation's δ_n
1	0.10	256.97°	16	0.98	21.07°
2	0.16	249.63°	17	0.75	50.69°
3	0.22	84.97°	18	0.96	129.84°
4	0.20	283.02°	19	0.79	57.24°
5	0.13	52.42°	20	0.84	165.36°
6	0.36	57.79°	21	0.69	171.06°
7	0.41	250.46°	22	0.44	300.49°
8	0.62	68.37°	23	0.66	171.96°
9	0.06	97.26°	24	0.32	270.03°
10	0.67	199.40°	25	0.37	276.13°
11	0.60	92.79°	26	0.15	198.63°
12	0.90	211.06°	27	0.26	324.32°
13	0.50	261.75°	28	0.21	315.22°
14	0.80	344.35°	29	0.20	69.04°
15	0.62	215.79°	30	0.11	331.71°

linear array with a SLL reduction to -19.4 dB, and Table 68.1 includes the amplitude excitation coefficients for all array elements. This SLL corresponds to an improvement of 33% in comparison to the SLL $= -13$ dB in [3]. Figure 68.2(c) shows the polar diagram of the radiation pattern of this linear array.

68.4 Conclusions

This paper shows that antenna array design and pattern synthesis can be modeled as multi-objective optimization problems, in which the optimization objectives are the maximization of multiple main beams and the minimization of the side lobe level. For the solution of these problems we propose an evolutionary optimization procedure based on two GAs. This approach uses a multi-objective GA in order to maximize main beams to various directions and a simple GA in order to reduce the side lobe level. Thus the radiation pattern of linear antenna arrays with uniform separations is computed efficiently, leading to better results.

Acknowledgments This work is co-funded by the European Social Fund (EKT) (75%) and National Resources (Greek Ministry of National Education and Religious Affairs) (25%) under the ΕΠΕΑΕΚ project "Archimedes – Support of Research Groups in TEI of Crete – 2.2.–7 – Smart antenna study and design using techniques of computational electromagnetics, and pilot development and operation of a digital audio broadcasting station at Chania (SMART-DAB)". Also, the authors would like to thank Professor John L. Tsalamengas and Associate Professor George Fikioris, of the National Technical University of Athens, for their helpful discussions and insightful comments.

References

1. Panduro MA, Covarrubias DH, Brizuela CA, Marante FR (2005) A multi-objective approach in the linear antenna array design. Int J Electron Commun 59:205–212
2. Haupt R (1995) An introduction to genetic algorithms for electromagnetics. IEEE Antenn Propag Mag 37:7–15
3. Marcano D, Duran F (2000) Synthesis of antenna arrays using genetic algorithms. IEEE Antenn Propag Mag 42:12–20
4. Weile DS, Michielssen E (1997) Genetic algorithm optimization applied to electromagnetics: a review. IEEE T Antenn Propag 45:343–353
5. Ares-Pena FJ, Rodriguez-Gonzalez JA, Villanueva-Lopez E, Rengarajan SR (1999) Genetic algorithms in the design and optimization of antenna array patterns. IEEE T Antenn Propag 47:506–510
6. Bray MG, Werner DH, Boeringer DW, Machuga DW (2002) Optimization of thinned aperiodic linear phased arrays using genetic algorithms to reduce grating lobes during scanning. IEEE T Antenn Propag 50:1732–1742
7. Haupt R (1994) Thinned arrays using genetic algorithms. IEEE T Antenn Propag 42:993–999
8. Yan KK, Lu Y (1997) Sidelobe reduction in array-pattern synthesis using genetic algorithm. IEEE T Antenn Propag 45:1117–112

Part VII
Computer Applications in Modern Medicine

The practice of modern medicine and biomedical research requires sophisticated information technologies with which to manage patient information, plan diagnostic procedures, interpret laboratory results, and carry out investigations. Medical informatics provides both a conceptual framework and a practical inspiration for this rapidly emerging scientific discipline at the intersection of computer science, decision science, information science, cognitive science, and biomedicine. This part is a useful reference work for individual readers needing to understand the role that computers can play in the provision of clinical services and the pursuit of biological questions. The part consists of several papers that explain basic concepts and illustrate them with specific systems and technologies. The papers consider an intention to adopt an e-health services system in a bureau of health, visualization and clustering of DNA sequences, the role of organizational innovation in an e-health service in Taiwan, an analysis of heart sounds with wavelet entropy, the detection of mitral regurgitation and normal heart sounds, wavelet entropy detection of heart sounds, the continuous wavelet transform analysis of heart sounds, limitations of lung segmentation techniques, segmentation of anatomical structures using volume definition tools, the shape analysis of heart rate Lorenz plots, and the human-readable rule induction in medical data mining.

Chapter 69
Intention to Adopt the E-Health Services System in a Bureau of Health

Hsieh-Hua Yang, Jui-Chen Yu, Hung-Jen Yang, and Wen-Hui Han

Abstract The objectives of this study are exploring the determinants of acceptance to use the e-health service system, and identifying the relative importance of the determinants. The participants are employees of the Health Bureaus and Health Stations located in the north of Taiwan. A total of 105 effective responses were received, a 95.4% response rate. The results showed that the original theory of reasoned action is robust enough in explaining behavioral intention. However, the importance of attitude is decreased when jointly perceived with behavioral control as predictors. It is concluded that behavioral intention is influenced by multiple factors; the relative importance of these factors is determined by volitional control. The implication for strategic planning is discussed.

69.1 Introduction

The government has made huge investments in electronic services in Taiwan. The Information System on Healthcare Administration was set up in 1981. A National Health Information Network began operating in 1989. A pilot project was tested out in the Hsinchu medical care region in 1991–1993 and then extended to other regions. A Taiwan health web site, http://www.doh.gov.tw was initiated in 1995. All of these systems are innovative for the public and government employees, especially for the employees of the Health Bureaus and Health Stations. In Taiwan, primary healthcare is provided by Health Bureaus and Health Stations. It is necessary to evaluate the adoption of the innovative systems by these employees.

Researchers [1] have investigated the user acceptance of the government services in Taiwan, but not their adoption by government employees. And

H.-H. Yang (✉)
Department of Health Care Administration, Oriental Institute of Technology,
58, sec. 2, Szechwan Rd., Banciao City, Taipei Country 220, Taiwan
e-mail: yansnow@gmail.com

N. Mastorakis et al. (eds.), *Proceedings of the European*
Computing Conference, Lecture Notes in Electrical Engineering 27,
DOI 10.1007/978-0-387-84814-3_69, © Springer Science+Business Media, LLC 2009

most of the researches exploring the use of information technology are grounded in the behavioral sciences, using theories and models to identity determinants of successful use. Kukafka et al. [2] made a systematic analysis of the literature on IT usage and found that most of the theories tested were used to study the relationship between attitudes and behaviors. It is suggested that a better understanding of the role of attitude can enhance the model's predictability about users' acceptance of information technology [3].

Numerous studies have applied the theory of planned behavior (TPB) to investigate behavior intention using attitudinal variables. The TPB is an extension of the theory of reasoned action (TRA), rather than an independent theory. The TRA and TPB are applied in this study to evaluate the employees' acceptance. The objectives of this study are exploring the determinants of acceptance to use the e-health service system, and identifying the relative importance of the determinants.

69.2 Literature Review

69.2.1 E-Health System in Taiwan

In the annual report of the Brown University Center for Public Policy, Taiwan was listed as having one of the top-ranked e-Government sites among the 198 investigated nations [4]. In Taiwan the e-government plan was formulated by the Research, Development, and Evaluation Commission (RDEC).

Health bureaus and health stations are subsidized and supervised in the procurement of computer facilities. There are plans for administrative information networks for primary healthcare. These include the planning for and establishment of standard networks in health bureaus and health stations; planning for the information systems and network applications; and training in knowledge and skills in the use of information in health bureaus and health stations. In 2003, the DOH continued to build health administration systems, promote an e-DOH, and install an e-document exchange center, in order to strengthen the performance of health administration [5].

The Internet Health Service Promotion Project and Health Bureau/Station Internet Public Services Project have been planned and implemented, and have been incorporated into the Digital Taiwan Plan, part of the Executive Yuan's Challenge 2008: National Priority Development Plan, with the aim of speeding up the health and medical care information infrastructure, upgrading the functions of the National Health Information Network, and establishing an information service environment of better quality and efficiency [5].

The Department of Health has created a system that enables the public to carry out application procedures online in the areas of health administration, drug administration, and food sanitation. Other functions of this system include downloading of application forms, application status inquiries, and pickup notification. The system was completed in December 2003 and went online in 2004, contributing toward the e-government goal of "single application point, total service" [6].

69.2.2 Theory of Reasoned Action

The theory of reasoned action (TRA), first introduced in 1967, is concerned with the relations between beliefs (behavioral and normative), attitudes, intentions, and behavior [7]. As its name implies, the theory of reasoned action is based on the assumption that human beings usually behave in a sensible manner; that they take account of available information and implicitly or explicitly consider the implications of their actions [8]. The TRA postulates that a person's intentions to perform (or not to perform) a behavior is the immediate determinant of that behavior. Intention is defined as the individual's decision to engage or not to engage in performing the action. In turn, the person's intentions are a function of the person's attitude toward the behavior, and his/her subjective norms. Attitude towards the behavior is defined as "the evaluative affect of the individual towards performing the behavior" [9]. Subjective norms refer to the social pressure exerted on the individual to perform or not perform the behavior. Figure 69.1 presents a visual overview of TRA. Impressively, these models have received robust support in numerous behavioral domains [10, 11].

69.2.3 Theory of Planned Behavior

The theory of planned behavior (TPB) is an extension of the TRA. Ajzen and colleagues [12–14] added perceived behavioral control to the TRA in an effort to account for factors outside the individual's control that may affect his intention and behavior. Figure 69.2 presents a visual overview of planned behavior theory. Ajzen [12] reviewed 16 studies of predicting behavioral intentions, and found that a considerable amount of variance in intentions can be

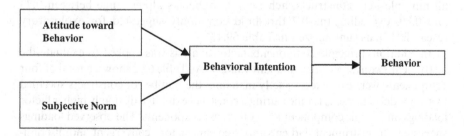

Fig. 69.1 Theory of reasoned action

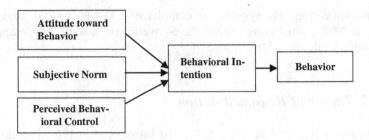

Fig. 69.2 Theory of planned behavior

accounted for by attitude, subjective norms, and perceived behavioral control [13]. The multiple correlations ranged from a low of 0.43 to a high of 0.94, with an average correlation of 0.71. The addition of perceived behavioral control to the model led to considerable improvements in the prediction of intentions. Most importantly, the regression coefficients of perceived behavioral control were significant in every study. However, Ajzen reminded us that the results for subjective norms were mixed, with no clearly discernible pattern.

69.3 Methods

Most of the constructs related to the model of planned behavior theory were measured with items adapted from prior research [1, 15]. Multiple item scales were developed for each of the constructs shown in Table 69.1. All items were measured using a five-point Likert-type scale with anchors ranging from "strongly disagree" to "strongly agree."

Behavioral intention was assessed with three items, attitude toward the behavior with two items, and perceived behavioral control with three items. The subjective norm was measured by normative beliefs and the motivation to comply. An individual's subjective norm is computed by multiplying her normative belief about each referent by her motivation to comply with that referent and then summing these product scores across all referents [14].

Regarding reliability, the measurement had strong internal consistency with all multiple-item constructs, achieving Cronbach's alpha range between 0.73 and 0.89, exceeding the 0.7 threshold commonly suggested for exploratory research [17], as summarized in Table 69.1.

In addition, principle components factor analysis was applied to evaluate the instrument's convergent/discriminant validity. As Table 69.2 shows, a total of four components were extracted, exactly matching the number of constructs specified by the model. The items for measuring a construct exhibited distinctly higher factor loadings on a single component than on other components. The observed loadings suggested our instrument had encompassed satisfactory convergent and discriminant validity. The validated instrument was used in this study.

Table 69.1 Reliability analysis using Cronbach's alpha values

Construct and item	Mean	SD	α
Behavioral intention			0.89
1 I intend to use this system	3.50	0.92	
2 It is likely that I will use this system	3.15	1.04	
3 I expect to use this system	3.46	0.93	
Attitude			0.73
1 Using this system would be a good idea	3.43	0.89	
2 Using this system would be a pleasant experience	3.34	0.95	
Subjective norm			0.89
My supervisor, whose opinion I value, supports me in using this system, and I would comply with him/her	15.37	5.57	
A colleague important to me supports my use of this system, and I would comply with him/her	15.17	5.60	
A friend who influences my behavior wants me to use this system, and I would comply with him/her	12.75	6.20	
Perceived behavioral control			0.86
1 I will be able to use this system	3.62	0.84	
2 Using this system is entirely within my control	3.17	1.04	
3 I have the resources, knowledge, and ability to use this system	3.50	0.86	

Table 69.2 Examination of convergent/discriminant validity

	Factor 1	Factor 2	Factor 3	Factor 4
Perceived behavioral control				
1	0.806	–	–	–
2	0.630	–	–	–
3	0.846	–	–	–
Subjective norm				
1	–	0.871	–	–
2	–	0.868	–	–
3	–	0.673	–	–
Behavioral intention				
1	–	–	0.570	–
2	–	–	0.798	–
3	–	–	0.851	–
Attitude				
1	–	–	–	0.758
2	–	–	–	0.775
Eigenvalue	2.568	2.454	2.424	1.753
% of variance	23.342	22.310	22.033	15.935

Extraction method: Principal component analysis
Rotation method: Varimax with Kaiser Normalization

All of the employees in the Bureau of Health were asked to complete the questionnaire. Of the 110 questionnaires distributed, a total of 105 effective responses were received, showing a 95.4% response rate.

69.4 Results

Analysis of the respondents showed an approximate 7 to 1 ratio in gender distribution in favor of females. Their characteristics are presented as Table 69.3. About 60% of the subjects were from the Health Bureau, and the other 39% from the Health Stations. About one third of the respondents had been employed for less than 5 years. Over a quarter of the respondents have been working in the Health Bureau over 16 years.

The analysis of determinants and the relative importance of these determinants are presented as below.

69.4.1 Determinants

Table 69.4 presents the results of stepwise regression analyses for the prediction of behavioral intention to use the e-health service system. For the prediction of intentions, attitude and subject norm were entered on the first step, and perceived behavioral control on the second step. The first step constitutes a test of the TRA. It can be seen that both attitude and subjective norm made significant

Table 69.3 Summary of participants' characteristics

Variables	N	%
Gender		
Male	13	12.4
Female	92	87.6
Department		
Bureau	63	60.0
Stations	41	39.0
Years of employment		
≤ 5	34	32.4
6–10	23	21.9
11–15	15	14.3
≥ 16	27	25.7
	Mean	SD
Behavioral intention	10.10	2.62
Attitude	6.77	1.63
Subjective norm	43.53	15.79
Perceived behavioral control	10.29	2.44

Table 69.4 Determinants of behavioral intention

Variable	Mode 1		Mode 2	
	B	t	B	t
Constant	2.841	3.619	1.017	1.348
A	0.689	4.835***	0.302	2.136*
SN	0.060	4.063***	0.038	2.882**
PBC			0.522	5.753***
R^2		0.501	0.624	
R^2 change			0.123***	

*p<0.05 **p<0.01 ***p<0.001
BI: behavioral intention
A: attitude
SN: subjective norm
PBC: perceived behavioral control

contributions to the prediction of intention, resulting in an R square of 0.501. This finding indicates that the original theory of reasoned action with its implication of attitude and subjective norm can influence behavioral intention.

The significant effects of attitude and subjective norm are still kept when jointly perceived with behavioral control as independent variables in a regression model to predict behavioral intention, as Table 69.4 shows. Model 2 tests the theory of planned behavior, with an R square of 0.624. The improvement of the model's prediction of behavior intention is 0.123, significant at the 0.001 level. Obviously, perceived behavioral control improves the model's explanatory power.

69.4.2 Relative Importance of Determinants

Standardized coefficients from regression analyses to test the model in explaining intention are presented in Table 69.5. The result of the first step demonstrates that the TRA model constructs predicted behavioral intention, with a multiple correlation of 0.708. And attitude is more important than subjective norm. On the second step, the addition of perceived behavioral control

Table 69.5 Model predictions of behavioral intention

	R
BI = 0.428*A + 0.360*SN	0.708
BI = 0.187*A + 0.232*SN + 0.485*PBC	0.790

BI: behavioral intention
A: attitude
SN: subjective norm
PBC: perceived behavioral control

significantly improved the explanatory power of the basic model, with a multiple correlation of 0.790. However, the importance of attitude is decreased. For these three independent variables, perceived behavioral control is the most important one, then subjective norm, and attitude the least one.

69.5 Discussion and Conclusion

The TRA and TPB have been widely applied and tested in the behavioral sciences. In this study, both theories are tested again. Nonetheless, there are some interesting findings. These findings will help us to understand the effect of individual factors on the behavior of adoption.

69.5.1 Multiple Factors of Using Behavior

The results showed that the adoption of the e-health service system was influenced by multiple factors. As indicated by Kukafka et al. [2], IT usage is influenced by multiple factors. In this study, *attitude* toward the behavior is the individual's positive or negative evaluation of adopting the technology [8]. It reflects personal interest, while *subjective norm* refers to the individuals' perceptions of social pressure to perform or not to perform the behavior under consideration [8]. It reflects social influence. The third factor is *perceived behavioral control* of internal and external factors. Of the two, the external factor refers to the situational or environmental factors external to the individual [8]. Thus individual, social, and situational features construct the multiple factors of adoption behaviour.

69.5.2 Volition of Participants

The theory of reasoned action was developed explicitly to deal with purely volitional behaviors [8]. In this context it has proved quite successful. However, if we try to apply the theory to behaviors that are not fully under volitional control, the degree of success will depend not only on one's desire or intention, but also on such partly nonmotivational factors as availability of requisite opportunities and resources. To the extent that people have the required opportunities and resources, and intend to perform the behavior, they should succeed in doing so.

This is the reason that Ajzen [12] added perceived behavioral control to the model, to overcome its limitations when "dealing with behaviours over which people have incomplete volitional control." In this situation, perceived control is expected to contribute to the prediction of intentions. In this study, the function of the e-health service system is to provide health services. The usage of this system is performing health services, not like other usages such as word processing, spreadsheets, or e-mail that have been studied. It is not surprising that the results showed that the influence of attitude on behavioral intention decreased when perceived behavioral control was added as a predictor.

The theory assumes that the relative importance of attitude toward the behavior and subjective norm depends in part on the intention under investigation. For some intentions attitudinal considerations are more important than normative considerations, while for other intentions normative considerations predominate [8]. Comparing the results of two regression models, subjective norm was more important than attitude in predicting behavior intention. Since the e-health service system is for providing health services, social pressure exerted by colleagues in the field should be considered more important than an individual's attitude. Furthermore, the subjects are employees of Health Bureaus and Health Stations. They have to comply with their supervisors. That's the reason why the adoption behavior cannot be entirely within the participants' volitional control, and subjective norm plays a more important role.

69.5.3 The Implication for Strategic Planning

Various factors internal to an individual can influence successful performance of an intended action. Some of these factors suggested by Ajzen [8] are skills, abilities, and information. These factors may present themselves as obstacles to behavioral control, but it is usually assumed that they can be overcome by training.

Examples of external factors are time, opportunity, and dependence of the behavior on the cooperation of other people [14]. These factors imply the extent to which circumstances facilitate or interfere with the performance of the behavior. This is the focus of strategic planning.

Information technology systems are changing the interface of healthcare delivery. The potential could be fully utilized if the role of individual factors in the development and implementation of such systems were better understood and managed. Successfully integrating technology into healthcare will require strategic planning and increased attention to the role of individual characteristics. The behavioral sciences will fulfill this duty.

References

1. Hung SY, Chang CM, Yu TJ Determinants of user acceptance of the e-government services: the case of online tax filing and payment system. Gov Inform Q 23:97–122
2. Kukafka R, Johnson SB, Linfante A, Allegrante JP (2003) Grounding a new information technology implementation framework in behavioral science: a systematic analysis of the literature on IT use. J Biomed Inform 36:218–227

3. Yang HD, Yoo Y (2004) It's about attitude: revisiting the technology acceptance model. Decis Support Syst 38:19–31
4. West DM Global e-Government survey. World Markets Research Centre. Brown University, Retrieved from http://www.insidepolitics.org/egovt06int. pdf
5. 2000Health_Information_Report.pdf (2000) Retrieved from http://www.doh.gov.tw/ufile/doc/2000Health_Information_Report.pdf
6. 2005Health_Information_Report.pdf (2005) Retrieved from http://www.doh.gov.tw/ufile/doc/2005Health_Information_Report.pdf
7. Fishbein M (ed) (1967) Readings in attitude theory and measurement. Wiley, New York
8. Ajzen I (1988) Attitudes, personality and behavior. The Dorsey Press, Chicago
9. Fishbein M, Ajzen I (1975) Beliefs, attitude, intention, and behavior: an introduction to theory and research. Addison-Wesley, Cambridge, MA
10. Ajzen I (2001) Nature and operation of attitudes. Annu Rev Psychol 52:27–58
11. Sheppard BH, Hartwick J, Warshaw PR (1988) The theory of reasoned action: a meta-analysis of past research with recommendations for modifications and future research. J Consum Res 15:325–343
12. Ajzen I (1991) The theory of planned behaviour. Organ Behav Hum Decis Process 50:179–211
13. Ajzen I, Driver BL (1991) Prediction of leisure participation from behavioral, normative, and control beliefs: an application of the theory of planned behaviour. Leisure Sci 13:185–204
14. Ajzen I, Madden TJ (1986) Prediction of goal-directed behavior: attitudes, intentions, and perceived behavioral control. J Exp Soc Psychol 22:453–474
15. Fang RJ, Yang HJ, Lin CC, Yang HH, Yu JC (2007) A comparison of applying the theory of reasoned action and the theory of planned behavior on predicting on-line KT intention. WSEAS Trans Math 6(2):432–438
16. Ajzen I, Fishbein M (1980) Understanding attitudes and predicting social behavior. Prentice Hall, Englewood Cliffs, NJ
17. Nunnally JC (1978) Psychometric theory. McGraw-Hill, New York

Chapter 70
Visualization and Clustering of DNA Sequences

Krista Rizman Žalik

Abstract Visual inspection can discover patterns that are quite difficult to discover with computers. We introduce a novel encoding algorithm providing graphical representation of DNA in order to characterize and compare long DNA sequences. The algorithm transforms a DNA sequence into four encoded sequences of natural numbers describing all the distances between each two the same and nearest nucleotides in a DNA sequence. Encoded DNA sequences of natural numbers are simpler for computer handling and numerical analysis, and can be employed for 2D graphical representations that offer visual analysis. This novel representation for processing DNA sequences is demonstrated using the coding sequences of the β-globin gene's exon I for eleven different species. Similarity analysis was done by means of 1D and 2D representation and numerical analysis with similarity/dissimilarity measures used for clustering.

70.1 Visualization of DNA Sequences

Visual data exploration is very useful when little is known about the data. The cluster analysis of 2D data by visual inspection is usually much faster and simpler than computational methods. The similarity analysis of visualized data uses human perception and brains to simultaneously process and compare more information. Visual data exploration and visual clustering have the following advantages over automatic data mining techniques, from statistics to clustering techniques:

- faster data exploration;
- good results, providing a higher degree of confidence;
- easy clustering of inhomogeneous and noisy data;
- understanding of complex mathematical or statistical algorithms is not required.

K. Rizman Žalik (✉)
Faculty of Natural Sciences and Mathematics, University of Maribor, Slovenia

N. Mastorakis et al. (eds.), *Proceedings of the European
Computing Conference*, Lecture Notes in Electrical Engineering 27,
DOI 10.1007/978-0-387-84814-3_70, © Springer Science+Business Media, LLC 2009

Because of the greater advantages of visual data mining, we visualize DNA sequences and their nucleotide content, and then perform similarity analysis and clustering from the graphical characteristics of the visualized DNA sequences. The algorithm considers the sequence as four distributions of distances between occurrences of one nucleotide (A, G, C, or T) in the DNA sequence and so takes into account also how well nucleotides are distributed in the sequence.

Similarity analysis of DNA sequences estimates the degree of similarity between finite sets of nucleic base strings (A, C, T, G). Similarity analysis by string comparison techniques is not suitable for use with DNA data, because DNA primary sequences consisting of only four bases vary in length. They can consist of over a hundred thousand bases or fewer than a hundred bases. Long sequences can be broken down into segments corresponding to exons or introns, but the segments belonging to different species may have different lengths. In Table 70.1 we show the first exon of beta-globin genes belonging

Table 70.1 The DNA sequences for exon I of the beta-globin gene for 11 species

Human 92:	ATGGTGCACCTGACTCCTGAGGAGAAGTCTGCCGTTACTGCCC TGTGG GGCAAGGTGAACGTGGATGAAGTTGGTGGTGAGGC CCTGGGCAG
Goat 81:	ATGGTGACTGCTGAGGAGAAGGCTGCCTGTCACCGGCTTCTGG GG CAAGGTGAAATGGATGTTGTCTGAGGCCCTGG GCAG
Opossum 92:	ATGGTGCACTTGACTTCTGAGGAGAAGAACTGCATCACTACCA TCTGGTCTAAG GTGCAGGTTGACCAGACTGGTGGTGAGGCC CTTGGCAG
Gallus 92:	ATGGTGCACTGGACTGCTGAGGAGAAGCAGCTCATCACCGGCC TCTGGGGCAAGGTCA ATGTGGCCGAATGTGGGGCCGAAGCC CTGGCCAG
Lemur 90:	ATGACTTTGCTGGTGCTGAGGAGAATGCTCATGTCACCCTCTGT GG GGCAAGGTGGATGTAGAGAAAGTTGGTGGCGAGGCCTT GGGCAG
Mouse 92:	ATGGTGCACCTGACTGATGCTGAGAAGGCTGCTGTCTCTTGCC TGTGGGGAAAGGTGAACTCCGATGAAGTTGGTGGTGAGGC CCTGGGCAGG
Rabbit 90:	ATGGTGCATCTGTCCAGTGAGGAGAAGTCTGCGGTCACTGCCC TGTGG GGCAAGGTGAATGTGGAAGAAGTTGGTGGTGAGGC CCTGGGC
Rat 92:	ATGGTGCACCTAACTGATGCTGAGAAGGCTACTGTTAGTGGCC TGTGGGGAAAGGTGAACCCTGATAATGTTGGCGCTGAGGCC CTGGGCAG
Gorilla 94:	ATGGTGCACCTGACTCCTGAGGAGAAGTCTGCCGTTACTGCCC TGTGGGGCAAGGTGAACGTGGATGAAGTTGGTGGTGAGGCC CTGGGCAGGA
Bovine 86:	ATGCTGACTGCTGAGGAGAAGGCTGCCGTCACCGCCTTTTGGG GC AAGGTGAAAGTGGATGAAGTTGGTGGTGAGGCCCTGGG CAG 86
Chimpanzee 105:	ATGGTGCACCTGACTCCTGAGGAGAAGTCTGCCGTTACTGCCC TGTGG GGCAAGGTGAACGTGGATGAAGTTGGTGGTGAGGC CCTGGG CAGGTTGGTATCAAGG

to eleven species, differing in length from 86 to 93 bases. The search for optimum correlation between sequences is rather difficult, and requires different operations, such as tracing (matching the same elements in the two sequences), or finding the smallest number of changes (deletions, insertions) that are necessary to match labels in two sequences [1].

Although directly comparing the sequences of four nucleic acid bases A, C, G, and T using computer programs is less straightforward because the sequences have different lengths, graphical representations of DNA sequences provide efficient simple comparison and visual inspection of data. They help when recognizing differences in similar DNA sequences, and finding similar DNA sequences or groups of similar DNA sequences.

The representation of large DNA sequences in low-dimensional graphs makes the data useful for biologists when making simple comparisons, similarity analyses, and clustering of DNA sequences. Many 1D, 2D, 3D, and even 4D graphical representations have been proposed to characterize DNA sequences. The reasons are twofold. DNA sequences, as strings of four nucleic acid bases, A, C, G, and T, do not provide useful or informative characterization. The second reason is the possibility of usefully extracting hidden information from complex data represented by low-dimensional graphical representations. 1D visualization of sequences by means of different colors was proposed ten years ago using the DNA view method [2]. In this method, a color code is used to represent DNA sequences. It uses color intensity proportional to the number of times the base is sequenced.

About twenty years ago, Hamori first used a three-dimensional H curve to represent a DNA sequence [3]. Gates [4] proposed a two-dimensional graphical representation that is simpler than the H curve degeneracy. The vectors representing A, G, C, and T were $(0,\pm1)$, $(1,0)$, $(\pm1,0)$, and $(0,1)$, respectively. However, Gates's graphical representation contains high degeneracy. For example, the sequences TGAC, TGACT, TGACT, etc., have the same graphical representation. In mathematical terms, sequence degeneracy is a result of repetitive closed loops or cycles in a DNA graph. In order to reduce this, Guo et al. [5] improved the representation by modifying the directions of the vectors assigned to the four bases.

High complexity and degeneracy have been major problems in many previous DNA graphical representations [5–7], limiting the application of DNA graphs. We view the DNA sequences as four sequences of natural numbers, representing the distances between pairs of the same nucleic acid bases, that offer simple representation by providing good visual analysis and clustering. Encoding must retain as much of the original information as possible. The suggested encoding of DNA sequences into sequences of natural numbers retains all the information. The results can be 1D or 2D graphical representations and possibly more, but graphs in three or more dimensions are less efficient for visual analysis. We suggest graphical representations in 1D with four colors in a pie chart, and in 2D with curves. The encoded DNA sequences can also be used for numerical characterization of DNA sequences. Finally, we obtain the results of examining similarities/dissimilarities among the coding sequences of the exon I of the beta-globin gene for 11 different species, each string consisting of about 100 basic amino acids, as suggested in the scheme.

70.2 A Visualization Algorithm

We suggest an algorithm that encodes and then visualizes four sequences of distances between the same bases in the original DNA sequence. For two equal bases (i.e., AA, GG, CC, or TT) near each other, we say that the distance between them is one. If there is another base between them, then the distance between these same bases is two. The distance of the first base in a sequence is computed as the distance from the beginning of the sequence. The length of a sequence is equal to the number of occurrences of the base in the DNA sequence.

The visualization algorithm uses the following recursion formula for calculating the $(i+1)^{th}$ point of the encoded sequence:

$$x_{(i+1,base1)} = x_{(i,base1)} + a \ distances_{(i,base1)} + b; x_{(0,base1)} = 0 \qquad (70.1)$$

where $x_{(i+1, base1)}$ is the $(i+1)^{th}$ element of the sequence of base *base1*. The ith unit of the distance sequence for base *base1* is written as distances (i, base1), with a and b as two coefficients. For a = 8 and b = 0, we get the following recursion function:

$$x_{(i+1,base1)} = x_{(i,base1)} + 8 \ distances_{(i,base1)} \qquad (70.2)$$

For the exon I of the beta-globin gene of a human (ATGGTGCACCTGAC TCCTGAGGAGAAGTCTGCCGTTACTGCCCTGTGGGGCAAGGTGAA CGTGGATGAAGTTGGTGGTGAGGCCCTGGGCAG), we get the distance sequence for base A: 1 7 5 7 3 2 1 11 15 1 5 1 6 3 1 11 11... and the encoded sequence using the encoding function (Eq. 70.2) for base A: 8 56 40 56 24...

70.2.1 1D Graphical Representation of DNA Sequences

If in the upper recursion formula *a* is the arc angle and *b* is the start angle and *b* = 0, then we can represent this sequence by a filled-in arch with the start angle being the sequence number and the arc angle being 8 degrees. One-dimensional graphical representations with colors, which form some kind of picture, can provide much more efficient representations as sequences of four characters denoting all four bases. We represent the DNA sequence by pie charts using one color for each of the four DNA bases. Figure 70.1 shows a 1D representation of an encoded sequence for humanin using encoding functions (Eq. 70.2).

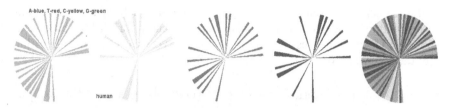

Fig. 70.1 Representation of DNA sequences of the exon I of the beta-globin gene by pie charts using encoding functions (Eq. 70.2). First green pie chart denotes spread of nucleotide G, second yellow C, third red T, and fourth blue shows the appearance of nucleotide A. The fifth pie chart shows the apperance of all four nucleotides

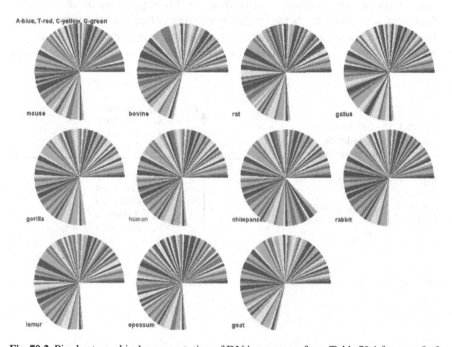

Fig. 70.2 Pie chart graphical representation of DNA sequences from Table 70.1 for exon I of the beta-globin gene for 11 species

Figure 70.2 shows pie chart graphical representations of DNA sequences from Table 70.1 for exon I of the beta-globin gene for 11 species using encoding functions (Eq. 70.2).

70.2.2 2D Representation of Encoded DNA Sequences by Base-Curves

For each of the four bases in the DNA sequences we obtain one encoded sequence, which can be represented graphically by curves to provide visual analysis. Each DNA sequence can be represented uniquely by four 2D graphs corresponding to the four encoded distance sequences for all four bases that appear in DNA sequences. We call them base curves. The encoding algorithm uses the following formula:

$$y_{(i+1,base)} = y_{(i,base)} + 2 \text{ distances}_{(I,base)} - 12; y_{(0,base1)} = 0;$$
$$x_{(i+1,base)} = x_{(i,base)} + 10; x_{(0,base1)} = 0; \tag{70.3}$$

It defines the unit vectors for different distances, as represented in Fig. 70.3. Figure 70.4 shows base curves for the eleven species from Table 70.1. The base curves of DNA sequences introduced in this paper have no degeneracy. They represent distances between the appearances of each of the four bases in the four base graphs and, therefore, provide a simple method for both computational scientists and molecular biologists when analyzing DNA sequences

Fig. 70.3 The unit vectors in the Cartesian coordinate plane for distances among bases. The vectors $X_1(10,-10)$, $X_2(10,-8)$, $X_3(10,-6)$, $X_4(10,-4)$, $X_5(10,-2)$, $X_6(10,0)$, ... are used to represent distances: 1, 2, 3, 4, ..., m

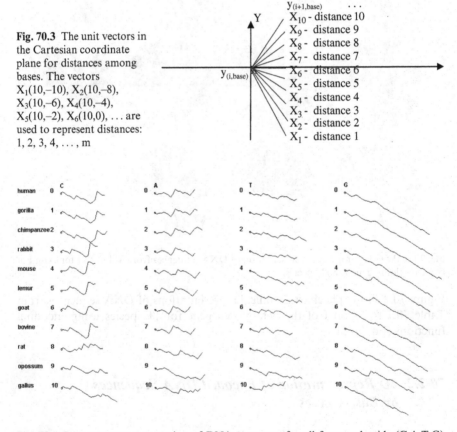

Fig. 70.4 Base curves representation of DNA sequences for all four nucleotides(C,A,T,G) from Table 70.1 for exon I of the beta-globin gene for 11 species

efficiently. When constructing a base curve, it can be seen that no circles can appear.

To prove this, let us denote the vectors used to represent distances as X_1, X_2, X_3, X_4, X_5, ..., X_m, as shown in Fig. 70.3. Let the number of X_1, X_2, X_3, X_4, X_5, ..., X_m vectors be x_1, x_2, x_3, x_4, x_5, ..., x_m respectively. The number of vectors can be null or a positive integer. If n is the length of the DNA sequence, then it follows that:

$$x_1 + 2x_2 + 3x_3 + 4x_4 + 5x_5 + ... + mx_m < = n0 < = x_1, x_2, x_3... < n \quad (70.4a)$$

Because X_1x_1, X_2x_2, X_3x_3, ... form a circuit, the following equations hold:

$$X_1x_1 + X_2x_2 + X_3x_3 + X_4x_4 + X_5x_5 + ... + X_mx_m = 0$$
$$x_1(10, - 10) + x_2(10, - 8) + x_3(10, - 6) + x_4(10, - 4) + ... = 0 \quad (70.4b)$$

$$- 10x_1 - 8x_2 - 6x_3 - 4x_4 + ... = 0 \quad (70.4c)$$

$$x_1 + x_2 + x_3 + x_4 + x_5 + ... + x_m = 0 \quad (70.4d)$$

Equations (70.4c) and (70.4d) hold if, and only if, $x_1 = x_2 = x_3 = x_4 = x_5 = ... = x_m = 0$.

Therefore, from Eq. (70.4a) n = 0, which means that either there can be a circuit for the length of sequence 0, or no circuit exists in this graphical representation. Base curves for 2D graphical representations of DNA sequences have no degeneracy because there is no overlapping and/or crossing itself by the curve representing DNA.

70.3 Examination of Similarities Among Encoded DNA Sequences

We used the suggested encoding sequences for graphical representation and visual investigation of similarities and dissimilarities of 11 species, as shown in Table 70.1. The underlying assumption is that if numbers in two sequences are similar, or if points forming two curves are spread similarly for all four bases, then the two DNA sequences represented by all four base curves are similar. Observing base curves of DNA sequences for all four nucleotides (C, A, T, G) from Table 70.1 for exon I of the beta-globin gene for 11 species in Fig. 70.4 shows that each species having one similar base curve also has others similar.

Encoded sequences can be easily saved and handled by computers. They can be used for computing similarity. Encoded sequences of distances can contain different numbers of elements n_{base} for different bases of the same DNA sequence. Therefore, any dissimilarity D(DNA1, DNA2) between two DNA sequences DNA1 and DNA2 can be examined using the formula below:

$$D(DNA1, DNA2) = \sum_{base=A;C;G;T} \sum_{i=1}^{n_{base}} |seq(i, DNA1, base)$$
$$- seq(i, DNA2, base)| \qquad (70.5)$$

where $seq(i,DNA1,base)$ denotes the ith element of the $DNA1$ encoded sequence for base $base$. Using dissimilarity D as a measure of the distance of two species, k-means can be used for clustering [8]. For our example of 11 species we compared the measure D. The smaller the sum D is, the more similar the two DNA sequences are. Figure 70.5 gives the dissimilarities for 11 coding sequences based on the measure D. From Fig. 70.5, we find that the most similar species are human-gorilla, human-chimpanzee, and gorilla-chimpanzee; while other species show considerable dissimilarity with the other eleven species. We also looked for groups of species having the same dissimilarity with all other species. Given a set of n species with a matrix of pair-wise similarity measures, we partitioned the species into clusters so that similar species were together and different ones apart. We used an upper dissimilarity matrix (Fig. 70.5) which is based on measure D

D	human	gorilla	chimp.	mouse	rabbit	lemur	goat	bovie	rat	gallus	opossum
human	0	5.13	30.1	36.6	25.4	35.6	44.6	28.0	44.3	42.8	71.2
gorilla		0	24.9	41.8	20.7	40.7	49.7	33.1	49.5	46.9	76.3
chipanze			0	62.1	45.7	62.9	74.7	58.1	73.3	71.9	98.5
mouse				0	35.1	15.4	57.7	44.7	42.8	62.5	76.1
rabbit					0	32.9	43.3	27.5	48.3	64.2	79.4
lemur						0	45.8	32.1	39.2	54.9	65.1
goat							0	26.5	40.5	45.2	65.2
bovie								0	43.2	38.9	77.3
rat									0	49.7	45.7
gallus										0	73.1
opossum											0

D1	human	gorilla	chimp.	mouse	rabbit	lemur	goat	bovie	rat	gallus	opossum
human	0	0.99	29.0	22.6	7.1	19.7	31.1	12.5	32.5	31.5	99.5
gorilla			22.1	25.5	6.8	23.9	35.0	16.0	37.3	34.1	101
chipanze				45.1	31.7	51.1	56.6	46.0	60.0	47.3	91.4
mouse				0	15.2	2.98	29.9	21.4	19.5	35.5	64.3
rabbit					0	12.8	23.8	8.65	28.4	31.2	90.8
lemur						0	22.8	13.6	17.0	31.8	69.6
goat							0	10.6	14.9	16.2	54.7
bovie								0	22.0	19.4	84.9
Rat									0	22.4	37.8
gallus										0	55.2
opossum											0

D+D1	human	gorilla	chimp.	mouse	rabbit	lemur	goat	bovie	rat	gallus	opossum
human	0	3.27	31.6	31.7	17.4	29.6	40.5	21.7	41.1	39.2	91.3
gorilla			25.1	36	14.7	34.6	45.3	26.3	46.4	43.3	95.1
chipanze				57.3	41.4	61.0	70.2	55.7	71.3	63.7	101
mouse					26.9	9.83	46.8	35.3	33.3	52.4	70.3
rabbit						24.4	35.9	19.3	41	45.6	91.0
lemur							36.7	24.4	30.1	46.4	72.0
goat								19.8	29.6	32.9	64.1
bovie									34.8	31.2	86.7
Rat										38.6	44.7
gallus											68.6
opossum											

Fig. 70.5 The dissimilarity matrix for the 11 coding sequences in Table 70.1 based on the measure of similarity D and D1 and D+D1

(Eq. 70.5) for clustering of species that had the same dissimilarity to all other species. For a symmetric matrix M of pair-wise dissimilarities, let M_k denote the kth row (or column) of M. For each pair of nodes i,j we calculate the difference $D1$, and if difference $||Mi-Mj||^2 <$ MAXDifference, then i and j are in the same cluster.

The algorithm finds any dissimilarity with other sequences by calculating the difference $D1(a,b)$ for each two DNA sequences (species) a and b from n species:

$$D1(a, b) = \sum_{i=1}^{n} |D(a, i) - D(b, i)|^2 \tag{70.6}$$

Clusters are obtained by upper formulas using the encoded sequences without any preprocessing like normalization, scaling, shifting, or further transformation. Although the distance sequences of each base for first axons of different species are of different length, we did not attempt to cut them to the uniform length but simply used the zero fill-up points for all of them. The three obtained clusters from dissimilarity measure D (Eq. 70.5) between two DNA sequences D are:

D/Dmax	cluster
1	human,gorilla
30	chimpanze
40	rabit

20	lemur, mouse

26	bovine, goat

>40   opossum, gallus, rat

The two obtained clusters from dissimilarity to all other DNA sequences D1 (Eq. 70.6) are:

D/Dmax	cluster
	human,gorilla
30	chimpanze

3	lemur, mouse
30	bovine, goat

>40   opossum, galus, rat

The three obtained clusters from the sum of both measures for dissimilarity D1 (Eq. 70.5) and D2 (Eq. 70.6) are:

D1+D2/Dmax

human,gorilla	
30	chimpanze

10	lemur, mouse

20	bovine, goat

> 40   opossum, gallus, rat

Human is in the same cluster as gorilla, chimpanzee, and rabbit, when using dissimilarity measure D. The second cluster contains bovine, goat, lemur, and mouse. Opossum, gallus, and rat show great dissimilarities with others among the eleven species. We can see that all measures give very similar results.

70.4 Conclusion

We show that pie charts and base curves based on encoded sequences are efficient for visual similarity analysis. Encoded DNA sequences are usable also for numerical analysis and computer handling and processing. The important conclusion we can draw is that the four generated sequences captured all the important features of the DNA sequences and are simpler for computer handling and numerical analysis. We used them for clustering. The simple algorithm for encoding DNA sequences proved to be very useful.

References

1. Kruskal J (1983) SIAM Rev 25:201
2. Singh G, Band KS (1995) DNA view: a quality assessment tool for the visualization of large sequenced regions. Comput Appl Biosci 11:317–319
3. Hamori E, Ruskin JH (1983) H curves, a novel method of representation of nucleotide series especially suited for long DNA sequences. J Biol Chem 258:1318–1327
4. Gates MA (1986) A simple way to look at DNA. J Theor Biol 119:319–328
5. Guo X, Nandy A (2003) Numerical characterization of DNA sequences in a 2-D graphical representation scheme of low degeneracy. Chem Phys Lett 361–366
7. Nandy A (1986) Two-dimensional graphical representation of DNA sequences and intron-exon discrimination in intron-rich sequences. Comput Appl Biosci 12:55–62
8. Rizman ŽK (2008) An efficient k-means clustering algorithm. Pattern Recogn Lett

Chapter 71
E-Health Service in Taiwan—The Role of Organizational Innovativeness

Hsieh-Hua Yang, Jui-Chen Yu, and Hung-Jen Yang

Abstract In Taiwan, the electronic information and online services provided by the health bureaus and the district health centers are still in their infancy. We argue that the inclusion of personal innovativeness and organizational innovativeness with respect to information technology would help us to further understand how perceptions are formed and the subsequent role they play in the formation of usage attitude. In this study, two comparable research models are tested. Findings reveal that compatibility has an effect on both perceived usefulness and perceived ease of use; and that personal innovativeness has a direct effect on perceived ease of use, but not on perceived usefulness; while organizational innovativeness has a direct effect on attitude. Some theoretical and practical implications are discussed.

71.1 Introduction

Because information technology is rapidly developing, and governments are making huge investments in electronic services, there is a variety of systems designed to provide innovative services. In Taiwan, the Department of Health has continued its efforts to computerize health administration operations. Taiwan's health web site, www.doh.gov.tw, was initiated in 1995. The Department of Health's Convenient Internet Services Plan was initiated in 2004 [10]. The current 25 health bureaus and 373 health stations are the medical and health service providers closest to citizens. They are also the places that people can experience the quality of government service with regard to medicine and health. But the electronic information and online services provided by the health bureaus and the district health centers are still in their infancy [11].

H.-H. Yang (✉)
Department of Health Care Administration, Oriental Institute of Technology,
58, sec. 2, Szechwan Rd., Banciao City, Taipei Country 220, Taiwan
e-mail: yansnow@gmail.com

N. Mastorakis et al. (eds.), *Proceedings of the European Computing Conference*, Lecture Notes in Electrical Engineering 27, DOI 10.1007/978-0-387-84814-3_71, © Springer Science+Business Media, LLC 2009

The adoption of innovative services needs innovative characteristics in individuals and an innovative climate in organizations. Studies pertaining to technological adoption have supported the importance of individual innovativeness; little attention has been paid to the organizational climate of innovation. We argue that the inclusion of personal innovativeness and organizational innovativeness, with respect to the e-health services system, would help us to further understand how perceptions are formed and the subsequent role they play in the formation of usage attitude. For the employees of the health bureaus and health stations, offering the e-health services system is equivalent to offering primary healthcare services. In this context, an essential acceptance criterion is whether or not the system fills the needs of individuals and provides the desired utility to their work. Thus compatibility was incorporated as an external variable.

71.2 Literature Review

The technology acceptance model (TAM) has garnered significant empirical support [8, 20] and has been appraised as a powerful and robust predictive model [14]. Research on technology adoption has shown that attitude becomes increasingly important, over time and with continuing experience, in determining technology usage behavior [5, 13, 20, 21]. These studies used attitude as the dependent variable, and tried to identify the direct and indirect effects of external variables. In this study, compatibility, personal innovativeness, and organizational innovativeness are incorporated as external variables.

71.2.1 Compatibility

Chau and Hu [5] included compatibility in their decomposed theory of planned behavior as an external variable to perceived usefulness and perceived ease of use, and the effects of compatibility were found to be significant only in relation to perceived usefulness [5]. In contrast, Agarwal et al. [3] found that relevant prior experience, mediated by general self-efficacy, affected perceived ease of use [3]. The role of experience in the technology acceptance model has sometimes been hypothesized as a moderating variable.

71.2.2 Personal Innovativeness

Personal innovativeness as a construct that is important to the study of individual behavior toward innovations has had a long-standing tradition in innovation diffusion research [19]. Agarwal and Prasad [2] adapted the concept to the domain of information technology and proposed a new instrument to measure personal innovativeness, which describes the extent to which the individual has an innate propensity toward adopting a new IT. Lewis et al. [16] found that personal innovativeness was a significant determinant of perceived ease of use. Yi et al. (2006) indicated that personal innovativeness is a distal determinant, achieving its influence indirectly through mediators such as perceived ease of use, subjective norm, and perceived behavioral control.

71.2.3 Organizational Innovativeness

Organizational variables were found to be particularly important in predicting technological innovation, especially innovations which produce changes in products and services [7]. Receptiveness of organizations to change has been found to be a significant factor for achieving success in technical innovations [24]. As TAM assumes that external variables predict usage only through their effect on perceived usefulness and perceived ease of use, organizational innovativeness has an indirect effect on attitude through perceived usefulness and perceived ease of use. The tested model is presented as Fig. 71.1.

However, Legris et al. [15] and Burton-Jones and Hubona [4] argued that external variables should be studied more systematically. Pijpers and Montfort (2006) categorized external variables into individual, organizational, task related, and IT resource characteristics to test their effect, and found variables that directly influence attitude. We argue that organizational innovativeness has a direct effect on attitude. As shown in Fig. 71.2, a direct relation should exist between organizational innovativeness and attitude.

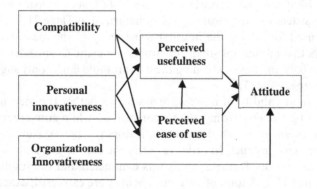

Fig. 71.1 Research model 1

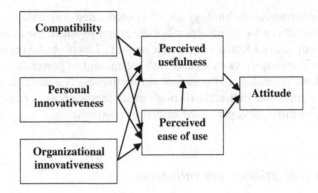

Fig. 71.2 Research model 2

71.3 Method

The system chosen as the innovation to be studied is appropriate, because it is still in the early stage. The 105 subjects in this study were employees of one health bureau and three health stations. Their mission is providing primary healthcare for the public.

Most of the constructs according to the technology acceptance model were measured with items adapted from prior research [8, 12, 22]. Multiple item scales were developed for each of the constructs shown in Table 71.1. All items were measured using a five-point Likert-type scale with anchors ranging from "strongly disagree" to "strongly agree." Attitude was assessed with two items, perceived usefulness with four items, perceived ease of use with three items, and compatibility with three items. Measurement of personal innovativeness was adapted from Agarwal and Prasad [2] and Hung et al. [12]. And measurement of organizational innovativeness was adapted from the Values in Organizational Culture scale [7].

The reliability analysis verified the precision of the survey instrument and the internal consistency of the measure. As summarized in Table 71.1, Cronbach's alpha was used for testing the internal consistency of the measurement. The result shows that all multiple-item constructs achieving Cronbach's alpha ranged between 0.79 and 0.98, exceeding the 0.7 threshold commonly suggested for exploratory research [17].

Discriminant validity of a measure assesses whether the measure is adequately distinguishable from related constructs, while convergent validity assesses the extent to which different indicators for the measure refer to the same conceptual construct. In order to verify the convergent and discriminant validity, confirmatory factor analysis was conducted and the results are presented in Table 71.2. A total of six components were extracted, exactly matching the number of constructs specified by the model. The items for measuring a construct exhibited distinctly higher factor loadings on a single component than

Table 71.1 Reliability analysis using Cronbach's alpha values

Construct and item	Mean	SD	α
Attitude			0.86
1 Using this system would be a good idea.	3.43	0.89	
2 Using this system would be a correct behavior.	3.34	0.91	
Perceived usefulness			0.93
1 Using this system would improve my performance.	3.53	0.95	
2 Using this system would improve my productivity.	3.37	0.98	
3 Using this system would enhance my effectiveness.	3.57	0.93	
4 Using this system would be useful to my work.	3.58	0.82	
Perceived ease of use			0.91
1 Learning to use this system would be easy for me.	3.39	0.99	
2 It would be easy for me to become skillful at using this system.	3.42	0.92	
3 I would find this system easy to use.	3.30	0.95	
Compatibility			0.91
1 Using this system will fit well with my work.	3.35	0.92	
2 The report form supported by this system will be compatible with my work.	3.32	1.01	
3 The setup of this system will be compatible with the way I work.	3.27	0.96	
Personal innovativeness			0.79
1 I am challenged by ambiguities and unsolved problems.	3.38	0.94	
2 I would enjoy myself doing original thinking and behavior.	3.64	0.80	
3 I find it stimulating to be original in my thinking.	3.57	0.86	
Organizational innovativeness			0.98
1 The service supplied by the organization is innovative.	3.74	0.65	
2 The organization is improving constantly.	3.78	0.65	
3 The organization is making breakthroughs by educating itself.	3.67	0.72	
4 The organization takes the leading role.	3.76	0.73	
5 The organization is seeking improvement.	3.72	0.73	
6 The organization has burning desire to struggle for victory.	3.69	0.70	
7 The organization supplies us an opportunity for growth and development.	3.75	0.69	

on other components. The observed loadings suggested that our instrument had encompassed satisfactory convergent and discriminant validity. The validated instrument was used in this study.

71.4 Results

Stepwise regression analyses were performed in order to determine the significance and strength of each effect posited. The results of the regression analysis are presented in Table 71.3. In testing mode 1, three steps were involved. First, stepwise regression was conducted for perceived usefulness and perceived ease of use against attitude. The R square value for this analysis is 0.416, and both variables have significant effects on attitude. Second, perceived ease of use,

Table 71.2 Examination of convergent/discriminant validity

Factor 1	Factor 2	Factor 3	Factor 4	Factor 5	Factor 6
Organizational innovativeness					
0.94	–	–	–	–	–
0.93	–	–	–	–	–
0.92	–	–	–	–	–
0.87	–	–	–	–	–
0.95	–	–	–	–	–
0.94	–	–	–	–	–
0.91	–	–	–	–	–
Perceived usefulness					
–	0.75	–	–	–	–
–	0.88	–	–	–	–
–	0.79	–	–	–	–
–	0.61	–	–	–	–
Perceived ease of use					
–	–	0.74	–	–	–
–	–	0.75	–	–	–
–	–	0.72	–	–	–
Personal innovativeness					
–	–	–	0.74	–	–
–	–	–	0.80	–	–
–	–	–	0.80	–	–
Compatibility					
–	–	–	–	0.60	–
–	–	–	–	0.87	–
–	–	–	–	0.74	–
Attitude					
–	–	–	–	–	0.76
–	–	–	–	–	0.84
Eigen value					
6.19	3.25	2.65	2.41	2.22	1.96
% of variance					
28.12	14.76	12.03	10.96	10.10	8.90
Cumulative % of variance					
28.12	42.88	54.91	65.87	75.98	84.88

compatibility, personal innovativeness, and organizational innovativeness were regressed against perceived usefulness. The R square value for this analysis is 0.618. Perceived ease of use and compatibility have significant effects on perceived usefulness. Third, compatibility, personal innovativeness, and organizational innovativeness were regressed against perceived ease of use. The R square value for this analysis is 0.595. Compatibility and personal innovativeness have significant effects on perceived ease of use.

Similarly, three steps were involved in testing model 2. First, stepwise regression was conducted for perceived usefulness, perceived ease of use, and

Table 71.3 Results of the regression analysis

	Model 1		Model 2	
	B	SE	B	SE
PU, PEOU, and OI against attitude				
Constant	2.137	0.565	0.496	0.825
PU	0.153	0.058*	0.146	0.056*
PEOU	0.246	0.074**	0.226	0.072**
OI			0.074	0.028**
R2	0.416		0.455	
R2 change			0.039**	
PEOU, Com, PI, and OI against PU				
Constant	3.092	1.467	2.886	1.063
PEOU	0.621	0.123***	0.621	0.123***
Com	0.396	0.119**	0.392	0.116**
PI	0.095	0.118	0.095	0.118
OI	−0.010	0.047		
R2	0.618		0.618	
R2 change			0.000	
Com, PI, and OI against PEOU				
Constant	0.748	1.180	0.802	0.853
Com	0.587	0.076***	0.588	0.073***
PI	0.326	0.089***	0.327	0.089***
OI	0.003	0.038		
R2	0.595		0.595	
R2 change			0.000	

*$p<0.05$ **$p<0.01$ ***$p<0.001$
A: attitude
PU: perceived usefulness
PEOU: perceived ease of use
Com: compatibility
PI: personal innovativeness
OI: organizational innovativeness

organizational innovativeness against attitude. The R square value for this analysis is 0.455 and all the three variables have significant effects on attitude. Compared with model 1, the improvement of the model's prediction of attitude is 0.039, significant at the .01 level. Obviously, organizational innovativeness improves the model's explanatory power. Second, perceived ease of use, compatibility, and personal innovativeness were regressed against perceived usefulness. The R square value for this analysis is 0.618, the same as model 1. Thirdly, compatibility and personal innovativeness were regressed against perceived ease of use. The R square value for this analysis is the same as model 1, 0.595. Clearly, organizational innovativeness did not have effect on perceived usefulness and perceived ease of use.

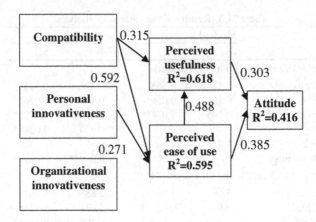

Fig. 71.3 Research result of model 1

Figures 71.3 and 71.4 portray the results of research models 1 and 2, respectively. The extended technology acceptance model incorporating the constructs of personal innovativeness and organizational innovativeness is proven to be relevant and applicable to the usage of the e-health services system. In both models, compatibility has a direct and strong influence on perceived usefulness and perceived ease of use; while personal innovativeness has a direct influence on perceived ease of use, and an indirect influence on perceived usefulness through perceived ease of use. The direct effect on attitude displayed by organizational innovativeness is supported in model 2. This means that organizational innovativeness has a direct effect on attitude jointly with perceived usefulness and perceived ease of use.

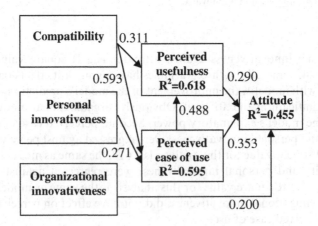

Fig. 71.4 Research result of model 2

71.5 Discussion and Conclusion

The objective of this study is testing the effects of compatibility, personal innovativeness, and organizational innovativeness on usage attitudes toward the e-health services system. The results provide support for the core concepts of the TAM model. Compatibility and personal innovativeness have effects on attitude via perceived usefulness and perceived ease of use. Nevertheless, the results support the idea that organizational innovativeness influences attitude directly.

Ssignificant evidence was found for the direct role of organizational innovativeness in building attitude. For successful adoption of the e-health services system, the climate within the health bureaus and health stations must be conducive to innovation. The organizational leaders' values, roles, and personalities are key factors in the adoption of technological innovations [7]. The involvement of the leaders should be encouraged, in order to foster the climate of innovation.

Within the context of the health bureaus and health stations, the acceptance of the e-health services system needs a process of implementation. During the implementation process, employees have to manage the original task and the new system at the same time. If the system is more compatible with the work, the employees will believe the system is useful and easy to implement, and a more innovative employee is more likely to perceive the new system as less difficult to implement.

As TAM proposes, both perceived usefulness and perceived ease of use are important in technology acceptance. However, Davis [9] found that usefulness dominated ease of use, whereas Adam et al. [1] found ease of use to be more influential than usefulness. The results showed that perceived ease of use is more important than perceived usefulness, contrary to the former and compatible with the latter study. Chau and Hu [5] indicated that professionals are found to be pragmatic, concentrating more on the usefulness of a technology and considering technology practice compatibility to be crucial. During the implementation process of e-health services systems, the quantity of tasks will be increased. For practical consideration, it is suggested that the reduction in effort is more important than usefulness during the adopting process. It is critical to examine whether the system is compatible and easy enough to allow employees to accomplish their tasks.

References

1. Adams DA, Nelson RR, Todd PA (1992) Perceived usefulness, ease of use, and usage of information technology: a replication. MIS Q 16(2):227–247
2. Agarwal R., Prasad J (1998) A conceptual and operational definition of personal innovativeness in the domain of information technology. Inform Syst Res 9(2):204–215
3. Agarwal R, Sambamurthy V, Stair RM (2000) Research report: the evolving relationship between general and specific computer self-efficacy—an empirical assessment. Inform Syst Res 11(4):418–430

4. Burton-Jones A, Hubona GS (2006) The mediation of external variables in the technology acceptance model. Inform Manag 43:706–717
5. Chau PYK, Hu PJH (2001) Information technology acceptance. Decis Sci 32(4):699–719
6. Cheng B-S (1990) A quantitive assessment of value in organizational culture. Chin J Psychol 32:31–49
7. Damanpour F (1987) The adoption of technological, administrative, and ancillary innovations: impact of organizational factors. J Manag 13(4):673–688
8. Davis FD, Bagozzi RP, Warshaw PR (1989) User acceptance of computer technology: a comparison of two theoretical models. Manag Sci 35(8):982–1003
9. Davis FD (1993) User acceptance of information technology: system characteristics, user perceptions and behavioral impacts. Int J Man-Mach Studies 38:475–487
10. Health Department Online Public Service Plan—2004 Report. Retrieved from www.doh.gov.tw
11. 2005Health_Information_Report.pdf. Retrieved from www.doh.gov.tw
12. Hung S-Y, Chang C-M, Yu T-J (2006) Determinants of user acceptance of the e-government services: the case of online TAC filing and payment system. Gov Inform Q 23:97–122
13. Karahanna E, Straub DW, Chervany NL (1999) Information technology adoption across time: a cross-sectional comparison of pre-adoption and post-adoption beliefs. MIS Q 23(2):183–213
14. King WR, He J (2006) A meta-analysis of the technology acceptance model. Inform Manag 43:740–755
15. Legris P, Ingham J, Collerette P (2003) Why do people use information technology? A critical review of the technology acceptance model. Inform Manag 40(3):191–204
16. Lewis W, Agarwal R, Sambamurthy V (2003) Sources of influence on beliefs about information technology use: an empirical study of knowledge workers. MIS Q 27(4):657–678
17. Nunnally JC (1978) Psychometric theory. McGraw-Hill, New York

Chapter 72
Analysis of Heart Sounds with Wavelet Entropy

S. Bunluechokchai and P. Tosaranon

Abstract It has been reported that heart sound detection is very useful in the diagnostic process of heart diseases, especially in patients with abnormality of the cardiac valves. Heart sounds originate from blood flow through the cardiac valves during heart contraction and relaxation, and they can be heard and recorded in the patient's chest. In this study, two groups of patients, those with normal hearts and those with aortic stenosis, were investigated. This paper presents the application of wavelet transform analysis to the heart sound signals of both groups. The continuous wavelet transform plot shows that the patients with aortic stenosis are likely to have irregular patterns of heart sounds in a time-scale representation, whereas normal heart sounds are likely to have relatively smooth surfaces. The wavelet entropy is then applied to the heart sounds. A disordered signal gives a high entropy value. Observations of the preliminary results in this study found that the heart sounds from patients with aortic stenosis tend to have higher wavelet entropy than those from patients with normal hearts.

72.1 Introduction

Heart sounds are generated from the mechanical functions of the heart, blood flow, and valve movement. They occur during each cardiac cycle. Although a new method of echocardiography for heart examination is now available, cardiac auscultation is still an important screening diagnostic method for early diagnosis of heart diseases. The necessity of early detection is to give a warning sign for further investigation before the serious pathological conditions of heart disease occur. In addition, detection of heart sounds can be performed by a technique known as phonocardiography. This technique has

S. Bunluechokchai (✉)
Department of Industrial Physics and Medical Instrumentation, Faculty of Applied Science, King Mongkut's Institute of Technology North Bangkok, 1518 Pibulsongkram Road, Bangsue, Bangkok 10800, Thailand

N. Mastorakis et al. (eds.), *Proceedings of the European Computing Conference*, Lecture Notes in Electrical Engineering 27, DOI 10.1007/978-0-387-84814-3_72, © Springer Science+Business Media, LLC 2009

many advantages over conventional auscultation due to the limited ability of human hearing and of the skill of physicians. The human ear is poorly suited for cardiac auscultation. Phonocardiography displays a graphical representation of heart sounds. It is easy to use and noninvasive. It provides diagnostic information for the detection of abnormal function of the cardiac valves in clinical practice. The heart sound signal is relatively easy to detect with a conventional sound sensor on the chest. Time-frequency analysis is one of the techniques to investigate nonstationary signals. As heart sounds are typically nonstationary signals, it is necessary to analyze both their time and frequency. Many researchers have studied signal processing techniques for analysis of heart sounds, such as the Wigner-Ville distribution, the short time Fourier transform, the wavelet transform (WT) [1, 4, 5, 7, 9–11], and heart sound segmentation [6, 8]. When compared to the other techniques, the main advantage of the wavelet transform is that it provides better time and frequency resolution for analysis of heart sounds. The normal heart sound signal consists of two main sounds: the first S1, and the second S2. The sound S1 is caused by the sudden closure of the mitral and tricuspid valves during ventricular contraction. The sound S2 is associated with the closure of the aortic and pulmonary valves and marks the end of ventricular systole. The period between S1 and S2 is called systole and the period between S2 and the next S1 is called diastole. Heart murmurs originate from turbulence in the blood flow. The conditions of blood flow turbulence are mainly valvular stenosis and valve insufficiency. Valvular stenosis may be due to calcium deposition preventing the valve from opening properly, thereby causing an obstruction of blood flow. Valve insufficiency occurs when the valve does not close completely, and there is a reverse flow direction or regurgitation of blood flow through a narrow opening. Murmurs are high-frequency, noise-like sounds. Systolic murmurs are caused during ventricular contraction by conditions such as aortic stenosis and mitral insufficiency. Two groups of patients, normal and aortic stenosis patients, were studied in this work. The application of continuous wavelet transform (CWT) analysis for the heart sounds was presented. Next, the computation of the wavelet entropy obtained from the CWT was performed for each patient group. This study investigates the possibility of using wavelet entropy for distinguishing between patients with normal hearts and those with aortic stenosis.

72.2 Materials and Methods

The wavelet transform has been considered a powerful tool of signal processing. It uses the wavelet function to analyze the signal of interest. The heart sound is a time-varying signal and the CWT has the ability to better analyze a nonstationary signal [2, 3]. Because the CWT is computed in terms of scale instead of frequency, the CWT plot of a signal is displayed as the time-scale domain rather than the time-frequency plane. It provides varying time-frequency resolution. A scaling parameter of the CWT is inversely related to the frequency. The selection of the mother wavelet and scaling parameter is crucial for the best detection of the signal. The mother wavelet used in this work is the Morlet wavelet, because a number of researchers have reported that it offers promising results for the analysis of heart sounds [7, 9, 11].

72.2.1 Data Acquisition

The heart sounds were recorded and saved on a hard disk for further analysis. They were obtained from subjects with a normal heart and from patients with aortic stenosis. They were digitized with 12-bit resolution at a sampling rate of 4096 Hz.

72.2.2 Data Analysis

The heart sound signals and electrocardiogram (ECG) signals were recorded simultaneously to correlate the heart sounds with the phase of the heart cycle in a time relationship. The heart sounds were processed by a low-pass filter with 300 Hz cutoff frequency. The first heart sound, S1, occurs at the onset of ventricular contraction, which corresponds to the R wave of the ECG signal. The R wave was taken to identify the start point of S1. The heart sounds then were segmented into each heart sound cycle consisting of S1 and S2, using the ECG as a reference. The CWT with scales from 10 to 200 was applied to one heart sound cycle for each patient group.

72.2.3 Wavelet Entropy

The concept of entropy has been widely known as a measure of the degree of disorder of a system. It involves computation of the probability density function. It is expected that the entropy measure can reveal statistically more of the signal information. The wavelet entropy computed from the wavelet transform has been applied to several biomedical signals. In this study, the wavelet entropy was calculated from the CWT. Computation of the heart sound energy (E_{ij}) in the time-scale plot was performed for each time i and scale j. The probability distribution of energy for each scale was derived as in Eq. (72.1).

$$P_{ij} = \frac{E_{ij}}{E_i} \tag{72.1}$$

where P_{ij} is the probability distribution at time i and scale j. E_{ij} is the energy at time i and scale j. Ei is the energy at time i. The wavelet entropy (W_i) is defined as in Eq. (72.2).

$$W_i = -\sum_{j=1}^{N} P_{ij} \times \log_2 P_{ij}$$ (72.2)

72.3 Results

Three consecutive heart sound cycles of one patient with a normal heart and those of one patient with aortic stenosis are plotted in Fig. 72.1a and b, respectively. It clearly shows the S1 and S2 heart sounds for each cycle (Fig. 72.1a), but it is not quite clear for the abnormal heart sounds (Fig. 72.1b).

The CWT with scales from 10 to 200 was applied to the heart sounds of Fig. 72.1a and b. The CWT contour plots are depicted in Fig. 72.2a and b, respectively. The CWT provides detailed information on the time-scale content of the heart sounds during the whole cardiac cycle. In Fig. 72.2a, the sounds of S1 and S2 can clearly be identified. Figure 72.2b shows the two major sounds of S1 and

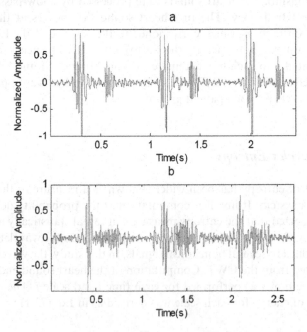

Fig. 72.1 Three consecutive heart sound cycles for the patient with normal heart (**a**), and the patient with aortic stenosis (**b**)

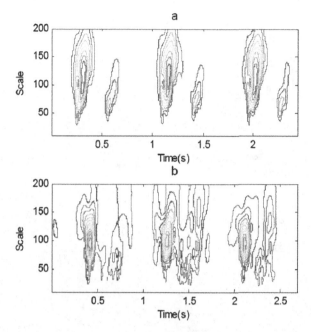

Fig. 72.2 The CWT contour plots for the normal patient (**a**), and the patient with aortic stenosis (**b**)

S2, and high-frequency information between them. The high-frequency content may be due to systolic murmurs; it does not appear in the normal heart sound (Fig. 72.2a).

The presence of high frequency murmurs was found at the smaller scales in the CWT time-scale plot. The CWT contour of the normal patient shows smooth curves for each cycle, whereas the abnormal patient gives discontinuous contours. In addition, the three-dimensional (3D) time-scale plots of one heart sound cycle for the patient with a normal heart and for the patient with aortic stenosis are illustrated in Fig. 72.3a and b, respectively. The difference in the time-scale plots between the two heart sound signals can be clearly observed. It was found that the normal heart sounds are likely to have relatively smooth surfaces, but the 3D surface of the abnormal heart sound seems to have more ragged edges. The abnormal heart sounds exhibit more irregular features in the time-scale plot. This would mean that the heart sound with aortic stenosis may be a disordered signal. With these findings, wavelet entropy was explored as a measure of a disordered pattern of the abnormal heart sounds. Computation of the wavelet entropy was processed, and the computed result is plotted in Fig. 72.4. It was observed that wavelet entropy in patients with aortic stenosis tends to be high, whereas the normal patients are likely to have low wavelet entropy.

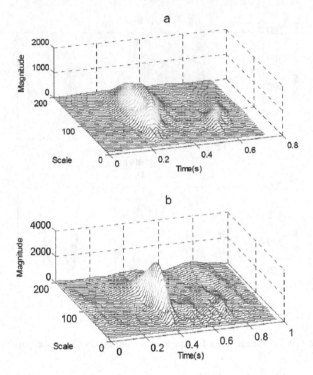

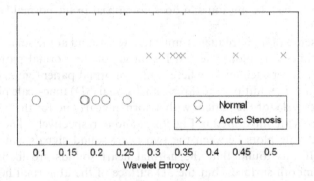

Fig. 72.3 The 3D continuous wavelet transform plots for the patient with normal heart (**a**), and the patient with aortic stenosis (**b**)

Fig. 72.4 The wavelet entropy for patients with normal heart and those with aortic stenosis

72.4 Discussion and Conclusion

Heart sounds give clinically useful information on the assessment of heart function, functional integrity of the cardiac valves, and hemodynamics. Because of the time-varying characteristics of heart sound signals, time-frequency representation is the most suitable technique to describe heart sounds. Two groups of patients were involved in this study: those with normal heart sounds and others with abnormal heart sounds (aortic stenosis). The CWT was applied to the heart sounds of both patient groups. It achieved good time resolution at high frequencies. The results of the CWT analysis (Figs. 72.2b and 72.3b) showed that the CWT can discover the high-frequency components of heart sounds in aortic stenosis, and that these components contain vital information for the diagnostic procedure. The condition of aortic stenosis caused by cardiac disease generates turbulence in flow, and high-frequency, noise-like murmurs are produced. These murmurs will have variations in frequency and intensity. It is thus assumed that abnormal heart sounds from patients with aortic stenosis are likely to be irregular in shape in the time-scale plot computed from the CWT (Figs. 72.2b and 72.3b). The normal heart produces periodically characteristic heart sounds, generally containing the S1 and S2 sounds. Thus the normal subject should show relatively smooth surfaces. This difference is very subjective for patient classification, and it is a qualitative result. A predictive indicator is thus needed to identify whether or not patients exhibit abnormal heart sounds for a quantitative analysis. This research has attempted to use wavelet entropy as a diagnostic measure for discrimination of patients. Heart diseases cause aberrations in the heart sounds. It is expected that abnormal heart sounds would contain additional information that is different from normal heart sounds. This extra information can be captured by a method of CWT-based entropy. The CWT-based entropy gives promising results. These results show that patients with aortic stenosis tend to have larger wavelet entropy. This would imply that abnormal heart sounds should be discontinuous events with nonstationary characteristics. Entropy is related to the degree of irregularity of a signal. The magnitude of CWT-based wavelet entropy would be used to quantify the different degrees of severity for heart diseases with aortic stenosis. In this study, an ECG reference was utilized to segment the heart sounds manually. Advanced studies in the automatic detection of heart sounds for each cardiac cycle without the ECG reference should be investigated. The proposed method of CWT-based entropy shows the potential capability of separating normal patients from those with aortic stenosis. It also may provide complementary clinical information.

References

1. Brusco M, Nazeran H (2004) Digital phonocardiography: a PDA-based approach. In: Proceedings of the 26th annual international conference of the IEEE EMBS, pp 2299–2302
2. Bunluechokchai S, English MJ (2003) Detection of wavelet transform-processed ventricular late potentials and approximate entropy. IEEE computers in cardiology, pp 549–552
3. Bunluechokchai S, English MJ (2003) Analysis of the high resolution ECG with the continuous wavelet transform. IEEE computers in cardiology, pp 553–556

4. Djebbari A, Reguig B (2002) Short-time Fourier transform analysis of the phonocardio-gram signal. In: The 7th IEEE international conference on electronics, circuits and systems, pp 844–847
5. Durand LG and Pibarot P (1995) Digital signal processing of the phonocardiogram: review of the most recent advancements. Crit Rev Biomed Eng163–219
6. Kumar P, Carvalho M, Antunes M, Gil P, Henriques J, Eugenio L (2006) A new algorithm for detection of S1 and S2 heart sounds. In: IEEE international conference on acoustic, speech, and signal processing ICASSP, pp 1180–1183
7. Mgdob HM, Torry JN, Vincent R, Al-Naami B (2003) Application of Morlet transform wavelet in the detection of paradoxical splitting of the second heart sound. IEEE com-puters in cardiology, pp 323–326.
8. Omran S, Tayel M (2004) A heart sound segmentation and feature extraction algorithm using wavelets. In: First international symposium on control, communications and signal processing, pp 235–238
9. Tovar-Corona B, Torry JN (1998) Time-frequency representation of systolic murmurs using wavelets. IEEE computers in cardiology, pp 601–604
10. Zhidong Z, Zhijin Z, Yuquan C (2005) Time-frequency analysis of heart sound based on HHT. In: IEEE international conference on communications, circuits and systems, pp 926–929
11. Zin ZM, Salleh SH, Daliman S, Sulaiman MD (2003) Analysis of heart sounds based on continuous wavelet transform. In: IEEE conference on research and development, pp 19–22

Chapter 73
Detection of Mitral Regurgitation and Normal Heart Sounds

S. Bunluechokchai and W. Ussawawongaraya

Abstract Detection of heart sounds has been considered to be useful in the diagnosis of heart diseases, especially in patients with abnormality of the cardiac valves. Heart sounds are generated from blood flow through the cardiac valves during heart contraction and relaxation. They can be listened to and recorded on the patient's chest. In this study, two groups of patients, some with normal hearts and others with mitral regurgitation, were investigated. The power spectrum of heart sound signals obtained from both patient groups was studied. It was discovered that the spectral power of the patients with mitral regurgitation was observed at frequencies above 500 Hz, whereas the normal heart sounds did not clearly occur. In addition, the continuous wavelet transform (CWT) was used to analyze the heart sounds of both groups. Results showed that patients with mitral regurgitation have high frequency components caused by murmurs at smaller scales in the time-scale plot. The CWT of heart sounds was further investigated by computing their energy. It was found that the heart sounds of patients with mitral regurgitation are likely to have higher energy than those of patients with normal hearts.

73.1 Introduction

Heart sounds originate from the contraction and relaxation of the heart, blood flow, and valve movements. They occur during each heart cycle. Heart diseases produce abnormal heart sounds. Although a new method of echocardiography for cardiac examination is now used, heart auscultation is still a simple and important tool for early diagnosis of heart diseases. In addition, heart sounds can be detected by a technique known as phonocardiography. This technique has many advantages over conventional auscultation due to the limitations of

S. Bunluechokchai (✉)
Department of Industrial Physics and Medical Instrumentation, Faculty of Applied Science, King Mongkut's Institute of Technology, North Bangkok,
1518 Pibulsongkram Road, Bangsue, Bangkok 10800, Thailand

N. Mastorakis et al. (eds.), *Proceedings of the European Computing Conference*, Lecture Notes in Electrical Engineering 27, DOI 10.1007/978-0-387-84814-3_73, © Springer Science+Business Media, LLC 2009

human hearing and physician skills. The human ear is not well suited to the frequency range of cardiac auscultation. Phonocardiography shows the graphical representation of heart sounds. It is easy to use and noninvasive. It provides diagnostic information for the detection of abnormal function of the cardiac valves in clinical practice. As heart sounds are typically nonstationary signals, it is necessary to analyze them for both time and frequency information. Various signal processing techniques have been developed to analyze heart sounds and murmurs, such as the Wigner-Ville distribution, the short time Fourier transform, the wavelet transform (WT) [1, 4, 5, 7, 9–11], and heart sound segmentation [6, 8]. Compared to the other techniques, the main advantage of the wavelet transform is that it provides better time and frequency resolution for the analysis of heart sounds. The normal heart sound signal is composed of two main parts: the first S1, and the second S2. The interval between S1 and S2 is called systole, and the interval between S2 and the next S1 is called diastole. The turbulence associated with the movement of blood through abnormal cardiac valves is responsible for the generation of heart murmurs. The conditions of blood flow turbulence are mainly valvular stenosis and valve insufficiency. Valvular stenosis may be due to calcium deposition preventing the valve from opening completely, thereby causing an obstruction of blood flow. Valve insufficiency occurs when the valve does not close completely, and there is a reverse flow direction or regurgitation of blood flow through a narrow opening. Murmurs are high-frequency, noise-like sounds. Systolic murmurs are caused during ventricular contraction by conditions such as aortic stenosis and mitral insufficiency. Two groups of patients, normal and mitral regurgitation patients, were studied in this research. The application of continuous wavelet transform (CWT) analysis for the heart sounds was described. Then the computation of energy obtained from the CWT was processed for each patient group to differentiate between patients with normal hearts and those with mitral regurgitation.

73.2 Materials and Method

The wavelet transform has been a powerful tool to describe the time-frequency representation of a signal. It has become popular for signal processing in recent years. It has also shown potential in biomedical signal processing. It offers varying time-frequency resolutions and it is defined as:

$$CWT(a,b) = \int_{-\infty}^{\infty} s(t)\psi_{a,b}(t)dt \qquad (73.1)$$

where $s(t)$ is the signal, $\psi(t)$ is the mother wavelet, a and b are the scaling and translation parameters, respectively, and t is the time. A wavelet family $\psi_{a,b}$ is the set of elemental functions obtained from dilations and translations of a mother wavelet ψ, defined as:

$$\psi_{a,b}(t) = \frac{1}{\sqrt{a}} \psi\left(\frac{t-b}{a}\right) \qquad (73.2)$$

The heart sound is a time-varying signal and the CWT has the capability to correctly perform analysis of a nonstationary signal with multi-resolution. The CWT also has the capability of analyzing both the high-frequency and low-frequency components of a signal. The wavelet is contracted at smaller scales, thus the wavelet transform can detect the high-frequency components of the signal. This property of the CWT is well suited to the analysis of high-frequency information [2, 3]. The CWT with smaller scales should be implemented to detect the high-frequency murmurs. As the CWT is computed in terms of scale instead of frequency, the CWT plot of a signal is represented as the time-scale domain rather than the time-frequency plot. A scale parameter of the CWT is inversely related to the frequency. It is of great importance in selecting the mother wavelet for the best detection of the signal. The mother wavelet used in this work is the Morlet wavelet, because a number of researchers have demonstrated that it provides satisfactory results in the detection of heart sounds [7, 9, 11].

73.2.1 Data Acquisition

The heart sounds were recorded and saved on a hard disk for further analysis. They were obtained from subjects with normal hearts and from patients with mitral regurgitation. They were digitized with 12-bit resolution at a sampling rate of 4096 Hz.

73.2.2 Data Analysis

The heart sounds were preprocessed by a high-pass filter with a 30 Hz cutoff frequency to remove low-frequency components from muscle or chest movements. The heart sound signals were recorded simultaneously with the electrocardiogram (ECG) signals to correlate the heart sounds with the phase of the heart cycle in the time relationship. The first heart sound, S1, occurs at the onset of ventricular contraction, which corresponds to the R wave of the ECG signal. The R wave was utilized to mark the start point of S1. The heart sounds then were segmented into each heart sound cycle, consisting of S1 and S2 and using the ECG as a reference. In this work, the detection of the R wave was done visually. The CWT with scales from 10 to 200 was applied to one heart sound cycle for each patient group.

73.3 Results

Figure 73.1a and b shows three consecutive heart sound cycles of one patient
with a normal heart and of one patient with mitral regurgitation, respectively. It
clearly shows the S1 and S2 of the heart sounds for each cycle, as shown in Fig.
73.1a, but it is not easy to locate the start point of S1 and S2 for the abnormal
sound, as shown in Fig. 73.1b.

The CWT with scales from 10 to 200 was applied to the heart sounds of
Fig. 73.1a and b. The CWT contour plots were then plotted in Fig. 73.2a and b,
respectively. The CWT gives detailed information on the time-scale content of
the heart sounds during the whole cardiac cycle. The S1 and S2 sounds are
clearly seen in Fig. 73.2a. Figure 73.2b shows the two main sounds S1 and S2,
and high-frequency information between them. The high-frequency components may be due to systolic murmurs; they do not occur in the normal heart
sound (Fig. 73.2a). The presence of high frequency murmurs was revealed at the
smaller scales in the CWT time-scale representation. In addition, power spectrum analysis of the heart sounds was computed by the Fourier transform to
discover the differences in the frequency components between normal and
abnormal heart sounds.

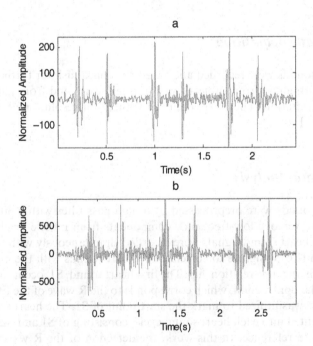

Fig. 73.1 Three consecutive heart sound cycles for the patient with normal heart (**a**), and the
patient with mitral regurgitation (**b**)

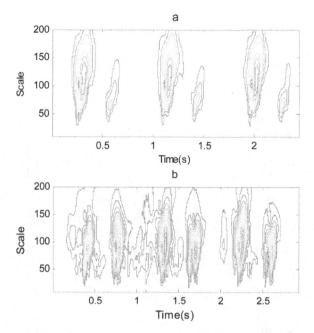

Fig. 73.2 The CWT contour plots for the normal patient (**a**), and the patient with mitral regurgitation (**b**)

Figure 73.3 illustrates the power spectrum for the patient with a normal heart (a) and the patient with mitral regurgitation (b). A noticeable difference between them can be found. The patient with mitral regurgitation displays the high-frequency content, whereas the high-frequency components of the normal heart are not observed. The main frequency components of the heart sounds from patients with mitral regurgitation occur at frequencies of approximately 550, 620, 1112, 1320, 1664, and 1876 Hz. In order to enhance the detection of the presence of the high-frequency parts, the heart sound signals of both patient groups were band-pass filtered at the cutoff frequency between 500 and 2000 Hz. The CWT was applied to the band-pass filtered heart sounds and then the CWT energy was computed for each patient. The computed energy is illustrated in Fig. 73.4.

73.4 Discussion and Conclusion

Phonocardiography is a method of recording heart sounds. It gives clinically useful information on the assessment of the cardiovascular system, the function of the heart, and the functional integrity of the cardiac valves. It displays heart sounds in time-amplitude plots. Due to the time-varying characteristics of heart

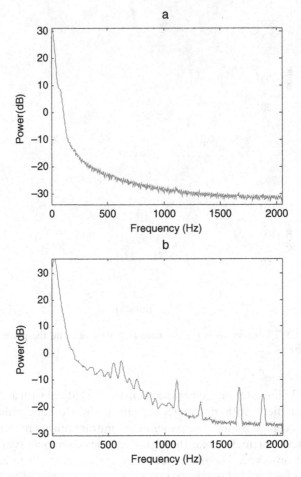

Fig. 73.3 The power spectrum for the patient with normal heart (**a**), and the patient with mitral regurgitation (**b**)

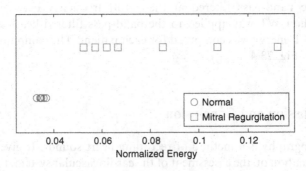

Fig. 73.4 The CWT energy for patients with normal hearts and those with mitral regurgitation

sound signals, time-frequency representation is the most appropriate technique to describe heart sounds. Two groups of patients were considered in this study: some with normal heart sounds and others with abnormal heart sounds (mitral regurgitation). When compared to the S1 and S2 of heart sounds, high-frequency murmurs are often of lower amplitude and they are generally masked by low-frequency components. It has been shown that the CWT has the potential ability to correctly localize the high-frequency signals. The CWT with smaller scales should be used for analysis of the high-frequency heart sounds. The CWT was applied to the heart sounds of both patient groups. Results of the CWT analysis (Fig. 73.2a and b) showed that the CWT can detect the high-frequency components of the heart sounds of patients with mitral regurgitation and these components will carry vital information for the diagnostic process. This will help in improving the diagnosis of abnormal heart sounds. Basic heart sound components appear mostly at frequencies below 500 Hz. It should be noted that there are high-frequency peaks above 500 Hz for the patient with mitral regurgitation. Analysis of these high-frequency contents was further performed by band-pass filtering the heart sounds; however, it is suggested that high-frequency components above 500 Hz should be further investigated. The CWT of the filtered heart sounds was performed and then the CWT energy of one cardiac cycle was computed for each patient. It was shown that the energy calculated from the patients with mitral regurgitation tends to be higher than the energy of those with normal hearts. It is difficult to accurately identify the onset and offset of S1 or the systolic murmurs from their heart sounds. In this research, the R wave as a reference was utilized to segment the heart sound cycle for computation of the CWT-based energy, and it was easily detected as compared to the heart sounds themselves. Heart auscultation is a very subjective technique to classify patients with abnormal heart sounds. Therefore, it is of great interest to develop an algorithm for diagnosis of the first stage of heart disease for assistance with clinical interpretation. Using the optimum reference for automated segmentation of heart sound signals, the CWT-based energy would provide guidance for the automatic classification between heart sounds and murmurs. It is possible to use that energy for differentiation between patients with normal hearts and those with mitral regurgitation. Clinical interpretation of heart sounds with CWT-based energy may provide complementary information in improving the diagnosis of abnormal heart sounds with better accuracy.

References

1. Brusco M, Nazeran H (2004) Digital phonocardiography: a PDA-based approach. In: Proceedings of the 26th annual international conference of the IEEE EMBS, pp 2299–2302
2. Bunluechokchai S, English MJ (2003) Detection of wavelet transform-processed ventricular late potentials and approximate entropy. IEEE computers in cardiology, pp 549–552

3. Bunluechokchai S, English MJ (2003) Analysis of the high resolution ECG with the continuous wavelet transform. IEEE computers in cardiology, pp 553–556
4. Djebbari A, Reguig B (2002) Short-time Fourier transform analysis of the phonocardiogram signal. In: The 7th IEEE international conference on electronics, circuits and systems, pp 844–847
5. Durand LG and Pibarot P (1995) Digital signal processing of the Phonocardiogram: review of the most recent advancements. Crit Rev Biomed Eng 163–219
6. Kumar P, Carvalho M, Antunes M, Gil P, Henriques J, Eugenio L (2006) A new algorithm for detection of S1 and S2 heart sounds. In: IEEE international conference on acoustic, speech, and signal processing ICASSP, pp 1180–1183
7. Mgdob HM, Torry JN, Vincent R, Al-Naami B (2003) Application of Morlet transform wavelet in the detection of paradoxical splitting of the second heart sound. IEEE computers in cardiology, pp 323–326
8. Omran S, Tayel M (2004) A heart sound segmentation and feature extraction algorithm using wavelets. In: First international symposium on control, communications and signal processing, pp 235–238
9. Tovar-Corona B, Torry JN (1998) Time-frequency representation of systolic murmurs using wavelets. IEEE computers in cardiology, pp 601–604
10. Zhidong Z, Zhijin Z, Yuquan C (2005) Time-frequency analysis of heart sound based on HHT. In: IEEE international conference on communications, circuits and systems, pp 926–929
11. Zin ZM, Salleh SH, Daliman S, Sulaiman MD (2003) Analysis of heart sounds based on continuous wavelet transform. In: IEEE conference on research and development, pp 19–22

Chapter 74
Wavelet Entropy Detection of Heart Sounds

T. Leeudomwong and P. Woraratsoontorn

Abstract Heart sound detection has been demonstrated to be important in the diagnosis of heart disease, especially in patients with abnormality of the cardiac valves. Heart sounds are produced from blood flow through the cardiac valves during heart contraction and relaxation, and they can be heard and recorded on the patient's chest. In this research, two groups of patients, those with normal hearts and those with mitral regurgitation, were studied. The application of wavelet transform analysis to heart sound signals is presented for both groups. The continuous wavelet transform plot shows that patients with mitral regurgitation are likely to have the irregular features of heart sounds in a three-dimensional diagram, whereas normal heart sounds seem to show relatively smooth surfaces. The wavelet entropy is then computed for the heart sounds. A disordered signal offers a high entropy value. Observations of the preliminary results in this research show that the heart sounds from the patients with mitral regurgitation tend to have higher wavelet entropy than those from the patients with normal hearts.

74.1 Introduction

Heart sounds originate from the mechanical functions of the heart, the flow of blood, and the movement of valves. It occurs in rhythmic manner during each cardiac cycle. The diagnosis of heart disease can be first achieved by listening to heart sounds. Although a new method of echocardiography for cardiac examination has currently been used, heart auscultation is still the important diagnostic method for early diagnosis of heart diseases. In addition, the detection of heart sounds can be achieved by a technique known as a phonocardiography. It is an easy and noninvasive technique that provides a graphical representation of

T. Leeudomwong (✉)
Department of Industrial Physics and Medical Instrumentation, Faculty of Applied
Science, King Mongkut's Institute of Technology, North Bangkok,
1518 Pibulsongkram Road, Bangsue, Bangkok 10800, Thailand

N. Mastorakis et al. (eds.), *Proceedings of the European*
Computing Conference, Lecture Notes in Electrical Engineering 27,
DOI 10.1007/978-0-387-84814-3_74, © Springer Science+Business Media, LLC 2009

the heart sound signals and offers the diagnostic information for the detection of abnormal function of the cardiac valves in clinical practice. The heart sound signals can be easily detected by placing a conventional sensor, generally a microphone, on the patient's chest. Because the heart sound is typically a nonstationary signal, it is necessary to analyze the heart sound signals using both time and frequency domain approaches. Many researchers have proposed various processing techniques for the analysis of heart sound signals, such as the Wigner-Ville distribution, the short time Fourier transform, the wavelet transform (WT) [1, 4, 5, 7, 9–11], and heart sound segmentation [6, 8]. When compared to the other techniques, the wavelet transform has the main advantage in that it provides better time and frequency resolution for analysis of heart sound signals. The normal heart sound signal is composed of two major sounds: the first S1, and the second S2. The sound S1 is generated by the sudden closure of the mitral and tricuspid valves during ventricular contraction. The sound S2 is correlated with the closure of the aortic and pulmonary valves and marks the end of ventricular systole. The period between S1 and S2 is called systole, and the period between S2 and the next S1 is called diastole. Murmurs are abnormal heart sounds that are produced by turbulent blood flow through the heart valves. The conditions of blood flow turbulence are mainly valvular stenosis and valve insufficiency. Valvular stenosis may be due to calcium deposition resulting in the valve not opening properly, and thereby causing an obstruction of blood flow. Valve insufficiency occurs when the valve does not close completely, and there is a reverse flow direction or regurgitation of blood flow through a narrow opening. Murmurs are high-frequency, noise-like sounds. Systolic murmurs are produced during ventricular contraction by conditions such as aortic stenosis and mitral insufficiency. Two groups of patients, normal and mitral regurgitation patients, were investigated in this study. The application of continuous wavelet transform (CWT) analysis for the heart sounds was described. Then the computation of the wavelet entropy obtained from the CWT was performed for each patient group. This study investigated the possibility of using wavelet entropy for discrimination between patients with normal hearts and those with mitral regurgitation.

74.2 Materials and Methods

The wavelet transform has been recognized as a powerful tool of signal processing. It utilizes the wavelet function to analyze the signal of interest. The heart sound is a time-varying signal and the CWT has the ability to better analyze a nonstationary signal [2, 3]. Because the CWT is computed in terms of scale instead of frequency, the CWT plot of a signal is described using the time-scale domain rather than the time-frequency plane. It gives varying time-frequency resolutions. A scaling parameter of the CWT is inversely related to the frequency. The selection of the mother wavelet and scaling parameter is very important for the best detection of the signal. The mother wavelet used in this work is the Morlet wavelet, because a number of researchers have found that it offers encouraging results for the analysis of heart sounds [7, 9, 11].

74.2.1 Data Acquisition

The heart sounds were recorded and saved on a hard disk for later analysis. They were obtained from subjects with normal hearts and from patients with mitral regurgitation. They were digitized with 12-bit resolution at a sampling rate of 4096 Hz.

74.2.2 Data Analysis

The heart sound and electrocardiogram (ECG) signals were recorded simultaneously to correlate the heart sounds with the phases of the heart cycle in the time relationship. The heart sounds were processed by a low-pass filter with a 300 Hz cutoff frequency. The first heart sound, S1, occurs at the onset of ventricular contraction, which corresponds to the R wave of the ECG signal. The R wave was employed to locate the start point of S1. The heart sounds then were segmented into cycles consisting of S1 and S2, using the ECG as a reference. The CWT with scales from 10 to 200 was applied to one heart sound cycle for each patient group.

74.2.3 Wavelet Entropy

The concept of entropy has been widely known as a measure of the uncertainty of a system. It involves computation of the probability density function. It is expected that entropy can reveal statistically more of the signal information. The wavelet entropy computed from the wavelet transform has been applied to many biomedical signals. In this paper, the wavelet entropy was calculated from the CWT. Computation of the heart sound energy (Eij) in the time-scale plot was obtained for each time i and scale j. The probability distribution of energy for each scale was calculated as in Eq.(74.1).

$$P_{ij} = \frac{E_{ij}}{E_i} \tag{74.1}$$

where P_{ij} is the probability distribution at time i and scale j. E_{ij} is the energy at time i and scale j. Ei is the energy at time i. The wavelet entropy (W_i) is defined as in Eq. (74.2).

$$W_i = -\sum_{j=1}^{N} P_{ij} \times \log_2 P_{ij} \tag{74.2}$$

74.3 Results

Three consecutive heart sound cycles of one patient with a normal heart and those of one patient with mitral regurgitation are shown in Fig. 74.1a and b, respectively. It clearly displays the S1 and S2 of heart sounds for each cycle, as shown in Fig. 74.1a, but it is not quite obvious for the abnormal heart sounds, as shown in Fig. 74.1b. The CWT with scales from 10 to 200 was applied to the heart sounds of Fig. 74.1a and b.

The CWT contour plots are then represented in Fig. 74.2a and b, respectively. The CWT shows detailed information on the time-scale content of the heart sounds during the whole cardiac cycle. In Fig. 74.2a, two major sounds of S1 and S2 can clearly be observed. Figure 74.2b shows the two main sounds of S1 and S2, and high-frequency information between them. The high-frequency content may be due to systolic murmurs; it does not occur in the normal heart sound (Fig. 74.2a).

The presence of high frequency murmurs was seen at the smaller scales in the CWT time-scale plot. The CWT contour of the normal patient displays the smooth curves for each cycle, whereas the abnormal patient shows discontinuous contours. In addition, three-dimensional (3D) time-scale plots of one heart sound cycle for the patient with a normal heart and one for the patient with mitral regurgitation are shown in Fig. 74.3a and b, respectively. The

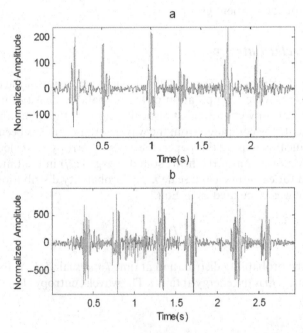

Fig. 74.1 Three consecutive heart sound cycles for the patient with normal heart (**a**), and the patient with mitral regurgitation (**b**)

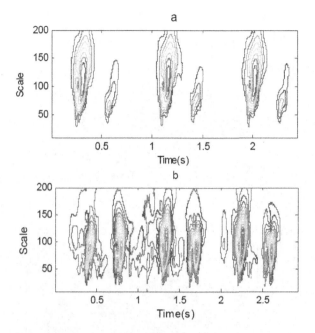

Fig. 74.2 The CWT contour plots for the normal patient (**a**), and the patient with mitral regurgitation (**b**)

noticeable difference between the two patients can be seen. It was observed that the normal heart sounds tend to have relatively smooth surfaces, whereas the 3D surface of the abnormal heart sounds seem likely to have more ragged edges. The abnormal heart sounds display more irregular features in the time-scale representation. It would mean that the heart sound with mitral regurgitation may be a disordered signal. With these findings, the wavelet entropy was explored as a measure of a disordered pattern of abnormal heart sounds. Computation of the wavelet entropy was performed, and the computed result is shown in Fig. 74.4. It was found that wavelet entropy in patients with mitral regurgitation tends to be high, whereas the normal patients are likely to have low wavelet entropy.

74.4 Discussion and Conclusion

Heart sounds and murmurs provide clinically important information on the assessment of the cardiovascular system, heart function, and functional integrity of the cardiac valves. Due to the nonstationary characteristics of heart sound signals, time-frequency representation is the most appropriate technique to study them. Two groups of patients were included in this study: some with normal heart sounds and others with abnormal heart sounds (mitral

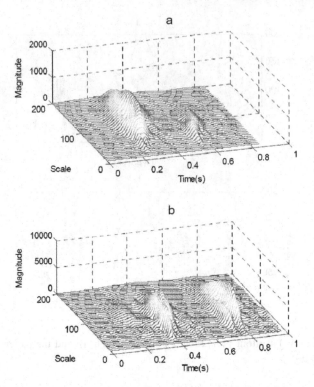

Fig. 74.3 The 3D Continuous wavelet transform plots for the patient with normal heart (**a**), and the patient with mitral regurgitation (**b**)

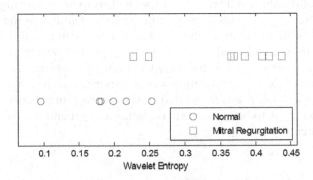

Fig. 74.4 The wavelet entropy for patients with a normal heart and for those with mitral regurgitation

regurgitation). The CWT was employed to investigate the time-varying frequency characteristics of heart sound signals. It was applied to the heart sounds of both patient groups. The CWT provides good time resolution at high frequencies. Results of the CWT analysis (Figs. 74.2b and 74.3b) showed that the CWT can reveal the high-frequency components of the heart sounds with mitral regurgitation, and these components contain vital information for the diagnostic process. The condition of mitral regurgitation caused by cardiac diseases produces turbulence in flow, and then the high-frequency, noise-like murmurs are observed. These murmurs have variation in frequency and intensity. It is thus expected that the abnormal heart sounds from patients with mitral regurgitation are likely to be irregular in shape in the time-scale plot computed from the CWT (Figs. 74.2b and 74.3b). The normal heart has characteristic periodical heart sounds, consisting typically of the S1 and S2 sounds. Then the normal subject should show relatively smooth surfaces. This difference is very subjective for patient discrimination and it is a qualitative result. The quantitative analysis of heart sounds is of great interest. A diagnostic indicator is thus required to identify whether or not the patients display the abnormal heart sounds. This study has investigated the use of wavelet entropy as a diagnostic measure for classification of patients. Results showed that patients with mitral regurgitation seem to have larger wavelet entropy. This would mean that abnormal heart sounds should be discontinuous events with nonstationary characteristics. Entropy is related to the degree of irregularity of a signal. Vital irregular information on heart murmurs would be detected by the CWT-based entropy. The magnitude of the CWT-based wavelet entropy value would be used to quantify the different degrees of severity for heart diseases with mitral regurgitation. In this study, an ECG reference was utilized to segment the heart sounds manually. The automated detection of heart sounds for each cardiac cycle without the ECG reference should be further investigated in future studies. It should be noted that the proposed method may provide complementary information. It would help in improving the diagnostic accuracy of abnormal heart sounds.

References

1. Brusco M, Nazeran H (2004) Digital phonocardiography: a PDA-based approach. In: Proceedings of the 26th annual international conference of the IEEE EMBS, pp 2299–2302
2. Bunluechokchai S, English MJ (2003) Detection of wavelet transform-processed ventricular late potentials and approximate entropy. IEEE computers in cardiology, pp 549–552
3. Bunluechokchai S, English MJ (2003) Analysis of the high resolution ECG with the continuous wavelet transform. IEEE computers in cardiology, pp 553–556
4. Djebbari A, Reguig B (2002) Short-time Fourier transform analysis of the phonocardiogram signal. In: The 7th IEEE international conference on electronics, circuits and systems, pp 844–847
5. Durand LG and Pibarot P (1995) Digital signal processing of the phonocardiogram: review of the most recent advancements. Crit Rev Biomed Eng163–219
6. Kumar P, Carvalho M, Antunes M, Gil P, Henriques J, Eugenio L (2006) A new algorithm for detection of S1 and S2 heart sounds. In: IEEE international conference on acoustic, speech, and signal processing ICASSP, pp 1180–1183

7. Mgdob HM, Torry JN, Vincent R, Al-Naami B (2003) Application of Morlet transform wavelet in the detection of paradoxical splitting of the second heart sound. IEEE computers in cardiology, pp 323–326
8. Omran S, Tayel M (2004) A heart sound segmentation and feature extraction algorithm using wavelets. In: First international symposium on control, communications and signal processing, pp 235–238
9. Tovar-Corona B, Torry JN (1998) Time-frequency representation of systolic murmurs using wavelets. IEEE computers in cardiology, pp 601–604
10. Zhidong Z, Zhijin Z, Yuquan C (2005) Time-frequency analysis of heart sound based on HHT. In: IEEE international conference on communications, circuits and systems, pp 926–929
11. Zin ZM, Salleh SH, Daliman S, Sulaiman MD (2003) Analysis of heart sounds based on continuous wavelet transform. In: IEEE conference on research and development, pp 19–22

Chapter 75
Continuous Wavelet Transform Analysis of Heart Sounds

P. Woraratsoontorn and T. Leeudomwong

Abstract Heart sound detection has been documented to be useful in the diagnosis of heart disease, especially in patients with abnormality of the cardiac valves. Blood flow through the cardiac valves during heart contraction and relaxation generates heart sounds which can be heard and recorded on the patient's chest. In this paper, two groups of patients, some with normal hearts and others with aortic stenosis, were studied. The heart sounds of both groups were analyzed for the power spectrum. It was found that the spectral power of the patients with aortic stenosis was observed at frequencies above 500 Hz, whereas the normal heart sounds did not clearly appear. In addition, the continuous wavelet transform (CWT) was applied to the heart sounds for both groups. Results showed that patients with aortic stenosis have high-frequency components caused by systolic murmur at lower scales in the time-scale plot. Theses high-frequency components were clearly seen between the S1 and S2 heart sounds. The CWT of heart sounds was further investigated for energy. It was shown that the heart sounds of patients with aortic stenosis seem to have higher energy than those of patients with healthy hearts.

75.1 Introduction

Heart sounds are produced from the contraction and relaxation of the heart, from blood flow, and from valve movements. They occur during the heart cycle. Acoustic analysis of heart sounds has been introduced. It offers important and useful information in the diagnosis and monitoring of cardiac valve diseases. Although a new method of echocardiography for cardiac examination is currently being used, heart auscultation is still the simplest and most important approach for early diagnosis of heart disease. In addition, detection of heart

P. Woraratsoontorn (✉)
Department of Industrial Physics and Medical Instrumentation, Faculty of Applied Science, King Mongkut's Institute of Technology, North Bangkok, 1518 Pibulsongkram Road, Bangsue, Bangkok 10800, Thailand

N. Mastorakis et al. (eds.), *Proceedings of the European Computing Conference*, Lecture Notes in Electrical Engineering 27, DOI 10.1007/978-0-387-84814-3_75, © Springer Science+Business Media, LLC 2009

sounds can be carried out by a technique known as phonocardiography. This technique has many advantages over conventional auscultation, due to the limitations of the human ear and the hearing experiences of physicians. The human ear is not well suited for cardiac auscultation. Phonocardiography displays a graphical representation of heart sounds in the time-amplitude plot. It is easy to use and noninvasive. It provides diagnostic information for the detection of abnormal functions of the cardiac valves in clinical practice. Time-frequency analysis is one of the techniques to study nonstationary signals. As heart sounds are typically nonstationary signals, it is necessary to analyze them for both time and frequency information. Several signal processing techniques have been proposed for the analysis of heart sounds, such as the Wigner-Ville distribution, the short time Fourier transform, the wavelet transform (WT) [1, 4, 5, 7, 9–11], and heart sound segmentation [6, 8]. When compared to the other techniques, the main advantage of the wavelet transform is that it offers better time and frequency resolution for the analysis of heart sounds. The normal heart sound signal is composed of two main parts: the first S1, and the second S2. The period between S1 and S2 is called systole, and the period between S2 and the following S1 is called diastole. The first heart sound is generally attributed to movement of blood into the ventricles, and the abrupt closure of the mitral and tricuspid valves during ventricular contraction. The second heart sound corresponds to the closure of the aortic and pulmonary valves. The turbulence associated with the movement of blood through abnormal cardiac valves is responsible for the generation of heart murmurs. The conditions of blood flow turbulence are mainly valvular stenosis and valve insufficiency. Valvular stenosis may be due to calcium deposition preventing the valve from opening properly, thereby causing an obstruction of blood flow. Valve insufficiency occurs when the valve does not close completely, and there is a reverse flow direction or regurgitation of blood flow through a narrow opening. Murmurs are high-frequency, noise-like sounds. Systolic murmurs during ventricular contraction are caused by conditions such as aortic stenosis and mitral insufficiency. Two groups of patients, normal and aortic stenosis patients, were studied in this paper. The application of continuous wavelet transform (CWT) analysis for the heart sounds was performed. Then the computation of energy obtained from the CWT was performed for each patient group to distinguish between patients with normal hearts and those with aortic stenosis.

75.2 Materials and Method

The wavelet transform is the most powerful tool among the time-frequency techniques. It has become popular for signal processing in recent years. It has also shown potential in biomedical signal processing. It provides varying time-frequency resolutions. It is defined as:

$$CWT(a,b) = \int\limits_{-\infty}^{\infty} s(t)\psi_{a,b}(t)dt \tag{75.1}$$

where $s(t)$ is the signal, $\psi(t)$ is the mother wavelet, a and b are the scaling and translation parameters, respectively, and t is the time. A wavelet family $\psi_{a,b}$ is the set of elemental functions obtained from dilations and translations of a mother wavelet ψ, defined as:

$$\psi_{a,b}(t) = \frac{1}{\sqrt{a}}\psi\left(\frac{t-b}{a}\right) \tag{75.2}$$

Heart sounds are time-varying signals, and the CWT has the capability to better analyze a nonstationary signal with multi-resolution. Also, the CWT has the capability of analyzing both the high-frequency and low-frequency components of a signal. The wavelet is contracted at smaller scales, thus the wavelet transform can detect the high-frequency components of the signal. This property of the CWT is well suited to the analysis of high-frequency information [2, 3]. The CWT with smaller scales should be implemented to detect the high-frequency murmurs. As the CWT is computed in terms of scale instead of frequency, the CWT plot of a signal is displayed as the time-scale domain rather than the time-frequency plane. A scale parameter of the CWT is inversely related to the frequency; it is of great importance in selecting the mother wavelet for the best detection of the signal. The mother wavelet used in this work is the Morlet wavelet, because a number of researchers have reported that this wavelet offers satisfactory results for the analysis of heart sounds [7, 9, 11].

75.2.1 Data Acquisition

The heart sounds were recorded and saved on a hard disk for later analysis. They were obtained from subjects with normal hearts and from patients with aortic stenosis. They were digitized with 12-bit resolution at a sampling rate of 4096 Hz.

75.2.2 Data Analysis

The heart sounds were pre-processed by a high-pass filter with a 30 Hz cutoff frequency to remove low-frequency components from muscle or chest movements. The heart sound signals were recorded simultaneously with an electrocardiogram (ECG) to correlate the heart sounds with the phases of the heart cycle in the time relationship. The first heart sound, S1, occurs at the onset of ventricular contraction, which corresponds to the R wave of the ECG signal. The R wave was used to locate the start point of S1. The heart sounds then were segmented into each heart sound cycle, consisting of S1 and S2, using the ECG as a reference. In this work, detection of the R wave was carried out visually. The CWT with scales from 10 to 200 was applied to one heart sound cycle for each patient group.

75.3 Results

Three consecutive heart sound cycles of one patient with a normal heart and one patient with aortic stenosis are plotted in Fig. 75.1a and b, respectively. It clearly shows the S1 and S2 of the heart sounds for each cycle, as shown in Fig. 75.1a, but it is not quite clear for the abnormal sound, as shown in Fig. 75.1b. The CWT with scales from 10 to 200 was applied to the heart sounds of Fig. 75.1a and b. The CWT contour plots were then depicted in Fig. 75.2a and b, respectively. The CWT provides detailed information on the time-scale content of the heart sounds during the whole cardiac cycle. In Fig. 75.2a, it clearly displays the S1 and S2. Figure 75.2b shows two major sounds of S1, S2, and high-frequency information between them. The high-frequency content may be due to systolic murmurs; it does not appear in the normal heart sound (Fig. 75.2a).

The presence of high frequency murmurs was found at the smaller scales in the CWT time-scale plot. In addition, power spectrum analysis of the heart

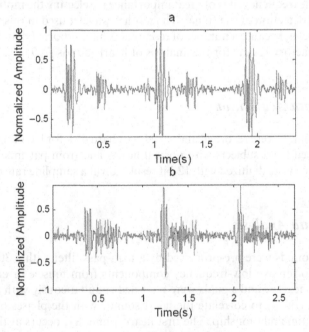

Fig. 75.1 Three consecutive heart sound cycles for the patient with normal heart (**a**), and the patient with aortic stenosis (**b**)

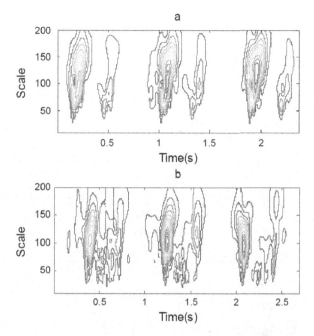

Fig. 75.2 The CWT contour plots for the normal patient (**a**), and the patient with aortic stenosis (**b**)

sounds was performed by the Fourier transform to discover the differences in the frequency components of normal and abnormal heart sounds. Figure 75.3 describes the power spectrum for the patient with a normal heart (a) and the patient with aortic stenosis (b). The difference between them can be observed. The patient with aortic stenosis exhibits the high-frequency components, whereas the high-frequency components of the normal heart are not seen. The dominant frequency components of heart sounds from patients with aortic stenosis appear at frequencies of approximately 555, 1110, 1665, and 1870 Hz.

In order to improve the detection of the presence of the high frequency parts, the heart sound signals of both patient groups were band-pass filtered between cutoff frequencies of 500 and 2000. The CWT was applied to the filtered heart sounds, and then the CWT energy was computed for each patient. The resulting energy is plotted in Fig. 75.4.

75.4 Conclusion

Phonocardiography gives clinically valuable information on the assessment of the cardiovascular system, the mechanisms of heart functions, and the functional integrity of the cardiac valves. It displays heart sounds in the time-amplitude plot. Due to the nonstationary characteristics of heart sound signals,

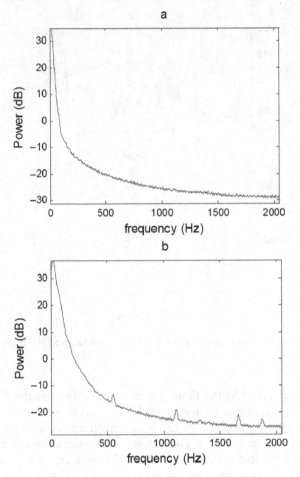

Fig. 75.3 The power spectrum for the patient with normal heart (**a**), and the patient with aortic stenosis (**b**)

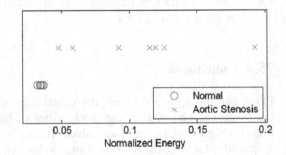

Fig. 75.4 The CWT energy for patients with normal heart and those with aortic stenosis

time-frequency representation is the most suitable technique to depict heart sounds. Two groups of patients were considered in this study: some with normal heart sounds and others with abnormal heart sounds (aortic stenosis). When compared to the S1 and S2 of heart sounds, heart frequency murmurs are often of lower amplitude and they are generally masked by low-frequency components. It has been reported that the CWT has the potential ability to correctly localize the high-frequency signals. The CWT with smaller scales should be utilized for analysis of the high-frequency heart sounds. The CWT was applied to the heart sounds of both patient groups. Results of the CWT analysis (Fig. 75.2a and b) showed that the CWT can reveal the high-frequency components of heart sounds with aortic stenosis and these components contain vital information for the diagnostic process. This can help in improving the diagnosis of abnormal heart sounds.

Most heart sound components are below the frequency of 500 Hz. It should be noted that there are high-frequency peaks above 500 Hz for the patient with aortic stenosis. Analysis of these high-frequency contents was further investigated by band-pass filtering the heart sounds; however, it is suggested that high-frequency components above 500 Hz should be further studied. The CWT of the filtered heart sounds was performed, and the CWT energy of one cardiac cycle was computed for each patient. It was shown that the energy calculated from the patients with aortic stenosis tends to be higher than the energy of those with normal hearts. It is difficult to accurately identify the onset and offset of S1 or the systolic murmurs from the heart sounds themselves. In this research, the R wave as a reference was utilized to segment the heart sound cycle for computation of energy, and it was easily detected compared to the heart sounds themselves. Heart auscultation is a very subjective method of identifying whether or not patients have abnormal heart sounds. It is thus of great importance to develop a method for diagnosing an initial condition of heart sound disease for the assistance of physicians. The objective of this study is to investigate the feasibility of the CWT-based energy for differentiation between patients with normal hearts and those with aortic stenosis. Using the R wave reference for automatic segmentation of heart sounds, it is possible to use the diagnostic measure of the CWT-based energy for automated discrimination between heart sounds and murmurs.

References

1. Brusco M, Nazeran H (2004) Digital phonocardiography: a PDA-based approach. In: Proceedings of the 26th annual international conference of the IEEE EMBS, pp 2299–2302
2. Bunluechokchai S, English MJ (2003) Detection of wavelet transform-processed ventricular late potentials and approximate entropy. IEEE computers in cardiology, pp 549–552
3. Bunluechokchai S, English MJ (2003) Analysis of the high resolution ECG with the continuous wavelet transform. IEEE computers in cardiology, pp 553–556

4. Djebbari A, Reguig B (2002) Short-time Fourier transform analysis of the phonocardio-gram signal. In: The 7th IEEE international conference on electronics, circuits and systems, pp 844–847
5. Durand LG and Pibarot P (1995) Digital signal processing of the Phonocardiogram: review of the most recent advancements. Crit Rev Biomed Eng 163–219
6. Kumar P, Carvalho M, Antunes M, Gil P, Henriques J, Eugenio L (2006) A new algorithm for detection of S1 and S2 heart sounds. In: IEEE international conference on acoustic, speech, and signal processing ICASSP, pp 1180–1183
7. Mgdob HM, Torry JN, Vincent R, Al-naami B (2003) Application of Morlet transform wavelet in the detection of paradoxical splitting of the second heart sound. IEEE computers in cardiology, pp 323–326
8. Omran S, Tayel M (2004) A heart sound segmentation and feature extraction algorithm using wavelets. In: First international symposium on control, communications and signal processing, pp 235–238
9. Tovar-Corona B, Torry JN (1998) Time-frequency representation of systolic murmurs using wavelets. IEEE computers in cardiology, pp 601–604
10. Zhidong Z, Zhijin Z, Yuquan C (2005) Time-frequency analysis of heart sound based on HHT. In: IEEE international conference on communications, circuits and systems, pp 926–929
11. Zin ZM, Salleh SH, Daliman S, Sulaiman MD (2003) Analysis of heart sounds based on continuous wavelet transform. In: IEEE conference on research and development, pp 19–22

Chapter 76
Limitations of Lung Segmentation Techniques

Nisar Ahmed Memon, Anwar Majid Mirza, and S.A.M. Gilani

Abstract High-resolution X-ray computed tomography (CT) is the standard for pulmonary imaging. Depending on the scanner hardware, CT can provide high spatial and high temporal resolution, excellent contrast resolution for the pulmonary structures and surrounding anatomy, and the ability to gather a complete three-dimensional (3D) volume of the human thorax in a single breath hold. Pulmonary CT images have been used for applications such as lung parenchyma density analysis, airway analysis, and lung and diaphragm mechanics analysis. A precursor to all of these quantitative analysis applications is lung segmentation. With the introduction of multi-slice spiral CT scanners, the number of volumetric studies of the lung is increasing, and it is critical to develop fast, accurate algorithms that require minimal to no human interaction to identify the precise boundaries of the lung. This paper presents the problem of inaccurate lung segmentation as observed in algorithms presented by researchers working in the area of medical image analysis. The different lung segmentation techniques have been tested using the data set of 9 patients consisting of a total of 413 CT scan slices. The slice thickness was 7.0 mm and the ages of patients varied from 30 to 73 years. We obtained data sets of patients from the Department of Radiology, Aga Khan Medical University Hospital, Karachi, Pakistan. After testing the algorithms against the data sets, the deficiencies of each algorithm have been highlighted.

76.1 Introduction

Computer-aided diagnosis using lung CT images has been a remarkable and revolutionary step in the early and premature detection of lung abnormalities. The computer-aided diagnosis (CAD) systems include systems for the

N. Ahmed Memon (✉)
Faculty of Computer Science and Engineering, Ghulam Ishaq Khan Institute
of Engineering Sciences and Technology, Topi, Pakistan
e-mail: memon_nisar@yahoo.com

N. Mastorakis et al. (eds.), *Proceedings of the European*
Computing Conference, Lecture Notes in Electrical Engineering 27,
DOI 10.1007/978-0-387-84814-3_76, © Springer Science+Business Media, LLC 2009

"automatic detection of abnormality nodules" and the "3D reconstruction of the lung," which assist radiologists in their final decisions. Image processing algorithms and techniques are applied to the images to clarify and enhance the image and then to separate the area of interest from the whole image. The separately obtained area is then analyzed for detection of nodules to diagnose the disease [1].

The accuracy and higher decision confidence value of any lung abnormality identification system depends on an efficient lung segmentation technique. It is therefore vital for the effective performance of any system that the entire lung part of the image is provided to it, and that no part present in the original image is eradicated.

76.1.1 Lung Cancer

The organs and tissues of the body are made up of tiny building blocks called cells. Cells in different parts of the body may look different and work differently, but most reproduce themselves in the same way. Cells are constantly becoming old and dying, and new cells are produced to replace them. Normally, the division and growth of cells is orderly and controlled; but if this process gets out of control for some reason, the cells will continue to divide and develop into a lump, which is called a tumour. Tumours can be either benign or malignant, as shown in Fig. 76.1 [2]. Cancer is the name given to a malignant tumour. A cancerous (malignant) tumour consists of cells which have the ability to spread beyond the original site. If left untreated, they may invade and destroy surrounding tissues. Sometimes cells break away from the original (primary) cancer and spread to other organs in the body by traveling in the bloodstream or lymphatic system. When these cells reach a new area of the body they may go on dividing and form a new tumour, often referred to as a "secondary tumour" or a "metastasis." It is important to realize that cancer is not a single disease with a single type of treatment; there are more than 200 different kinds of cancer, each with its own name and treatment.

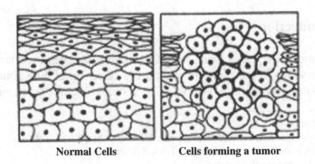

Normal Cells   **Cells forming a tumor**

Fig. 76.1 Normal and benign cells

Lung cancer is a malignancy of the lungs that is the leading cause of cancer deaths for both men and women globally. About 90 percent of all lung cancer occurs in current or former smokers. The American Cancer Society estimates that 164,000 new cases of lung cancer are diagnosed annually in the United States, and an estimated 157,000 people die from the disease each year [3]. According to the Canadian Cancer Society, 20,600 new cases of lung disease are diagnosed in Canada annually, and the disease causes 17,700 deaths a year. Most cases are diagnosed between the ages of 55 and 65. The number of yearly deaths almost equals the number of new cases. Symptoms do not appear until the disease is quite advanced.

76.1.2 Medical Image Segmentation

Image segmentation refers to the process of partitioning an image into distinct regions by grouping together neighboring pixels based on some predefined criterion of similarity. This criterion can be determined using specific properties of pixels representing objects in the image. In other words, segmentation is a pixel classification technique that allows the formation of regions of similarities in the image [4].

Segmentation has remained an important tool in medical image processing, and it has been useful in many applications. These applications include detection of the coronary border in angiograms, multiple sclerosis lesion quantification, surgery simulations, surgical planning, measuring tumor volume and its response to therapy, functional mapping, automated classification of blood cells, studying brain development, detection of microcalcification on mammograms, image registration, atlas matching, heart image extraction from cardiac cineangiograms, detection of tumours, etc. [5].

In medical imaging, segmentation is important for feature extraction, image measurements, and image display. In some applications it may be useful to classify image pixels into anatomical regions, such as bones, muscles, and blood vessels; while in others, into pathological regions, such as cancer, tissue deformities, and multiple sclerosis lesions. In some studies the goal is to divide the entire image into subregions such as the white matter, gray matter, and cerebrospinal fluid spaces of the brain; while in others one specific structure has to be extracted, for example breast cancer from magnetic resonance images.

76.1.3 Segmentation Techniques

Segmentation refers to the differentiation of the objects of interest in an image from the background. For intensity images, i.e., images which are represented by point-wise intensity levels, four popular approaches are used: (a) threshold techniques, (b) edge-based methods, (c) region-based techniques, and (d) connectivity-preserving relaxation methods.

76.1.3.1 Threshold Techniques

Based on image characteristics and grey-level pixel information, threshold techniques work best when the objects have a range of levels which squarely falls outside the range of levels in the background. The method does not work well with blurred regions, since spatial information is ignored. Thus, if blurred images do exist in the case, thresholding should be avoided.

76.1.3.2 Edge-Based Methods

Edge-based methods primarily emphasize the shapes and contours of the objects under consideration by highlighting their boundaries. The inherent weakness of these methods is the connecting of broken contour lines, which makes them prone to failure in the presence of blurring.

76.1.3.3 Region-Based Methods

A region-based method usually groups neighboring pixels of similar intensity levels into connected regions. Adjacent regions are then merged under some criterion involving, perhaps, homogeneity or sharpness of region boundaries. Very strict merging criteria create fragmentation; lenient ones overlook blurred boundaries and over-merge. Hybrid techniques using a mix of the methods above are also popular.

76.1.3.4 Connectivity-Preserving Relaxation Methods

The active contour model is the connectivity-preserving relaxation-based segmentation method which uses spline curves to approximate shapes. The main idea is to start with some initial boundary shape represented in the form of spline curves, and iteratively modify it by applying various shrink/expansion operations according to some energy function. Although the energy-minimizing model is not new, coupling it with the maintenance of an "elastic" contour model gives it an interesting new twist. The dilemma is that the method can converge to local minima, which must be avoided.

76.2 Literature Survey

Samuel et al. [6] have introduced the use of the rolling ball algorithm for the segmentation of lungs. At the first stage, each CT-image is grey level thresholded to segment the thorax from the background, and then the lungs from the thorax. In the next step, the rolling ball algorithm is applied to the lung segmentation contours to avoid the loss of juxtapleural nodules. The authors have used a 17-case database for verifying the results with the help of radiologists. Kerr [7] has introduced another method of segmentation which he has called the TRACE method for the segmentation of lungs. The size, shape, and texture of lungs vary considerably between patients, and among the images of a single patient, because of the possible presence of various disease processes, and the change in anatomy with the vertical position. Consequently, the boundary between the lung and the surrounding tissues can vary from a smooth-edged, sharp-intensity transition to irregularly jagged edges with a less distinct intensity transition. The TRACE algorithm has been developed with the idea in mind of a nonapproximating technique for edge detection and representation. Shiying et al. [8] have presented a fully automatic method for identifying lungs in 3D pulmonary X-ray CT images. The method is divided into three main steps: 1) the lung region is extracted from a CT-scan image by grey-level thresholding, 2) the left and right lungs are separated by identifying the anterior and posterior junctions by dynamic programming, and 3) a sequence of morphological operations is used to smooth the irregular boundary along the mediastinum in order to obtain results consistent with those obtained by manual analysis, in which only the most central pulmonary arteries are excluded from the lung region. Boscolo et al. [9] have used a novel segmentation technique that combines a knowledge-based segmentation system with a sophisticated active contour model. This approach uses the guidance of a higher-level process to robustly perform the segmentation of various anatomic structures. The user needs to provide the initial contour placement, and the high-level process carries out the required parameter optimization automatically. El-Baz et al. [10] have developed a fully automatic computer-assisted diagnosis (CAD) system for lung cancer screening using chest spiral CT scans. This paper presents the first phase of an image analysis system for 3D reconstruction of the lungs and trachea, detection of lung abnormalities, identification or classification of these abnormalities with respect to a specific diagnosis, and distributed visualization of the results over computer networks. Binsheng et al. [11] have used the method of selecting the threshold by analyzing the histogram. The threshold is then used to initially separate the lung parenchyma from the other anatomical structures on the CT images. As the apparent densities of voxels and bronchial walls in the lungs differ, structures with higher densities, including some higher density nodules, could be grouped into soft tissues and bones, leading to an incomplete extraction of the lung mask. To obtain a complete hollow-free lung mask, morphological closing is applied. A spherical shape of the structural element is chosen for the morphological operator and the filter size is approximately determined. With the help of a 3D mask, the lungs can be readily extracted from the original chest CT images. El-Baz et al. [12] have used optimal grey-level thresholding for the extraction of the thorax area. Once the threshold has been selected and applied, region growing and connectivity analysis are used to extract the exact cavity region with accuracy. This scheme gives promising results for all CT sections which have unique intensities of lung parenchyma.

Both [8, 10] and [12] have used iterative grey-level thresholding for the extraction of the thorax and lungs from the CT scan images. For these techniques, and a few more like region growing [13] and the active contour method [13], it is necessary for their accurate and precise performance that there should be a distinct outer black region, which should enclose the lung parenchyma and the adjoining features. However, it may happen that the lung tissue and cavities may not be enclosed in a completely black outline, and thus these methodologies may only work partially.

76.3 Implementation

We have implemented two algorithms. The algorithms are: (1) the thresholding and morphology algorithm proposed by Zhao et al. [11], and (2) the thresholding and rolling ball algorithm proposed by Armato et al. [6]. We have assigned the terms "Scheme-I" and "Scheme-II" to these two algorithms, respectively.

76.3.1 Scheme-I (Thresholding + Morphology)

In Scheme-I, Zhao et al. [11] have used optimal thresholds after analyzing the histogram of the chest CT scan image in order to segment the thorax area from each CT scan image. The morphological closing is then applied to the resultant images. The overall flow chart of the scheme is shown in Fig. 76.2. The iterative method has been used for threshold selection. Once the threshold is calculated, it is applied to each CT scan image, which gives the binary image as the result. The voxels having a density lower than the threshold value are recognized as lung candidates, assigned value "1," and appear white; whereas other voxels are assigned value "0" and appear black, as shown in Fig. 76.3(b).

Because of their low densities, both lung parenchyma and background will be classified as the "lung" on the resultant binary images. Because the lung parenchyma is usually completely isolated from the background by the chest wall, it can be readily determined by labeling the 3D-connected components (i.e., grouping geometrically connected voxels that have value "1" and assigning an identical number to the voxels in each individual group) and selecting the largest component that does not touch any margin of the images. The resultant image is shown in Fig. 76.3(c). As the apparent densities of the vessels and bronchial walls in the lung differ, structures with higher densities, including some higher density nodules, could be grouped into soft tissues and bones, leading to an incomplete extraction of the lung mask as shown in Fig. 76.3(c). To obtain a complete, hollow-free lung mask, morphological closing is applied. Figure 76.3(d) shows the result. A spherical shape of the structural element is

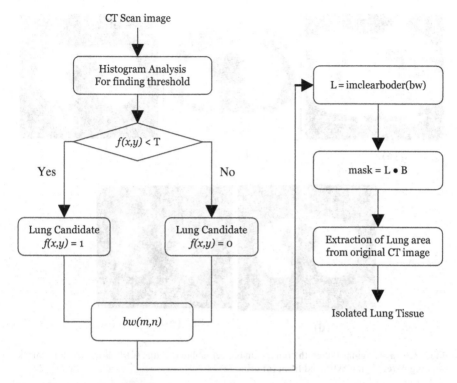

Fig. 76.2 Flow chart of scheme-I

chosen for the morphological operator. We chose a ball of radius 10 pixels for the closing operation. The final lung parenchyma has been extracted with the help of the lung mask. The final result is shown in Fig. 76.3(e). After implementing the algorithm of scheme-I, we applied this scheme on the images received from Aga Khan University Hospital, Pakistan. The results are shown in Fig. 76.4.

76.3.2 Scheme-II (Thresholding + Ball Algorithm)

Armato et al. [6] have used the grey-level thresholding methods for each CT scan to segment the thorax from the background, and then the lungs from the thorax. A rolling ball algorithm is then applied to the lung segmentation contours to avoid the loss of juxtapleural nodules. The flow chart of the algorithm is shown in Fig. 76.5. In order to implement the algorithm, we first used the method of grey-level thresholding. After the computation of the threshold, it was applied to the corresponding CT scan image. The original CT scan image is shown in Fig. 76.6(a). After thesholding we have a binary

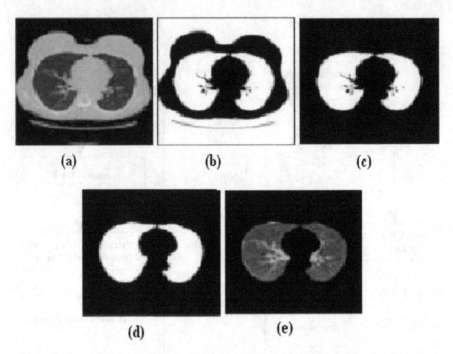

Fig. 76.3 (**a**) Original image (**b**) binary image (**c**) isolated lungs (**d**) hollow-free lung mask (**e**) lung parenchyma extracted from (**a**) using (**c**)

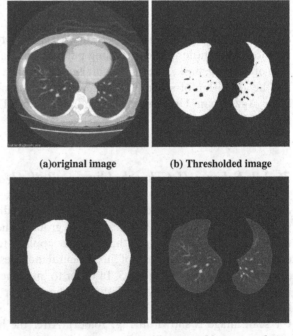

Fig. 76.4 Results of scheme-I

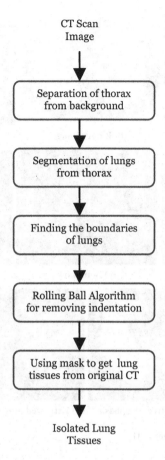

Fig. 76.5 Flow chart of Scheme-II

image. The voxels having a density lower than the threshold value are recognized as lung candidates and assigned the value "1," whereas other voxels are assigned the value "0," as shown in Fig. 76.6(b). Because the lung parenchyma is usually completely isolated from the background by the chest wall, it is readily determined by labeling the 3D-connected components and selecting the largest component that does not touch any margin of the image. The result is shown in Fig. 76.6(c). The gaps in the lung parenchyma due to other anatomical structures are filled in order to get the hollow-free lung mask, as shown in Fig. 76.6(d). Because of nodules present at the lung wall, parts of the lungs have been eradicated and indentations are produced, as shown in Fig. 76.6(e). So in order to overcome this loss of juxtapleural nodules, the rolling ball algorithm is applied to fill the indentations. A spherical shape structuring element of radius 10 pixels has been used. Figure 76.6(f) shows the result. Again holes are filled to get the complete lung mask as shown in Fig. 76.6(g). Finally, the lung parenchyma has been extracted from the original CT scan image with the help of the mask. Fig. 76.6(h) depicts the result. After implementing the algorithm of scheme-II, we

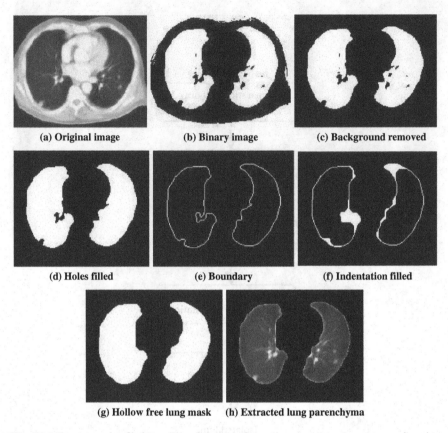

(a) Original image   (b) Binary image   (c) Background removed

(d) Holes filled   (e) Boundary   (f) Indentation filled

(g) Hollow free lung mask   (h) Extracted lung parenchyma

Fig. 76.6 Process cycle of scheme-II

used it as a test on the image database. The results are shown in Fig. 76.7. They show that the algorithm gives the promised results. It fills the indentations produced by the juxtapleural nodules.

76.4 Results

After implementing the above two schemes, we found some limitations of each, and have described them below:

76.4.1 Limitations of Scheme-I

While testing the database slice by slice it was found that the spherical structuring element selected for morphological opening to get the hollow-free lung mask worked for almost the entire database of the patient. However, for a few slices it did not give promising results. Those results are shown in Fig. 76.8.

Fig. 76.7 Results of scheme-II

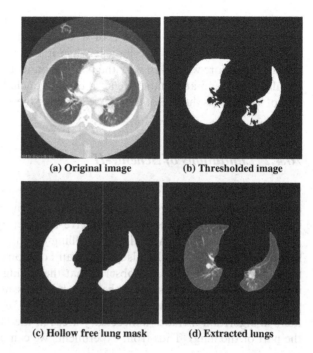

(a) Original image   (b) Thresholded image

(c) Hollow free lung mask   (d) Extracted lungs

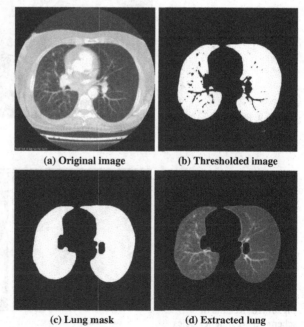

(a) Original image   (b) Thresholded image

Fig. 76.8 Results of scheme-I with deficiencies

(c) Lung mask   (d) Extracted lung

Thus, it was observed that the ball size selected for morphological closing did not work for the entire database of a single patient. Sometimes a ball of specified size worked for one patient while it was not sufficient for another patient. Therefore we had to vary the ball size patient by patient, which makes the algorithm semi-automatic. Hence it can be concluded that the main limitation of scheme-I is the size of the ball selected for the morphological operation.

76.4.2 Limitations of Scheme-II

While testing the database of each patient slice by slice for scheme-II, we observed that the ball algorithm was almost working for each patient's database, but for a few slices it did not give satisfactory results, as shown in Figures 76.9 and 76.10. Applying this rolling ball algorithm to overcome the loss of juxtapleural nodules also creates an additional processing time overhead. From Fig. 76.9 we can observe that the rolling ball algorithm fills the indentations on the left side of the lung but does not completely fill those produced on the right side of the lung. From Fig. 76.10 we can observe that the ball algorithm includes unnecessary areas in the lung region (not actually the part of the lung). Thus from experiments we can conclude that scheme-II

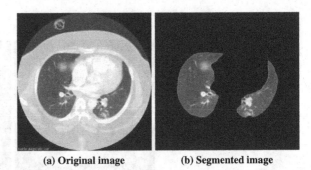

Fig. 76.9 Results of scheme-II with deficiencies   **(a) Original image**   **(b) Segmented image**

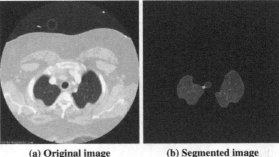

Fig. 76.10 Results of scheme-II with deficiencies   **(a) Original image**   **(b) Segmented image**

has the following very serious shortcomings: (1) a fixed ball size, (2) processing time overhead, and (3) the inclusion of unnecessary areas as lung regions.

76.5 Discussion

In this work, we have studied the performance of the different segmentation techniques that are used in computer-aided diagnosis (CAD) systems using thorax CT scans. These methods for segmentation give good results on test databases of reasonable size. There is still much room to work on computer-aided diagnosis in CT imaging. CT lung density is influenced by factors such as subject tissue volume, air volume, image acquisition protocol, physical material properties of lung parenchyma and transpulmonary pressure. These factors make the selection of a grey-level segmentation threshold difficult, because different thresholds are likely required for different subjects. The selection of the optimal threshold by the iterative method is one possible solution. But again, because of different densities of the anatomical structures on each slide, as discussed above, it is necessary to compute the threshold for each slide, which is costly from a computational point of view. Thus there is a need for a sophisticated algorithm which can calculate a single threshold for the entire database of a single patient. Also, the algorithms reported so far have used databases with slice thicknesses usually in the range of 5–10 mm. But with the advent of modern MSCT scanners and their capability to acquire submillimeter slice data over the whole thorax with a single breath-hold, the CAD software algorithms for the diagnosis of lung cancer must be improved, using databases having a slice thickness of 1 mm or less.

Acknowledgment   The authors are very grateful to the Faculty of Computer Science and Engineering of the Ghulam Ishaq Khan Institute, for providing the research resources for this study. The work has been financed by the Quaid-e-Awam University of Engineering and Technology, Nawabshah, Pakistan, under the faculty development programme.

References

1. Khawaja MA, Aziz MZ, Iqbal N (2004) Effectual lung segmentation for CAD systems using CT scan images. In: Proceedings of IEEE, INMIC conference, FAST, Lahore, Pakistan
2. Strickland RN (2002) Image processing techniques for tumor detection. Marcel Dekker, New York
3. Wicmker R, Rogalla P, Blaffert T (2005) Aspects of computer-aided detection (CAD) and volumetry of pulmonary nodules using multislice CT. Brit J Radiol 78:46–56
4. Dhawan AP (2003) Medical image analysis. IEEE press series in biomedical engineering. Wiley, New York
5. Kogowska J (2000) Overview and fundamentals of medical image segmentation. Handbook of medical imaging. Academic Press, San Diego, CA, pp 69–85

6. Armato SG III, Giger ML, Moran CJ (1999) Computerized detection of pulmonary nodules on CT scans. Radiographics 19:1303–1311
7. Kerr J (2000) The TRACE method for segmentation of lungs from chest CT images by deterministic edge linking. University of New South Wales, Department of Artificial Intelligence, Australia
8. Hu S, Huffman EA, Reinhardt JM (2001) Automatic lung segmentation for accurate quantification of volumetric x-ray CT images. IEEE T Med Imaging 20(6)
9. Boscolo R, Brown MS, McNitt-Gray MF (2002) Medical image segmentation with knowledge-guided robust active contours. Radiographics 22:437–448
10. El-Baz A, Farag AA, Falk R, La Rocca R (2002) Detection, visualization, and identification of lung abnormalities in chest spiral CT scans: phase 1. International conference on biomedical engineering, Cairo, Egypt
11. Zhao B, Gamsu G, Ginsberg MS (2003) Automatic detection of small lung nodules on CT utilizing a local density maximum algorithm. J Appl Clin Med Phys 4(3)
12. El-Baz A, Farag AA, Falk R, La Rocc R (2003) A unified approach for detection, visualization, and identification of lung abnormalities in chest spiral CT scans. Proceedings of the conference on computer-assisted radiology and surgery, London
13. Wan S-Y, Higgins WE (2003) Symmetric region growing. IEEE T Image Process 12(8)

Chapter 77
Segmentation of Anatomical Structures Using Volume Definition Tools

S. Zimeras and G. Karangelis

Abstract The definition of structures and the extraction of an organ's shape are essential parts of medical imaging applications. These might be applications like diagnostic imaging, image-guided surgery, or radiation therapy. The aim of the volume definition process is to delineate the specific shape of an organ on a digital image as accurately as possible, especially for 3D rendering, radiation therapy, and surgery planning. This can be done either through manual user interaction or by applying imaging processing techniques for the automatic detection of specific structures in the image using segmentation techniques. Segmentation is the process that separates an image into its important features (primitives) so that each of them can be addressed separately. This converts the planar pixel of the image into a distinguishable number of individual organs or tumors that can be clearly identified and manipulated. The segmentation process might involve complicated structures, and in this case usually only an expert can perform the task of identification manually on a slice-by-slice basis. Humans can perform this task using complex analysis of shape, intensity, position, texture, and proximity to surrounding structures. In this study we present a set of tools that are implemented in several computer-based medical applications. The central focus of this work is on the techniques used to improve the time and interaction needed by a user when defining one or more structures. These techniques involve interpolation methods for manual volume definition as well as methods for semi-automatic organ shape extraction. Finally, we will investigate volume segmentation aspects within the framework of radiation therapy.

S. Zimeras (✉)
Department of Statistics and Actuarial-Financial Mathematics, University of the Aegean, G.R.832 00, Karlovassi, Samos, Greece
e-mail: zimste@aegean.gr

N. Mastorakis et al. (eds.), *Proceedings of the European Computing Conference*, Lecture Notes in Electrical Engineering 27, DOI 10.1007/978-0-387-84814-3_77, © Springer Science+Business Media, LLC 2009

767

77.1 Introduction

Volume definition aims to localise areas of interest as well the surrounding tissues, for use in several medical imaging applications. Methods for performing segmentations vary widely, depending on the specific application, the imaging modality, and other factors. This might involve treatment (e.g., cancer treatment, image-guided surgery, and invasive techniques) or diagnostic imaging [1–3]. In general, there is currently no single segmentation method that yields acceptable results for every kind of medical imaging. Nowadays medical imaging devices give us the capability to acquire large amounts of digital data, providing high-resolution information in a short time. Although this improves visualisation results, it also increases the work effort during volume definition because the user has to go through a large number of images step-by-step. Traditionally this is done on the original acquired image plane. This means that on each original volume plane a number of contours are defined, relative to the number of structures we want to have. The user frequently has the capability to modify and edit the contour or each contour point. Eventually, the sequence of the contoured objects is triangulated, creating surface objects that can be reconstructed in three dimensions. The target definition can be used for calculations and measurements or as an orientation indicator. For example, in the case of diagnostic imaging, the physician needs to know the size of a structure on a specific or arbitrary plane, the volume it occupies, and its 3D shape. In this study we will present a number of methods that improve the time and interaction one needs to define one or more structures. These techniques involve interpolation methods for the manual volume definition and methods for the semi-automatic organ shape extraction (segmentation techniques). Once an accurate segmentation is obtained, this information may be used by doctors to compare the volume and morphology characteristics of each region against known anatomical norms, other regions in the same image set, and corresponding regions in related image sets. The above tools are already in use and involve applications of the InViVo family. InViVo is a 3D visualization system of medical volume data that has been developed over several years by Fraunhofer-IGD, and applied in ultrasound diagnostic imaging, interstitial brachytherapy, and virtual simulation of external beam radiotherapy.

77.2 Manual Volume Definition

The manual volume definition is the most common and traditional way to define contours. Usually the user has the capability to create a contour line using either discrete or continuous points. The manual process is used mainly by the physician to define the target volume or volume of interest in applications like surgery or cancer treatment. With the manual definition the user has under his/her control the whole process, and can achieve, according to her/his judgement, the required accuracy. It is difficult to replace the manual process with an automatic one, since the physician is the one that will decide about the final result.

Depending on the application, the volume data set, the available editing tools, and the doctor's experience, the target and structure volume delineation might take from 10 min to 45 min. In most systems, the volume definition tools can be applied only in the original, usually axial direction for the CT and MR data acquisitions, although images on orthogonal reconstructed planes are provided as well. These views usually are the coronal and the sagittal. The reconstructed views provide an excellent tool for navigation and structure investigation. In this study, the implemented volume definition tools are capable of drawing in any reconstructed viewing direction.

77.2.1 Linear Interpolation

Linear interpolation between contours is the first approach used to provide an acceleration tool for manual contouring (Fig. 77.1). The mechanism of the linear interpolation is applied when at least one slice exists between the key contours. By key contours we mean the original contours the user has drawn. To perform the linear interpolation we create triangles between the contour points of the key contours, as described in [1]. For this operation, both contour's points must be rotated towards the same direction. The interpolated contour points are created after calculating the intersection of each triangle side with the intermediate slice.

In some cases organs of interest create branches, for example when the air canal divides in two sections for the left and right lung, and the vein tree in the liver. Although we see only parts of the tree body and its branches, the whole tree must be considered a complete structure. To achieve this in practice, the system should allow the design of bisections (or bifurcations). When the user draws a subdivided contour into at least two contours without leaving any

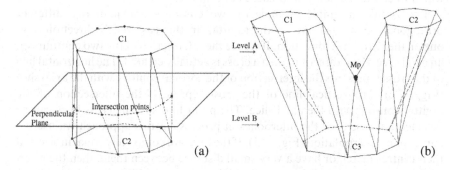

Fig. 77.1 On the *left side,* a simple case where an interpolated contour is created from the plane intersection with the triangles. On the *right side,* an example of contour bisection

empty slice between the key contours, the decision for branching the structure is rather simple. A more complicated case is created when the user leaves an empty slice between the key contours. In this case, a more sophisticated interpolation has to be performed. The approach we use to handle this problem is again based on triangle connections between the key contours, similar to the single contour linear interpolation.

To make a simple example: We subdivide one contour into two that have a theoretical number of slices between them. The key factor to this process is the three-dimensional position of the point (Mp) between the originally drawn contours. Assuming that the original contours are drawn in the axial slice direction, the horizontal (Xc) and vertical (Yc) position is equal to the centre of the bounding box of the two contours that are on the same slice level. The (Zc), which defines the longitudinal position of the Mp, is the mean value of the distance between the bounding box z-value and the z-value of the original contour C3. This point indicates the connection point of the branch. Most points of the two original contours that are located near their bounding box will connect through this point for the triangle generation (Fig. 77.1b). If an axial plane cuts the triangulated branched object at level A, then the intersection points will result in two interpolated contours. If we perform the same cut at level B, the result will be a merger of the two contours on that level.

77.2.2 Orthogonal Contour Interpolation

The orthogonal contour interpolation serves to create a volume combining and interpolating orthogonally drawn contours. The idea for this tool came from the fact that in some cases there are structures in the human body that can be visualised and investigated in a better way on orthogonal reconstructed slices, since they are located longitudinally in the human body. Principally, the algorithm needs at least two orthogonal contours to work. The plane perpendicular to these two contours creates intersection points that are the key points on which to create the new interpolated contour.

Let us take a simple example of two contours drawn in two different orthogonal directions: one on the sagittal, in this case the X direction, and one on the coronal, Y direction. If we see the intersection of the two contours on an axial slice, the Z direction, then a cross is actually created. The horizontal line of the cross represents the intersection of the coronal contour with the axial slice (Fig. 77.2). The vertical line of the cross represents the intersection of the sagittal contour with the axial slice. The point limits of the cross lines can be detected and used for the interpolation process. In this approach we use the cubic spline interpolation (Fig. 77.3). If the lines are completely equal in size and their centres match or have a very small distance between them, then the result of the interpolation will be a circle. If the two vertical lines are unequal, the result will be an ellipse. The same method is applied to every axial contour

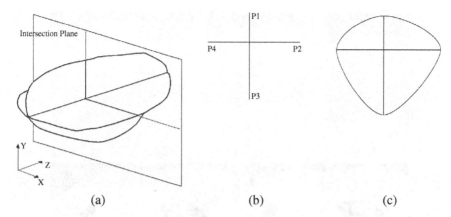

Fig. 77.2 The orthogonal contour interpolation. In (**a**), the two drawn contours and the intersection plane are in the Z direction. The result of the intersection is shown in (**b**). After applying the cubic spline interpolation, a new contour is created (**c**)

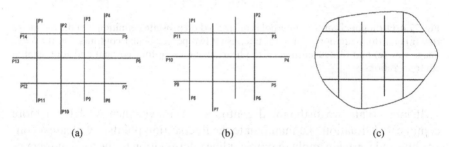

Fig. 77.3 Here we see a more complicated case for the orthogonal contour interpolation. In (**a**), we see the intersections of the sagittal and coronal drawn contours. After applying the convex hull algorithm (**b**), the number of points has been reduced. The final contour is created after applying the cubic spline interpolation (**c**)

separately. The final result is to create a volume that on the axial direction has the same size as the maximum of the originally drawn perpendicular contours (Fig. 77.2).

We should make a special decision in circumstances when the two contours have different longitudinal (Z) length, as in our example, or when the centres of their intersections are at a big distance, which usually happens at the end sides of the contours. In the first case there are two alternatives: either expanding the shorter contour until both longitudinal lengths are the same, or using the longer intersection to create a new drawing which the user will modify later. This can be a circle with the same diameter as the intersection length. If the two centres of the intersection lines are relatively far apart, then these intersections are ignored.

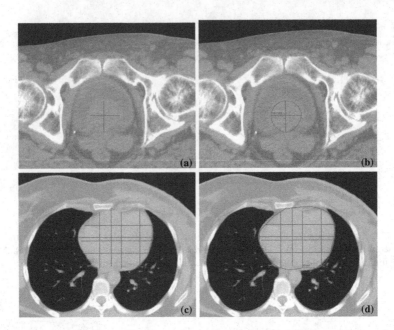

Fig. 77.4 Orthogonal contour interpolation applied on prostate definition. On the *top* image row (**a**) and (**b**): definition of the prostate, using two perpendicular contours. On the *lower* image row (**c**) and (**d**): generation of the heart shape using four coronal and four sagittal contour sections

If more than two orthogonal contours are drawn, then we have a more complicated situation. To manipulate the intersection points, we create a convex hull and we again apply the cubic spline interpolation to the total number of convex hull points (Fig. 77.4). The described tool offers the user the ability to quickly define volumes that have an elliptical or spherical shape. The main disadvantage of this technique is that it cannot handle complicated structures like concave objects.

77.3 Volume Interpolation From Non-Parallel Contours

The interpolation techniques described above are usually applied only on a single slice level (2D). The use of high resolution CT data allows the use of multiplanar reconstructions (MPR) for the sagittal and coronal directions, in relation to the patient's anatomy. These two images are orthogonal to each other and perpendicular with the axial plane. Navigation through these images helps in the observation of complex anatomy. The sagittal and coronal views often offer a better overview of organs' 3D shapes. Defining volumes in these directions could be of benefit, since several organs are aligned along the longitudinal body axis. Although this can be just a visual trick, because physicians are used to working with X-ray images and observing structures in relation to the rest of the anatomy, they regard the multiplanar reconstructed images as providing additional information about the shape of the organ. The MPR techniques are used in MRI imaging, in which the image reconstruction plane can be placed in an oblique orientation for the investigation of the complex anatomy of brain tumors.

The problem we have to solve in our case is the generation of a surface and contours from structured, closed, non-parallel contours. The data points represent the contour points as they are generated by the user on the different levels of the axial slices and/or on the MPRs. The problem of the non-parallel contours could be also formulated as the generation of surfaces from scattered data, which are very common in industrial applications [4, 5]. There is a large variety of solutions to that problem. Gibson [6] extracted smooth surface models by treating the existing binary data as a constraining element in an energy-minimizing deformable surface system. The resulting data structure could either be used to create Euclidean distance maps for direct volume rendering, or be employed directly as a polygonal surface model [6]. Whitaker [7] presented a modified version of the constrained deformable surface model. His constrained level-set model approach creates smooth models while bypassing the need for separate surface representation. However, while these methods generate smooth surface representations, both the level-set model and the surface net remain discretely sampled and retain the problem that they are non-scale invariant.

The main restriction of this method is the relatively small number of samples that can be handled. This drawback makes the above approach unsuitable with applications for which a large number of sampling points are needed.

In this study, we demonstrate the use of implicit function interpolation to reconstruct 3D organ shapes from closed planar parallel and non-parallel contours that have been defined selectively by the user. The total number of contour points will be used as the input data to the implicit surface algorithm with arbitrary order. The number of these sampling points will not exceed the level of a few hundred, and therefore the calculation times will be in acceptable ranges despite the complexity of the algorithm. The output result of the reconstruction algorithm is provided in two forms: as a triangulated mesh, or as multiple parallel contours extracted in arbitrary plane directions. We can create smooth functions for a given 2D or 3D shape using implicit functions. Our interest is the use of 2D planar contours as input, and to generate a smooth 3D shape as output. The boundary constraints refer to the on-surface points. The question now becomes how to generate the off-surface points. Since in this approach we use closed planar 2D curves, we have to specify the number of points that should be interior and exterior to the curve. Paired with each boundary constraint is a normal constraint, which is a location at which the implicit function is required to take any non-zero positive and negative value (off-surface points). The locations of the normal constraints should be towards the interior of the desired curve, and the line passing through the normal constraint and its paired boundary constraint should be parallel to the desired normal to the curve. The function that is returned is an implicit function that describes our curve and the interpolation curve.

The most powerful advantage provided from the implicit function interpolation is the capability to interpolate constraints located in 3D space. Investigating anatomical structures in these images very often provides more information than the standard cross-section (axial) images. Probably the greatest advantage is that organ length and shape can be better appreciated. For example, if one is looking for the length of the kidney through the axial slices, then more than 20 slices with 5 mm voxel thickness in the z direction have to be investigated until the positive and negative z direction limit is found. If, of course, the slice thickness is smaller, then the number of slices increases and even more slices have to be investigated. In the framework of 3D simulation, powerful and efficient segmentation tools must be present. Hence, we adopted a fully 3D manual contour definition for the organ and tumor segmentation with the help of the implicit function interpolation.

To realize this concept, the system allows the user to define closed planar contours on any of these planes. An even bigger degree of organ definition could be offered by allowing the segmentation of structures on any arbitrary cutting plane of the volume. The limitation of such an approach will be the difficulty in orienting the cutting plane in the volume. The definition of anatomical structures in the sagittal and coronal images offers a better description of the organ end shape, providing more information than the axial contours. A common example is the round-off shape of organs, which is omitted when using axial planar contours, as shown in the previous example. However, this important detail must be included and taken into account for visualization and calculation purposes. The most obvious problem one might face during nonparallel contour definition is the possibility of having different contour sizes. Usually anatomical structures in images have a different appearance. This is due to image quality factors such as CT acquisition parameters, slice thickness, and internal organ movements. During manual contour definition, it is often the case that the sagittal or the coronal contours have unequal slice lengths at the organ extremities. Considering this effect on the global surface shape we are about to reconstruct, it is rather less important. However, this problem becomes more noticeable when getting near those regions where the non-parallel contour constraints are close to each other (organ extremities) and the implicit function is forced to pass though all sample points.

The implicit surface interpolation algorithm can be applied even if we consider a large number of contours in all three main planes. This will increase the number of sample points and will make the topological relation among them more complex. As already mentioned, defining structures on the basis of the sagittal and coronal reconstructions presumes that the image quality used is good enough to discriminate the necessary structures. The example in Fig. 77.5 illustrates the definition of the femur head, which is considered a complex

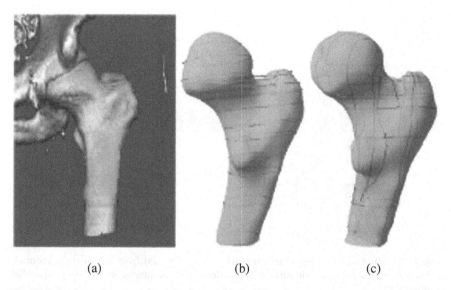

(a) (b) (c)

Fig. 77.5 Surface reconstruction of a femur: (**a**) the 3D volume reconstruction of the femur bone;.(**b**) an implicit surface reconstruction model of the femur using only parallel contours. (**c**) a combination of different non-parallel contours and the final implicit surface of the femur

structure with a rather irregular concave shape. To reconstruct the femur shape using parallel contours (axial direction) we needed to place more than 13 contours, as shown in Fig. 77.5(b). The total area defined within the axial contours was 245.5 cm^2 and the number of sample points used was 315. For the same shape we needed to place nine non-parallel contours, as Fig. 77.5(c) illustrates; five axial, two coronal, and two sagittal, which covered 282.3 cm^2 with 320 sample points. An example of non-parallel contour definition over the CT slices is given in Fig. 77.6. Both implicit surface shapes have been reconstructed with small topological differences from the original shapes. One can notice the difference in the femur head between the parallel and the non-parallel implicit surfaces.

An interesting issue to notice when reconstructing implicit surfaces from non-parallel contours is the inaccuracies of the surface at those points where contours from different directions are intersecting. From the algorithmic point of view, this effect illustrates one important property of the implicit surface interpolation. On the other hand, it introduces topological errors in the shape of the structure we are attempting to reconstruct. To solve this problem we introduce the averaging of the two neighboring sample points. The result can replace the two or more original sample points with a new one decreasing the surface inconsistency. Probably the best way to prevent this effect is to provide the necessary tools to the user in order to be able to adjust mismatches of the sample points interactively, and also to be able to see the result in real time 2D and 3D views.

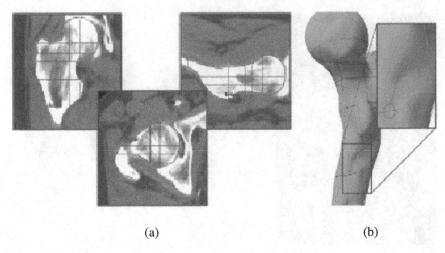

(a)   (b)

Fig. 77.6 Definition of non-parallel contours on CT slices: (**a**) from *left* to *right:* coronal, axial, and sagittal contours; (**b**) surface inconsistencies at the intersections of the nonparallel contours

77.4 Conclusions

In this study, different volume definition techniques have been presented. In the first part we presented interpolation techniques aimed to accelerate the manual contouring process, including the well-known linear interpolation technique implemented for simple bisection organ shapes. In addition, we introduced a new interpolation method, the orthogonal interpolation, which enables the user to define a volume by drawing only a small number of contours. The main drawback of this method is its weakness in handling complex organ shapes. Because of the high order of continuity of the complexity, the interpolation scheme can reconstruct a surface from 3D sample points that describes very accurately the shape of different anatomical structures. Even more advanced is the capability of the algorithm to handle the nonparallel contours, either perpendicular or even arbitrarily placed relative to each other. This has been a new concept for radiotherapy imaging applications. The conventional 2D segmentation tools for axial cross sections have been the standard methods used for many years now. The advantage of the new concept is the flexibility and freedom given to clinicians to investigate new ways of defining anatomical structures, taking advantage of the full 3D information available on the multiplanar reconstructed slices, such as sagittal and coronal. As a result, the segmentation process improves in terms of speed and selectivity. In addition, the time needed to segment structures can be dramatically reduced, since fewer contours are required to describe the shapes of structures.

Acknowledgments The authors would like to thank Professor George Sakas and Professor Nikos Zamboglou for their useful scientific help and comments about the progress of this work. Also, many thanks to MedCom Company and Städtisches Klinikum Offenbach, which provided equipment and medical data sets for the implementation of the above work. This work was supported partially by a Marie Curie Industry Host Fellowship Grant no: HPMI-CT-1999–00005; and the first author is an MCFA member.

References

1. Strassman GJ, Kolotas C, Heyd R (2000) Navigation system for interstitial brachytherapy. Radiother Oncol 56:49–57
2. Hoegemann D, Stamm G (1999) Volumetric evaluation and 3D-visualization of the liver before living-related donation. CARS'99, France, pp 259–252
3. Meyer JL, Purdy JA (eds) (1996) 3D conformal radiotherapy. Front Radiat Ther Oncol 29:31–42
4. Hoppe H, DeRose T, Duchamp T, McDonald J, Stuetzle W (1992) Surface reconstruction from unorganised points. Computer graphics, SIGGRAPH' 1992, pp 71–78
5. Amenta N, Bern M, Kamvysselis M (1998) A new Voronoi-based surface reconstruction algorithm. Computer Graphics, SIGGRAPH' 1998, pp 415–421
6. Gibson S (1998) Using distance maps for accurate surface representation in sampled volumes. In: Proceedings of the symposium on volume visualization, ACM SIGGRAPH, pp 23–30
7. Whitaker R (2000) Reducing aliasing artifacts in iso-surfaces of binary volumes. In: Volume visualization and graphics symposium 2000, ACM SIGGGRAPH, pp 23–32

Chapter 78
Shape Analysis of Heart Rate Lorenz Plots

France Sevšek

Abstract A quantitative geometrical method to analyse heart rate variability is presented. From the heart rate data a Lorentz plot was constructed and converted into a grey-scale image. Imaging techniques were applied to it to determine the outline of the attractor area. Its shape was described in terms of Fourier coefficients. The method was applied to the RR-interval data collected from 15 healthy male subjects (30 ± 4 years) during a step test (YMCA protocol) and the consequent 10 min relaxation.

78.1 Introduction

Under normal physiological conditions, heart rate is not a periodic oscillator—the time interval between heart beats is constantly changing because of both the fluctuating inputs to the system and the dynamic responses of the cardiovascular regulatory mechanisms [1]. Although heart rate variability has been known about for more than a century, its analysis and interpretation are still an active research field, mainly because of new developments in computational and digital signal-processing techniques, as well as new understandings of nonlinear systems [2]. The standard procedure is to deduce the time intervals between R-peaks from measured electrocardiograms. These intervals may be Fourier analysed, but recently, methods of nonlinear dynamics, including the phase space representation, have often been applied [3]. From the heart rate data, a standard Lorenz (or Poincaré) plot is constructed by plotting each RR time interval as a function of the immediately preceding one. These plots give a visual representation of the RR data, but their shapes are also used to classify the data. The shapes of the attractor regions are also determined quantitatively by approximating them by ellipsoids and calculating their principal axes [4]. It was thus of interest to develop a more general method for quantitative representation of the shapes of heart rate

F. Sevšek (✉)
University of Ljubljana, University College of Health Studies, Poljanska 26a,
1000, Ljubljana, Slovenia
e-mail: france.sevsek@vsz.uni-lj.si

N. Mastorakis et al. (eds.), *Proceedings of the European
Computing Conference*, Lecture Notes in Electrical Engineering 27,
DOI 10.1007/978-0-387-84814-3_78, © Springer Science+Business Media, LLC 2009

attractor regions. Here we describe a method based on computer imaging and determination of Fourier coefficients of the graph outline. This work is based on our previous research in shape analysis of phospholipid vesicles [5] and has been briefly introduced elsewhere [6].

The developed method was applied to the data collected during a standard step test procedure. This is a simple method used to estimate the maximal oxygen capacity and thus physical fitness of subjects with submaximal exercise testing [7]. It consists of subjects performing work at a well defined power level long enough for the heart rate to reach the steady state value from which the maximal oxygen consumption may be estimated. In a step test the work is done by stepping on and off a bench at a given pace (usually imposed by a metronome). The work done by the subjects was additionally regulated by changing the mass they were carrying in a backpack.

78.2 Methods

Data analysis was done mostly on a Pentium IV computer, with 512 MB memory, running under the Linux operating system. Computer programmes were specially developed for this purpose—they consisted of routines written in the C language and shell procedures. Heart rate data were collected from 15 healthy male subjects, 30 ± 4 years old, 83 ± 9 kg, with body mass index 26 ± 2 kgm^{-2}. They performed the standard YMCA step test procedure: 24 steps per minute on a 30 cm high bench for 3 min [8]. Each test was repeated four times: the normal test and three tests with additional loads of 5, 10, and 15 kg carried by the subjects in a rucksack. Heart RR intervals were measured by Polar Vantage NV (Polar Electro, Oy, Finland) heart rate monitors during the stepping phase as well as during the following 10 min relaxation period. Prior to the test all subjects signed an informed consent as required by the National Committee for Medical Ethics, which approved this research.

78.3 Image Processing

From the RR time interval data, a standard Lorenz plot was constructed by plotting each RR time interval as a function of the immediately preceding one (Fig. 78.1). Beside these standard Lorenz plots, we also considered difference plots. They were constructed in the same way, but the differences between the RR time interval and the preceding one were plotted on the vertical axes (Fig. 78.2).

All the plots were centred at the average RR value. Data were plotted on a 512×512 byte array of cells where the value in each cell was proportional to the count of the graph points corresponding to it. The resulting array was interpreted as an image with 256 grey levels, and standard image analysis techniques were applied to it.

First, it was normalised to ensure that the full range of 256 values was used, and then smoothed using a $\frac{1}{8}(010/141/010)$filter. Usually the resulting images

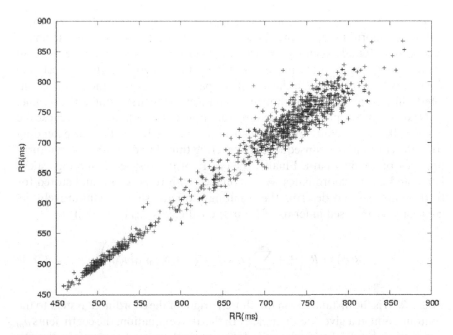

Fig. 78.1 A typical Lorenz plot for RR intervals: the values of RR time intervals are plotted as functions of the previous ones

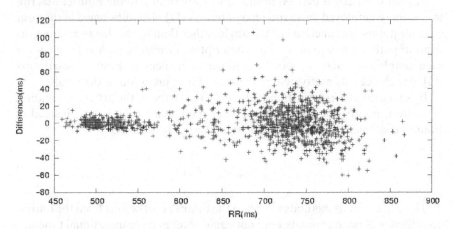

Fig. 78.2 A typical difference plot for RR intervals: the values of the differences of RR time intervals are plotted as a function of RR intervals

consisted of scattered, not connected, points, as a consequence mainly of the short measurement time. On these images the normalisation and smoothing cycles were repeated up to one hundred times until the compact attractor region was obtained. The image was then binarised by thresholding at half of the maximal height.

The outline of the resulting black region was determined by a contour-following algorithm [5]. In the chosen region of interest, a point on the image contour was found. Then a maze-walking algorithm was used to determine all the remaining points of the attractor outline. At each step, it starts to investigate the points from the right to the left of the path and moves to the first encountered white point. This algorithm always returns to the first point of the outline.

Next, a contour-pruning procedure was applied. It was necessary because the maze-walking procedure often yielded some one-pixel-wide threads extending from the contour as a consequence of entering thin closed channels and returning back on its own path. Finally, the centre of the vesicle outline was calculated, and all the coordinates were expressed with respect to it and stored for further analysis. To describe the resulting attractor shape quantitatively, its contour was analysed in terms of Fourier coefficients (a_m ψ and b_m):

$$R(\phi) = R_0 \left(1 + \sum_{m=1}^{m_{max}} [a_m \cos(m\phi) + b_m \sin(m\phi)] \right) \quad (78.1)$$

where $R(\phi)$ is the distance from the chosen origin of the coordinate system to the contour point at a given polar angle ϕ. In the above equation, the coefficients a_m and b_m have been defined relative to (in units of) the average contour radius R_0. Thus, the coefficients depend only on the contour's shape and not on its size.

Fourier coefficients defined in such a way are similar to the Fourier descriptors usually employed in shape recognition [9–11]. The difference is that our contour points are functions of the angle rather than of the distance along the contour path. Although other shape description measures, such as moments or even simple compactness, were sometimes equivalent to Fourier descriptors [12], our choice was motivated by the ease of interpretation of the results.

In analysing the experimental data, only the shape of the attractor region is of interest, not its orientation. It was thus more convenient to use, instead of Fourier coefficients a_m and b_m, their squares (u_m^2) defined as:

$$u_m^2 = \frac{1}{2} [a_m^2 + b_m^2] \quad (78.2)$$

There are various methods to obtain the Fourier coefficients from the outline coordinates. Since our points were not equidistant and computational time was not crucial, we used the most straightforward method—least square fitting of Eq. (78.1) to the experimentally determined contour points. The sum of the squares of the differences between the measured and the calculated vesicle outline points was minimised. The resulting normal equations gave a linear system that could be easily solved. For this purpose the method of LU decomposition [13] was used. It decomposes the matrix into the product of a lower and an upper triangular one, from which the solutions can be calculated by a simple substitution.

78.4 Results

To test the experimental procedure four sessions of 15 heart rate measurements recorded during 3 min step tests and the following 10 min relaxation intervals were analysed. Allowing for two missing measurements, this resulted in 58 heart rate data series. At the beginning of the test the RR intervals quickly decreased and in less than 2 min settled to nearly steady state value. After 3 min, when exercising was finished and the subject lay down, the RR interval quickly increased. Since the heart rate attractor regions were expected to be different during the exercising and resting periods, the two activities were analysed separately.

Because the measuring time was quite short, the resulting graphs were not compact on the 256 × 256 grid. About 100 smoothing and normalising cycles were usually required to obtain graphs with a compact central region. These were then analysed, as described above. It was found that no more than the first ten Fourier coefficients were needed to sufficiently describe the shape of the obtained smoothed attractor region. The quality of fit was also monitored by plotting the contours, as calculated from the resulting Fourier coefficients, over the attractor image (Fig. 78.3).

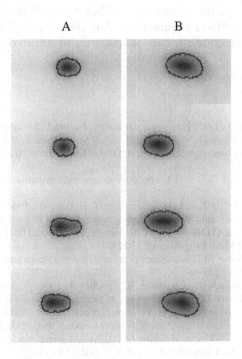

Fig. 78.3 Samples of difference plots for the step test (**A**) and the following relaxation (**B**). Graphs represent measurements with 0, 5, 10, and 15 kg external loads from *top* to *bottom*, respectively

78.5 Discussion

The main purpose of this study was to develop a new method to quantitatively characterise the shapes of the heart rate attractor. The method was tested on data from healthy subjects during exertion and the following relaxation. The reported procedure proved to be efficient at describing heart rate variability data. Visual inspection of the resulting images indicated the quality of the recorded data. The ones with multiple regions were mainly related to changing heart rate regimes during the experiment or to recording problems. But most of the standard Lorenz plots were of the well-known ellipsoidal or club-like shapes, centred along the image diagonal. Although Lorenz plots are more generally accepted to represent the heart rate phase space, the difference plots were preferred, since their interpretation was much more straightforward— they reflect typical variations of HR at any given heart rate value. For shape analysis, the difference plots also proved to be more convenient, as they usually consisted of a symmetric horizontally-extended attractor region. Their shape could be quantitatively determined by the first few Fourier coefficients of the region outline.

This method enables us to qualitatively describe the shapes of heart rate attractors. Because of the problems with the eventual small number of data points and the resulting averaging, it is believed that this method is useful mainly with long period measurements. And the physiological interpretations of the resulting heart rate attractor shapes are also still an open issue.

References

1. Malik M, Camm A (1995) Heart rate variability. Futura Publishing Co., Amonk, NY
2. Stefanovska A, Haken H, McClintock P, Hožič M, Bajrovič F, Ribarič S (2000) Reversible transitions between synchronisation states of the cardiorespiratory system. Phys Rev Lett 85:4831–4834
3. Schmidt G, Morfill G (1994) Complexity diagnostics in cardiology: methods. PACE 17:2336–2340
4. Tulppo M, Mäkikallio T, Takala T, Seppänen T, Huikuri H (1996) Quantitative beat-to-beat analysis of heart rate dynamics during exercise. Am J Physiol 271:H244–H252
5. Sevšek F, Gomišček G (2004) Shape determination of attached fluctuating phospholipids vesicles. Comput Meth Prog Biomed 73:189–194
6. Sevšek F, Jakovljević M (2002) Heart rate variability—a shape analysis of Lorenz plots. Cell Mol Biol Lett 7:159–161
7. Fleg J, Pina I, Balady G, Chaitman B, Fletcher B, Lavie C, Limacher M, Stein R, Williams M, Bazzarre T (2000) Assessment of functional capacity in clinical and research applications: an advisory from the committee on exercise, rehabilitation, and prevention, council on clinical cardiology, American heart association. Circulation 102:1591–1597
8. American College of Sports Medicine (2000) ACSM guidelines for exercise testing and prescription. 6th ed. Lippincott Williams & Wilkins, Philadelphia
9. Bankman I, Spisz T, Pavlopoulos S (2000) Handbook of medical imaging, processing and analysis. Chap. 14. Academic Press, Cleveland
10. Gonzalez R, Woods R (1992) Digital image processing. Addison-Wesley, Boston

11. Sanchez-Marin F (2000) Automatic recognition of biological shapes with and without representation of shape. Artif Intell Med 8:173–186
12. Rangayyan R, El-Faramawy N, Desautels J, Alim O (1997) Measures of acutance and shape classification of breast tumors. IEEE T Med Imaging 16:799–810
13. Press W, Teukolsky S, Vetterling W, Flannery B (1992) Numerical recipes in C. Cambridge University Press, Cambridge

Chapter 79
Human Readable Rule Induction in Medical Data Mining

Nor Ridzuan Daud and David Wolfe Corne

Abstract In general, the human-readable rule refers to data shown in a format easily read by most humans—normally this is in the form of IF...THEN rules. This is the most convenient way for physicians to express their knowledge in medical diagnosis. In particular, if learned diagnostic rules can be presented in such a form, physicians are much more likely to trust and believe the consequent diagnoses. This paper investigates the performances of existing state-of-the-art classification algorithms, mainly rule induction and tree algorithms, on benchmark problems in medical data mining. The findings indicate that certain algorithms are better for generating rules that are both accurate and short; these algorithms are recommended for further research towards the goal of improved accuracy and readability in medical data mining.

79.1 Introduction

In general, the human-readable rule refers to data shown in a format easily read by most humans—normally in the form of IF...THEN rules. This kind of knowledge representation is the most convenient way for physicians to express their knowledge in medical diagnosis. For example, IF (patient_body = weak) and (body_temperature > 101) THEN (diagnosis = fever). Some of the benefits of IF...THEN rules are that they are modular, each defining a relatively small and, at least in principle, independent piece of knowledge. To discover this kind of rule, data mining methods are normally used to extract high-level knowledge from real-world medical data sets. We focus on the data mining task of classification, which is the task addressed by most rule induction algorithms that extract high-level knowledge from data [6]. Rule, or decision tree, induction is the most established and effective medical data mining technology in use today.

N.R. Daud (✉)
Faculty of Computer Science and Information Technology, University of Malaya,
50603 Kuala Lumpur, Malaysia

N. Mastorakis et al. (eds.), *Proceedings of the European*
Computing Conference, Lecture Notes in Electrical Engineering 27,
DOI 10.1007/978-0-387-84814-3_79, © Springer Science+Business Media, LLC 2009

There have been several comparative performance analyses done between rule induction algorithms, but these have been focused mainly on the predictive accuracy aspect of the algorithms [15, 3]. We present a comparative analysis with a focus not only on accuracy but also on the number of rules produced by these algorithms. Normally, a diagnosis is based on a collection of rules generated by an algorithm; hence, if the algorithm produces a smaller collection, we consider it a more human-readable result. Our aim is to find potential rule induction algorithms which can discover human-readable rules from large medical data sets. The candidate algorithms not only must be able to have high predictive accuracy but also must contain as few rules as possible. The medical data sets we are using for this comparative analysis are described further in the next section. We perform our comparative analysis using the WEKA platform as described later in Section 79.3.

79.2 The Medical Data Sets

The data sets used in this investigation were obtained from the UCI Machine Learning Repository at the University of California (UCI). For the investigation, 10 medical data sets (see Table 79.1) have been incorporated in the benchmark analysis because it will be used for further study in data mining to produce interesting human-readable rules, particularly in medical data sets. The size of the data sets ranges from 132 instances to 768 instances. The input data set is in the format of the Waikato Environment for Knowledge Analysis (WEKA) ARFF file format. All the medical data sets are commonly used for supervised learning studies. They contain at least 5–30 nominal and numeric attributes. The classes of the data sets range from 2 to 19 nominal classes. Most of the data types are integer and real-valued attributes [16, 17].

Table 79.1 UCI medical data sets

No	Problem	Instances	Attributes (continuous/nominal)	Classes
1	breast-cancer	286	9 (8/1)	2
2	breast-cancer (Wisconsin)	569	30 (30/0)	2
3	echocardiogram	132	9 (9/0)	2
4	heart-disease	303	13 (13/0)	5
5	hepatitis	155	19 (19/0)	2
6	lymphography	148	18 (18/0)	4
7	pima-diabetes	768	8 (8/0)	2
8	primary-tumor	339	18 (18/0)	19
9	thyroid-gland	215	5 (5/0)	3
10	horse-colic	300	23 (23/0)	2

79.3 Rule Induction Algorithms

The investigation was done using the Waikato Environment for Knowledge Analysis (WEKA) open source software environment. WEKA (version 3.5.3) is a collection of machine learning algorithms for data mining tasks. WEKA provides numerous classification algorithms, but only tree- and rule-based algorithms are used in this investigation because we are interested only in rule induction methods (a decision tree is equivalent to a set of rules). We have identified and tested six rule and five decision-tree algorithms inside WEKA for their suitability in handling nominal class prediction, which is the common type of class attribute for medical data sets.

79.3.1 WEKA Classifier Algorithms

Inside the WEKA system, there exist many classification algorithms which can be divided into two types: rule induction and decision-tree algorithms [18]. Rule induction algorithms generate a model as a set of rules. The rules are in the standard form of IF...THEN rules. Meanwhile, decision-tree algorithms generate a model by constructing a decision tree in which each internal node is a feature or attribute. The leaf nodes are class outputs. Table 79.2 describes the six rule and five decision-tree algorithms chosen for this investigation. Other WEKA classifier algorithms, such as Prism, M5Rules, NNge, LMT, NBTree, RandomForest, ID3, and M5P, are not included because of their inability to handle both nominal and numeric data types, which is needed when considering the typical nature of medical data sets.

Table 79.2 Selected WEKA classifier algorithms

No.	Algorithms	Acronym	Classifier types
1	OneR	OneR	Rule
2	Ridor	Ridor	Rule
3	PART	PART	Rule
4	JRip	JRip	Rule
5	DecisionTable	DT	Rule
6	ConjunctiveRule	CR	Rule
7	J48 (C4.5)	J48	Tree
8	ADTree	ADT	Tree
9	DecisionStump	DS	Tree
10	RandomTree	RT	Tree
11	REPTree	RTree	Tree

79.3.2 Rule and Decision Tree Induction Algorithms Summary

OneR or "One Rule" is a simple algorithm proposed by Holt [10]. The OneR algorithm builds one rule for each attribute in the training data and then selects the rule with the smallest error rate as its "one rule." The Ridor algorithm is the implementation of a RIpple-DOwn Rule learner proposed by Brian and Compton [2]. It generates a default rule first and then the exceptions for the default rule with the least (weighted) error rate. Next it generates the "best" exceptions for each exception and iterates until pure. IREP or Incremental Reduced Error Pruning [9] is used to generate the exceptions.

PART is a separate-and-conquer rule learner proposed by Witten and Frank [18]. The algorithm produces sets of rules called "decision lists," which are ordered sets of rules. PART builds a partial C4.5 decision tree in each iteration and makes the "best" leaf into a rule. The algorithm is a combination of C4.5 [12] and RIPPER [3] rule learning. JRip implements a propositional rule learner, Repeated Incremental Pruning to Produce Error Reduction (RIPPER), which was proposed by Cohen [3, 4], as an optimized version of IREP [9]. RIPPER builds a rule set by repeatedly adding rules to an empty rule set until all positive examples are covered. Rules are formed by greedily adding conditions to the antecedent of a rule (starting with an empty antecendent) until no negative examples are covered. After a rule set is constructed, an optimization postpass massages the rule set so as to reduce its size and improve its fit to the training data.

The DecisionTable algorithm builds a decision rule using a simple decision table majority classifier as proposed by Kohavi [11]. It summarizes the data set with a "decision table" which contains the same number of attributes as the original data set. The ConjuctiveRule algorithm implements a single conjunctive rule learner that can predict numeric and nominal class labels, as described in Conjuctive Rule [5]. A rule consists of antecedents "AND"ed together and the consequent (class value) for the classification/regression. In this case, the consequent is the distribution of the available classes (or mean for a numeric value) in the data set.

The J48 algorithm is the WEKA implementation of the C4.5 top-down decision tree learner proposed by Quinlan [12]. The algorithm uses the greedy technique and is a variant of ID3, which determines at each step the most predictive attribute, and splits a node based on this attribute. It deals with numeric attributes by determining where thresholds for decision splits should be placed. The Alternating Decision Trees (ADTree) algorithm is a generalization of decision trees, voted decision trees, and voted decision stumps [8]. The algorithm applies boosting procedures to decision tree algorithms to produce accurate classifiers. The classifiers are in the form of a majority vote over a number of decision trees, but having smaller and easier to understand classification rules.

The DecisionStump algorithm builds simple binary decision "stumps" (one-level decision tress) for both numeric and nominal classification problems [19]. DecisionStump is usually used in conjunction with a boosting algorithm such as LogitBoost. It does regression (based on mean-squared error) or classification (based on entropy). RandomTree is an algorithm for constructing a tree that considers K randomly chosen attributes at each node, as described in Random Tree [13]. The REPTree algorithm is a fast decision tree learner, as described in REPTree [14]. It builds a decision/regression tree using information gain/variance, and prunes it using reduced-error pruning (with back-fitting).

79.4 Experimental Setup

For the experimental setup, all the original medical data sets are converted to ARFF (Attribute Relation File Format), as this is the input file format for the WEKA system. Next, all the identified algorithms are tested with each medical data set with the option of using 10-fold cross-validation. 10-cross validation (10CV) is a standard way of predicting the error rate [1]. Also, note that for the data testing, we use the standard default setting inside the WEKA system (version 3.5.3) without any modification. For the experimental results, we are interested in the percentage of correctly classified instances of the algorithms (accuracy percentage) and the number of rules or size of trees produced by the classifiers.

79.4.1 Summary of Experimental Results

Table 79.3 below shows a summary of the number of rules produced by the six rule algorithms. It also shows a summary of the size of the trees produced by the five tree algorithms. For the rule algorithms, it is noted that ConjuctiveRule always produces only a single rule, while the other algorithms output between 2 and 434 rules. This is followed by the OneR algorithm which had an average of 4.1 rules. The third algorithm which produced reasonably few and short rules was JRip with an average of 4.5 rules. The Ridor algorithm was the one that produced the longest and most rules compared with other algorithms, outputting an average of 49.6 rules. We can also observe that algorithms such as OneR, JRip, and ConjuctiveRule are among the more stable algorithms in producing fewer rules, with an average of less than 8 rules. However, among the six rule algorithms, the most unstable algorithm is DecisionTable, which is highly variable in the number of rules produced, even for two-class problems.

For the tree algorithms, we note that the DecisionStump algorithm always produces a fixed size of tree (with three nodes), while the other algorithms output trees with between 3 and 551 nodes. This is followed by the REPTree algorithm, which produced trees with an average size of 14.6

Table 79.3 Number of rules or size of trees produced by the algorithms

No	Name	OneR	Ridor	PART	JRip	DT	CR	J48	ADT	DS	RT	RTree
1	pima-diabetes	8	4	13	5	32	1	39	31	3	551	33
2	breast-cancer	7	3	18	3	11	1	14	31	3	505	13
3	echocardiogram	3	2	6	2	3	1	3	31	3	73	7
4	hepatitis	3	2	8	3	46	1	21	31	3	151	7
5	horse-colic	3	5	15	4	24	1	11	31	3	533	9
6	breast-cancer (Wisconsin)	4	5	7	5	88	1	25	31	3	103	9
7	thyroid-gland	3	3	4	5	21	1	17	0	3	41	7
8	lymphography	3	7	11	8	30	1	25	0	3	141	11
9	heart-disease	4	31	49	3	8	0	67	0	3	349	17
10	primary-tumor	3	434	46	7	53	1	81	31	3	517	37
	Average	4.1	49.6	17.7	4.5	31.6	1	30.3	31	3	296.4	14.6

nodes. The third algorithm which produced reasonably small trees was J48, with an average of 30.3 nodes. It is also noted that ADTree was consistent in producing trees with 31 nodes in each test. However, ADTree was unable to handle data sets having more than two prediction classes. RandomTree was the algorithm that produced the largest trees compared with the other algorithms, with an average size of 296.4 nodes. We can also notice that algorithms such as DecisionStump, REPTree, ADTree, and J48 are among the stable algorithms in producing small trees with an average of less than 19.7 nodes. However, among the five tree algorithms, the most unstable algorithm in terms of tree size is RandomTree, even when only two-class problems were considered.

Table 79.4 above shows a summary of the accuracy (percentage of correctly classified instances) produced by the rule and tree algorithms. For the rule algorithms, it was observed that the DecisionTable algorithm was the most accurate classifier (having an average of 74.3%) compared with the other algorithms, which output an average accuracy range between 69.6 and 74.2%. This is followed by the Ridor and PART algorithms, which shared an average accuracy of 74.2%. The third algorithm which produced higher accuracy was JRip, with an average accuracy of 73.5%. ConjunctiveRule had the lowest accuracy compared with the other algorithms, with an average of 69.6%. We can also notice from the results that the overall accuracies of the rule algorithms vary considerably, depending on factors such as the number of attributes and the number of classes of the medical data sets. None of the rule algorithms has consistently good accuracy over the all data sets. We can also observe that the average accuracies are low when the numbers of classes are increasing (higher than four classes), with an average of below 55% for all the rule algorithms.

For the tree algorithms, it was observed that the ADTree algorithm was the most accurate classifier (having an average of 80.5%) compared with the other algorithms, which output an average accuracy range between 69.1 and 74.6%. However, the average accuracy of the ADTree algorithm was based on only six data sets, because the algorithm was unable to handle the processing of data sets with more than two prediction classes. This was followed by the J48 algorithm, which had an average accuracy of 74.2%. The third algorithm which produced higher accuracy was REPTree, with an average of 73.5%. The DecisionStump algorithm had the lowest accuracy compared with the other algorithms by output, an average of 69.1%. Here we can also note from the results that the overall accuracies of the tree algorithms are inconsistent, depending on factors such as the number of attributes and the number of classes of the medical data sets. None of the tree algorithms has consistent accuracy over all data sets. Again, we can also observe that the average accuracies are low when the numbers of classes are increasing (more than four classes), with an average of below 57% for all the tree algorithms.

Table 79.4 Accuracy produced by rule and tree algorithms (%)

No	Name	OneR	Ridor	PART	JRip	DT	CR	J48	ADT	DS	RT	RTree
1	pima-diabetes	71.3	73.4	73	75.6	73.4	73	71.4	74.6	70.4	65.6	75.2
2	breast-cancer	69.5	71.6	70.9	72.3	75.5	71.3	75.1	71.6	72	70.9	71.3
3	echocardiogram	92.3	91.6	87.7	90	92.3	92.3	90.8	90	92.3	78.6	90.8
4	hepatitis	81.9	78	80.6	73.5	73.5	76.7	78.7	80	77.4	77.4	78
5	horse-colic	69.3	67	69	73	70.6	64.6	72	72.6	73	67.6	73.3
6	breast-cancer (Wisconsin)	89.2	93.1	93.4	92.7	92.4	90.1	93.1	94	88.9	90.1	92.4
7	thyroid-gland	92	92.5	93.9	93.4	92	77.6	92	0	77.2	91.1	71.3
8	lymphography	75.6	82.4	79	75	77.7	67.5	78.3	0	60.8	74.3	75
9	heart-disease	52.4	54.4	55.4	51.8	57	55.1	52.4	0	51.4	46.2	56.7
10	primary-tumor	26.8	38.3	39.5	37.7	38.9	27.4	41.8	0	27.1	33.6	41
	Average (%)	72.0	74.2	74.2	73.5	74.3	69.6	74.6	80.5	69.1	69.5	72.5

79.4.2 The Tradeoffs Between Readability and Accuracy

Figure 79.1 above illustrates a summary of results showing the trade-off between readability (sorted according to number of rules or size of tree) and accuracy, for both the rule induction and decision-tree induction algorithms. It can be observed that ConjunctiveRule stands out in terms of having a small number of rules; however its accuracy is not competitive. High accuracy results are dominated by the ADTree algorithm (80.5%). Meanwhile, although the ADTree algorithm had the highest accuracy, its tree sizes are quite large (8th worst in terms of average size). On the other hand, the DecisionStump algorithm, despite having the smallest trees, ranks as the least accurate of all the tested algorithms. In 3rd place in terms of readability is the OneR algorithm, but despite having an average of 4.1 rules, it only managed to get 8th place in the overall accuracy ranking. The JRrip algorithm, although producing an average of 4.5 rules, managed to get the 6th place (73.5%) in the overall accuracy ranking. The last position in the readability ranking is held by the RandomTree algorithm, which also has the second-to-last position in the ranking of overall accuracy. From these results, we can see the predictable and expected tradeoff between accuracy and readability. However, it is perhaps surprising and instructive to see the "shape" of this tradeoff. Generally, the algorithms show a trend in which reduced readability (more rules or larger trees) leads to slightly better accuracy. For example, J48 shows a slightly better overall average accuracy then JRip, but this is at a considerable cost to readability. It therefore seems to be quite challenging to find algorithms that provide significant

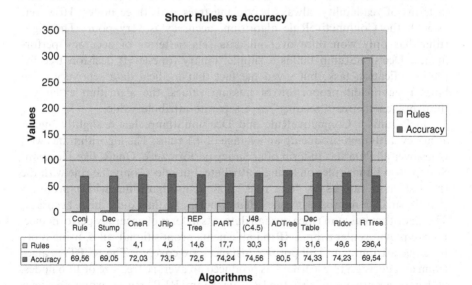

	Conj Rule	Dec Stump	OneR	JRip	REP Tree	PART	J48 (C4.5)	ADTree	Dec Table	Ridor	R Tree
Rules	1	3	4,1	4,5	14,6	17,7	30,3	31	31,6	49,6	296,4
Accuracy	69,56	69,05	72,03	73,5	72,5	74,24	74,56	80,5	74,33	74,23	69,54

Fig. 79.1 Short rules vs. accuracy in all algorithms

improvements in accuracy without sacrificing readability. In this study we hope to identify the better algorithms on which to base future research, with the goal of trying to meet that challenge.

79.5 Discussion

In these experiments, we applied 11 classifier algorithms to the UCI medical data sets to study their characteristics in producing accurate but "short" rules. Here, we are interested only in the algorithms that possess the ability to generate both accurate and readable rules. In this case, "readable" means that the number of rules is small, or that the tree size is small (e.g., less than 20 rules, or trees with less than 20 nodes—the number of nodes in a tree is roughly equivalent to the size of the corresponding rule set). Based on the results so far, we have focused on six algorithms: ConjuctiveRule, DecisionStump, OneR, JRip, REPTree, and PART, because of their ability to generate good results with a readability level of 20 rules/nodes.

ConjunctiveRule is the winner among all the algorithms in terms of the number of rules, but the algorithm is lacking. For example, in the case of the heart-disease data set (Table 79.3), the algorithm only managed to output the majority class as its rule because none of the instances is covered by its candidate rule. This is not the case for other algorithms, which are able to produce a number of rules on the heart-disease data set. We also noted that this algorithm has no wins at all over other data sets in term of its accuracy performance. The DecisionStump algorithm is the second best in terms of readability, always producing trees with three nodes. However, as with the ConjunctiveRule algorithm, accuracy is very poor. The algorithm has only won once over all data sets in terms of accuracy performance. DecisionStump builds a simple, binary (one-level) decision tree for the classification task, but given the fact that medical data sets very often have a significant proportion of missing values, the algorithm extends a third branch to treat "missing" as a separate attribute value.

OneR, unlike ConjuctiveRule and DecisionStump, has a slightly higher accuracy, although producing an average of 4.1 rules. The algorithm has won twice over all data sets in term of accuracy performance. OneR, like Decision-Stump, also builds rules on a single attribute, but selects only the rule with the smallest error rate as its "one rule." The JRip algorithm, which produces slightly smaller numbers of rules than OneR, also has quite high accuracy. The algorithm has won twice over all data sets in term of accuracy performance. Unlike previous algorithms, JRip implements propositional rules, using more than one attribute until all positive instances are covered. The REPTree algorithm as a decision tree learner has produced an average tree size of 14.6 nodes, with good accuracy similar to the JRip algorithm. REPTree has won three times over all data sets in term of accuracy. The PART algorithm has the highest accuracy among the six rule induction algorithms. It produces sets of rules called "decision lists." However, it produces the most rules, with an average of 17.7. The algorithm has won twice over all data sets in terms of accuracy.

79.6 Conclusions

In conclusion, based on the analysis of results and after study of the characteristics and behaviours of the six candidate rule induction algorithms, we have singled out for further research two promising algorithms for combining human readability and accuracy: JRip and PART. However, we prefer JRip over PART, since the readability of its results is better. This finding is justified below.

JRip and PART both have high accuracy. Despite PART producing more rules than some of the other algorithms, its accuracy is the highest among them all. Meanwhile, JRip has both the ability and the potential to produce accurate but readable rules. The algorithm also is stable and has been shown to generate fewer than eight rules on the medical data sets tested. PART is less consistent in terms of its readability. DecisionStump and OneR, although producing fewer rules, provide accuracy that is too low. These algorithms also are not able to induce rules which associate more than one attribute. In knowledge discovery, it is of course crucial to investigate interaction between attributes in order to induce interesting and useful prediction rules. There is no single one of the tested algorithms which has prevailed on the tested medical data sets in terms of both accuracy and readability. Some algorithms are good at handling certain data sets which have different combinations of numeric and nominal attributes. Some algorithms performed poorly when the number of classes to predict was above three or four. We end by recommending the JRip algorithm as a foundation for further research into finding a better tradeoff between readability and accuracy than the current state-of-the-art algorithms. We have begun an investigation in which we are examining modifications to the JRip algorithm; specifically, we are hybridising it with an evolutionary algorithm. Our results so far are encouraging, suggesting that building on the JRip approach will lead to new data mining algorithms that are particularly well suited to medical data sets and the needs of physicians and consultants for accurate, understandable, and trustworthy diagnosis.

References

1. Boukaert RR (2003) Choosing between two learning algorithms based on calibrated tests. In: Machine learning, pp 51–58
2. Brian RG, Compton P (1995) Induction of ripple-down rules applied to modeling large databases. J Intell Inf Syst 5(3):211–228
3. Cohen WW (1995) Fast effective rule induction. In: Machine learning. Lake Tahoe, California

4. Cohen WW (1996) Learning rules that classify e-mail. In: Machine learning in information access, pp 18–25
5. Conjuctive Rule class (2007) Weka website, http://www.dbs.informatik.uni-muenchen.de/Lehre/KDD_Praktikum/weka/doc/weka/classifiers/rules
6. Fayyad U, Shapiro GP, Smyth P (1996) From data mining to knowledge discovery in databases. AI Mag 17(3):37–54
7. Frank E, WittenIH (1998) Generating accurate rule sets without global optimization. In: Machine learning, pp 144–151
8. Freund Y, Mason L (1999) The alternating decision tree learning algorithm. In: Machine learning, pp 124–133
9. Furnkranz J, Widmer G (1994) Incremental reduced error pruning. In: Machine learning, Morgan Kaufmann, New Jersey, pp 70–77
10. Holt RC (1993) Very simple classification rules perform well on most commonly used data sets. In: Machine learning, 11:63–91
11. Kohavi R (1995) The power of decision tables. In: Machine learning, pp 174–189
12. Quinlan R (1993) C4.5: Programs for machine learning. Morgan Kaufmann, San Mateo, CA
13. RandomTree class (2007) WEKA website, http://www.lsi.upc.es/~bejar/apren/docum/doc/weka/classifiers/trees/RandomTree
14. REPTree class (2007) WEKA website, http://www.dbs.informatik.uni-muenchen.de/Lehre/KDD_Praktikum/weka/doc/weka/classifiers/trees
15. Sandholm T et al (1996) Comparison of regression methods, symbolic induction methods and neural networks in morbidity diagnosis and mortality prediction in equine gastrointestinal colic. http://citeseer.ist.psu.edu/196017.html
16. UCI Machine Learning Repository (2007), http://www.ics.uci.edu/~mlearn/MLRepository.html
17. WEKA web site (2007), http://www.cs.waikato.ac.nz/~ml/weka
18. Witten IH, Frank E (2005) Data mining: practical machine learning tools and techniques. Morgan Kaufmann, San Francisco
19. Witten IH, Frank E, Trigg L, Hall M, Holmes G, Cunningham SJ (1999) WEKA: practical machine learning tools and techniques with java implementations. In: Emerging knowledge engineering and connectionist-based information systems, pp 192–196

Author Index

Subject Index

Lecture Notes in Electrical Engineering

Printed in the United States
By Bookmasters